Handbook of Research on Emerging Applications of Fuzzy Algebraic Structures

Chiranjibe Jana
Vidyasagar University, India

Tapan Senapati
Padima Janakalyan Banipith (H.S.), India

Madhumangal Pal
Vidyasagar University, India

IGI Global
DISSEMINATOR OF KNOWLEDGE

A volume in the Advances in Computer and Electrical Engineering (ACEE) Book Series

Published in the United States of America by
IGI Global
Engineering Science Reference (an imprint of IGI Global)
701 E. Chocolate Avenue
Hershey PA, USA 17033
Tel: 717-533-8845
Fax: 717-533-8661
E-mail: cust@igi-global.com
Web site: http://www.igi-global.com

Library of Congress Cataloging-in-Publication Data

Names: Jana, Chiranjibe, 1984- editor. | Senapati, Tapan, 1985- editor. | Pal, Madhumangal, 1967- editor.
Title: Handbook of research on emerging applications of fuzzy algebraic structures / Chiranjibe Jana, Tapan Senapati, and Madhumangal Pal, editors.
Description: Hershey, PA : Engineering Science Reference, [2020] | Includes bibliographical references. | Summary: "This book examines the use of fuzzy algebraic structures in various applications"-- Provided by publisher.
Identifiers: LCCN 2019018176 | ISBN 9781799801900 (hardcover) | ISBN 9781799801924 (ebook)
Subjects: LCSH: Fuzzy sets. | Fuzzy mathematics. | Algebra.
Classification: LCC QA248.5 .H35 2020 | DDC 511.3/223--dc23
LC record available at https://lccn.loc.gov/2019018176

This book is published in the IGI Global book series Advances in Computer and Electrical Engineering (ACEE) (ISSN: 2327-039X; eISSN: 2327-0403)

British Cataloguing in Publication Data
A Cataloguing in Publication record for this book is available from the British Library.

For electronic access to this publication, please contact: eresources@igi-global.com.

Advances in Computer and Electrical Engineering (ACEE) Book Series

Srikanta Patnaik
SOA University, India

ISSN:2327-039X
EISSN:2327-0403

Mission

The fields of computer engineering and electrical engineering encompass a broad range of interdisciplinary topics allowing for expansive research developments across multiple fields. Research in these areas continues to develop and become increasingly important as computer and electrical systems have become an integral part of everyday life.

The **Advances in Computer and Electrical Engineering (ACEE) Book Series** aims to publish research on diverse topics pertaining to computer engineering and electrical engineering. **ACEE** encourages scholarly discourse on the latest applications, tools, and methodologies being implemented in the field for the design and development of computer and electrical systems.

Coverage

- Circuit Analysis
- Programming
- Computer Hardware
- Microprocessor Design
- Digital Electronics
- Chip Design
- Optical Electronics
- Computer Science
- Analog Electronics
- Electrical Power Conversion

IGI Global is currently accepting manuscripts for publication within this series. To submit a proposal for a volume in this series, please contact our Acquisition Editors at Acquisitions@igi-global.com or visit: http://www.igi-global.com/publish/.

The Advances in Computer and Electrical Engineering (ACEE) Book Series (ISSN 2327-039X) is published by IGI Global, 701 E. Chocolate Avenue, Hershey, PA 17033-1240, USA, www.igi-global.com. This series is composed of titles available for purchase individually; each title is edited to be contextually exclusive from any other title within the series. For pricing and ordering information please visit http://www.igi-global.com/book-series/advances-computer-electrical-engineering/73675. Postmaster: Send all address changes to above address. ©

Titles in this Series

For a list of additional titles in this series, please visit:
https://www.igi-global.com/book-series/advances-computer-electrical-engineering/73675

Novel Approaches to Information Systms Design
Naveen Prakash (Indraprastha Institute of Information Technology, Delhi, India) and Deepika Prakash (NIIT University, India)
Engineering Science Reference • © 2020 • 300pp • H/C (ISBN: 9781799829751) • US $215.00

IoT Architectures, Models, and Platforms for Smart City Applications
Bhawani Shankar Chowdhry (Mehran University of Engineering and Technology, Pakistan) Faisal Karim Shaikh (Mehran University of Engineering and Technology, Pakistan) and Naeem Ahmed Mahoto (Mehran University of Engineering and Technology, Pakistan)
Engineering Science Reference • © 2020 • 291pp • H/C (ISBN: 9781799812531) • US $245.00

Nature-Inspired Computing Applications in Advanced Communication Networks
Govind P. Gupta (National Institute of Technology, Raipur, India)
Engineering Science Reference • © 2020 • 319pp • H/C (ISBN: 9781799816263) • US $225.00

Pattern Recognition Applications in Engineering
Diego Alexander Tibaduiza Burgos (Universidad Nacional de Colombia, Colombia) Maribel Anaya Vejar (Universidad Sergio Arboleda, Colombia) and Francesc Pozo (Universitat Politècnica de Catalunya, Spain)
Engineering Science Reference • © 2020 • 300pp • H/C (ISBN: 9781799818397) • US $215.00

Tools and Technologies for the Development of Cyber-Physical Systems
Sergey Balandin (FRUCT Oy, Finland) and Ekaterina Balandina (Tampere University, Finland)
Engineering Science Reference • © 2020 • 300pp • H/C (ISBN: 9781799819745) • US $235.00

Handbook of Research on New Solutions and Technologies in Electrical Distribution Networks
Baseem Khan (Hawassa University, Hawassa, Ethiopia) Hassan Haes Alhelou (Tishreen University, Syria) and Ghassan Hayek (Tishreen University, Syria)
Engineering Science Reference • © 2020 • 439pp • H/C (ISBN: 9781799812302) • US $270.00

Major Applications of Carbon Nanotube Field-Effect Transistors (CNTFET)
Balwinder Raj (National Institute of Technical Teachers Training and Research, Chandigarh, India) Mamta Khosla (Dr. B. R. Ambedkar National Institute of Technology, Jalandhar, India) and Amandeep Singh (National Institute of Technology, Srinagar, India)
Engineering Science Reference • © 2020 • 255pp • H/C (ISBN: 9781799813934) • US $185.00

IGI Global
DISSEMINATOR OF KNOWLEDGE

701 East Chocolate Avenue, Hershey, PA 17033, USA
Tel: 717-533-8845 x100 • Fax: 717-533-8661
E-Mail: cust@igi-global.com • www.igi-global.com

List of Contributors

Table of Contents

Detailed Table of Contents

Fuzzy matrices are successfully used when fuzzy uncertainty occurs in a problem. Fuzzy matrices have become popular in the last two decades. In this chapter, two new binary fuzzy operators, $\oplus$ and $\odot$, for fuzzy matrices are introduced. Several properties on $\oplus$ and $\odot$ are presented here. Also, some results on existing operators along with these new operators are presented. In this chapter, Hamming and Euclidean distances are introduced for fuzzy matrix and investigated a number of properties. Here, one application is mentioned to recognize a printed character automatically with the help of Hamming distance. The proposed method is applicable when the characters are printed with same font, same size, and same shape. In the case of hand-written characters, the proposed method is also equally effective if it is modified. A lot of works have been done in fuzzy matrices, but there is a huge scope to work with these matrices.

This chapter deals with solution methodology of fuzzy system of linear equations (FSLEs). In fuzzy set theory, finding solutions of FLSEs has long been a well-known problem to the researchers. In this chapter, the fuzzy number has been converted into interval number, and the authors have solved the interval system of linear equation for finding the fuzzy valued solution. Here, a fuzzy valued linear system has been introduced and a numerical example has been solved and presented for illustration of purpose.

The notion of fuzzy multigroups is an application of fuzzy multisets or fuzzy bags to classical group theory. This chapter explores the notions of fuzzy comultisets, characteristic fuzzy submultigroups, and direct product of fuzzy multigroups as extensions of cosets, characteristic subgroups, and direct product of groups. The relationship between fuzzy comultisets of a fuzzy multigroup and the cosets of a group is established. Some results on the concept of fuzzy comultisets are deduced. A number of properties of characteristic fuzzy submultigroups of fuzzy multigroups are outlined, and some related results are obtained. Also, the author presents some properties of direct product of fuzzy multigroups and establish some results.

This chapter considers k-nn classification of objects that assume values of imprecise nature with respect to attributes being considered. In order to handle imprecise data values, two approaches of crisp conversion of fuzzy data sets are considered. The approaches considered in this chapter are borrowed from Credibility Theory of Liu. A comparative study on the choice of approach for crisp conversion of fuzzy data has been carried out with the help of certain multivariate simulated data sets. Conclusions drawn from the study are presented.

This chapter explores the various algebraic structures that exist in the fuzzy set environment in a multiset context. The authors have also established analogs of groups, rings, and ideals of sets in multiset environment. The conditions under which a multiset that is derived from a group becomes a multiset group (shortly mset group) and that derived from a ring to become a multiset ring (shortly mset ring) have been mentioned. The distinguishing between multiset subring and multiset ideals has been discussed on in the chapter. The authors have also delved into properties such as intersection and union of mset groups, mset rings, and ideals.

The basic notions of soft sets theory are introduced by Molodtsov to deal with uncertainties when solving problems in practice as in engineering, social science, environment, and economics. This notion is convenient and easy to apply as it is free from the difficulties that appear when using other mathematical tools as theory of theory of fuzzy sets, rough sets, and theory of vague sets. The soft set theory has recently gaining significance for finding rational and logical solutions to various real-life problems, which involve uncertainty, impreciseness, and vagueness. The concepts of intuitionistic fuzzy soft left almost semigroups and the intuitionistic fuzzy soft ideal are introduced in this chapter, and some of their basic properties are studied.

In a lattice $\mathfrak{L}$, the authors used the concept of belongingness and quasi-coincidence of fuzzy point to a fuzzy set, and by this notion, $(\in,\in\vee q)$-fuzzy sublattice, $(\in,\in\vee q)$-fuzzy ideal, cartesian product of $(\in,\in\vee q)$-fuzzy sublattice, $(\in,\in\vee q)$-fuzzy complemented sublattice, and cartesian product of $(\in,\in\vee q)$-

fuzzy complemented sublattice are introduced, and their properties are briefly studied. The relationship between fuzzy sublattice and (∈,∈∨q)-fuzzy sublattice, fuzzy ideal and (∈,∈∨q)-fuzzy ideal of L are established. The authors prove that the cartesian product of two (∈,∈∨q)-fuzzy ideals of a lattice is not necessarily a fuzzy ideal of a lattice. The theory of image and inverse image of an (∈,∈∨q)-fuzzy sublattice and (∈,∈∨q)-fuzzy ideal, an (∈,∈∨q)-fuzzy complemented sublattice, and (∈,∈∨q)-fuzzy complemented ideal of 𝔏 on the basis of homomorphism of lattices are also significantly established.;

In this chapter, the author studies the uni-hesitant fuzzy set-theoretical approach to the ideals of BCK-algebras. For a hesitant fuzzy set H on S and a subset of [0,1], the set L(H;ʎ):={x∈S|xH⊆ʎ}, is called the uni-hesitant level set of H. Moreover, the author discusses the relations between uni-hesitant fuzzy commutative ideals and uni-hesitant fuzzy ideals. Further, he considered the characterizations of uni-hesitant fuzzy commutative ideals in BCK-algebras. Finally, he proved some conditions for a uni-hesitant fuzzy ideal to be a uni-hesitant fuzzy commutative ideal.

In this chapter, the authors introduce the notion of a fuzzy bi-interior ideal of semiring and they characterize the regular semiring in terms of fuzzy bi-interior ideal of semiring. And also they proved Let I be a non-empty subset of a semiring M and χ_I be the characteristic function of I. Then I is a bi interior ideal of semiring if and only if χ_I is a fuzzy bi-interior ideal of semiring M.

In this chapter, the authors introduce the concepts of (∈,∈∨qkδ)-fuzzy interior ideal in semigroup, which is a generalisation of (∈,∈∨qk)-fuzzy interior ideal and (∈,∈∨q)-fuzzy interior ideal in semigroup, and investigate some of its properties. They give example to show that (∈,∈∨qkδ)-fuzzy interior ideal in semigroup is not an (∈,∈∨qk)-fuzzy interior ideal and (∈,∈∨q)-fuzzy interior ideal in semigroup. They show that every (∈,∈∨qkδ)-fuzzy ideal is an (∈,∈∨qkδ)-fuzzy interior ideal, but the converses are not true in general.

In this chapter, the concepts of bipolar fuzzy H-ideals of BCI-algebras are introduced and their natures are investigated. Relations between bipolar fuzzy subalgebras, bipolar fuzzy ideals, and bipolar fuzzy H-ideals are discussed. Conditions for a bipolar fuzzy ideal to be a bipolar fuzzy H-ideal are provided. Some characterization theorems of bipolar fuzzy H-ideals are established. A bipolar fuzzy H-ideal is established by using a finite collection of H-ideals. The authors have shown that if every bipolar fuzzy

H-ideal has the finite image, then every descending chain of H-ideals terminates at finite step.

The authors studied the notion of (α,β)-intuitionistic fuzzy BCK/BCI-subalgebras by applying the Lukasiewicz 3-valued implication operator, where $\alpha,\beta\in\{\in,q,\in\wedge q,\in\vee q\}$ for $\alpha\neq\in\wedge q$. In this chapter, an intuitionistic fuzzy set A is an $(\in,\in)$ (or $(\in\wedge q,\in)$ or $(\in,\in\vee q)$)-intuitionistic fuzzy subalgebras if and only if for any $p\in(0,1]$ (or $p\in(0,0.5]$ or $p\in(0.5,1]$), then Ap is a fuzzy subalgebras of X respectively. Next, the authors defined intuitionistic fuzzy subalgebras with thresholds (s,t) and then provided intuitionistic fuzzy subalgebras with thresholds (0,1) (or (0,0.5) or (0.5,1)) by the concept of quasi-coincidence of fuzzy point respectively. Also, A is an intuitionistic fuzzy subalgebras with thresholds (s,t) if and only if for any $p\in(s,t]$, then cut set A_p is a fuzzy subalgebras.

The theory of interval-valued intuitionistic fuzzy sets is a generalization of both interval-valued fuzzy sets and intuitionistic fuzzy sets. In this chapter, the notion of interval-valued intuitionistic fuzzy subnear-ring is introduced, and some interesting properties are discussed. Some relations on the family of all interval-valued intuitionistic fuzzy subnear-ring are presented, and some related properties are investigated. Also, the authors represent upper and lower level set of interval-valued intuitionistic fuzzy set.

This chapter focuses on the basics of a theory for such an interval-valued fuzzy set becoming doubt interval-valued fuzzy subalgebra and doubt interval-valued fuzzy ideal in BF-algebras. After a detailed study of its properties, the authors come to the conclusion that in BF-algebras an interval-valued fuzzy set in BF-algebras is a doubt interval-valued fuzzy subalgebra and a doubt interval-valued fuzzy ideal if and only if the complement of this interval-valued fuzzy set is an interval-valued fuzzy subalgebra and an interval-valued fuzzy ideal.

The notion of interval valued fuzzy set was first introduced by Zadeh as a generalization of fuzzy sets. Using interval valued fuzzy set, various algebraic structures and related topics were discussed. This chapter

This chapter provides an elaborate discussion on the use of fuzzy preview control designed with the help of look up tables and granular computing for the studies on under actuated systems which are dynamic systems which have lesser degrees of freedom than the actuating input signals. It is an important design consideration for most systems to be categorized under this section which makes them the essential component for analysis. The chapter provides an in depth analysis taking the case studies of various control system test beds such as the family of the inverted pendulum systems which make the prototype models for validation before applying the developed logic to control of high end complex dynamic systems in terms of stability, robustness, speed of operation and more importantly to see the fuzzy structure that controls these systems in terms of lookup tables.

Similarity is an ambiguous term that can be interpreted differently depending on the context of use. In this chapter, the authors review some of its uses before focusing on decision making. Ignoring the uncertainty of human knowledge would be denying a major attribute. Thus, they linked it to the fuzzy context. However, even taking this aspect into consideration, the opinion itself must be relevant. Therefore, the more a decision is similar to other opinions, the more it is coherent. Hence, there is a need to measure the similarity between each couple of expressed opinions.

Fuzzy logic has been serving the industry for decades by resolving the ambiguities that appear as a result of imprecise environment. High-stake decision-making processes require inputs from various stakeholders to incorporate. If the risk is high, as in the case of high investment decision making, a robust system of incorporating opinions from multiple stakeholders must be set in place in order to avoid any inconsistency or bad decisions. Fuzzy matrices and arithmetic can play a rescuer in such situations. In this chapter, the authors demonstrate a decision-making framework incorporating the use of fuzzy numbers and arithmetic to make critical decisions in strategic marketing and new product development. Forecasting in the domains of new products is an utmost complex and critical process because no relevant history is available owing to the product's 'one-of-its-kind' nature. In such cases, computation via analogy is an interesting paradigm, which is also discussed in the chapter.

Preface

Due to the advancement of science and technology our modern society becomes very complex and as a result our decision processes have become uncertain and difficult to analyze. The important of fuzziness and untruth play important role in decision making, but until 1965 we did not know how the fuzziness arising from subjectivity can model and analyzed. In 1965, Zadeh introduced a novel concept well known as fuzzy set theory. This invention provides a mathematical framework for generalizing some notions of classical algebraic structures. After the development of fuzzy set theory many researchers carried out their research on the generalizations of fuzzy sets and several other mathematical structures. And these researches are successful applied in artificial intelligence, computer science, expert system, control engineering, knowledge discovery, robotics, decision sciences, automata theory, medicine, graph theory, finite state machine, logics and many other branches of pure and applied mathematics. In 1971, Rosenfield introduced the concept of fuzzy subgroups. This seminal work was the first fuzzification of any algebraic structure. This development opened a new direction, new revolution, new path of thinking to mathematicians, computer scientists, engineers, physicists, social scientists, chemists and many others. After that, several important discrete structures have been fuzzified by many researchers in various ways. It is true that not all of these fuzzifications are applied in real domain, but we are sure that many of these fuzzifications of algebraic structures will be very useful tools for applications in different areas in near feature. On the other hand, several extensions of fuzzy set theory are growing on to deal with imprecision and uncertainty in a wider sense, e.g. intuitionsistic fuzzy sets (IFS), interval-valued intuitionistic fuzzy sets (IVIFS), type-2 fuzzy sets, bipolar intuitionistic fuzzy sets and Zadeh's generalized theory of uncertainty.

This book has cover interesting co-related topics including fuzzy matrices, fuzzy decision-making, fuzzy similarity relations, solution of fuzzy linear equation, fuzzy logic based applications, neighborhood based classification, soft ideals, fuzzy multigroup notions, multi-set approach algebraic structures, fuzzy sublattice and complemented fuzzy sublattice structures based on fuzzy point, Uni-hesitant fuzzy BCKalgebra, fuzzy bi-interior ideals in semiring, $(\in k, \in k \vee qk)$ -fuzzy interior ideals in semigroups, bipolar fuzzy H-ideals in BCI-algebras, Lukuswize 3-valued logic applied on BCK/BCI-algebras, IVIF subnear ring, doubt fuzzy translations on BF algebra, fuzzy H-ideals on β-algebra, ideals in Q1-algebra, regular and intra-regular neutrosophics left almost semihypergroups, picture fuzzy aggregation operators and applications, and fuzzy preview control with granular computing.

This book is a description of such updated algebraic theories and applications of fuzzy structures. The summaries of all 21 chapters are presented below.

Chapter 1

The fuzzy matrix theory is an extension of the classical matrix algebra. This extension is successfully used to model several uncertain problems when fuzziness occurs in the problem. Fuzzy matrices have become popular for last four decades. In this chapter, two new binary fuzzy operators $\oplus$ and $\odot$ for fuzzy matrices are introduced. Several properties on new peoposed operators $\oplus$ and $\odot$ are presented here. Also, some results on existing operators along with these new operators are presented. In this chapter, Hamming and Euclidean distances are introduced for fuzzy matrix and investigated a number of properties. Here, one application is mentioned to recognize a printed character automatically with the help of Hamming distance. The proposed method is applicable when the characters are printed with same font, same size and shape. In the case of hand written character, the proposed method is also equally effective if it is modified. A lot of works have been done in fuzzy matrices, even there is a huge scope to work with these matrices.

Chapter 2

This chapter deals with solution methodology of fuzzy system of linear equations (FSLEs). In fuzzy set theory, finding solutions of FLSEs has long been a well-known problem to the researchers. In this chapter, the fuzzy number has been converted into interval number and we have solved the interval system of linear equation for finding the fuzzy valued solution. Here, a fuzzy valued linear system has been introduced and a numerical example has been presented for illustration purpose.

Chapter 3

The notion of fuzzy multigroups is an application of fuzzy multisets or fuzzy bags to classical group theory. This chapter explores the notions of fuzzy comultisets, characteristic fuzzy submultigroups and direct product of fuzzy multigroups as extensions of cosets, characteristic subgroups and direct product of groups. The relationship between fuzzy comultisets of a fuzzy multigroup and the cosets of a group is established. Some results on the concept of fuzzy comultisets are deduced. A number of properties of characteristic fuzzy submultigroups of fuzzy multigroups are outlined and some related results are obtained. Also, we present some properties of direct product of fuzzy multigroups and establish some results.

Chapter 4

This chapter considers k-nn classification of objects which assume values of imprecise nature with respect to attributes being considered. In order to handle imprecise data values two approaches of crisp conversion of fuzzy data sets is considered. The approaches considered in this work are borrowed from Credibility Theory of Liu (2007). A comparative study on the choice of approach for crisp conversion of fuzzy data has been carried out with the help of certain multivariate simulated data sets. Conclusions drawn from the study are presented.

Chapter 5

This chapter explores the various algebraic structures that exist in the fuzzy set environment in a multiset context. The authors have also established analogs of groups, rings and ideals of sets in multiset environment. The conditions under which a multiset that is derived from a group becomes a multiset group (shortly mset group) and that derived from a ring to become a multiset ring (shortly mset ring) have been mentioned. The distinguishing between multiset subring and multiset ideals has been discussed on in the chapter. The authors have also delved into properties such as intersection and union of mset groups, mset rings and ideals. The authors noted that many applications of sets in fields such as cryptography, automation, machine learning, artificial intelligence, neural networking and so on could be vastly improved if the usage context were to be switched to multisets, which was a deciding factor in the determining of the paper.

Chapter 6

The basic notions of soft sets theory are introduced by Molodtsov to deal with uncertainties when solving problems in practice as in engineering, social science, environment, and economics. This notion is convenient and easy to apply as it is free from the difficulties that appear when using other mathematical tools as theory of fuzzy sets, rough sets and theory of vague sets, etc. The soft set theory has recently gaining significance for finding rational and logical solutions to various real life problems, which involve uncertainty, impreciseness, and vagueness. The concepts of intuitionistic fuzzy soft left almost semigroups and the intuitionistic fuzzy soft ideal are introduced in this chapter, and some of their basic properties are studied.

Chapter 7

In a lattice $\mathfrak{L}$, the concept of belongingness and quasi-coincidence of fuzzy point to a fuzzy set, by this notion $(\in, \in\vee q)$-fuzzy sublattice, $(\in, \in\vee q)$-fuzzy ideal, cartesian product of $(\in, \in\vee q)$-fuzzy sublattice, $(\in, \in\vee q)$-fuzzy complemented sublattice and cartesian product of $(\in, \in\vee q)$-fuzzy complemented sublattice are introduced and briefly studied their properties. The relationship between fuzzy sublattice and $(\in, \in\vee q)$-fuzzy sublattice, fuzzy ideal and $(\in, \in\vee q)$-fuzzy ideal of $\mathfrak{L}$ are established. Also, it is proved that the Cartesian product of two $(\in, \in\vee q)$-fuzzy ideals of a lattice is not necessarily a fuzzy ideal of a lattice. It is also significantly established the theory of image and inverse image of an $(\in, \in\vee q)$-fuzzy sublattice and $(\in, \in\vee q)$-fuzzy ideal, an $(\in, \in\vee q)$-fuzzy complemented sublattice and $(\in, \in\vee q)$-fuzzy complemented ideal of $\mathfrak{L}$ on the basis of homomorphism of lattices.

Chapter 8

In this chapter, the authors studied the uni-hesitant fuzzy set theoretical approach to the ideals of B*CK*-algebras. For a hesitant fuzzy set $\mathcal{H}$ on S and a subset of $[0,1]$, the set $L(\mathcal{H};\lambda) := \{x \in S \mid x\mathcal{H} \subseteq \lambda\}$, is called the uni-hesitant level set of $\mathcal{H}$. Moreover, the relations between uni-hesitant fuzzy commutative ideals and uni-hesitant fuzzy ideals are discussed. Further, the characterizations of uni-hesitant fuzzy commutative ideals in *BCK*-algebras is considered. Finally, some conditions for a uni-hesitant fuzzy ideal to be a uni-hesitant fuzzy commutative ideal are proved.

Chapter 9

In this chapter, the notion of a fuzzy bi-interior ideal of simmering is introduced. Several properties are presented along with examples. Also, the regular semiring is characterized in terms of fuzzy bi-interior ideal of semiring.

Chapter 10

In this chapter, the concepts of $\left(\in,\in \vee q_k^{\delta}\right)$-fuzzy interior ideal in semigroup is introduced, which is a generalisation of $(\in, \in\vee q_k)$ -fuzzy interior ideal and $(\in,\in\vee q)$-fuzzy interior ideal in semigroup, and investigated some of its properties. Also, an example is provided to show that $\left(\in,\in \vee q_k^{\delta}\right)$-fuzzy interior ideal in semigroup is not an $(\in, \in\vee q_k)$-fuzzy interior ideal and $(\in,\in\vee q)$-fuzzy interior ideal in semigroup. It is also shown that every $\left(\in,\in \vee q_k^{\delta}\right)$-fuzzy ideal is an $\left(\in,\in \vee q_k^{\delta}\right)$-fuzzy interior ideal, but the converses are not true in general.

Chapter 11

In this chapter, the concepts of bipolar fuzzy H-ideals of BCI-algebras is introduced and their natures are investigated. Relations between bipolar fuzzy subalgebras, bipolar fuzzy ideals and bipolar fuzzy H-ideals are discussed. Conditions for a bipolar fuzzy ideal to be a bipolar fuzzy H-ideal are provided. Some characterization theorems of bipolar fuzzy H-ideals are established. A bipolar fuzzy H-ideal is established by using a finite collection of H-ideals. Also, it is shown that if every bipolar fuzzy H-ideal has the finite image, then every descending chain of H-ideals of X terminates at finite step.

Chapter 12

The notion of (α,β) -intuitionistic fuzzy *BCK/BCI*-subalgebras are studied by applying the Lukasiewicz 3-valued implication operator, where $\alpha,\beta\in\{\in,q,\in\wedge q,\in\vee q\}$ for $\alpha\neq\in\wedge q$. In this chapter, an intuitionistic fuzzy set A is an $(\in,\in)$(or $(\in\wedge q,\in)$ or $(\in,\in\vee q)$)-intuitionistic fuzzy subalgebras if and only if for any $p\in(0,1]$ (or $p\in(0,0.5]$ or $p\in(0.5,1]$), then A_p is a fuzzy subalgebras of X respectively. Next, defined intuitionistic fuzzy subalgebras with thresholds (s,t) and then provided intuitionistic fuzzy subalgebras with thresholds (0,1) (or (0,0.5) or (0.5,1)) by the concept of quasi-coincidence of fuzzy point respectively. Also, A is an intuitionistic fuzzy subalgebras with thresholds (s,t) if and only if for any $p\in(s,t]$, then cut set A_p is a fuzzy subalgebras.

Chapter 13

The theory of interval-valued intuitionistic fuzzy sets are generalization of both interval-valued fuzzy sets and intuitionistic fuzzy sets. In this chapter, the notion of interval-valued intuitionistic fuzzy subnear-ring is introduced and some imprtant properties are discussed. Some relations on the family of all interval-valued intuitionistic fuzzy subnear-ring are presented and investigate some related result.

Chapter 14

This chapter focuses on the basics of a theory for such an interval-valued fuzzy set becoming doubt interval-valued fuzzy subalgebra and doubt interval-valued fuzzy ideal in BF-algebras. After a detailed study of its properties, the authors come to this conclusion that in BF-algebras an interval-valued fuzzy set in BF-algebras is a doubt interval-valued fuzzy subalgebra and a doubt interval-valued fuzzy ideal if and only if the complement of this interval-valued fuzzy set is an interval-valued fuzzy subalgebra and an interval-valued fuzzy ideal. It is illustrated with suitable example that the union of two doubt interval-valued fuzzy subalgebras / two doubt interval-valued fuzzy ideals is also doubt interval-valued fuzzy subalgebra/ doubt interval-valued fuzzy ideal. Also the Cartesian product of doubt interval-valued fuzzy ideals in BF algebras are introduced. Another unique inclusion that the authors have made in this chapter fuzzy-translation, fuzzy-multiplication of a doubt interval-valued fuzzy subalgebra and doubt interval-valued fuzzy ideal in a BF-algebra as an extension of doubt interval-valued fuzzy subalgebra and doubt interval-valued fuzzy ideal.

Chapter 15

The notions of interval valued fuzzy set were first introduced by Zadeh as a generalization of fuzzy sets. Using interval valued fuzzy set, various algebraic structures and related topics were discussed. This paper extends fuzzy H-ideal into interval valued fuzzy H-ideals of β–algebra and some related results.

Chapter 16

In this chapter, the notions of implicative ideal, fantastic ideal and normal ideal in a right distributive QI-algebra are introduced and proved that these notions are equivalent.

Chapter 17

In this chapter, regular and intra-regular neutrosophic LA-semihypergroups are characterized by giving some examples with different neutrosophic identities and prove a various important result about it with the help of different neutrosophic identities. The main aim in this paper is to characterize the idea of regular and intra-regular neutrosophic LA-semihypergroups, (Weakly, Strongly, Completely) regular neutrosophic LA-semihypergroups using neutrosophuic sets. Also different substructures of regular neutrosophic LA-semihypergroup namely (weakly, left, right, strongly, completely) regular neutrosophic LA-semihypergroup are defined with some basic results. Further, invertible neutrosophic LA-semihypergroup are also defined in the part of the paper.

Chapter 18

The main objective of proposed work is to introduce a series of picture fuzzy weighted averaging and geometric aggregation operators by using t-norm and t-conorm. In this chapter, the generalized form of weighted averaging and geometric aggregation operators for picture fuzzy information are discussed. Further, the proposed aggregation operators of picture fuzzy number are applied to multi-attribute group decision making problems. To implement the proposed models, some numerical applications of group decision making problems are provided. Also, by comparing with previous model, it is concluded that the proposed technique is more effective and reliable.

Chapter 19

In this chapter, authors provided an elaborate discussion on the use of fuzzy preview control designed with the help of look up tables and granular computing for the studies on under actuated systems which are dynamic systems which have lesser degrees of freedom than the actuating input signals. It is an important design consideration for most systems to be categorized under this section which makes them the essential component for analysis. The chapter provides an in depth analysis taking the case studies of various control system test beds such as the family of the inverted pendulum systems which make the prototype models for validation before applying the developed logic to control of high end complex dynamic systems in terms of stability, robustness, speed of operation and more importantly to see the fuzzy structure that controls these systems in terms of lookup tables.

Chapter 20

Due to the multiplicity of definitions, interpretations and uses, similarity concept remains challenging to surround. The difficulty is all the more persistent in the presence of uncertainty. In this work, a review of similarity development especially in fuzzy context is presented, then extend and compare a coherence measure that merges similarity and dissimilarity in a decision-making context.

Chapter 21

Fuzzy logic has been serving the industry since decades by resolving the ambiguities appear as a result of imprecise environment. High stake decision making processes require inputs from various stakeholders to incorporate. If the risk is high, in case of high investment decision making, a robust system of incorporating opinions from multiple stakeholders must be set in place in order to avoid any inconsistency or bad decision. Fuzzy matrices and arithmetic can play a rescuer in such situations. In this chapter, a decision making framework is demonstrated incorporating the use of fuzzy numbers and arithmetic to make critical decisions in strategic marketing, and new product development. Forecasting in the domains of new products is an utmost complex and critical process, because no relevant history is available owing to the product's 'one of its kind' nature. In such cases, computation via analogy is an interesting paradigm, which will also be discussed in the present chapter.

This book provides relevant theoretical frameworks and the latest empirical research findings in the area of research on fuzzy algebras and fuzzy matrices in a large number of real-life applications. The target audience of this book is composed of academics, practitioners, upper-level students, and professionals interested in an innovative overview of fuzzy algebras and matrices. Moreover, the book will provide insights and support from practitioners and academia in order to highlight the most debatable aspects in the field.

We are highly grateful to all the contributors for writing the chapters by spending their valuable time and expertise. Also, we are very thankful to the reviewers for their time and suggestion to review the chapters within the stipulated period. Without the support of contributors and reviewers, it is not possible for us to bring this book on fuzzy algebraic structure. We are also thankful to our colleagues and friends for their continuous support and encouragement to prepare the book. We shall feel great to receive constructive criticisms for the improvement of the book from the experts as well as the learners. We thank the IGI Global Publisher for their sincere care and suggestion in the publication of the book.

Chiranjibe Jana
Vidyasagar University, India

Tapan Senapati
Padima Janakalyan Banipith (H.S.), India

Madhumangal Pal
Vidyasagar University, India

Chapter 1
An Introduction to Fuzzy Matrices

Madhumangal Pal
Vidyasagar University, India

ABSTRACT

Fuzzy matrices are successfully used when fuzzy uncertainty occurs in a problem. Fuzzy matrices have become popular in the last two decades. In this chapter, two new binary fuzzy operators, $\oplus$ and $\odot$, for fuzzy matrices are introduced. Several properties on $\oplus$ and $\odot$ are presented here. Also, some results on existing operators along with these new operators are presented. In this chapter, Hamming and Euclidean distances are introduced for fuzzy matrix and investigated a number of properties. Here, one application is mentioned to recognize a printed character automatically with the help of Hamming distance. The proposed method is applicable when the characters are printed with same font, same size, and same shape. In the case of hand-written characters, the proposed method is also equally effective if it is modified. A lot of works have been done in fuzzy matrices, but there is a huge scope to work with these matrices.

1. INTRODUCTION

Fuzzy matrices arise in many applications, one of which is as adjacency matrices of fuzzy relations and fuzzy relational equations have important applications in pattern classification and in handling fuzziness in knowledge based systems. Several authors presented a number of results on fuzzy matrices. Kim and Roush (1980), studied the canonical form of an idempotent matrix. Hashimoto (1983) studied the canonical form of a transitive matrix. Kolodziejczyk (1987) presented the canonical form of a strongly transitive matrix. Xin (1992, 1994) studied controllable fuzzy matrices. Hemasinha et al. (1993) investigated iterates of fuzzy circulates matrices. Ragab and Emam (1995) presented some properties of the min-max composition of fuzzy matrices. Thomason (1977) and Kim (1988) defined the adjoint of a square fuzzy matrix. Kim et al. (1989), studied determinant of square fuzzy matrices. Ragab and Emam (1994) presented some properties on determinant and adjoint of a square fuzzy matrix. Mondal and Pal (2013a) defined soft matrix and presented several results. Pal (2015a, 2015b) defined new kind of fuzzy matrix whose rows and columns are uncertain. Mondal and Pal (2016a, 2019) studied on bipolar fuzzy

DOI: 10.4018/978-1-7998-0190-0.ch001

matrix, eigenvalue of bipolar fuzzy matrix and their convergence. Pal et al. (2013b, 2014) studied on intuitionistic fuzzy matrix. Pandey et al. (2014) has applied fuzzy matrix in decision making problems.

In this chapter, we introduce two new binary operators on fuzzy matrices which are denoted by the symbols $\oplus$.and $\odot$. Also some properties of the fuzzy matrices over these new operations and some pre-defined operations are presented. After that, the distance between two fuzzy matrices are defined and investigated relationship among the distances. Two real life problems have been solved based on Hamming and Euclidean distances.

The first problem is "recognition of character" using Hamming distance. In the age of automation, men having the natural propensity of performing all types of work automatically instead of human interference, the character recognition is one of the most successful and epoch-making application of automatic pattern recognition. The objective of character recognition system is to recognize a character automatically by machine without typing or human interaction. There are various types of optical character recognition system in different languages. Several works on optical character recognition system are available in literature. Here we discuss and analyze a new system of character recognition. In this process recognition of character is possible by measuring the Hamming distance (Shyamal & Pal, 2005).

2 PRELIMINARIES

The definition of a fuzzy matrix is given below.

Definition 2.1 Fuzzy Matrix. A fuzzy matrix (FM) A of order m×n.is defined as $A = [\langle a_{ij}, a_{ij\mu} \rangle]_{m\times n}$ where $a_{ij\mu}$.is the membership value of the element a_{ij}.in A. For simplicity, we write A as $A=[a_{ij\mu}]$.

A special type of fuzzy matrix called Boolean fuzzy matrix is defined below.

Definition 2.2 Boolean Fuzzy Matrix. A fuzzy matrix $A = [a_{ij}]_{m\times n}$ .is said to be a Boolean fuzzy matrix of order m×n if all the elements of A are either 0 or 1.

In classical matrix $A=[a_{ij}]$ the elements are either any real numbers or complex numbers. The basic arithmetic operators are +,-,*,/ and some mathematical functions. The classical matrices are defined only on real and complex fields. No geometric representation is possible as the elements may be very large and small also.

Fuzzy matrix is the extension of classical matrix. Following issues are new in fuzzy matrix and these issues lead fuzzy matrix a new area of fuzzy mathematics.

1. Many new operators are defined
2. The elements are taken from different fields
3. Can be represented by geometrically
4. Several verities of fuzzy matrices are available, like fuzzy sets

The elements of matrices are taken from different fields. They may be membership values, i.e. they lie between 0 and 1 including them. May be intuitionistic fuzzy number, interval-valued fuzzy number, interval numbers, triangular fuzzy numbers, trapezoidal fuzzy numbers, pentagonal fuzzy numbers, picture fuzzy number, Pythagorean fuzzy number, bipolar fuzzy number, etc. Based on the fuzzy numbers different types of fuzzy matrices are defined listed below:

1. Fuzzy matrix whose elements are membership number, lies between 0 and 1, including 0 and 1
2. Matrix whose elements are interval numbers
3. Matrix whose elements are triangular fuzzy numbers
4. Picture fuzzy matrix
5. Pythagorean fuzzy matrix
6. Intuitionistic fuzzy matrix
7. Interval-valued fuzzy matrix
8. Interval-valued intuitionstic fuzzy matrix
9. Bipolar fuzzy matrix
10. Type-2 fuzzy matrix
11. Soft matrix, etc.

In this chapter mainly we will discuss about the fuzzy matrix of first type listed above, i.e. fuzzy matrix whose elements are taken from the closed interval [0,1].

Fuzzy matrices arise in many applications, one of which is as adjacency matrices of fuzzy relations and fuzzy relational equations have important applications in pattern classification, handling fuzziness in knowledge based systems and many other fields.

Let I.be the set of all real numbers between 0 and 1, i.e., $I=\{x: 0\leq x\leq 1\}$. For all $x,y,\alpha\in I$. the following operations are defined.

(i) $x\vee y = \max(x,y)$

(ii) $x\wedge y = \min(x,y)$.

(iii) $x \ominus y = \begin{cases} x, & \text{if } x > y \\ 0, & \text{if } x \leq y \end{cases}$

(iv) $x^{(\alpha)}(\text{upper } \alpha - \text{cut}) = \begin{cases} 1, & \text{if } x \geq \alpha \\ 0, & \text{if } x < \alpha \end{cases}$

(v) $x_{(\alpha)}(\text{lower } \alpha - \text{cut}) = \begin{cases} x, & \text{if } x \geq \alpha \\ 0, & \text{if } x < \alpha \end{cases}$

(vi) x^c (complement) = $1-x$

Now, we define two new operators $\oplus$.and $\odot$.as follows:

(vii) $x\oplus y = x+y-xy$ and

(viii) $x\odot y = xy$

where the operations '+', '–', and '.' are ordinary addition, subtraction and multiplication respectively.

It may be noted that the values of $x\vee y$, $x\wedge y$, $x\ominus y$, $x\oplus y$, $x\odot y$, $x^{(\alpha)}$, $x_{(\alpha)}$ and x^c belong to the set I.

It is obvious that (i) $1\oplus x=1$. (ii) $1\odot x=x$. (iii) $0\oplus x=x$.and (iv) $0\odot x=0$.We define the following operators for any two matrices $A=[a_{ij}]$ .and $B=[b_{ij}]$ of order $m\times n$.

(i) $A \vee B = [a_{ij} \vee b_{ij}]$.
(ii) $A \wedge B = [a_{ij} \wedge b_{ij}]$
(iii) $A \ominus B = [a_{ij} \ominus b_{ij}]$
(iv) $A^{[k+1]} = A^{[k]} \odot A$, $A^{[1]} = A$, $k=1,2,\ldots$
(v) $[k+1]A = [k]A \oplus A$, $[1]A = A$, $k=1,2,\ldots$
(vi) $A @ B = \left[\dfrac{a_{ij} + b_{ij}}{2}\right]$
(vii) $A\$B = \left[\sqrt{a_{ij}.b_{ij}}\right]$
(viii) $A^{(\alpha)} = \left[a_{ij}^{(\alpha)}\right]$ (upper α-cut fuzzy matrix)
(ix) $A_{(\alpha)} = [a_{ij(\alpha)}]$ (lower α-cut fuzzy matrix)
(x) $A' = \left[a_{ji}\right]$ (the transpose of A)
(xi) $A^c = [1 - a_{ij}]$ (the complement of A)
(xii) $A \leq B$ if and only if $a_{ij} \leq b_{ij}$ for all i,j
(xiii) For any two fuzzy matrices A and B, $\min(A,B) = \begin{cases} A, & \text{if } A \leq B \\ B, & \text{if } B \leq A \end{cases}$

The two new operations $\odot$ and $\oplus$ on the FMs are defined as

(xiv) $A \oplus B = [a_{ij} + b_{ij} - a_{ij} b_{ij}]$
(xv) $A \odot B = [a_{ij} b_{ij}]$

Every fuzzy matrix can be visualized as a three-dimensional figure. But, this representation is not possible for classical matrix without any proper scaling. To illustrate this fact, we consider two fuzzy matrices A.and B.as follows:

The 3D representation of the matrices A.and B.are shown in Figures 1 and 2.

In the following we define some special types of matrices. Let $R=[r_{ij}]$ .be an $n \times n$ fuzzy matrix. Then,

Figure 1. Geometrical representation of the matrix A

$$A = \begin{bmatrix} 0.5 & 0.1 & 0.7 & 0.5 \\ 0.3 & 0.8 & 0.1 & 0.6 \\ 0.6 & 0.4 & 0.9 & 0.8 \\ 0.2 & 0.7 & 0.3 & 0.4 \end{bmatrix}$$

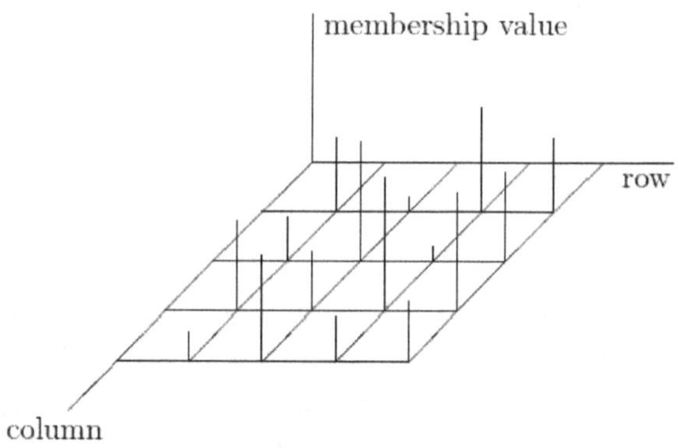

Figure 2. Geometrical representation of the matrix B

$$B = \begin{bmatrix} 0.3 & 0.8 & 0.6 & 0.4 \\ 0.2 & 0.7 & 0.5 & 0.3 \\ 0.7 & 0.5 & 0.3 & 0.9 \\ 0.9 & 0.4 & 0.1 & 0.6 \end{bmatrix}$$

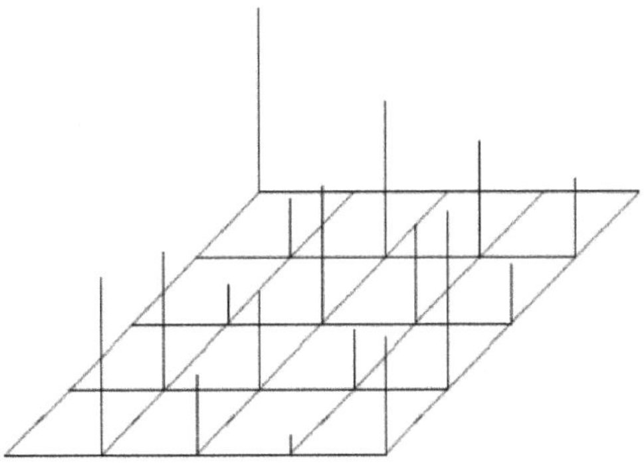

Figure 3. Geometrical representation of the matrix A⊕B

$$A \oplus B = \begin{bmatrix} 0.65 & 0.82 & 0.88 & 0.70 \\ 0.44 & 0.94 & 0.55 & 0.72 \\ 0.88 & 0.70 & 0.93 & 0.98 \\ 0.92 & 0.82 & 0.37 & 0.76 \end{bmatrix}$$

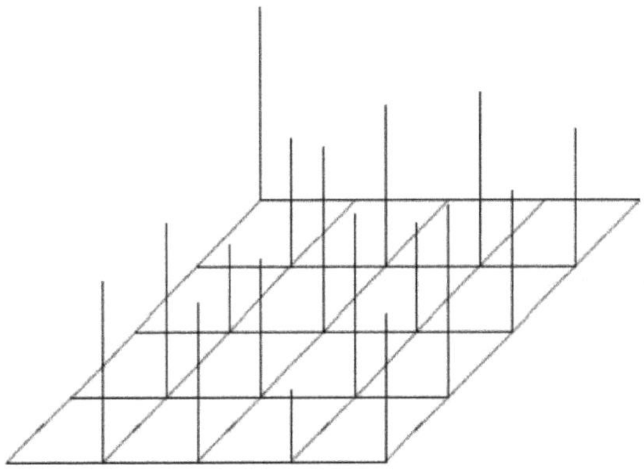

(i) *R* is reflexive if and only if $r_{ii}=1$ for all $i=1,2,\dots,n$
(ii) *R* is irreflexive if and only if $r_{ii}=0$ for all $i=1,2,\dots,n$
(iii) *R* is nearly irreflexive if and only if $r_{ii}\leq r_{ij}$ for all $i,j=1,2,\dots,n$
(iv) *R* is symmetric if and only if $R' = R$.
(v) *R* is constant if and only if $r_{ij}=r_{kj}$ for all $i,j,k=1,2,\dots,n$
(vi) *R* is identity if and only if $r_{ii}=1$ and $r_{ij}=0$ $(i\neq j)$ for all i,j.
The identity matrix of order $n\times n$ is generally denoted by I_n.
(vii) *R* is weakly reflexive if $r_{ii}\geq r_{ij}$ for all i,j
(viii) *R* is diagonal if $r_{ii}\geq$ and $r_{ij}=0$ $(i\neq j)$ for all i,j.

If all the entries of a matrix are 0 (respectively 1) then we denote it by *O* (respectively *U*). Throughout the paper we assume that $A=[a_{ij}]$, $B=[b_{ij}]$, $C=[c_{ij}]$ and $D=[d_{ij}]$.

3 SOME RESULTS

The matrices $A^{[k]}$ .and $[k]A$ converge to null matrix O and the matrix U respectively. These results are proved in the following.

Property 3.1 *Let $A=[a_{ij}]$ be a fuzzy matrix of order n×n.*

(i) If $a_{ij}<1$ for all i,j, then $\lim_{k\to\infty} A^{[k]} = O$,

(ii) If $a_{ij}>0$ for all i,j then $\lim_{k\to\infty} [k]A = U$..

Proof. For the fuzzy matrices $A=[a_{ij}]$ and $B=[b_{ij}]$ we have, $A\odot B=[a_{ij}b_{ij}]$. Therefore, $A\odot A = A^{[2]} = [a_{ij}^2]$, $A^{[3]} = A^{[2]}\odot A = [a_{ij}^3]$. In general, for any positive integer k. $A^{[k]} = [a_{ij}^k]$. Hence, $\lim_{k\to\infty} A^{[k]} = O$. Again, $[2]A=A\oplus A=[2a_{ij}- a_{ij}^2]=[1-(1-a_{ij})^2]$. Also, $1-a_{ij}\leq 1$. Therefore, for positive integer k, $[k]A=[1-(1-a_{ij})^k]$ and hence $\lim_{k\to\infty}[k]A = U$ ■

For $0\leq a,b\leq 1$ $a,b\leq a$ and $a,b\leq b$ imply $a,b\leq \min(a,b)$. Therefore, $A\odot B\leq \min\{A,B\}$.

Property 3.2 *Let A and B be two fuzzy matrices, then*

(i) $A\oplus B\geq A\odot B$
(ii) If A and B are symmetric, then $A\oplus B$ and $A\odot B$ are also symmetric
(iii) If A and B are nearly irreflexive, then $A\oplus B$ and $A\odot B$ are nearly irreflexive.

Proof. (i) The ij h element of $A\oplus B$ is $a_{ij}+b_{ij}-a_{ij}b_{ij}$ and that of $A\odot B$.is $a_{ij}b_{ij}$. Assume that $a_{ij}+b_{ij}-a_{ij}b_{ij}\geq a_{ij}b_{ij}$.i.e, $a_{ij}(1-b_{ij})+b_{ij}(1-a_{ij})\geq 0$ which is true as $0\leq a_{ij}\leq 1$.and $0\leq b_{ij}\leq 1$. Hence $A\oplus B \geq A\odot B$.

(ii) Let $A=[a_{ij}]$ and $B=[b_{ij}]$ be two symmetric fuzzy matrices such that $A\oplus B$ and $A\odot B$ are defined. Therefore $a_{ij}=a_{ji}$ and $b_{ij}=b_{ji}$. Let c_{ij} be the ij h element of $A\oplus B$. Then $c_{ij}=a_{ij}+b_{ij}-a_{ij}b_{ij}=a_{ji}+b_{ji}-a_{ji}b_{ji}=c_{ji}$. Hence $A\oplus B$.is symmetric. Again let d_{ij} be the ij h element of $A\odot B$. Then $d_{ij}=a_{ij}b_{ij}=a_{ji}b_{ji}=d_{ji}$. Hence $A\odot B$.is symmetric.
(iii) Since A and B are nearly irreflexive, $a_{ii}\leq a_{ij}$ and $b_{ii}\leq b_{ij}$ for all i,j. Let c_{ij}.and d_{ij} be the ij h elements of $A\oplus B$.and $A\odot B$.respectively. Then $c_{ij}-c_{ii}= (a_{ij}+b_{ij}-a_{ij}b_{ij}) - (a_{ii}+b_{ii}-a_{ii}b_{ii}) = (1-a_{ii})(1-b_{ii}) - (1-a_{ij})(1-b_{ij})\geq 0$ as $1-a_{ii}\geq 1-a_{ij}$.and $1-b_{ii}\geq 1-b_{ij}$ i.e. $c_{ij}\geq c_{ii}$. Hence $A\oplus B$.is irreflexive. Now $d_{ij}-d_{ii} = a_{ij}b_{ij}-a_{ii}b_{ii}\geq 0$ i.e. $d_{ij}\geq d_{ii}$. Hence $A\odot B$ is irreflexive. ■

The operator $\oplus$.is expanding operator while the operator $\odot$.is contracting operator, that is, if the operator $\oplus$.is used between the same matrices then the resultant matrix is larger than the original matrix. The fact is opposite for the operator $\odot$. This fact is justified in the following property.

Property 3.3 *For any fuzzy matrix A.*

(i) $A\oplus B\geq A$ and

(ii) $A \odot A \leq A$.

Proof. (i) The *ij* h element of $A \oplus B$.is $2a_{ij} - a_{ij}^2 = a_{ij} + a_{ij}.(1 - a_{ij}) \geq a_{ij}$. Therefore, $A \oplus A \geq A$.

(ii) The *ij* h element a_{ij}^2 of $A \odot A$ is less than a_{ij}. Therefore, $A \odot A \leq A$. ■

The following results are obvious. The operators $\oplus$ and $\odot$ are commutative as well as associative.

Property 3.4 *Let A. B and C be any three fuzzy matrices. Then*

(i) $A \oplus B = B \oplus A$,
(ii) $(A \oplus B) \oplus C = A \oplus (B \oplus C)$,
(iii) $A \odot B = B \odot A$,
(iv) $(A \odot B) \odot C = A \odot (B \odot C)$,

De Morgan's laws (over transpose) for the operators * and ∘ are

(i) $(A*B)' = A' \circ B'$ and
(ii) $(A \circ B)' = A'*B'$,

where A' is the transpose of A. The operators $\oplus$.and $\odot$ do not obey the De Morgan's laws over transpose.

Property 3.5 *Let A, B, and C be three fuzzy matrices.*

(i) $(A \oplus B)' = A' \oplus B'$,
(ii) $(A \odot B)' = A' \odot B'$,
(iii) If $A \leq B$, then $A \oplus C \leq B \oplus C$ and $A \odot C \leq B \odot C$

Proof. (i) Let c_{ij}.and d_{ij} be the *ij* h elements of $A \oplus B$ and $A' \oplus B'$.respectively. Therefore, $e_{ij}=c_{ji}$ is the *ij* h element of $(A \oplus B)'$. Then $c_{ij}=a_{ij}+b_{ij}-a_{ij}b_{ij}$, $d_{ij}=a_{ji}+b_{ji}-a_{ji}b_{ji}$, $e_{ij}=a_{ji}+b_{ji}-a_{ji}b_{ji}=d_{ij}$. Therefore, $e_{ij}=d_{ij}$.for all *i,j*. Hence $(A \oplus B)' = A' \oplus B'$.

(ii) Let c_{ij} and d_{ij} be the *ij* h elements of $A \odot B$ and $A' \odot B'$ respectively. Naturally, $e_{ij}=c_{ji}$ is the *ij* h element of $(A \odot B)'$. Then $c_{ij}=a_{ij}b_{ij}$. Thus $e_{ij}=a_{ji}b_{ji}$ and $d_{ij}=a_{ji}b_{ji}$. Therefore, $d_{ij}=e_{ij}$ for all *i,j*. Hence $(A \odot B)' = A' \odot B'$.

(iii) Let d_{ij}. e_{ij},f_{ij} and g_{ij} be the *ij*.h elements of $A \oplus C$, $B \oplus C$, $A \odot C$, and $B \odot C$ respectively. Then $d_{ij}= a_{ij}+c_{ij}-a_{ij}c_{ij}$, $e_{ij}= b_{ij}+c_{ij}-b_{ij}c_{ij}$, $f_{ij}= a_{ij}c_{ij}$, $g_{ij}= b_{ij}c_{ij}$. Since, $A \leq B$. $a_{ij} \leq b_{ij}$. Therefore, $a_{ij}(1-c_{ij}) \leq b_{ij}(1-c_{ij})$ or $a_{ij}+c_{ij}-a_{ij}c_{ij} \leq b_{ij}+c_{ij}-b_{ij}c_{ij}$. That is, $d_{ij} \leq e_{ij}$.for all *i,j*. Hence $A \oplus C \leq B \oplus C$. Again, $a_{ij}c_{ij} \leq b_{ij}c_{ij}$. i.e., $f_{ij} \leq g_{ij}$.for all *i,j*. Hence $A \odot C \leq B \odot C$. ■.

Property 3.6 *For any n×n fuzzy matrix A.*

(i) $I_n \oplus (A \oplus A')$ is reflexive and symmetric,
(ii) $A \ominus I_n$ is irreflexive,
(iii) $A \oplus A'$ is nearly irreflexive and symmetric,
(iv) $I_n \oplus (A \oplus A') = I_n \vee (A \oplus A')$.

Proof. (i) $A \oplus A' = [a_{ij} + a_{ji} - a_{ij}.a_{ji}]$ and $I_n \oplus (A \oplus A') = [r_{ij}]$, where r_{ij}=1 and $r_{ij}=a_{ij}+a_{ji}-a_{ij}a_{ji}$ for $i \neq j$. Now, $r_{ji}=a_{ji}+a_{ij}-a_{ij}a_{ji}=r_{ij}$. That is, each diagonal element of $I_n \oplus (A \oplus A')$ is 1 and all non–diagonal elements are $a_{ij}+a_{ji}-a_{ij}a_{ji}$. Therefore, $I_n \oplus (A \oplus A')$.is reflexive and also symmetric.

(ii) The diagonal elements of $A \ominus I_n$ are 0 because $a_{ii} \leq 1$. Hence $A \ominus I_n$ is irreflexive.
(iii) Let $R = A \oplus A'$. i.e., $r_{ij}=a_{ij}+a_{ji}-a_{ij}a_{ji}=r_{ji}$. Therefore, R.is symmetric. Again, $r_{ii} = 2a_{ii} - a_{ii}^2$..Since A.is nearly irreflexive, $a_{ii} \leq a_{ij}$. Therefore, $1-a_{ii} \geq 1-a_{ij}$. Now, $r_{ij}-r_{ii}$ = {1-(1-a_{ij}).(1-a_{ji})} – {1-(1-a_{ii}).(1-a_{ii})} 2em = (1-a_{ii}).(1-a_{ii}) – (1-a_{ij}).(1-a_{ji})≥0. Therefore, $A \oplus A'$.is nearly irreflexive and symmetric.
(iv) As max{1, $a_{ij}+a_{ji}-a_{ij}a_{ji}$}=1 and max{0, $a_{ij}+a_{ji}-a_{ij}a_{ji}$}=$a_{ij}+a_{ji}-a_{ij}a_{ji}$. $I_n \vee (A \oplus A') = [r_{ij}]$ where r_{ii}=1 and $r_{ij}=a_{ij}+a_{ji}-a_{ij}a_{ji}$, for $i \neq j$. Again $I_n \oplus (A \oplus A') = [r_{ij}]$. where r_{ii}=1 and $r_{ij}=a_{ij}+a_{ji}-a_{ij}a_{ji}$, for $i \neq j$. Hence the result. ■.

Property 3.7 *Let A.and B.be two fuzzy matrices, then A⊕B ≥ A∨B ≥ A⊖B.*
Proof. Let c_{ij}. d_{ij} and e_{ij} be the *ij*.h elements of the fuzzy matrices *A⊕B*. *A∨B* and *A⊖B* respectively. Now,

$$c_{ij} = a_{ij} + b_{ij} - a_{ij}.b_{ij} = \begin{cases} a_{ij} + b_{ij}.(1 - a_{ij}) \geq a_{ij} \\ b_{ij} + a_{ij}.(1 - b_{ij}) \geq b_{ij} \end{cases}.$$

d_{ij}=max(a_{ij},b_{ij}) ≤ $a_{ij}+b_{ij}-a_{ij}b_{ij}=c_{ij}$. Thus $c_{ij} \geq d_{ij}$ for all *i,j*. Hence *A⊕B≥A∨B*. Again,

$$e_{ij} = a_{ij} \ominus b_{ij} = \begin{cases} a_{ij}, & a_{ij} > b_{ij} \\ 0, & a_{ij} \leq b_{ij} \end{cases}.$$

That is, $e_{ij} \leq a_{ij} \leq$max(a_{ij},b_{ij}) for all *i,j*. Thus *A⊖B ≤ A∨B*. Finally, *A⊕B ≥ A∨B ≥ A⊖B*. ■.

Property 3.8 *For any two fuzzy matrices A.and B.*

(i) (*A∨B*)∨(*A⊖B*) = *A∨B*,
(ii) (*A∨B*)⊖(*A⊖B*) ≤ *B*,
(iii) (*A⊕B*) ≥ (*A∨B*)∨(*A⊖B*),
(iv) *A⊕B*≥(*A∨B*)⊖(*A⊖B*).

Proof. (i) Let c_{ij}, d_{ij}, and e_{ij} be the *ij*.h elements of *A*∨*B*, *A*⊖*B*, and (*A*∨*B*)∨(*A*⊖*B*) respectively. Then

$$c_{ij} = \max(a_{ij}, b_{ij}), \quad 2emd_{ij} = \begin{cases} a_{ij}, & \text{if } a_{ij} > b_{ij} \\ 0, & \text{if } a_{ij} \le b_{ij} \end{cases}.$$

$$e_{ij} = \begin{cases} max\left(max\left(a_{ij}, b_{ij}\right), a_{ij}\right), & if \ a_{ij} > b_{ij} \\ max\left(max\left(a_{ij}, b_{ij}\right), 0\right), & if \ a_{ij} \le b_{ij} \end{cases} = \begin{cases} a_{ij}, & if \ a_{ij} > b_{ij} \\ b_{ij}, & if \ a_{ij} \le b_{ij}. \end{cases}..$$

That is, *ij*.h element e_{ij} of (*A*∨*B*)∨(*A*⊖*B*) is either a_{ij} or b_{ij} according as a_{ij}> or ≤b_{ij}. Also, the *ij*.h element of c_{ij} is either a_{ij} or b_{ij} according as a_{ij}> or ≤b_{ij}. Therefore, c_{ij}=e_{ij} for all *i,j*. Hence (*A*∨*B*)∨(*A*⊖*B*) = *A*∨*B*.

(ii) Let f_{ij} be the *ij*.h element of (*A*∨*B*)⊖(*A*⊖*B*). Then *ij*.h element c_{ij} of *A*∨*B* is

$$c_{ij} \max = (a_{ij}, b_{ij}) = \begin{cases} a_{ij}, & \text{if } a_{ij} > b_{ij} \\ b_{ij}, & \text{if } a_{ij} \le b_{ij} \end{cases}$$

and the *ij*.h element of *A*⊖*B* is

$$d_{ij} = \begin{cases} a_{ij}, & \text{if } a_{ij} > b_{ij} \\ 0, & \text{if } a_{ij} \le b_{ij} \end{cases}.$$

Therefore,

$$f_{ij} = \begin{cases} 0, & \text{if } a_{ij} > b_{ij} \\ b_{ij}, & \text{if } a_{ij} \le b_{ij} \end{cases}.$$

That is the elements of (*A*∨*B*)⊖(*A*⊖*B*) are either 0 or b_{ij}. Hence (*A*∨*B*)⊖(*A*⊖*B*)≤*B*.

(iii) (*A*∨*B*)∨(*A*⊖*B*) = *A*∨*B*≤*A*⊕*B* (by property 2.3.7).
(iv) It is obvious that *B*≤*A*∨*B*.

Hence (*A*∨*B*)⊖(*A*⊖*B*)≤ *B*≤*A*∨*B*≤*A*⊕*B*. ■.

Property 3.9 *Let A, B, and C be three fuzzy matrices, then*

(i) *A*⊕(*B*∨*C*) = (*A*⊕*B*)∨(*A*⊕*C*),
(ii) *A*⊕(*B*⊖*C*) ≥ (*A*⊕*B*)⊖(*A*⊕*C*),
(iii) *A*⊖(*B*⊕*C*) ≤ (*A*⊖*B*)⊕(*A*⊖*C*),
(iv) *A*⊖(*B*∨*C*) ≤ (*A*⊖*B*)∨(*A*⊖*C*),

(v) $A\vee(B\oplus C) \leq (A\vee B)\oplus(A\vee C)$,
(vi) $A\vee(B\ominus C) \geq (A\vee B)\ominus(A\vee C)$.

4 RESULTS ON α CUT OF FUZZY MATRIX

The upper α cut fuzzy matrix is basically a Boolean fuzzy matrix. It represents only two states 0 and 1. But, lower α cut fuzzy matrix is a multi-graded fuzzy matrix. When the elements of this matrix are less than α then lower α cut fuzzy matrix represents same state 0 and in all other cases it represents the actual states.

Some results on α cuts of matrices are presented below:

Property 4.1 *For any two fuzzy matrices A and B.*

(i) $(A\ominus B)^{(\alpha)}\geq A^{(\alpha)}\ominus B^{(\alpha)}$,
(ii) $(A\vee B)^{(\alpha)}=A^{(\alpha)}\vee B^{(\alpha)}$,
(iii) $(A\oplus B)^{(\alpha)}\geq A^{(\alpha)}\oplus B^{(\alpha)}$.

Proof. (i) Let $C=A\ominus B$, $E=A^{(\alpha)}\ominus B^{(\alpha)}$ and $D=C^{(\alpha)}$.

Case 1. $a_{ij}\geq b_{ij}$. Therefore, $c_{ij}=a_{ij}$ and $d_{ij} = a_{ij}^{(\alpha)}$.

Subcase 1.1. $a_{ij}\geq b_{ij}\geq\alpha$.Then $d_{ij}=1$ and $a_{ij}^{(\alpha)} = 1, b_{ij}^{(\alpha)} = 1$ so $e_{ij}=0$. That is, $d_{ij}>e_{ij}$.

Subcase 1.2. $a_{ij}\geq\alpha\geq b_{ij}$. Also, in this case $d_{ij} = 1, a_{ij}^{(\alpha)} = 1, b_{ij}^{(\alpha)} = 0$ so $e_{ij}=1$. Therefore, $d_{ij}=e_{ij}$.

Subcase 1.3. $\alpha\geq a_{ij}\geq b_{ij}$. Here, $d_{ij} = 0, a_{ij}^{(\alpha)} = b_{ij}^{(\alpha)} = 0$ and hence $e_{ij}=0$. That is, $d_{ij}=e_{ij}$. Hence, $d_{ij}\geq e_{ij}$ for all i,j. Thus, $(A\oplus B)^{(\alpha)}\geq A^{(\alpha)}\ominus B^{(\alpha)}$.

Case 2. $b_{ij}>a_{ij}$. In this case, $c_{ij}=0$ and $d_{ij}=0$.

Subcase 2.1. $b_{ij}\geq a_{ij}\geq\alpha$. In this case, $b_{ij}^{(\alpha)} = 1$ and $a_{ij}^{(\alpha)} = 1$, so $e_{ij}=0$.

Subcase 2.2. $b_{ij}\geq\alpha\geq a_{ij}$. Then, $b_{ij}^{(\alpha)} = 1$ and $a_{ij}^{(\alpha)} = 0$. Therefore, $e_{ij}=0$.

Subcase 2.3. $\alpha\geq b_{ij}\geq a_{ij}$. Here, $a_{ij}^{(\alpha)} = b_{ij}^{(\alpha)} = 0$, $e_{ij}=0$.

Therefore, $d_{ij}=e_{ij}$ when $b_{ij}\geq a_{ij}$ and hence $(A\ominus B)^{(\alpha)}=A^{(\alpha)}\ominus B^{(\alpha)}$. Finally, $(A\ominus B)^{(\alpha)}\geq A^{(\alpha)}\ominus B^{(\alpha)}$. Other proofs are similar.

Property 4.2 *For the fuzzy matrices A and B.*

(i) $(A\oplus B)_{(\alpha)}\geq A_{(\alpha)}\oplus B_{(\alpha)}$,
(ii) $(A\ominus B)_{(\alpha)}=A_{(\alpha)}\ominus B_{(\alpha)}$,
(iii) $(A\vee B)_{(\alpha)}=A_{(\alpha)}\vee B_{(\alpha)}$.

Proof. (i) $(A\oplus B)_{(\alpha)}=[(a_{ij}+b_{ij}-a_{ij}b_{ij})_{(\alpha)}]$ and $A_{(\alpha)}\oplus B_{(\alpha)}=\left[a_{ij(\alpha)} + b_{ij(\alpha)} - a_{ij(\alpha)}.b_{ij(\alpha)}\right]$..

Case 1. $a_{ij}, b_{ij} \geq \alpha$. Therefore

$$(a_{ij} \oplus b_{ij})_{(\alpha)} = (a_{ij} + b_{ij} - a_{ij}.b_{ij})_{(\alpha)} = a_{ij} + b_{ij} - a_{ij}.b_{ij}.$$

as

$$a_{ij} + b_{ij} - a_{ij}.b_{ij} = a_{ij} + b_{ij}.(1 - a_{ij}) \geq a_{ij} \geq \alpha.$$

Again,

$$a_{ij(\alpha)} \oplus b_{ij(\alpha)} = a_{ij(\alpha)} + b_{ij(\alpha)} - a_{ij(\alpha)}.b_{ij(\alpha)} = a_{ij} + b_{ij} - a_{ij}.b_{ij}$$

as $a_{ij}, b_{ij} \geq \alpha$. Thus in this case $(A \oplus B)_{(\alpha)} = A_{(\alpha)} \oplus B_{(\alpha)}$.

Case 2. $a_{ij}, b_{ij} < \alpha$. Here,

$$(a_{ij} \oplus b_{ij})_{(\alpha)} = (a_{ij} + b_{ij} - a_{ij}.b_{ij})_{(\alpha)} = a_{ij} + b_{ij} - a_{ij}.b_{ij}$$

or 0 according as $a_{ij} + b_{ij} - a_{ij}b_{ij} \geq \alpha$ or $< \alpha$. But, $a_{ij(\alpha)} \oplus b_{ij(\alpha)} = 0 \oplus 0 = 0$. Thus $(A \oplus B)_{(\alpha)} \geq A_{(\alpha)} \oplus B_{(\alpha)}$.

Case 3. $a_{ij} < \alpha$ and $b_{ij} \geq \alpha$. Then

$$(a_{ij} \oplus b_{ij})_{(\alpha)} = (a_{ij} + b_{ij} - a_{ij}.b_{ij})_{(\alpha)} = a_{ij} + b_{ij} - a_{ij}.b_{ij},.$$

because,

$$a_{ij} + b_{ij} - a_{ij}.b_{ij} = b_{ij} + a_{ij}.(1 - b_{ij}) \geq b_{ij} \geq \alpha.$$

Again,

$$a_{ij(\alpha)} \oplus b_{ij(\alpha)} = a_{ij(\alpha)} + b_{ij(\alpha)} - a_{ij(\alpha)}.b_{ij(\alpha)} = 0 + b_{ij} - 0 = b_{ij}.$$

Therefore, $(A \oplus B)_{(\alpha)} \geq A_{(\alpha)} \oplus B_{(\alpha)}$. Hence finally, $(A \oplus B)_{(\alpha)} \geq A_{(\alpha)} \oplus B_{(\alpha)}$..Other proofs are similar.

5 RESULTS ON COMPLEMENT OF FUZZY MATRIX

The complement of a fuzzy matrix is used to analyse the complement nature of any system. For example, if A represents the crowdness of a network at a particular time period then its complement A^c represents the clearness of this network at the same time period. Using the following results, we can study the complement nature of a system with the help of original fuzzy matrix.

The operator complement obey the De Morgan's laws for the operator $\oplus$ and $\odot$.

Property 5.1 *For the fuzzy matrices A and B.*

(i) $(A\oplus B)^c = A^c\odot B^c$,
(ii) $(A\odot B)^c = A^c\oplus B^c$,
(iii) $(A\oplus B)^c \leq A^c\oplus B^c$,
(iv) $(A\odot B)^c \geq A^c\odot B^c$,

Proof. (i) $A^c\oplus B^c = [1\text{-}a_{ij}b_{ij}]$ and $A\odot B = [a_{ij}b_{ij}]$. Then $(A\odot B)^c = [1\text{-}a_{ij}b_{ij}]$. Therefore, $(A\odot B)^c = A^c\oplus B^c$.

(ii) $(A\oplus B)^c = [1\text{-}(a_{ij}+b_{ij}\text{-}a_{ij}b_{ij})]$, $A^c=[1\text{-}a_{ij}]$, and $B^c=[1\text{-}b_{ij}]$. Now, $A^c\odot B^c = [1\text{-}(a_{ij}+b_{ij}\text{-}a_{ij}b_{ij})]$. Hence, $(A\oplus B)^c = A^c\odot B^c$.
(iii) From property 2.3.2(i), $A\oplus B\geq A\odot B$, we have $(A\oplus B)^c \leq (A\odot B)^c\ A^c\oplus B^c$.
(iv) $(A\odot B)^c \geq (A\oplus B)^c = A^c\odot B^c$. ■.

DISTANCE BETWEEN FUZZY MATRICES

In this section, we define two basic distances between FMs. Let M be the set of fuzzy matrices and R be the set of real numbers. The distance δ between FM A, $B\in M$ is a mapping from the set M to the set R i.e., δ: $M\times M\rightarrow R$. The two basic distances – Hamming distance and Euclidean distance are investigated here. The definitions of these two distances are given below.

6.1 Hamming Distance

The Hamming distance between two FMs A and B of order $m\times n$ is

$$H(A,B) = \sum_{i=1}^{m}\sum_{j=1}^{n}\left|a_{ij} - b_{ij}\right|..$$

It is obvious that $0\leq H(A,B)\leq mn$.

Lemma 6.1 *If $a_{ij}=1$ and $b_{ij}=0$ for all i,j or $a_{ij}=0$ and $b_{ij}=1$ for all i,j then $H(A,B)=mn$.*

Lemma 6.2 *If $a_{ij}=b_{ij}$ for all i,j then $H(A,B)=0$.*

The Hamming distance H: $M \times M \rightarrow R$ satisfies the following conditions:

(i) $H(A,B) \geq 0$ for all $A,B \in M$ (non-negativity),
(ii) $H(A,B)=0$ iff $A=B$ for all $A,B \in M$,
(iii) $H(A,B)=H(B,A)$, $A,B \in M$ (symmetry),
(iv) $H(A,B) \leq H(A,C)+H(C,B)$ for all $A,B,C \in M$ (triangular inequality).

Thus the distance function H is a metric on M.

Normalized Hamming Distance

The normalized Hamming distance between two FMs A and B is denoted by $H^*(A,B)$ and is defined as

$$H^*\left(A,B\right) = \frac{H\left(A,B\right)}{m.n}..$$

It may be noted that $0 \leq H^*(A,B) \leq 1$.

6.2 Euclidean Distance

The Euclidean distance between two FMs A and B of order $m \times n$ is defined as

$$E\left(A,B\right) = \sqrt{\sum_{i=1}^{m}\sum_{i=1}^{n}(a_{ij} - b_{ij})^2}\ .$$

It is also obvious that $0 \leq E\left(A,B\right) \leq \sqrt{m.n}$. The Euclidean distance E is also a metric on M.

Normalized Euclidean Distance

The Normalized Euclidean distance between two FMs A and B is denoted by $E^*(A,B)$.and is defined as

$$E^*\left(A,B\right) = \frac{E\left(A,B\right)}{\sqrt{m.n}}..$$

Here also $0 \leq E^*(A,B) \leq 1$. The maximum Hamming distance between two Boolean fuzzy matrices A and B of order $m \times n$ is mn. It is possible only when one of the following conditions hold, $a_{ij}=1$ and $b_{ij}=0$ or $a_{ij}=0$ and $b_{ij}=1$. Again if $a_{ij}=b_{ij}$ then distance will be zero, it is the minimum distance between two matrices.

Depending on the values of $H^*(A,B)$ or $E^*(A,B)$ we can define the following terms.

(i) If $H^*(A,B)=1$ or $E^*(A,B)=1$ then A and B are totally opposite.

(ii) If $H^*(A,B)=0.5$ or $E^*(A,B)=0.5$ then A and B are nearly equal.
(iii) If $H^*(A,B)=0.25$ or $E^*(A,B)=0.25$ then A and B are almost equal.
(iv) If $H^*(A,B)=0$ or $E^*(A,B)=0$ then A and B are exactly equal.

7 PROPERTIES OF DISTANCE

A large number of properties for Hamming distance and Euclidean distance are investigated in this section, which are presented below.

Property 7.1 *For any fuzzy matrix A,*

(i) $H(A,A^{[2]})\leq H(A,A^{[k]})$; $k=2,3,\ldots$
(ii) $H(A,[2]A)\leq H(A,[k]A)$; $k=2,3,\ldots$
(iii) $E(A,A^{[2]})\leq E(A,A^{[k]})$; $k=2,3,\ldots$
(iv) $E(A,[2]A)\leq E(A,[k]A)$; $k=2,3,\ldots$
(v) $H(A,A^{[2]})= H(A,[2]A)$
(vi) $E(A,A^{[2]})= E(A,[2]A)$

Proof.(i) Let $A=[a_{ij}]$. Then, $A^{[2]} = A \odot A = \left[a_{ij}^2\right]$. For any positive integer k. $A^{[k]} = \left[a_{ij}^k\right]$. Since $0\leq a_{ij}\leq 1$ then, $a_{ij}^2 \geq a_{ij}^3 \geq \ldots \geq a_{ij}^k$ for any positive integer k. Hence, $H(A,A^{[2]})\leq H(A,A^{[k]})$; $k=2,3,\ldots$.

(ii) Let $A=[a_{ij}]$.Then, $[2]A = A \oplus A =\left[2a_{ij} - a_{ij}^2\right]=\left[1-\left(1-a_{ij}\right)^2\right]$. For any positive integer k. we have $[k]A=[1-(1-a_{ij})^k]$. Now,

$$H\left(A,[2]A\right) = \sum_{i=1}^{n}\sum_{j=1}^{n} | a_{ij} - 1 + (1-a_{ij})^2 |$$

and

$$H\left(A,[k]A\right) = \sum_{i=1}^{n}\sum_{j=1}^{n} | a_{ij} - 1 + (1-a_{ij})^k |.$$

Since, $1-a-(1-a)^2\leq 1-a-(1-a)^k$ for $k\geq 2$. Therefore, $|a-1+(1-a)^2|\geq|a-1+(1-a)^k|$. Since, $0\leq a_{ij}\leq 1$ we have, $(1-a_{ij})^2 \geq (1-a_{ij})^k$. for any positive integer k. Hence, $H(A,[2]A)\leq H(A,[k]A)$.

Proofs of (iii), (iv) are similar to (i).

(v) $H\left(A, A^{[2]}\right)=\sum_{i=1}^{n}\sum_{j=1}^{n}\left|a_{ij}-a_{ij}^{2}\right|$ and $H\left(A,[2]A\right)=\sum_{i=1}^{n}\sum_{j=1}^{n}\mid a_{ij}-1+(1-a_{ij})^{2}\mid$. We have, $(a-1)+(1-a)^2=a(a-1)$ and $a-a^2=a(1-a)$.Therefore, $\mid a_{ij}-a_{ij}^{2}\mid=\mid a_{ij}-1+(1-a_{ij})^{2}\mid$. Hence, $H(A,A^{[2]})=H(A,[2]A)$.

Proof of (vi) is similar to (v). ■.

Property 7.2 *Let I_n be the identity fuzzy matrix of order n×n and $A=[a_{ij}]$ be any FM of the same order. Then,*

(i) I_n^c and $A\ominus I_n$ are irreflexive *FM*s.
(ii) $I_n \vee A$ and $I_n \oplus A$ are reflexive *FM*s.
(iii) $I_n \wedge A$, $I_n \odot A$ and $I_n \$ A$ are diagonal *FM*s.
(iv) $I_n @ A$ is weakly reflexive *FM*.
(v) $I_n \oplus \left(A \oplus A'\right)$ is reflexive and symmetric *FM*.

Property 7.3 *Let I_n be the identity FM of order n×n and $A=[a_{ij}]$ be any FM of the same order. Then,*

(i) $H(I_n \odot A, I_n \wedge A) = 0$.
(ii) $H(I_n \oplus A, I_n \vee A) = 0$.
(iii) $H(I_n \odot A, I_n \$ A) \leq n$.
(iv) $H(I_n, I_n \odot A) \leq n$.
(v) $H(I_n, I_n \wedge A) \leq n$.
(vi) $H(I_n, I_n \$ A) \leq n$.
(vii) $H\left(I_n, I_n^c\right) = n^2$.
(viii) $H\left(I_n \oplus \left(A \oplus A'\right), I_n \oplus A\right) \leq n^2 - n$.
(ix) $H\left(I_n^c, A \ominus I_n\right) \leq n^2 - n$.
(x) $H(I_n, I_n \oplus A) \leq n^2 - n$.
(xi) $H(I_n, I_n \vee A) \leq n^2 - n$.

Proof. Proofs of (i) and (ii) are trivial.

(iii) Since both the FMs $I_n \odot A$ and $I_n \$ A$ are diagonal, then all the non-diagonal elements of both the FMs are same, in particular they are zero. The diagonal elements of $I_n \odot A$ and $I_n \$ A$ are different. Here the number of diagonal elements is *n*. Hence, $H(I_n \odot A, I_n \$ A) \leq n$. Again if all the diagonal elements of *A* are either 1 or 0, then all the diagonal elements of $I_n \odot A$ and $I_n \$ A$ are same, they are either 1 or 0 respectively. Therefore, $H(I_n \odot A, I_n \$ A)=0$.

The proofs of (iv), (v) and (vi) are similar to (iii).

(vii) In this case I_n is identity FM of order $n \times n$ then it is obvious that all the diagonal elements of I_n are 1 and all the non-diagonal elements are 0. I_n^c is irreflexive FM of the same order whose all the diagonal elements are 0 all the non-diagonal elements are 1. Hence, $H\left(I_n, I_n^c\right) = n^2$.

(viii) Here all the diagonal elements ('n in number) of $I_n \oplus \left(A \oplus A'\right)$ and $I_n \bigoplus A$ are same, they are 1, since both the FMs $I_n \oplus \left(A \oplus A'\right)$ and $I_n \bigoplus A$ are reflexive FM of order $n \times n$. But the non-diagonal elements of both the FM are different from each other. The number of non-diagonal elements of FM $I_n \oplus \left(A \oplus A'\right)$ and $I_n \bigoplus A$ are n^2-n.

Therefore, $H\left(I_n \oplus \left(A \oplus A'\right), I_n \oplus A\right) \le n^2 - n$. The proofs of (ix), (x) and (xi) are similar to (viii).

■

Property 7.4 *For any two fuzzy matrices A=[a_{ij}] and B=[b_{ij}] of the same order,*

(i) $H(A,A \wedge B) = H(B,A \vee B)$.
(ii) $H(A,A \vee B) = H(B,A \wedge B)$.
(iii) $H(A,A \bigoplus B) = H(B,A \bigodot B)$.
(iv) $H(A,A \bigodot B) = H(B,A \bigoplus B)$.
(v) $H(A \bigoplus B, A \vee B) = H(A \bigodot B, A \wedge B)$.
(vi) $H(A,B) = H(A \vee B, A \wedge B)$.
(vii) $H(A \bigodot B, A \vee B) = H(A \bigoplus B, A \wedge B)$.
(viii) $H(A,A@B) = H(B,A@B)$.
(ix) $H(A@B, A \vee B) = H(A@B, A \wedge B)$.
(x) $H(A \bigoplus B, A@B) = H(A \bigodot B, A@B)$.
(xi) $H(A, A@B) = H(A@B, A \wedge B)$.
(xii) $H(A^c \bigoplus B^c, (A \bigoplus B)^c) = H(A^c \bigodot B^c, (A \bigodot B)^c)$.
(xiii) $H(A^c \bigoplus B^c, (A \bigodot B)^c) = H(A^c \bigoplus B^c, (A \bigodot B)^c)$.

Proof. (vii) We have,

$$H\left(A \odot B, A \vee B\right) = \sum_{i=1}^{n}\sum_{j=1}^{n}\left|a_{ij}.b_{ij} - max\left\{a_{ij}, b_{ij}\right\}\right|$$

and

$$H\left(A \oplus B, A \wedge B\right) = \sum_{i=1}^{n}\sum_{j=1}^{n}\left|a_{ij} + b_{ij} - a_{ij}.b_{ij} - max\left\{a_{ij}, b_{ij}\right\}\right|.$$

Two cases will arise:

Case I. When $a_{ij} \geq b_{ij}$ then

$$H\left(A \odot B, A \vee B\right) = \sum_{i=1}^{n}\sum_{j=1}^{n}\left|a_{ij}.b_{ij} - a_{ij}\right| \text{ and } H\left(A \oplus B, A \wedge B\right) = \sum_{i=1}^{n}\sum_{j=1}^{n}\left|a_{ij} - a_{ij}.b_{ij}\right|.$$

Case II. When $b_{ij} \geq a_{ij}$ then

$$H\left(A \odot B, A \vee B\right) = \sum_{i=1}^{n}\sum_{j=1}^{n}\left|a_{ij}.b_{ij} - b_{ij}\right| \text{ and } H\left(A \oplus B, A \wedge B\right) = \sum_{i=1}^{n}\sum_{j=1}^{n}\left|b_{ij} - a_{ij}.b_{ij}\right|.$$

Hence, from both the cases we have, $H(A \odot B, A \vee B) = H(A \oplus B, A \wedge B)$.

(xiii) (xiii) We have $A^c=[1-a_{ij}]$ and $B^c=[1-b_{ij}]$. Similarly, $(A \odot B)^c=[1-a_{ij}b_{ij}]$ and $(A \oplus B)^c= [1-(a_{ij}+b_{ij}-a_{ij}b_{ij})]$. Now

$$\begin{aligned} H(A^c \oplus B^c, \left(A \odot B\right)^c\right) &= \sum_{i=1}^{n}\sum_{j=1}^{n}\left|1 - a_{ij} + 1 - b_{ij} - \left(1 - a_{ij}\right).\left(1 - b_{ij}\right) - 1 + a_{ij}.b_{ij}\right| \\ &= \sum_{i=1}^{n}\sum_{j=1}^{n}\left|1 - \left(a_{ij} + b_{ij} - a_{ij}.b_{ij}\right) - \left(1 - a_{ij}\right).\left(1 - b_{ij}\right)\right|. \end{aligned}$$

Again,

$$H(\left(A \oplus B\right)^c, A^c \odot B^c\right) = \sum_{i=1}^{n}\sum_{j=1}^{n}\left|1 - \left(a_{ij} + b_{ij} - a_{ij}.b_{ij}\right) - \left(1 - a_{ij}\right).\left(1 - b_{ij}\right)\right|.$$

Hence, $H(A^c \oplus B^c, (A \odot B)^c) = H(A^c \oplus B^c, (A \odot B)^c)$. Remaining proofs are same as (vii). ■

Property 7.5 *For any two fuzzy matrices $A=[a_{ij}]$ and $B=[b_{ij}]$ of the same order,*

(i) $E(A,A \wedge B) = E(B,A \vee B)$.
(ii) $E(A,A \vee B) = E(B,A \wedge B)$.
(iii) $E(A,A \oplus B) = E(B,A \odot B)$.
(iv) $E(A,A \odot B) = E(B,A \oplus B)$.
(v) $E(A \oplus B, A \vee B) = E(A \odot B, A \wedge B)$.
(vi) $E(A,B) = E(A \vee B, A \wedge B)$.
(vii) $E(A \odot B, A \vee B) = E(A \oplus B, A \wedge B)$.
(viii) $E(A,A@B) = E(B,A@B)$.
(ix) $E(A@B, A \vee B) = E(A@B, A \wedge B)$.
(x) $E(A \oplus B, A@B) = E(A \odot B, A@B)$.
(xi) $E(A, A@B) = E(A@B, A \wedge B)$.

(xii) $E(A^c \bigoplus B^c, (A \bigoplus B)^c) = E(A^c \bigodot B^c, (A \bigodot B)^c)$.
(xiii) $E(A^c \bigoplus B^c, (A \bigodot B)^c) = E(A^c \bigodot B^c, (A \bigodot B)^c)$.

Property 7.6 *For any two fuzzy matrices* $A=[a_{ij}]$ *and* $B=[b_{ij}]$ *of same order,*

(i) $H(A, A \bigodot B) \geq H(A, A \wedge B)$.
(ii) $H(A, A \bigodot B) \geq H(A \bigoplus B, A \vee B)$.
(iii) $H(A, A \bigoplus B) \geq H(A, A \vee B)$.
(iv) $H(A, A \bigoplus B) \geq H(A \bigoplus B, A \vee B)$.
(v) $H(A \bigodot B, A \vee B) \geq H(A \vee B, A \wedge B)$.
(vi) $H(A \vee B, A \wedge B) \geq H(A @ B, A \vee B)$.
(vii) $H(A \bigoplus B, A @ B) \geq H(A @ B, A \vee B)$.
(viii) $H(A \bigodot B, A \vee B) \geq H(A \bigoplus B, A @ B)$.
(ix) $H(A \bigoplus B, A \bigodot B) \geq H(A \bigoplus B, A @ B)$.
(x) $H(A \bigoplus B, A \wedge B) \geq H(A \bigoplus B, A \vee B)$.
(xi) $H(A \bigoplus B, A \bigodot B) \geq H(A \bigoplus B, A \vee B)$.
(xii) $H(A \bigodot B, A \vee B) \geq H(A \bigoplus B, A \vee B)$.
(xiii) $H(A \bigoplus B, A \wedge B) \geq H(A \vee B, A @ B)$.
(xiv) $H(A \vee B, A \$ B) \geq H(A \vee B, A \wedge B)$.
(xv) $H(A \vee B, A \$ B) \geq H(A \wedge B, A \$ B)$.
(xvi) $H(A \bigoplus B, A \$ B) \geq H(A \vee B, A \$ B)$.
(xvii) $H(A \bigoplus B, A \wedge B) \geq H(A \vee B, A \wedge B)$.
(xviii) $H(A \bigodot B, A \bigoplus B) \geq H(A \bigodot B, A \vee B)$.

Proof. Proofs of (i) to (iv) are trivial.

(v) We have,

$$H\left(A \odot B, A \vee B\right) = \sum_{i=1}^{n}\sum_{j=1}^{n}\left|a_{ij}.b_{ij} - max\left\{a_{ij}, b_{ij}\right\}\right|$$

and

$$H\left(A \vee B, A \wedge B\right) = \sum_{i=1}^{n}\sum_{j=1}^{n}\left|max\left\{a_{ij}, b_{ij}\right\} - min\left\{a_{ij}, b_{ij}\right\}\right|.$$

Two cases will arise.

Case I. When $a_{ij} \geq b_{ij}$ then

$$H\left(A \odot B, A \vee B\right) = \sum_{i=1}^{n}\sum_{j=1}^{n}\left|a_{ij}.b_{ij} - a_{ij}\right| \text{ and } H\left(A \vee B, A \wedge B\right) = \sum_{i=1}^{n}\sum_{j=1}^{n}\left|a_{ij} - b_{ij}\right|.$$

Since $b_{ij} \geq a_{ij}b_{ij}$ as $0 \leq a_{ij}, b_{ij} \leq 1$ therefore $|a_{ij}b_{ij} - a_{ij}| \geq |a_{ij} - b_{ij}|$. Hence, $H(A \odot B, A \vee B) \geq H(A \vee B, A \wedge B)$.

Case II. When $b_{ij} \geq a_{ij}$ we have

$$H\left(A \odot B, A \vee B\right) = \sum_{i=1}^{n}\sum_{j=1}^{n}\left|a_{ij}.b_{ij} - b_{ij}\right| \text{ and } H\left(A \vee B, A \wedge B\right) = \sum_{i=1}^{n}\sum_{j=1}^{n}\left|b_{ij} - a_{ij}\right|.$$

Hence, from both the cases we have, $H(A \odot B, A \vee B) \geq H(A \vee B, A \wedge B)$. All the remaining proofs are similar to (v). ■

Property 7.7 *For any two fuzzy matrices* $A=[a_{ij}]$ *and* $B=[b_{ij}]$ *of same order,*

(i) $E(A, A \odot B) \geq E(A, A \wedge B)$.
(ii) $E(A, A \odot B) \geq E(A \oplus B, A \vee B)$.
(iii) $E(A, A \oplus B) \geq E(A, A \vee B)$.
(iv) $E(A, A \oplus B) \geq E(A \oplus B, A \vee B)$.
(v) $E(A \odot B, A \vee B) \geq E(A \vee B, A \wedge B)$.
(vi) $E(A \vee B, A \wedge B) \geq E(A @ B, A \vee B)$.
(vii) $E(A \oplus B, A @ B) \geq E(A @ B, A \vee B)$.
(viii) $E(A \odot B, A \vee B) \geq E(A \oplus B, A @ B)$.
(ix) $E(A \oplus B, A \odot B) \geq E(A \oplus B, A @ B)$.
(x) $E(A \oplus B, A \wedge B) \geq E(A \oplus B, A \vee B)$.
(xi) $E(A \oplus B, A \odot B) \geq E(A \oplus B, A \vee B)$.
(xii) $E(A \odot B, A \vee B) \geq E(A \oplus B, A \vee B)$.
(xiii) $E(A \oplus B, A \wedge B) \geq E(A \vee B, A @ B)$.
(xiv) $E(A \vee B, A \$ B) \geq E(A \vee B, A \wedge B)$.
(xv) $E(A \vee B, A \$ B) \geq E(A \wedge B, A \$ B)$.
(xvi) $E(A \oplus B, A \$ B) \geq E(A \vee B, A \$ B)$.
(xvii) $E(A \oplus B, A \wedge B) \geq E(A \vee B, A \wedge B)$.
(xviii) $E(A \odot B, A \oplus B) \geq E(A \odot B, A \vee B)$.

Property 7.8 *For any two fuzzy matrices* $A=[a_{ij}]$ *and* $B=[b_{ij}]$ *of the same order,*

(i) $H(\left(A \wedge B\right) \wedge \left(A \ominus B\right), \left(A \wedge B, \vee \left(A \ominus B\right)\right) \geq H\left(\left(A \wedge B\right) \wedge \left(A \ominus B\right), \left(A \vee B\right) \vee \left(A \ominus B\right)\right)$.
(ii) $H(\left(A \wedge B\right) \wedge \left(A \ominus B\right), \left(A \vee B, \vee \left(A \ominus B\right)\right) \geq H\left(\left(A \wedge B\right) \wedge \left(A \ominus B\right), \left(A \vee B\right) \wedge \left(A \ominus B\right)\right)$.
(iii) $H(\left(A \wedge B\right) \wedge \left(A \ominus B\right), \left(A \vee B, \vee \left(A \ominus B\right)\right) \geq H\left(\left(A \vee B\right) \wedge \left(A \ominus B\right), \left(A \wedge B\right) \vee \left(A \ominus B\right)\right)$.

(iv) $H((A \vee B) \vee (A \ominus B), (A \vee B, \wedge (A \ominus B)) \geq H((A \vee B) \wedge (A \ominus B), (A \wedge B) \vee (A \ominus B))$.

(v) $H((A \vee B) \vee (A \ominus B), (A \vee B, \wedge (A \ominus B)) \geq H((A \vee B) \vee (A \ominus B), (A \wedge B) \vee (A \ominus B))$.

Property 7.9 *For any two fuzzy matrices A=*$[a_{ij}]$ *and B=*$[b_{ij}]$ *of the same order,*

(i) $E(A \wedge B) \wedge (A \ominus B), (A \wedge B, \vee (A \ominus B)) \geq E((A \wedge B) \wedge (A \ominus B), (A \vee B) \vee (A \ominus B))$.

(ii) $E((A \wedge B) \wedge (A \ominus B), (A \vee B, \vee (A \ominus B)) \geq E((A \wedge B) \wedge (A \ominus B), (A \vee B) \wedge (A \ominus B))$.

(iii) $E((A \wedge B) \wedge (A \ominus B), (A \vee B, \vee (A \ominus B)) \geq E((A \vee B) \wedge (A \ominus B), (A \wedge B) \vee (A \ominus B))$.

(iv) $E((A \vee B) \vee (A \ominus B), (A \vee B, \wedge (A \ominus B)) \geq E((A \vee B) \wedge (A \ominus B), (A \wedge B) \vee (A \ominus B))$.

(v) $E((A \vee B) \vee (A \ominus B), (A \vee B, \wedge (A \ominus B)) \geq E((A \vee B) \vee (A \ominus B), (A \wedge B) \vee (A \ominus B))$.

8 APPLICATION TO CHARACTER RECOGNITION

Several conventional methods are available to recognize the characters. Here we discuss about the recognition of a character (printed only) by the help of Hamming distance. The recognition of hand written characters is a challenging task for computer scientist due to the large variability of style of the hand written characters. Specially, in the case of hand written characters variation of shape occurs due to writing habit, style, mood, etc., which differs from man to man and also depends on some other factors such as the writing instruments, writing surface, etc. Now we shall illustrate how character recognition is possible by using distance function. There are 52 fixed printed English alphabets in binary matrix form including upper case and lower case alphabets. Now a printed manuscript of same font, same size and shape (length, width etc.) is taken to serve the purpose. A scanner is placed on the top of the printed manuscript and scanner scans each of the character. As a scanning process goes on, the scanner will draw a horizontal scan line when it will find black pixels of the character at the time of its top-down movement from the top of the manuscript. It will draw another horizontal scan line i.e., parallel to the first horizontal line at the fag-end of that character. After completion of downward movement, the scanner will start its horizontal movement and it will draw two parallel vertical scan lines touching at least one black pixel in the left and right side of the same character. Thus we form a minimum sized rectangular box called *bounding box* which is needed to enclose the character. The black pixels of the character within the bounding box are filled up by 1 and white pixels are filled up by 0. It is nothing but forming the binary matrix of the character. Now we measure the Hamming distance between the obtained binary matrix and the previous stored binary matrix representations of 52 given characters. For which specific binary matrix of 52 former given characters the Hamming distance is zero, the scanned character will be that specific character. In this case, there are some limitations. The characters would have same font, same size and shape. Here, for example, we consider two printed characters, say, 'h' and 'g'. Every pixel of both the characters (tracing path of the character) is filled up by 1 and the other pixels (background of the character) is filled up by 0. The Boolean matrices of these two characters are shown in Figure 4. The Boolean matrices for the characters 'h' and 'g' are respectively denoted by X and Y.

Figure 4. Binary matrices (a) for 'h' and (b) for 'g'

(a)	(b)
111100000000000000	000001111111000111
011100000000000000	000111111111100111
011100000000000000	001111111111110111
011100000000000000	011111100000111111
011100000000000000	011110000000001111
011100000000000000	111110000000001111
011100000000000000	111100000000001111
011100000000000000	111100000000000111
011100000000000000	111100000000000111
011100000000000000	111100000000000111
011100011110000000	111100000000000111
011101111111111000	111100000000001111
011111111111111100	011110000000001111
011111000001111100	011110000000011111
011110000000011110	011111000000111111
011100000000001110	001111111111110111
011100000000001110	000011111111100111
011100000000001110	000000111100000111
011100000000001110	000000000000000111
011100000000001110	000000000000000111
011100000000001110	011100000000000111
011100000000001110	011100000000000111
011100000000001110	011100000000001110
011100000000001110	001110000000011110
011100000000001110	001111000000111110

The Hamming distance between the matrices X and Y is, $H(X,Y)=215$. This implies that the characters 'h' and 'g' for the matrices X and Y are not same, i.e., they are different from each other. But $H(X,X)=0$ and $H(Y,Y)=0$. Here Hamming distance in both the cases are zero. These imply that both the characters are same in each of the cases. Same characters being of same font, same size and shape, the Boolean matrix representation will be exactly same. Hence always Hamming distance will be zero, if we measure the distance between the same characters.

REFERENCES

Atanassov, K., & Gargov, G. (1989). Interval valued intuitionistic fuzzy sets. *Fuzzy Sets and Systems, 31*(3), 343–349. doi:10.1016/0165-0114(89)90205-4

Atanassov, K. T. (1986). Intuitionistic fuzzy sets. *Fuzzy Sets and Systems, 20*(1), 87–96.

Bourke, M. M., & Fisher, D. G. (1996). Convergence, eigen fuzzy sets and stability analysis of relational matrices. *Fuzzy Sets and Systems, 81*(2), 227–234. doi:10.1016/0165-0114(95)00212-X

Buckley, J. J., & Qu, Y. (1991). Solving systems of linear fuzzy equations. *Fuzzy Sets and Systems, 43*(1), 33–43. doi:10.1016/0165-0114(91)90019-M

Dubois, D., & Prade, H. (1997). *Fuzzy sets and systems: Theory and applications*. Orlando, FL: Academic Press.

Elsner, L., & van den Driessche, P. (1999). On the power method in max algebra. *Linear Algebra and Its Applications, 302–303*, 17–32. doi:10.1016/S0024-3795(98)10171-4

Fan, Z.-T. (1999). A note on the power sequence of a fuzzy matrix. *Fuzzy Sets and Systems, 102*(2), 281–286. doi:10.1016/S0165-0114(97)00118-8

Fan, Z. T., & Liu, D.-F. (1997). Convergence of the power sequence of a nearly monotone increasing fuzzy matrix. *Fuzzy Sets and Systems, 88*(3), 363–372. doi:10.1016/S0165-0114(96)00040-1

Fan, Z.-T., & Liu, D.-F. (1998a). A detailed study on the power sequence of matrices of commonly used types. *Fuzzy Sets and Systems, 99*(2), 197–203. doi:10.1016/S0165-0114(97)00025-0

Fan, Z.-T., & Liu, D.-F. (1998b). On the oscillating power sequence of a fuzzy matrix. *Fuzzy Sets and Systems, 93*(1), 75–85. doi:10.1016/S0165-0114(96)00194-7

Goguen, J. A. (1985). Fuzzy Sets and Systems: Theory and Applications (Didier Dubois and Henri Prade). *SIAM Review, 27*(2), 270–274. doi:10.1137/1027081

Guu, S.-M., Chen, H.-H., & Pang, C.-T. (2001). Convergence of products of fuzzy matrices. *Fuzzy Sets and Systems, 121*(2), 203–207. doi:10.1016/S0165-0114(00)00006-3

Hansen, E. (1980). Global optimization using interval analysis? the multi-dimensional case. *Numerische Mathematik, 34*(3), 247–270. doi:10.1007/BF01396702

Hashimoto, H. (1983a). Canonical form of a transitive fuzzy matrix. *Fuzzy Sets and Systems, 11*(1–3), 157–162. doi:10.1016/S0165-0114(83)80076-1

Hashimoto, H. (1983b). Convergence of powers of a fuzzy transitive matrix. *Fuzzy Sets and Systems, 9*(1–3), 153–160. doi:10.1016/S0165-0114(83)80015-3

Hashimoto, H. (1984). Subinverses of fuzzy matrices. *Fuzzy Sets and Systems, 12*(2), 155–168. doi:10.1016/0165-0114(84)90035-6

Hashimoto, H. (1985). Transitivity of generalized fuzzy matrices. *Fuzzy Sets and Systems, 17*(1), 83–90. doi:10.1016/0165-0114(85)90009-0

Hemasinha, R., Pal, N. R., & Bezdek, J. C. (1993). Iterates of fuzzy circulant matrices. *Fuzzy Sets and Systems*, *60*(2), 199–206. doi:10.1016/0165-0114(93)90346-J

Imai, H., Okahara, Y., & Miyakoshi, M. (2000). The period of powers of a fuzzy matrix. *Fuzzy Sets and Systems*, *109*(3), 405–408. doi:10.1016/S0165-0114(98)00080-3

Jian-Xin, L. (1992). Controllable fuzzy matrices. *Fuzzy Sets and Systems*, *45*(3), 313–319. doi:10.1016/0165-0114(92)90149-X

Jian-Xin, L. (1994). Convergence of powers of controllable fuzzy matrices. *Fuzzy Sets and Systems*, *62*(1), 83–88. doi:10.1016/0165-0114(94)90074-4

Kauffmann, A. (1965). *Introduction to the Theory of Fuzzy Subsets*. New York: Academic Press.

Kim, J. B. (1988). Determinant theory for fuzzy and boolean matrices. *Congressus Numerantium*, 273–276.

Kim, J. B., Baartmans, A., & Sahadin, N. S. (1989). Determinant theory for fuzzy matrices. *Fuzzy Sets and Systems*, *29*(3), 349–356. doi:10.1016/0165-0114(89)90046-8

Kim, K. H. (1982). *Boolean matrix theory and applications*. New York: Dekker.

Kim, K. H., & Roush, F. W. (1980). Generalized fuzzy matrices. *Fuzzy Sets and Systems*, *4*(3), 293–315. doi:10.1016/0165-0114(80)90016-0

Kim, K. H., & Roush, F. W. (1987). Fuzzy Matrix Theory. In *Analysis of Fuzzy Information*. CRC Press.

Kołodziejczyk, W. (1987). Canonical form of a strongly transitive fuzzy matrix. *Fuzzy Sets and Systems*, *22*(3), 297–302. doi:10.1016/0165-0114(87)90073-X

Kołodziejczyk, W. (1988). Convergence of powers of s-transitive fuzzy matrices. *Fuzzy Sets and Systems*, *26*(1), 127–130. doi:10.1016/0165-0114(88)90011-5

Lee, H.-S. (2001). An optimal algorithm for computing the max–min transitive closure of a fuzzy similarity matrix. *Fuzzy Sets and Systems*, *123*(1), 129–136. doi:10.1016/S0165-0114(00)00062-2

Li, J.-X. (1989). Some results on the nilpotent fuzzy matrices. *Fuzzy Systems Math.*, *1*, 52–55.

Li, J.-X. (1992). Controllable fuzzy matrices. *Fuzzy Sets and Systems*, *45*(3), 313–319. doi:10.1016/0165-0114(92)90149-X

Li, J.-X. (1994). Convergence of powers of controllable fuzzy matrices. *Fuzzy Sets and Systems*, *62*(1), 83–88. doi:10.1016/0165-0114(94)90074-4

Liu, W., & Ji, Z. (2002). The periodicity of square fuzzy matrices based on minimal strong components. *Fuzzy Sets and Systems*, *126*(2), 233–240. doi:10.1016/S0165-0114(01)00033-1

Mishref, M. A., & Emam, E. G. (1992). Transitivity and subinverses in fuzzy matrices. *Fuzzy Sets and Systems*, *52*(3), 337–343. doi:10.1016/0165-0114(92)90241-U

Mondal, S., & Pal, M. (2013a). Similarity Relations, Invertibility and Eigenvalues of Intuitoinistic Fuzzy Matrix. *Fuzzy Information and Engineering*, *5*(4), 431–443. doi:10.100712543-013-0156-y

Mondal, S., & Pal, M. (2013b). Soft matrices. *Journal of Uncertain Systems*, *7*(4), 254–264.

Mondal, S., & Pal, M. (2014). Intuitionistic fuzzy incline matrix and determinant. *Annals of Fuzzy Mathematics and Informatics*, *8*(1), 19–32.

Mondal, S., & Pal, M. (2016a). Rank of interval-valued fuzzy matrices. *Afrika Matematika*, *27*(1–2), 97–114. doi:10.100713370-015-0325-8

Mondal, S., & Pal, M. (2016b). Similarity relations, eigenvalues and eigenvectors of bipolar fuzzy matrix. *Journal of Intelligent & Fuzzy Systems*, *30*(4), 2297–2307. doi:10.3233/IFS-152000

Moore, R. E., & Bierbaum, F. (1979). *Methods and applications of interval analysis*. Philadelphia: Siam.

Pal, Khan, & Shyamal. (2002). Intuitionistic fuzzy matrices. *Notes on Intuitionistic Fuzzy Sets*, *8*(2), 51–56.

Pal, M. (2001). Intuitionistic fuzzy determinant. *V. U. J. Physical Sciences*, *7*, 87–93.

Pal, M. (2015a). Fuzzy matrices with fuzzy rows and columns. *Journal of Intelligent & Fuzzy Systems*, *30*(1), 561–573. doi:10.3233/IFS-151780

Pal, M. (2015b). Interval-valued Fuzzy Matrices with Interval-valued Fuzzy Rows and Columns. *Fuzzy Information and Engineering*, *7*(3), 335–368. doi:10.1016/j.fiae.2015.09.006

Pal, M., & Mondal, S. (2019). Bipolar fuzzy matrices. *Soft Computing*. doi:10.100700500-019-03912-9

Pandey, P., Kumar, S., & Shrivastava, S. (2014). A unified strategy for forecasting of a new product. *Decision (Washington, D.C.)*, *41*(4), 411–424.

Pradhan, R., & Pal, M. (2013). Convergence of maxgeneralized mean-min generalized mean powers of intuitionistic fuzzy matrices. *The Journal of Fuzzy Mathematics*, *22*(2), 477–492.

Ragab, M. Z., & Emam, E. G. (1995). On the min-max composition of fuzzy matrices. *Fuzzy Sets and Systems*, *75*(1), 83–92. doi:10.1016/0165-0114(94)00332-2

Ragab, M. Z., & Eman, E. G. (1994). The determinant and adjoint of a square fuzzy matrix. *Fuzzy Sets and Systems*, *61*(3), 297–307. doi:10.1016/0165-0114(94)90172-4

Sengupta, A., & Pal, T. K. (2000). On comparing interval numbers. *European Journal of Operational Research*, *127*(1), 28–43.

Shyamal, A. K., & M. Pal. (2001). Distance between interval-valued intuitionistic fuzzy sets. *Busefal*, *86*(1).

Shyamal, A. K., & Pal, M. (2002a). Distances between intuitionistics fuzzy matrices. *V. U. J. Physical Sciences*, *8*, 81–91.

Shyamal, A. K., & Pal, M. (2002b). Interval-valued fuzzy matrices. In *Proc. International Conference on 'Analysis and Discrete Structures'*, (pp. 22–24). Kharagpur, India: Indian Institute of Technology.

Shyamal, A. K., & Pal, M. (2004). Two new operators on fuzzy matrices. *Journal of Applied Mathematics and Computing*, *15*(1–2), 91–107. doi:10.1007/BF02935748

Shyamal, A. K., & Pal, M. (2004a). Matrices of interval numbers with applications. In *Proc. of National Seminar on ``Recent Advances in Applied Mathematics''*, (pp. 26–31). Midnapore, India: Vidyasagar University.

Shyamal, A. K., & Pal, M. (2004b). Triangular fuzzy matrices. In *Proc. `Fuzzy Logic and Its Applications in Technology and Management'*. Kharagpur, India: Indian Institute of Technology.

Shyamal, A. K., & Pal, M. (2005). Distance between fuzzy matrices and its applications-I. *Journal of Natural and Physical Sciences*, *19*(1), 38–51.

Szmidt, E., & Kacprzyk, J. (2000). Distances between intuitionistic fuzzy sets. *Fuzzy Sets and Systems*, *114*(3), 505–518. doi:10.1016/S0165-0114(98)00244-9

Thomason, M. G. (1977). Convergence of powers of a fuzzy matrix. *Journal of Mathematical Analysis and Applications*, *57*(2), 476–480. doi:10.1016/0022-247X(77)90274-8

Xuecheng, L. (1992). Entropy, distance measure and similarity measure of fuzzy sets and their relations. *Fuzzy Sets and Systems*, *52*(3), 305–318.

Zadeh, Fu, Tanaka, & Shimura. (Eds.). (1975). Fuzzy sets and their applications to cognitive and decision processes: Papers. New York: Academic Press.

Zadeh, L. A. (1965). Fuzzy sets. *Information and Control*, *8*(5), 338–353.

Zadeh, L. A. (1971). Similarity relations and fuzzy orderings. *Information Sciences*, *3*, 177–200.

Zhu, R. (1984). Convergence of powers of a fuzzy matrix. *Fuzzy Math.*, *3*, 45–49.

Chapter 2
Solutions of Fuzzy System of Linear Equations

Laxminarayan Sahoo
Raniganj Girls' College, India

ABSTRACT

This chapter deals with solution methodology of fuzzy system of linear equations (FSLEs). In fuzzy set theory, finding solutions of FLSEs has long been a well-known problem to the researchers. In this chapter, the fuzzy number has been converted into interval number, and the authors have solved the interval system of linear equation for finding the fuzzy valued solution. Here, a fuzzy valued linear system has been introduced and a numerical example has been solved and presented for illustration of purpose.

1 INTRODUCTION

System of linear equations plays an important role in many areas such as economics, physics, engineering, statistics and management science also. Recently, in many application oriented problems some or all of the system parameters and their measures are considered using imprecise valued rather than precise valued (Rahgooy, Yazdi, & Monsefi, 2009). To overcome the problem with imprecise measured generally stochastic, fuzzy and fuzzy-stochastic approaches are used. In stochastic approach, the parameters are assumed as random variables with known probability distribution. In fuzzy approach, the parameters are considered as fuzzy sets (Zadeh, 1965) with known membership functions or fuzzy numbers (Sahoo, Bhunia, &. Kapur, 2012). On the other hand, in fuzzy-stochastic approach, some parameters are viewed as fuzzy sets and other as random variables. Here, we have represented imprecise parameters as a fuzzy number and fuzzy number has been converted to interval number as interval representation is the best representation among others (Hansen, & Walster, 2004, Moore, 1979, Moore, 2009). In this chapter, we have considered fuzzy system of linear equations (Jahantigh, Khezerloo, & Khezerloo, 2010, Otadi, & Mosleh, 2012, Rahgooy, Yazdi, & Monsefi, 2009). For this purpose, we have introduced interval system of linear equations for solving fuzzy system of linear equations and its solution methodology has been presented. This proposed method has been illustrated with a numerical example.

DOI: 10.4018/978-1-7998-0190-0.ch002

2 SOME MATHEMATICAL BACKGROUND

In this section some definitions and basic concepts related to fuzzy sets, fuzzy numbers and interval numbers used in this chapter are described briefly.

2.1 Preliminary Definitions About Fuzzy Set

A fuzzy set $\tilde{A}$ is defined by a membership function $\mu_{\tilde{A}}(x)$, which corresponds to each element x in X to a real number in the interval $0 \leq x \leq 1$. The function $\mu_{\tilde{A}}(x)$ represents the grade of membership of x in $\tilde{A}$.

Definition 2.1 The α-cut (Zimmermann, 2001) of a fuzzy set $\tilde{A}$ is a crisp subset of X and is denoted by $\tilde{A}_\alpha = \{x \in X : \mu_{\tilde{A}}(x) \geq \alpha\}$, where $\mu_{\tilde{A}}(x)$ is the membership function of $\tilde{A}$ and $0 \leq \alpha \leq 1$.

Definition 2.2 A fuzzy set $\tilde{A}$ is called a normal fuzzy set if there exists at least one $x \in X$ such that $\mu_{\tilde{A}}(x) = 1$.

Definition 2.3 A fuzzy set $\tilde{A}$ is called convex iff for every pair of $x_1, x_2 \in X$, the membership function of $\tilde{A}$ satisfies $\mu_{\tilde{A}}(\lambda x_1 + (1-\lambda)x_2) \geq \min\{\mu_{\tilde{A}}(x_1), \mu_{\tilde{A}}(x_2)\}$, where $0 \leq \lambda \leq 1$.

Definition 2.4 A fuzzy number $\tilde{A}$ is a fuzzy set which is both convex and normal.

Definition 2.5 The triangular fuzzy number (TFN) is a normal fuzzy number denoted as $\tilde{A} = (a_1, a_2, a_3)$ where $a_1 \leq a_2 \leq a_3$ and its membership function $\mu_{\tilde{A}}(x): X \to [0,1]$ is defined by

$$\mu_{\tilde{A}}(x) = \begin{cases} \dfrac{x - a_1}{a_2 - a_1} & \text{if } a_1 \leq x \leq a_2 \\ 1 & \text{if } x = a_2 \\ \dfrac{a_3 - x}{a_3 - a_2} & \text{if } a_2 \leq x \leq a_3 \end{cases}$$

2. 2. α Level Interval of Fuzzy Number

Let $\tilde{A}$ be a fuzzy number, then α Level Interval of Fuzzy number or α-cut is $[A^-(\alpha), A^+(\alpha)]$ and the nearest interval approximation (Grzegorzewski, 2002) of $\tilde{A}$ is $[\int_0^1 A^-(\alpha)d\alpha, \int_0^1 A^+(\alpha)d\alpha]$.

Definition 2.6 Let $\tilde{A} = (a_1, a_2, a_3)$ is a triangular fuzzy number. The α-level interval of $\tilde{A}$ is $[a_1 + (a_2 - a_1)\alpha,\ a_3 - (a_3 - a_2)\alpha]$.

2.3. Finite Interval Arithmetic

Interval arithmetic was introduced by Moore (Moore, 1979, Moore, Kearfott, & Cloud, 2009). In interval arithmetic, an imprecise variable a is represented as an interval number denoted by $\bar{A} = [a^-, a^+]$ and is defined by $\bar{A} = [a^-, a^+] = \{a : a^- \leq a \leq a^+, a \in \mathbb{R}\}$, where a^- and a^+ are the lower and upper limits respectively and $\mathbb{R}$ is the set of all real numbers. Actually, every real number can be treated as an interval, such as for all $a \in \mathbb{R}$, a can be written as an interval with the form $[a,a]$. The basic arithmetical operations of interval numbers $\bar{A} = [a^-,a^+]$ and $\bar{B} = [b^-,b^+]$ are defined as

$$\bar{A} + \bar{B} = [a^- + b^-, a^+ + b^+], \quad \bar{A} - \bar{B} = [a^- - b^+, a^+ - b^-]$$

For any scalar λ,

$$\lambda\bar{A} = \begin{cases} [\lambda a^-, \lambda a^+] & \text{for } \lambda \geq 0 \\ [\lambda a^+, \lambda a^-] & \text{for } \lambda < 0 \end{cases},$$

$$\bar{A} \times \bar{B} = [\min(a^- b^-, a^- b^+, a^+ b^-, a^+ b^+), \ \max(a^- b^-, a^- b^+, a^+ b^-, a^+ b^+)],$$

$$\frac{\bar{B}}{\bar{A}} = \bar{B} \times \frac{1}{\bar{A}} = [b^-, b^+] \times [\frac{1}{a^+}, \frac{1}{a^-}],$$

provided $0 \notin [a^-,a^+]$.

For more details one may refer to the works of Hansen and Walster (Hansen, & Walster, 2004).

3. PROPOSED METHOD FOR SOLVING FUZZY SYSTEM OF LINEAR EQUATIONS (FSLE)

Let us considered a linear system as follows:

$$AX = \bar{b} \tag{1}$$

where

$$A = \begin{pmatrix} a_{11} & a_{12} & \cdots & a_{1n} \\ a_{21} & a_{22} & \cdots & a_{2n} \\ \vdots & \vdots & \vdots & \vdots \\ a_{n1} & a_{n2} & \cdots & a_{nn} \end{pmatrix}$$

is a matrix of order $n \times n$ whose elements are crisp/precise valued and

$$\bar{b} = \begin{pmatrix} [b_1^-, b_1^+] \\ [b_2^-, b_2^+] \\ \vdots \\ [b_n^-, b_n^+] \end{pmatrix}$$

is a matrix of order $n \times 1$ whose elements are interval valued.

Then equation (1) is called an interval linear system where A is crisp matrix and $\bar{b}$ is an interval matrix. Obviously $X = \bar{X}$ is also an interval matrix and our aim is to find out such $\bar{X}^*$ which satisfy the equation (1).

Now, transform equation (1) into $2n \times 2n$ matrix as follows:

$$\begin{aligned} &c_{11}x_1^- + \ldots + c_{1n}x_n^- + c_{1(n+1)}(-x_1^+) + \ldots + c_{1(n+n)}(-x_n^+) = b_1^- \\ &\vdots \\ &c_{n1}x_1^- + \ldots + c_{nn}x_n^- + c_{n(n+1)}(-x_1^+) + \ldots + c_{n(n+n)}(-x_n^+) = b_n^- \\ &c_{(n+1)1}x_1^- + \ldots + c_{(n+1)n}x_n^- + c_{(n+1)(n+1)}(-x_1^+) + \ldots + c_{(n+1)(n+n)}(-x_n^+) = -b_1^+ \\ &\vdots \\ &c_{(n+n)1}x_1^- + \ldots + c_{(n+n)n}x_n^- + c_{(n+n)(n+1)}(-x_1^+) + \ldots + c_{(n+n)(n+n)}(-x_n^+) = -b_n^+ \end{aligned} \tag{2}$$

where

$$c_{ij} = c_{(i+n)(j+n)} = a_{ij}, \quad \text{if } a_{ij} \geq 0 \text{ and } c_{i(j+n)} = c_{(i+n)j} = -a_{ij}, \quad \text{if } a_{ij} < 0 \tag{3}$$

and any c_{ij} which is not determined by equation (3) is set as zero. Using matrix notation, equation (2) can be written as

$$CX_{new} = b_{new} \tag{4}$$

where

$$C = \begin{pmatrix} c_{11} & c_{12} & \cdots & c_{1(n+n)} \\ c_{21} & c_{22} & \cdots & c_{2(n+n)} \\ \vdots & \vdots & \vdots & \vdots \\ c_{(n+n)1} & c_{n2} & \cdots & c_{(n+n)(n+n)} \end{pmatrix}, \; X_{new} = \begin{pmatrix} x_1^- \\ x_2^- \\ \vdots \\ x_n^- \\ -x_1^+ \\ -x_2^+ \\ \vdots \\ -x_n^+ \end{pmatrix} \text{ and } b_{new} = \begin{pmatrix} b_1^- \\ b_2^- \\ \vdots \\ b_n^- \\ -b_1^+ \\ -b_2^+ \\ \vdots \\ -b_n^+ \end{pmatrix}$$

The structure of C in equation (4) indicates that c_{ij}, $1 \leq i, j \leq n$, and that

$$C = \begin{pmatrix} U & V \\ V & U \end{pmatrix}$$

where U contains the positive entries of A, V the absolute values of the negative entries of A and $A = U - V$. Now we find out C^{-1} (if exist) and obtain the following:

$$X_{new} = C^{-1} b_{new} \tag{5}$$

Theorem 3.1 If C^{-1} exists it must have the same structure as C, i.e.

$$C^{-1} = \begin{pmatrix} E & F \\ F & E \end{pmatrix}$$

Theorem 3.2 The unique solution X^*_{new} of interval linear system (4) is a interval vector iff the inverse matrix of C exists and nonnegative.

4. NUMERICAL EXAMPLE AND ILLUSTRATION

To illustrate the proposed method we have considered the following system of linear equations.

$$10\tilde{x}_1 - 4\tilde{x}_2 = (20, 22, 25) - 4\tilde{x}_1 + 16\tilde{x}_2 = (25, 28, 30) \tag{6}$$

Now, equation (6) can be written in interval form using α-cut as follows:

$$10[x_1^-, x_1^+] - 4[x_2^-, x_2^+] = [20 + 2\alpha, 25 - 3\alpha] - 4[x_1^-, x_1^+] + 16[x_2^-, x_2^+] = [25 + 3\alpha, 30 - 2\alpha] \tag{7}$$

To solve equation (7), we have used equation (3) and we obtained the following:

$$\begin{pmatrix} 10 & 0 & 0 & 4 \\ 0 & 16 & 4 & 0 \\ 0 & 4 & 10 & 0 \\ 4 & 0 & 0 & 16 \end{pmatrix} \begin{pmatrix} x_1^- \\ x_2^- \\ -x_1^+ \\ -x_2^+ \end{pmatrix} = \begin{pmatrix} 20 + 2\alpha \\ 25 + 3\alpha \\ -25 + 3\alpha \\ -30 + 2\alpha \end{pmatrix} \tag{8}$$

Solving equation (8) we get

Table 1. Solutions of fuzzy system of linear equation for different values of α

α	$\tilde{x}_1 = [x_1^-, x_1^+]$	$\tilde{x}_2 = [x_2^-, x_2^+]$
0.0	[3.055556, 3.472222]	[2.430556, 2.638889]
0.1	[3.072222, 3.447222]	[2.443056, 2.630556]
0.2	[3.088889, 3.422222]	[2.455556, 2.622222]
0.3	[3.105556, 3.397222]	[2.468056, 2.613889]
0.4	[3.122222, 3.372222]	[2.480556, 2.605556]
0.5	[3.138889, 3.347222]	[2.493056, 2.597222]
0.6	[3.155556, 3.322222]	[2.505556, 2.588889]
0.7	[3.172222, 3.297222]	[2.518056, 2.580556]
0.8	[3.188889, 3.272222]	[2.530556, 2.572222]
0.9	[3.205556, 3.247222]	[2.543056, 2.563889]
1.0	[3.222222, 3.222222]	[2.555556, 2.555556]

$$x_1^- = \frac{110 + 6\alpha}{36},\ x_1^+ = \frac{125 - 9\alpha}{36},\ x_2^- = \frac{350 + 18\alpha}{144} \text{ and } x_2^+ = \frac{380 - 12\alpha}{144}.$$

So,

$$\tilde{x}_1 = [\frac{110 + 6\alpha}{36}, \frac{125 - 9\alpha}{36}] \text{ and } \tilde{x}_2 = [\frac{350 + 18\alpha}{144}, \frac{380 - 12\alpha}{144}].$$

For different values of α the solutions has been shown in Table 1. From Table 1, it follows that in case of α=1, the optimum solution is précised/crisp valued. To realize the case, we have also plotted the graphs and these have been shown in Figure 1 and Figure 2.

From these two graphs, it is seen that when α=1, we get crisp solutions and for other values we get fuzzy solutions in term of interval numbers.

5. CONCLUSION

In this chapter, we have solved fuzzy system of linear equations and its solution methodology has been presented. For this purpose, we have introduced fuzzy system of linear equations and its solution methodology has been discussed with a numerical example. From numerical result, it is clear that the optimum result is fuzzy valued in terms of interval numbers. Finally it can be concluded that the entire methodologies attempted in this chapter well to solve any problems arising in mathematics, engineering, management science etc.with fuzzy/imprecise parameters.

Figure 1. Different values of $\tilde{x}_1$ for different values of α

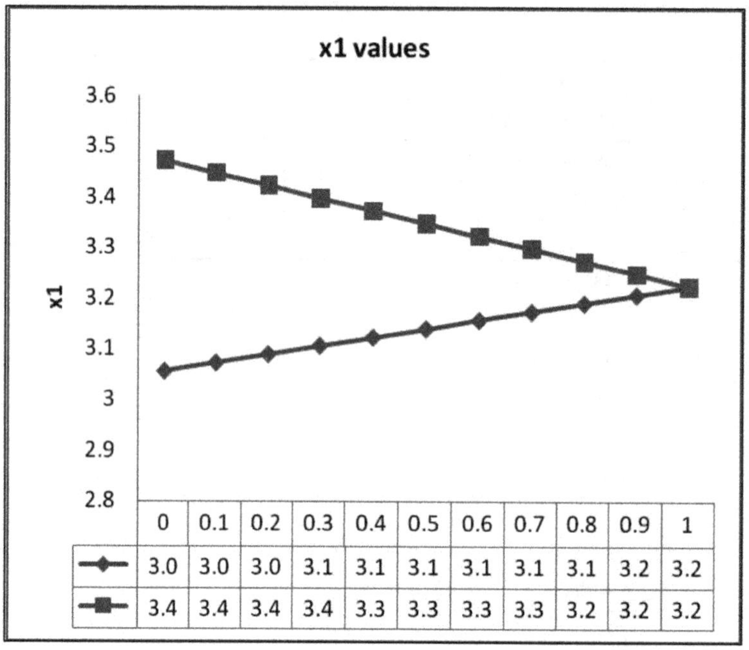

	0	0.1	0.2	0.3	0.4	0.5	0.6	0.7	0.8	0.9	1
◆	3.0	3.0	3.0	3.1	3.1	3.1	3.1	3.1	3.1	3.2	3.2
■	3.4	3.4	3.4	3.4	3.3	3.3	3.3	3.3	3.2	3.2	3.2

Figure 2. Different values of $\tilde{x}_2$ for different values of α

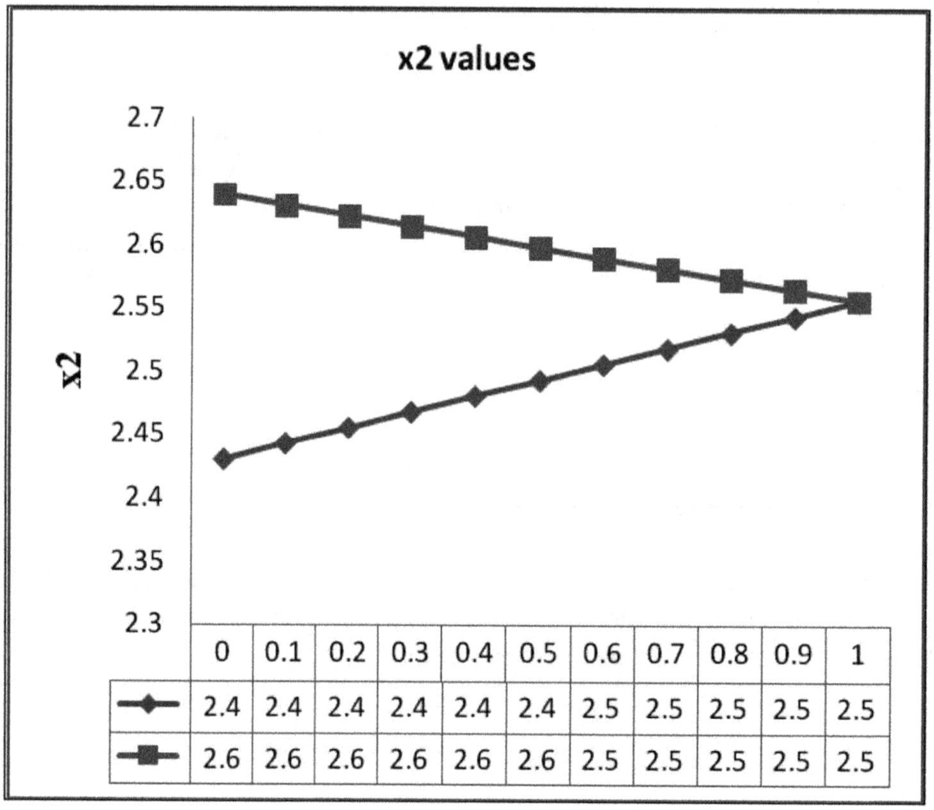

	0	0.1	0.2	0.3	0.4	0.5	0.6	0.7	0.8	0.9	1
◆	2.4	2.4	2.4	2.4	2.4	2.4	2.5	2.5	2.5	2.5	2.5
■	2.6	2.6	2.6	2.6	2.6	2.6	2.5	2.5	2.5	2.5	2.5

REFERENCES

Grzegorzewski, P. (2002). Nearest interval approximation of a fuzzy number. *Fuzzy Sets and Systems*, *1320*, 321–330.

Hansen, E. R., & Walster, G. W. (2004). *Global optimization using interval analysis*. New York: Marcel Dekker Inc.

Jahantigh, M.A.,Khezerloo, S., & Khezerloo, M. (2010). Complex fuzzy Linear Systems. *Int. Journal of Industrial Mathematics*, *2*(1), 21-28.

Moore, R. E. (1979). *Methods and applications of interval analysis*. Philadelphia: SIAM.

Moore, R. E., Kearfott, R. B., & Cloud, M. J. (2009). *Introduction to interval analysis*. Philadelphia: SIAM.

Otadi, M., & Mosleh, M. (2012). Solution of fuzzy matrix equation System. *International Journal of Mathematics and Mathematical Sciences*, *2012*, 1–8.

Rahgooy, T., Yazdi, H.S., & Monsefi, R. (2009). Fuzzy complex system of linear equations applied to circuit Analysis. *International Journal of Computer and Electrical Engineering*, *1*(5), 535-541.

Sahoo, L., Bhunia, A. K., & Kapur, P. K. (2012). Genetic algorithm based multi-objective reliability optimization in interval environment. *Computers & Industrial Engineering*, *62*(1), 152–160.

Zadeh, L. A. (1965). Fuzzy sets. *Information and Control*, *8*(3), 338–352.

Zimmermann, H. J. (2001). Fuzzy Set Theory-and its Applications (4th ed.). Springer Science+ Business Media, LLC.

Chapter 3
Some Group Theoretic Notions in Fuzzy Multigroup Context

Paul Augustine Ejegwa
University of Agriculture, Nigeria

ABSTRACT

The notion of fuzzy multigroups is an application of fuzzy multisets or fuzzy bags to classical group theory. This chapter explores the notions of fuzzy comultisets, characteristic fuzzy submultigroups, and direct product of fuzzy multigroups as extensions of cosets, characteristic subgroups, and direct product of groups. The relationship between fuzzy comultisets of a fuzzy multigroup and the cosets of a group is established. Some results on the concept of fuzzy comultisets are deduced. A number of properties of characteristic fuzzy submultigroups of fuzzy multigroups are outlined, and some related results are obtained. Also, the author presents some properties of direct product of fuzzy multigroups and establish some results.

1 INTRODUCTION

The concept of fuzzy sets was proposed by Zadeh (1965) as a method for representing imprecision in real-world situations. The concept of fuzzy set has grown stupendously over the years giving birth to fuzzy group proposed in (Rosenfeld, 1971). Some analogous concepts and results which were conventional in group theory have been established in fuzzy groups since inception. See (Ajmal and Prajapati, 1992; Bhattacharya and Mukherjee, 1987; Kim, 2003; Kumar, 1992; Mordeson *et al.,* 2005; Mukherjee and Bhattacharya, 1984; Mukherjee and Bhattacharya, 1986; Onasanya and Ilori, 2013; Ray, 1999; Seselja and Tepavcevic, 1997; Sidky and Mishref, 1991) for details.

Motivated by Zadeh (1965), the idea of fuzzy multisets was introduced in (Yager, 1986) as a generalization of fuzzy sets in multisets framework. See (Biswas, 1999; Ejegwa, 2015; Miyamoto, 1996; Miyamoto and Mizutani, 2004; Mizutani *et al.,* 2008; Syropoulos, 2012) for some details on fuzzy multisets. Recently, Shinoj *et al.* (2015) followed the foot steps of Rosenfeld (1971) and introduced a non-classical group called fuzzy multigroup. The concept of fuzzy multigroups constitutes an application of fuzzy multisets to the notion of group.

DOI: 10.4018/978-1-7998-0190-0.ch003

Baby *et al.* (2015) continued the algebraic study of fuzzy multisets by proposing the idea of abelian fuzzy multigroups. The work in (Ejegwa, 2018a), which was built on (Shinoj *et al.*, 2015), introduced the concept of fuzzy multigroupoids and presented the idea of fuzzy submultigroups with a number of results. Some properties of abelian fuzzy multigroups were explicated in (Ejegwa, 2018b) with some results. The concept of normal fuzzy submultigroups of fuzzy multigroups was proposed in (Ejegwa, 2018c) with some number of results. In (Ejegwa, 2018d), the notion of homomorphism in the context of fuzzy multigroups was defined and some homomorphic properties of fuzzy multigroups were presented.

Since the concept of fuzzy multigroups is relatively noveltic, we seek to extend some notions hitherto established in group theory to fuzzy multigroups for the enrichment of fuzzy multigroup theory. Sequel, we explore the notions of fuzzy comultisets, characteristic fuzzy submultigroups and direct product of fuzzy multigroups. By organization, the chapter is thus presented: Section 2 provides some preliminaries on fuzzy multisets and fuzzy multigroups. In Section 3, we study the idea of fuzzy comultisets and discuss some of its properties. Section 4 discusses characteristic fuzzy submultigroups of fuzzy multigroups. In Section 5, we explicate the concept of direct product of fuzzy multigroups with some results. Finally, Section 6 concludes the paper and provides direction for future studies.

2 PRELIMINARIES

In this section, we review some existing definitions and results for the sake of completeness and reference.

Definition 2.1 Assume X is a set of elements. Then a fuzzy bag/multiset A drwan from X can be characterized by a count membership function CM_A such that $CM_A: X \to Q$, where Q is the set of all crisp bags or multisets from the unit interval $I=[0,1]$.

A fuzzy multiset can also be characterized by a high-order function. In particular, a fuzzy multiset A can be characterized by a function $CM_A: X \to N^I$ or $CM_A: X \to [0,1] \to N$, where $I=[0,1]$ and $N = \mathbb{N} \cup \{0\}$.

It implies that $CM_A(x)$ for $x \in X$ is given as

$$CM_A(x) = \{\mu_A^1(x), \mu_A^2(x), \ldots, \mu_A^n(x), \ldots\},$$

where

$$\mu_A^1(x), \mu_A^2(x), \ldots, \mu_A^n(x), \ldots \in [0,1]$$

such that

$$\mu_A^1(x) \geq \mu_A^2(x) \geq \ldots \geq \mu_A^n(x) \geq \ldots,$$

whereas in a finite case, we write

$$CM_A(x) = \{\mu_A^1(x), \mu_A^2(x), ..., \mu_A^n(x)\},$$

for $\mu_A^1(x) \geq \mu_A^2(x) \geq ... \geq \mu_A^n(x)$.

A fuzzy multiset A can be represented in the form

$$A = \{\langle \frac{CM_A(x)}{x} \rangle \mid x \in X\} \text{ or } A = \{\langle x, CM_A(x) \rangle \mid x \in X\}.$$

In a simple term, a fuzzy multiset A of X is characterized by the count membership function $CM_A(x)$ for $x \in X$, that takes the value of a multiset of a unit interval $I=[0,1]$.

For example, let $X=\{a,b,c\}$ be a set. Then, a fuzzy multiset of X is given as

$$A = \left\{ \langle \frac{0.5, 0.4, 0.3}{a} \rangle, \langle \frac{0.6, 0.4, 0.4}{b} \rangle, \langle \frac{0.7, 0.4, 0.2}{c} \rangle \right\}.$$

We denote the set of all fuzzy multisets by $FMS(X)$.

Definition 2.2 Let $A,B \in FMS(X)$. Then A is called a fuzzy submultiset of B written as $A \subseteq B$ if $CM_A(x) \leq CM_B(x)$ $\forall x \in X$. Also, if $A \subseteq B$ and $A \neq B$, then A is called a proper fuzzy submultiset of B and denoted as $A \subset B$.

Suppose for $X=\{a,b,c\}$,

$$A = \left\{ \langle \frac{0.5, 0.4, 0.3}{a} \rangle, \langle \frac{0.6, 0.4, 0.4}{b} \rangle, \langle \frac{0.7, 0.4, 0.2}{c} \rangle \right\}$$

and

$$B = \left\{ \langle \frac{0.6, 0.6, 0.4}{a} \rangle, \langle \frac{0.6, 0.5, 0.45}{b} \rangle, \langle \frac{0.7, 0.5, 0.4}{c} \rangle \right\}$$

are fuzzy multisets of X. Then it is easy to see that $A \subseteq B$.

Definition 2.3 Let $A,B \in FMS(X)$. Then the intersection and union of A and B, denoted by $A \cap B$ and $A \cup B$, respectively, are defined by the rules that for any object $x \in X$,

1. $CM_{A \cap B}(x) = CM_A(x) \wedge CM_B(x)$,
2. $CM_{A \cup B}(x) = CM_A(x) \vee CM_B(x)$,

where $\wedge$ and $\vee$ denote minimum and maximum respectively.

Definition 2.4 Let $A,B\in FMS(X)$. A and B are comparable to each other if $A\subseteq B$ or $B\subseteq A$, and $A=B$ if $CM_A(x)\leq CM_B(x)\ \forall x\in X$.

Definition 2.5 Let $A\in FMS(X)$. Then the cardinality of A denoted by $|A|$ is the length of the membership sequence $CM_A(x)=\mu_A^1(x),\mu_A^2(x),...,\mu_A^m(x)$. We define the length $L(x,A)$, that is, the length of $\mu_A^i(x), i=1,...,m$ as

$$L(x;A)=\vee\{i\mid\mu_A^i(x)\neq 0\},$$

where $\vee$ stands for maximum.

The cardinality between two fuzzy multisets, say A and B of X, is the lengths of the membership sequences

$$CM_A(x)=\mu_A^1(x),\mu_A^2(x),...,\mu_A^m(x) \text{ and } CM_B(x)=\mu_B^1(x),\mu_B^2(x),...,\mu_B^n(x)$$

defined as $L(x;A,B) = \vee\{L(x;A),L(x;B)\}$. Where no ambiguity arises, we write $L(x) = L(x;A,B)$ for simplicity.

For example, let

$$A=\left\{\langle\frac{0.3,0.2}{x}\rangle,\langle\frac{1,0.5,0.5}{y}\rangle\right\} \ and\ B=\left\{\langle\frac{0.7,0.1}{w}\rangle,\langle\frac{0.6}{x}\rangle,\langle\frac{0.8,0.6}{y}\rangle\right\}.$$

Then $L(w;A)=0$, $L(x,A)=2$, $L(y;A)=3$, $L(w;B)=2$, $L(x;B)=1$ and $L(y;B)=2$. Also, $L(w)=2$, $L(x)=2$ and $L(y)=3$. Then $|A|=3$ and $|B|=2$. We can rewrite A and B as

$$A=\left\{\langle\frac{0,0,0}{w}\rangle,\langle\frac{0.3,0.2,0}{x}\rangle,\langle\frac{1,0.5,0.5}{y}\rangle\right\} \ and\ B=\left\{\langle\frac{0.7,0.1}{w}\rangle,\langle\frac{0.6,0}{x}\rangle,\langle\frac{0.8,0.6}{y}\rangle\right\}$$

by completing the membership sequences.

Definition 2.6 A fuzzy multiset B of a set X is said to have sup-property if for any subset $W\subset X\ \exists\ w_0\in W$ such that

$$CM_B(w_0)=\underset{w\in W}{\vee}\{CM_B(w)\}.$$

Definition 2.7 Let X be a group. A fuzzy multiset A of X is said to be a fuzzy multigroupoid of X if $CM_A(xy)\geq CM_A(x)\wedge CM_A(y)\forall x,y\in X$.

For example, a fuzzy multiset

$$A=\left\{\langle\frac{1,0.9}{0}\rangle,\langle\frac{0.7,0.5}{1}\rangle,\langle\frac{0.8,0.6}{2}\rangle\right\}$$

of a group $X=\{0,1,2\}$ of modulo 3 is a fuzzy multigroupoid.

Definition 2.8 Let X be a group. A fuzzy multigroupoid A of X is called a fuzzy multigroup of X if $CM_A(x^{-1})\geq CM_A(x)\forall x\in X$. It follows immediately that, $CM_A(x^{-1})=CM_A(x)$, $\forall x\in X$ since

$$CM_A(x)=CM_A\left((x^{-1})^{-1}\right)\geq CM_A\left(x^{-1}\right).$$

It can be easily verified that if A is a fuzzy multigroup of X, then

$$CM_A(e)=\bigvee_{x\in X} CM_A(x)\quad \forall x\in X.$$

That is, $CM_A(e)$ is the tip of A, where e is the identity element of X.

We notice that, every fuzzy multigroupoid and fuzzy multigroup are fuzzy multisets, but the converse are not true. Again, every fuzzy multigroup is fuzzy multigroupoid, but the converse is not true.

The set of all fuzzy multigroups of X is denoted by $FMG(X)$.

Remark 2.1 We notice that, a fuzzy multiset A of a group X is a fuzzy multigroup if $\forall x,y\in X$, $CM_A(x^{-1})\geq CM_A(x)\wedge CM(y)$ holds.

Example 2.1 Let $X=\{e,a,b,c\}$ be a Klein 4-group such that $ab=c$, $ac=b$, $bc-a$, $a^2=b^2=c^2=e$.

Again, let

$$A=\left\{\langle\frac{1,0.9}{e}\rangle,\langle\frac{0.7,0.5}{a}\rangle,\langle\frac{0.8,0.6}{b}\rangle,\langle\frac{0.7,0.5}{c}\rangle\right\}$$

be a fuzzy multiset of X. We investigate whether $A\in FMG(X)$.

$CM_A(ea)=CM_A(a)=0.7,0.5 \geq CM_A(e)\wedge CM_A(a)=0.7,0.5$
$CM_A(eb)=CM_A(b)=0.8,0.6 \geq CM_A(e)\wedge CM_A(b)=0.8,0.6$
$CM_A(ec)=CM_A(c)=0.7,0.5 \geq CM_A(e)\wedge CM_A(c)=0.7,0.5$
$CM_A(ab)=CM_A(c)=0.7,0.5 \geq CM_A(a)\wedge CM_A(b)=0.7,0.5$
$CM_A(ac)=CM_A(b)=0.8,0.6 \geq CM_A(a)\wedge CM_A(c)=0.8,0.6$
$CM_A(bc)=CM_A(a)=0.7,0.5 \geq CM_A(b)\wedge CM_A(c)=0.7,0.5$
$CM_A(aa)=CM_A(e)=1,0.9 \geq CM_A(a)\wedge CM_A(a)=0.7,0.5$
$CM_A(bb)=CM_A(e)=1,0.9 \geq CM_A(b)\wedge CM_A(b)=0.8,0.6$
$CM_A(cc)=CM_A(e)=1,0.9 \geq CM_A(c)\wedge CM_A(c)=0.7,0.5$
$CM_A(ee)=CM_A(e)=1,0.9 \geq CM_A(e)\wedge CM_A(e)=1,0.9$
$CM_A(a^{-1})=CM_A(a)=0.7,0.5$; $CM_A(b^{-1})=CM_A(b)=0.8,0.6$

$CM_A(c^{-1})=CM_A(c)=0.7,0.5$; $CM_A(e^{-1})=CM_A(e)=1,0.9$

It follows that A is a fuzzy multigroup of X.

Clearly, a fuzzy multigroup is a fuzzy group that admits repetition of membership values. That is, a fuzzy multigroup collapses into a fuzzy group whenever repetition of membership values is ignored.

Definition 2.9 Let $\{A_i\}_{i\in I}$, $I=1,\dots,n$ be an arbitrary family of fuzzy multigroups of X. Then

$$CM_{\bigcap_{i\in I} A_i}(x) = \wedge_{i\in I} CM_{A_i}(x)\ \forall x \in X \ \ and\ \ CM_{\bigcup_{i\in I} A_i}(x) = \vee_{i\in I} CM_{A_i}(x)\ \forall x \in X.$$

The family of fuzzy multigroups $\{A_i\}_{i\in I}$ of X is said to have inf or sup assuming chain if either $A_1\subseteq A_2\subseteq\dots\subseteq A_n$ or $A_1\supseteq A_2\supseteq\dots\supseteq A_n$, respectively.

Definition 2.10 Let A be a fuzzy multigroup of a group X. Then A^{-1} is defined by $CM_{A^{-1}}(x) = CM_A(x^{-1})$ $\forall x\in X$.

By Definition 2.8, we get $CM_{A^{-1}}(x) = CM_A(x^{-1}) = CM_A(x)$. That is, $A^{-1}=A$. Thus $A\in FMG(X)$ $\Leftrightarrow A^{-1}\in FMG(X)$.

Definition 2.11 Let $A,B\in FMG(X)$. Then, the product $A\circ B$ of A and B is defined to be a fuzzy multiset of X as follows:

$$CM_{A\circ B}(x) = \begin{cases} \vee_{x=yz} \{CM_A(y) \wedge CM_B(z)\}, & if\ \exists y,z \in X\ such\ that\ x = yz \\ 0, & otherwise. \end{cases}$$

Definition 2.12 Let $A\in FMG(X)$. A fuzzy submultiset B of A is called a fuzzy submultigroup of A denoted by $B\sqsubseteq A$ if B form a fuzzy multigroup. A fuzzy submultigroup B of A is a proper fuzzy submultigroup denoted by $B\subseteq A$, if $B\sqsubseteq A$ and $A\neq B$.

Remark 2.2 If $A\in FMG(X)$ and $B\sqsubseteq A$, then $B\in FMG(X)$. Again, suppose $C\in FMS(X)$ and $C\subseteq B$. Then $C\sqsubseteq A$ if and only if $C\sqsubseteq B$.

Theorem 2.1 *Let X be a finite group and A be a submultigroup of $B\in MG(X)$.* Define $H=\{g\in X \mid C_A(g)=C_A(e)\}$, $K=\{x\in X \mid C_{Ax}(y)=C_{Ae}(y)\}$, where e denotes the identity element of X. Then H and K are subgroups of X. Again, $H=K$.

Definition 2.13 Let $A\in FMG(X)$. Then A is said to be commutative if for all $x,y\in X$, $CM_A(xy) = CM_A(yx)$

Definition 2.14 Let $A\in FMG(X)$. Then the sets $A_{[\alpha]}$ and $A_{(\alpha)}$ defined as

1. $A_{[\alpha]}=\{x\in X \mid CM_A(x)\geq\alpha\}$ and
2. $A_{(\alpha)}=\{x\in X \mid CM_A(x)>\alpha\}$

are called strong upper alpha-cut and weak upper alpha-cut of A, where $\alpha\in[0,1]$.

Definition 2.15 Let $A\in FMG(X)$. Then the sets $A^{[\alpha]}$ and $A^{(\alpha)}$ defined as

1. $A^{[\alpha]}=\{x\in X|CM_A(x)\leq\alpha\}$ and
2. $A^{(\alpha)}=\{x\in X|CM_A(x)<\alpha\}$

are called strong lower alpha-cut and weak lower alpha-cut of A, where $\alpha\in[0,1]$.

Theorem 2.2 *Let $A\in FMG(X)$. Then $A_{[\alpha]}$ is a subgroup of X for all $\alpha\leq CM_A(e)$ and $A^{[\alpha]}$ is a subgroup of X for all $\alpha\geq CM_A(e)$, where e is the identity element of X and $\alpha\in[0,1]$.*

Proposition 2.1 *Let A be a fuzzy multigroup of X and let $x\in X$. Then $CM_A(x)=\alpha$ if and only if $x\in A_{[\alpha]}$ and $x\notin A_{[\beta]}$ $\forall\beta>\alpha$.*

Definition 2.16 Let $A\in FMG(X)$. Then, the sets A_* and A^* are defined as

1. $A_*=\{x\in X|CM_A(x)>0\}$ and
2. $A^*=\{x\in X|CM_A(x)+CM_A(e)\}$,

where e is the identity element of X.

Proposition 2.2 *Let $A\in FMG(X)$. Then A_* and A^* are subgroups of X.*

Definition 2.17 Let $A,B\in FMG(X)$ such that $A\subseteq B$. Then, A is called a normal fuzzy submultigroup of B if for all $x,y\in X$, it satisfies

$CM_A(xyx^{-1})\geq CM_A(y)$.

Proposition 2.3 *Let $A,B\in FMG(X)$. Then, the following statements are equivalent:*

1. A is a normal fuzzy submultigroup of B.
2. $CM_A(xyx^{-1})=CM_A(y)\forall x,y\in X$.
3. $CM_A(xy)=CM_A(yx)\forall x,y\in X$.

Definition 2.18 Two fuzzy multigroups A and B of X are conjugate to each other if for all $x,y\in X$,

$CM_A(x)=CM_B(yxy^{-1})$ and $CM_B(y)=CM_A(xyx^{-1})$.

Definition 2.19 Let X and Y be groups and let f: $X\to Y$ be a homomorphism. Suppose A and B are fuzzy multigroups of X and Y, respectively. Then f induces a homomorphism from A to B which satisfies

1. $CM_A\left(f^{-1}(y_1y_2)\right)\geq CM_A\left(f^{-1}(y_1)\right)\wedge CM_A\left(f^{-1}(y_2)\right)\ \forall y_1,y_2\in Y$,
2. $CM_B\left(f(x_1x_2)\right)\geq CM_B\left(f(x_1)\right)\wedge CM_B\left(f(x_2)\right)\ \forall x_1,x_2\in X$,

where

1. the image of A under f, denoted by $f(A)$, is a fuzzy multiset over Y defined by

$$CM_{f(A)}(y)=\begin{cases}\bigvee_{x\in f^{-1}(y)} CM_A(x), & f^{-1}(y)\neq\varnothing\\ 0, & otherwise\end{cases}$$

for each $y \in Y$.

2. the inverse image of B under f, denoted by $f^{-1}(B)$, is a fuzzy multiset over X defined by

$$CM_{f^{-1}(B)}(x)=CM_B(f(x))\,\forall x\in X.$$

It follows that, a homomorphism f of X onto Y is called an automorphism of A onto A if f is both injective and surjective, that is, bijective.

3 FUZZY COMULTISETS

In this section, we define fuzzy comultiset with examples, and explicate some of its properties.

Definition 3.1 Let X be a group. For any fuzzy submultigroup A of a fuzzy multigroup B of X, the fuzzy submultiset yA of B for $y \in X$ defined by

$$CM_{yA}(x)=CM_A(y^{-1}x)\forall x\in X$$

is called the left fuzzy comultiset of A. Similarly, the submultiset Ay of B for $y \in X$ defined by

$$CM_{Ay}(x)=CM_A(xy^{-1})\forall x\in X$$

is called the right fuzzy comultiset of A.

Example 3.1 Let $X=\{\rho_0,\rho_1,\rho_2,\rho_3,\rho_4,\rho_5\}$ be a permutation group on the set $S=\{1,2,3\}$ such that

$\rho_0=(1)$, $\rho_1=(123)$, $\rho_2=(132)$, $\rho_3=(23)$, $\rho_4=(13)$, $\rho_5=(12)$.

Suppose

$$B=\left\{\langle\frac{1,1}{\rho_0}\rangle,\langle\frac{0.9,0.8}{\rho_1}\rangle,\langle\frac{0.9,0.8}{\rho_2}\rangle,\langle\frac{0.8,0.7}{\rho_3}\rangle,\langle\frac{0.8,0.7}{\rho_4}\rangle,\langle\frac{0.8,0.7}{\rho_5}\rangle\right\}$$

is a fuzzy multigroup of *X*. Then

$$A=\left\{\langle\frac{1,1}{\rho_0}\rangle,\langle\frac{0.8,0.7}{\rho_1}\rangle,\langle\frac{0.8,0.7}{\rho_2}\rangle,\langle\frac{0.7,0.6}{\rho_3}\rangle,\langle\frac{0.7,0.6}{\rho_4}\rangle,\langle\frac{0.7,0.6}{\rho_5}\rangle\right\}$$

is a fuzzy submultigroup of *B*.

The left fuzzy comultisets of *A* in *B* for $\rho_0,\rho_1,\rho_2,\rho_3,\rho_4,\rho_5\in X$ are:

$$\rho_0A=\left\{\langle\frac{1,1}{\rho_0}\rangle,\langle\frac{0.8,0.7}{\rho_1}\rangle,\langle\frac{0.8,0.7}{\rho_2}\rangle,\langle\frac{0.7,0.6}{\rho_3}\rangle,\langle\frac{0.7,0.6}{\rho_4}\rangle,\langle\frac{0.7,0.6}{\rho_5}\rangle\right\},$$

$$\rho_1A=\left\{\langle\frac{0.8,0.7}{\rho_2}\rangle,\langle\frac{1,1}{\rho_0}\rangle,\langle\frac{0.8,0.7}{\rho_1}\rangle,\langle\frac{0.7,0.6}{\rho_5}\rangle,\langle\frac{0.7,0.6}{\rho_3}\rangle,\langle\frac{0.7,0.6}{\rho_4}\rangle\right\},$$

$$\rho_2A=\left\{\langle\frac{0.8,0.7}{\rho_1}\rangle,\langle\frac{0.8,0.7}{\rho_2}\rangle,\langle\frac{1,1}{\rho_0}\rangle,\langle\frac{0.7,0.6}{\rho_4}\rangle,\langle\frac{0.7,0.6}{\rho_5}\rangle,\langle\frac{0.7,0.6}{\rho_3}\rangle\right\},$$

$$\rho_3A=\left\{\langle\frac{0.7,0.6}{\rho_3}\rangle,\langle\frac{0.7,0.6}{\rho_5}\rangle,\langle\frac{0.7,0.6}{\rho_4}\rangle,\langle\frac{1,1}{\rho_0}\rangle,\langle\frac{0.8,0.7}{\rho_2}\rangle,\langle\frac{0.8,0.7}{\rho_1}\rangle\right\},$$

$$\rho_4A=\left\{\langle\frac{0.7,0.6}{\rho_4}\rangle,\langle\frac{0.7,0.6}{\rho_3}\rangle,\langle\frac{0.7,0.6}{\rho_5}\rangle,\langle\frac{0.8,0.7}{\rho_1}\rangle,\langle\frac{1,1}{\rho_0}\rangle,\langle\frac{0.8,0.7}{\rho_2}\rangle\right\},$$

$$\rho_5A=\left\{\langle\frac{0.7,0.6}{\rho_5}\rangle,\langle\frac{0.7,0.6}{\rho_4}\rangle,\langle\frac{0.7,0.6}{\rho_3}\rangle,\langle\frac{0.8,0.7}{\rho_2}\rangle,\langle\frac{0.8,0.7}{\rho_1}\rangle,\langle\frac{1,1}{\rho_0}\rangle\right\}.$$

The right fuzzy comultisets can be computed similarly.

Remark 3.1 Since a fuzzy multiset is unordered, it follows that $xA=Ax$ $\forall x\in X$. It is easily seemed that $xA=yA=zA=A$ $\forall x,y,x\in X$ for the same reason. These are not so in the classical case.

The following result proves that the right and left fuzzy comultisets of a fuzzy submultigroup in a fuzzy multigroup are equal.

Proposition 3.1 *If A is a fuzzy submultigroup of B∈FMG(X), then the right and left fuzzy comultisets of A in B are identical.*

Proof. Assume *A* is a fuzzy submultigroup of *B*. Then $\forall x\in X$ we have

$$\begin{aligned}CM_{yA}(x)&=CM_A(y^{-1}x)\geq CM_A(y)\wedge CM_A(x)\\&=CM_A(x)\wedge CM_A(y)\\&=CM_A(x)\wedge CM_A(y^{-1}).\end{aligned}$$

Suppose by hypothesis, $CM_A(x)\wedge CM_A(y)=CM_A(xy)$. Then we have

$$CM_{yA}(x)\geq CM_{Ay}(x).$$

Again,

$$\begin{aligned}CM_{Ay}(x)&=CM_A(yx^{-1})\geq CM_A(x)\wedge CM_A(y)\\&=CM_A(y)\wedge CM_A(x)\\&=CM_A(y^{-1})\wedge CM_A(x).\end{aligned}$$

By the same hypothesis, we get

$$CM_{Ay}(x)\geq CM_{yA}(x).$$

Hence, $CM_{yA}(x)=CM_{Ay}(x)\Rightarrow yA=Ay$.

Theorem 3.1 *Let A be a fuzzy submultigroup of $B\in FMG(X)$. Then $gA=hA$ for $g,h\in X$ if and only if*

$$CM_A(g^{-1}h)=CM_A(h^{-1}g)=CM_A(e).$$

Proof. Let $gA=hA$. Then $CM_{gA}(g)=CM_{hA}(g)$ and $CM_{gA}(h)=CM_{hA}(h)$ $\forall g,h\in X$. Hence

$$CM_A(g^{-1}h)=CM_A(h^{-1}g)=CM_A(e).$$

Conversely, let $CM_A(g^{-1}h)=CM_A(h^{-1}g)$ $\forall g,h\in X$. For every $x\in X$, we have

$$\begin{aligned}CM_{gA}(x)&=CM_A(g^{-1}x)=CM_A(g^{-1}hh^{-1}x)\\&\geq CM_A(g^{-1}h)\wedge CM_A(h^{-1}x)\\&=CM_A(h^{-1}x)\\&=CM_{hA}(x).\end{aligned}$$

Similarly,

$$\begin{aligned}CM_{hA}(x)&=CM_A(h^{-1}x)=CM_A(h^{-1}gg^{-1}x)\\&\geq CM_A(h^{-1}g)\wedge CM_A(g^{-1}x)\\&=CM_A(g^{-1}x)\\&=CM_{gA}(x).\end{aligned}$$

Hence $CM_{gA}(x)=CM_{hA}(x)\Rightarrow gA=hA$.

Corollary 3.1 *Let A be a fuzzy submultigroup of B∈FMG(X). Then Ag=Ah for g,h∈X if and only if* $CM_A(gh^{-1})=CM_A(hg^{-1})=CM_A(e)$.

Proof. Straightforward from Theorem 3.1.

Proposition 3.2 *Let A,B∈FMG(X) such that A⊆B. If gA=hA, then* $CM_A(g)=CM_A(h)$ $\forall g,h\in X$.

Proof. Let X be a group and $g\in X$. Suppose $gA=hA$, then we have

$$\begin{aligned} CM_{gA}(g)=CM_{ha}(g) &\Rightarrow CM_A(g^{-1}g)=CM_A(h^{-1}g)\\ &\Rightarrow CM_A(e)=CM_A(h^{-1}g) \end{aligned}$$

$\forall g,h\in X$. Since $CM_A(e)=CM_A(h^{-1}g)\Rightarrow CM_A(h)=CM_A(g)$, the result follows.

Alternatively, suppose $z\in X$, we get

$$\begin{aligned} &CM_{gA}(z)=CM_{ha}(z)\Rightarrow CM_A(g^{-1}z)=CM_A(h^{-1}z)\\ &\Rightarrow CM_{Az^{-1}}(g)=CM_{Az^{-1}}(h)\\ &CM_A(g)=CM_A(h) \end{aligned}$$

since $Az^{-1}=A$.

Theorem 3.2 *Let B∈FMG(X). Any fuzzy submultigroup A of B and for any z∈X, the fuzzy submultiset* zAz^{-1}, *where* $CM_{zAz^{-1}}(x)=CM_A(z^{-1}xz)$ *for each x∈X is a fuzzy submultigroup of B.*

Proof. Let $x,y\in X$ and $A\sqsubseteq B$. We prove that $zAz^{-1}\sqsubseteq B$ $\forall z\in X$. Now

$$\begin{aligned} CM_{zAz^{-1}}(xy^{-1}) &= CM_A(z^{-1}xy^{-1}z)\\ &= CM_A(z^{-1}xzz^{-1}y^{-1}z)\\ &\geq CM_A(z^{-1}xz)\wedge CM_A(z^{-1}y^{-1}z)\\ &= CM_{zAz^{-1}}(x)\wedge CM_{zAz^{-1}}(y^{-1})\\ &= CM_{zAz^{-1}}(x)\wedge CM_{zAz^{-1}}(y)\quad \forall z\in X. \end{aligned}$$

Thus $CM_{zAz^{-1}}(xy^{-1})\geq CM_{zAz^{-1}}(x)\wedge CM_{zAz^{-1}}(y)$. Hence zAz^{-1} is a fuzzy submultigroup of B.

Corollary 3.2 *Let* $\{A_i\}_{i\in I}\in FMG(X)$, *then*

1. $\bigcap_{i\in I} zA_iz^{-1}\in FMG(X)\quad \forall z\in X$,
2. $\bigcup_{i\in I} zA_iz^{-1}\in FMG(X)\quad \forall z\in X$

provided $\{A_i\}_{i \in I}$ have sup/inf assuming chain.

Proof. The results follow from Theorem 3.2.

The following results are the application of product of fuzzy multigroups to the idea of fuzzy comultisets.

Proposition 3.3 *Let A be a fuzzy submultigroup of B∈FMG(X) and for all g,h∈X, then the following statements hold;*

1. $Ag \circ Ag = Ag$.
2. $Ag \circ Ah = Ah \circ Ag$.
3. $(Ag \circ Ah)^{-1} = (Ah)^{-1} \circ (Ag)^{-1}$.
4. $(Ag \circ Ah)^{-1} = Ag \circ Ah$.

Proof. Let $g,h \in X$.

(i) From Definition 2.11, we have

$$\begin{aligned} CM_{Ag \circ Ag}(x) &= \vee_{x=yz}\left(CM_{Ag}(y) \wedge CM_{Ag}(z)\right) \ \forall y,z \in X \\ &= \vee_{y \in X}\left(CM_{Ag}(xy^{-1}) \wedge CM_{Ag}(y)\right) \ \forall x \in X. \end{aligned}$$

Hence $Ag \circ Ag = Ag$.

(ii) We get

$$\begin{aligned} CM_{Ag \circ Ah}(x) &= \vee_{x=yz}\left(CM_{Ag}(y) \wedge CM_{Ah}(z)\right) \ \forall y,z \in X \\ &= \vee_{x=yz}\left(CM_{Ah}(z) \wedge CM_{Ag}(y)\right) \ \forall y,z \in X \\ &= CM_{Ah \circ Ag}(x). \end{aligned}$$

Hence $Ag \circ Ah = Ah \circ Ag$.

(iii) We show that, the left and right hand sides are equal.

$$CM_{(Ag \circ Ah)^{-1}}(x) = CM_{Ag \circ Ah}(x^{-1}) = CM_{Ag \circ Ah}(x).$$

Again, from the right hand side we get

$$CM_{(Ah)^{-1}\circ(Ag)^{-1}}(x) = \bigvee_{y\in X}\left(CM_{(Ah)^{-1}}(y^{-1}) \wedge CM_{(Ag)^{-1}}(yx)\right) \ \forall x \in X$$
$$= \bigvee_{y\in X}\left(CM_{Ah}(y^{-1}) \wedge CM_{Ag}(yx)\right) \ \forall x \in X$$
$$= CM_{Ah\circ Ag}(x) = CM_{Ag\circ Ah}(x).$$

Hence $(Ag\circ Ah)^{-1}=(Ah)^{-1}\circ(Ag)^{-1}$.

(iv) Straightforward from (iii).

Remark 3.2 Proposition 3.3 also holds for left fuzzy comultisets.

Proposition 3.4 *Let X be a group. If A is a fuzzy submultigroup of a commutative fuzzy multigroup B of X, then*

1. $Ay\circ Az=Ayz \ \forall y,z\in X$,
2. $yA\circ zA=yzA \ \forall y,z\in X$.

Proof. (i) Let $A\in FMG(X)$ and $x\in X$, then we have

$$CM_{Ay\circ Az}(x) = \bigvee_{x=zy}\left(CM_{Ay}(z) \wedge CM_{Az}(y)\right) \ \forall y,z \in X$$
$$= \bigvee_{x=zy}\left(CM_{A}(zy^{-1}) \wedge CM_{A}(yz^{-1})\right) \ \forall y,z \in X$$
$$= \bigvee_{x=zy} CM_{A\cap A}\left((zy^{-1})(yz^{-1})\right) \ \forall y,z \in X$$
$$= CM_{A}(xz^{-1}y^{-1}), x = zy, \ \forall y,z \in X$$
$$= CM_{Ayz}(x), \forall y,z \in X.$$

Hence $Ay\circ Az=Ayz$.

(ii) Similar to (i).

Corollary 3.3 *Let A be a fuzzy submultigroup of a commutative fuzzy multigroup B of a group X and $y,z\in X$. Then the following statements are equivalent.*

1. $(Ay\circ Az)^{-1}=Ay\circ Az$,
2. $Ay\circ Az=Ayz$.

Proof. By implication, $A\in FMG(X)$ and it is commutative.

(i)⇒ (ii). $CM_{(Ay\circ Az)^{-1}}(x) = CM_{Ay\circ Az}(x) = CM_{Ayz}(x) \ \forall x\in X$ by Proposition 3.4.

(ii)⇒ (i). $CM_{Ayz}(x)=CM_{Ay\circ Az}(x)=CM_{Ay\circ Az}(x^{-1})=CM_{(Ay\circ Az)^{-1}}(x)$ $\forall x\in X$.

Remark 3.3 If $(Ay\circ Az)^{-1}=Ay\circ Az$ and $Ay\circ Az=Ayz$, then $(Ay\circ Az)^{-1}=Ayz$.

Theorem 3.3 *Let A be a fuzzy submultigroup of B∈FMG(X) such that B is commutative, then* $Ag\circ Ah=Agh$ *if and only if* $gA\circ hA=ghA$ $\forall$ $g,h\in X$. *Consequently,* $Agh=ghA$.

Proof. Let $A\in FMG(X)$ and $g,h\in X$. Suppose $Ag\circ Ah=Agh$. By Definition 2.11, we get

$$\begin{aligned} CM_{Agh}(x) &= CM_{Ag\circ Ah}(x)=\vee_{y\in X}\left(CM_{Ag}(y)\wedge CM_{Ah}(y^{-1}x)\right)\\ &= \vee_{y\in X}\left(CM_{A}(yg^{-1})\wedge CM_{A}(y^{-1}xh^{-1})\right)\\ &= \vee_{y\in X}\left(CM_{A}(g^{-1}y)\wedge CM_{A}(h^{-1}y^{-1}x)\right)\\ &= \vee_{y\in X}\left(CM_{gA}(y)\wedge CM_{hA}(y^{-1}x)\right)\\ &= CM_{gA\circ hA}(x)=CM_{ghA}(x)\\ &\Rightarrow gA\circ hA=ghA \end{aligned}$$

Conversely, let $gA\circ hA=ghA$. Then we have

$$\begin{aligned} CM_{ghA}(x) &= CM_{gA\circ hA}(x)=\vee_{y\in X}\left(CM_{gA}(y)\wedge CM_{hA}(y^{-1}x)\right)\\ &= \vee_{y\in X}\left(CM_{A}(g^{-1}y)\wedge CM_{A}(h^{-1}y^{-1}x)\right)\\ &= \vee_{y\in X}\left(CM_{A}(yg^{-1})\wedge CM_{A}(y^{-1}xh^{-1})\right)\\ &= \vee_{y\in X}\left(CM_{Ag}(y)\wedge CM_{Ah}(y^{-1}x)\right)\\ &= CM_{Ag\circ Ah}(x)=CM_{Agh}(x)\\ &\Rightarrow Ag\circ Ah=Agh \end{aligned}$$

Hence the result follow.

Theorem 3.4 *Let X be a finite group and A be a fuzzy submultigroup of B∈FMG(X). Define* $H=\{g\in X \mid CM_A(g)=CM_A(e)\}$.

Then $Hx=Hy\Leftrightarrow Ax=Ay$ $\forall x,y\in X$. Similarly, $xH=yH\Leftrightarrow xA=yA$.

Proof. This result gives a relationship between fuzzy comultisets of a fuzzy submultigroup of a fuzzy multigroup and the cosets of a subgroup of a given group.

By Theorem 2.1, we know that H is a subgroup of X and $H=\{x\in X \mid CM_{Ax}(z)=CM_{Ae}(z)\}$.

Now, suppose that $Hx=Hy$. Then $xy^{-1}\in H$. Thus

$$CM_{Axy^{-1}}(z)=CM_{Ae}(z)\ \ \forall z\in X \ and\ so\ CM_A(zyx^{-1})=CM_A(z).$$

Put $z=zy^{-1}$, we get

$$CM_A(zy^{-1}yx^{-1})=CM_A(zy^{-1})\Rightarrow CM_A(zx^{-1})=CM_A(zy^{-1})\Rightarrow CM_{Ax}(z)=CM_{Ay}(z)$$

and so, $Ax=Ay$.

Conversely, suppose that $Ax=Ay$, that is $CM_{Ax}(z)=CM_{Ay}(z)$ $\forall z\in X$. This implies that

$CM_A(zx^{-1})=CM_A(zy^{-1})$.

Put $z=y$, we get $CM_A(yx^{-1})=CM_A(e)$. So $yx^{-1}\in H$. Thus $Hx=Hy$.

The proof of $xH=yH\Leftrightarrow zA=yA$ is similar.

Corollary 3.4 *Let X be a group. If A is a fuzzy submultigroup of a fuzzy multigroup B of X and x,y∈X. Then xA=yA and Ax=Ay if and only if* $CM_A(y^{-1}x)=CM_A(yx^{-1})=CM_A(e)$.

Proof. Let $x,y\in X$, and recall that $H=\{x\in X|CM_A(x)CM_A(e)\}$. Suppose $xA=yA$ and $Ax=Ay$. Then, $y^{-1}x, yx^{-1}\in H$ as in Theorem 3.4. So, $CM_A(y^{-1}x)=CM_A(e)=CM_A(yx^{-1})$.

Conversely, assume $CM_A(y^{-1}x)=CM_A(e)=CM_A(yx^{-1})$. This implies that,

$CM_A(y^{-1}x)=CM_A(x^{-1}x)$ and $CM_A(yx^{-1})=CM_A(yy^{-1})$

and hence,

$CM_{yA}(x)=CM_{xA}(x)$ and $CM_{Ax}(y)=CM_{Ay}(y)$ $\forall x,y\in X$.

Thus $xA=yA$ and $Ax=Ay$.

4 CHARACTERISTIC FUZZY SUBMULTIGROUPS OF FUZZY MULTIGROUPS

Recall that, a subgroup Y of a group X is a characteristic subgroup if $Y^{\theta}=Y$ for every automorphism, θ of X, where Y^{θ} denotes $\theta(Y)$. Now, we define the analogue of this notion in fuzzy multigroup setting.

Definition 4.1 Let A be a fuzzy submultigroup of $B\in FMG(X)$. Then, A is called a characteristic (f-invariant) fuzzy submultigroup of B if

$$CM_{A^{\theta}}(x)=CM_A(x)\ \ \forall x\in X$$

for every automorphism, θ of X. That is, $\theta(A)\subseteq A$ for every $\theta\in Aut(X)$.

Definition 4.2 Let A be a fuzzy multigroup of a group X and θ a function from X into itself. Define the multiset A^θ of X by

$$CM_{A^\theta}(x) = CM_A(x^\theta), \ \ where\ x^\theta = \theta(x) = x \ \ \forall x \in X.$$

Proposition 4.1 *Let $A\in FMG(X)$ and $g\in X$. If θ is an automorphism of X defined by $\theta(x)=gxg^{-1}$ $\forall x\in X$, then $A^g=A^\theta$.*

Proof. Let $A\in FMG(X)$ and $g\in X$. Suppose $\theta: X\to X$ is defined by $\theta(x)=gxg^{-1}$ $\forall x\in X$. By Definition 2.18, we get

$$CM_{A^g}(x) = CM_A(gxg^{-1}) = CM_A(\theta(x)) = CM_{A^\theta}(x).$$

Thus the result follows.

Proposition 4.2 *Let $A\in FMG(X)$. If θ is a homomorphism of X into itself, then A^θ is a fuzzy multigroup of X.*

Proof. Let $x,y\in X$. Then $CM_{A^\theta}(xy) = CM_A((xy)^\theta) = CM_A(x^\theta y^\theta)$ since θ is a homomorphism. Since A is a fuzzy multigroup of X, we have

$$CM_A(x^\theta y^\theta) \geq CM_A(x^\theta) \wedge CM_A(y^\theta) = CM_{A^\theta}(x) \wedge CM_{A^\theta}(y).$$

Thus

$$CM_{A^\theta}(xy) \geq CM_{A^\theta}(x) \wedge CM_{A^\theta}(y).$$

Also,

$$CM_{A^\theta}(x^{-1}) = CM_A((x^{-1})^\theta) = CM_A((x^\theta)^{-1}) = CM_A(x^\theta) = CM_{A^\theta}(x).$$

Hence A^θ is a fuzzy multigroup of X.

Theorem 4.1 *Let $\theta: X\to X$ be an automorphism and $A\in FMS(X)$. Then $A^\theta\in FMG(X)$ if and only if $A\in FMG(X)$.*

Proof. Suppose $A\in FMG(X)$. Then using the same logic in the proof of Proposition 4.2, it follows that $A^\theta\in FMG(X)$.

Conversely, assume A^θ is a fuzzy multigroup of X. That is,

$$CM_{A^{\theta}}(xy) \geq CM_{A^{\theta}}(x) \wedge CM_{A^{\theta}}(y) \quad and \quad CM_{A^{\theta}}(x^{-1}) = CM_{A^{\theta}}(x)$$

∀*x,y*∈*X* and for every θ∈*Aut*(*X*). Thus

$$\begin{aligned} CM_{A^{\theta}}(xy) &= CM_{A}((xy)^{\theta}) = CM_{A}(\theta(xy)) \\ &= CM_{A}(xy), \\ &\Rightarrow CM_{A}(xy) \geq CM_{A}(x) \wedge CM_{A}(y) \;\; \forall x, y \in X. \end{aligned}$$

Also,

$$\begin{aligned} CM_{A^{\theta}}(x^{-1}) &= CM_{A}((x^{-1})^{\theta}) = CM_{A}((x^{\theta})^{-1}) \\ &= CM_{A}((\theta(x))^{-1}) = CM_{A}(x^{-1}), \\ &\Rightarrow CM_{A}(x^{-1}) = CM_{A}(x) \;\; \forall x \in X. \end{aligned}$$

Hence *A*∈*FMG*(*X*).

Proposition 4.3 *Let B∈FMG(X). Then every characteristic submultigroup of B is a normal fuzzy submultigroup.*

Proof. Let *x,y*∈*X* and let *A* is a characteristic fuzzy submultigroup of *B*. To prove that *A* is a normal fuzzy submultigroup of *B*, we have to show that $CM_A(xy)=CM_A(yx)$ ∀*x,y*∈*X*.

Let θ be the automorphism of *X* defined by $\theta(y)=x^{-1}gx$ ∀*g*∈*X*.
Now since *A* is a characteristic fuzzy submultigroup of *B*, $A^{\theta}=A$. So we have

$$\begin{aligned} CM_{A}(xy) &= CM_{A^{\theta}}((xy)) = CM_{A}((xy)^{\theta}) \\ &= CM_{A}(x^{-1}(xy)x) \\ &= CM_{A}(yx). \end{aligned}$$

by definition of θ
Hence A^{θ} is normal by Proposition 2.3.

Remark 4.1 Let *A,B,C*∈*FMG*(*X*) such that *C*⊆*B*⊆*A*.

1. If *C* is a characteristic fuzzy submultigroup of *B* and *B* is a characteristic fuzzy submultigroup of *A*, then *C* is a characteristic fuzzy submultigroup of *A*.
2. If *C* is a characteristic fuzzy submultigroup of *B* and *B* is a normal fuzzy submultigroup of *A*, then *C* is a normal fuzzy submultigroup of *A*.

Proposition 4.4 *If $B\in FMG(X)$ and A is a characteristic fuzzy submultigroup of B. Then A^* and A_* are characteristic subgroups of X. Also, A^* is a characteristic subgroup of B^* and A_* is a characteristic subgroup of B_*.*

Proof. We know that A_* and A^* are subgroups of X by Proposition 2.16. Now, we proof that A_* and A^* are characteristic subgroups of X. It is sufficient to show that $\theta(A^*)\subseteq A^*$ $\forall\theta\in Aut(X)$.

Let $\theta\in Aut(X)$. Then $CM_{A^\theta}(x)=CM_A(x)$, since A is a characteristic fuzzy submultigroup of B. Let $x\in A^*$, then $CM_A(x)=CM_A(e)$, which implies that

$$CM_{A^\theta}(x)=CM_A(\theta(x))=CM_A(x)=CM_A(e),$$

so $\theta(x)\in A^*$. Hence $\theta(A^*)\subseteq A^*$. This completes the proof.

Similarly, the proof of the fact that A_* is a characteristic subgroup of X follows.

Recall that, A is a characteristic fuzzy submultigroup of B, and A^* and A_* are characteristic subgroups of X. Synthesizing these, it implies that A^* is a characteristic subgroup of B^* and A_* is a characteristic subgroup of B_*.

Theorem 4.2 *Let X be a finite group and A be a characteristic fuzzy submultigroup of $B\in FMG(X)$. Then $A_{[\alpha]}$ is a characteristic subgroup of X $\forall$ $\alpha\leq CM_A(e)$ and $A^{[\alpha]}$ is a characteristic subgroup of X $\forall$ $\alpha\geq CM_A(e)$, where e is the identity element of X and $\alpha\in[0,1]$.*

Proof. It implies from Theorem 2.2 that, $A_{[\alpha]}$ is a subgroup of X $\forall$ $\alpha\leq CM_A(e)$ and $A^{[\alpha]}$ is a subgroup of X $\forall$ $\alpha\geq CM_A(e)$, where $\alpha\in[0,1]$. Now, we proof first that $A_{[\alpha]}$ is a characteristic subgroups of X.

Whenever we show that $\theta(A_{[\alpha]})\subseteq A_{[\alpha]}$ $\forall\theta\in Aut(X)$ we are done. Let $\theta\in Aut(X)$. Then $CM_{A^\theta}(x)=CM_A(x)$, since A is a characteristic fuzzy submultigroup of B. Let $x\in A_{[\alpha]}$, then $CM_A(x)\geq\alpha$, which implies that

$$CM_{A^\theta}(x)=CM_A(\theta(x))=CM_A(x)\geq\alpha,$$

so $\theta(x)\in A_{[\alpha]}$. Thus $\theta(A_{[\alpha]})\subseteq A_{[\alpha]}$. Hence $A_{[\alpha]}$ is a characteristic subgroups of X. Similarly, $A^{[\alpha]}$ is a characteristic subgroups of X.

Remark 4.2 Since A is a characteristic fuzzy submultigroup of B, and $A_{[\alpha]}$ and $A^{[\alpha]}$ are characteristic subgroups of X. Synthesizing these, it happens that $A_{[\alpha]}$ is a characteristic subgroup of $B_{[\alpha]}$ and $A^{[\alpha]}$ is a characteristic subgroup of $B^{[\alpha]}$.

Now, we give a statement of the converse of Theorem 4.2.

Theorem 4.3 *Let X be a finite group and A be a fuzzy submultigroup of $B\in FMG(X)$. If $A_{[\alpha]}$, for $\alpha\in[0,1]$ (also $A^{[\alpha]}$) is a characteristic subgroup of X, then A is a characteristic fuzzy submultigroup of B.*

Proof. Since X is finite, $|A|<\infty$. Let $Im(A)=\{\alpha_0,\dots,\alpha_n\}$ with $\alpha_0>\dots>\alpha_n$. By hypothesis, it follows that $A_{[\alpha]}=\{x\in X \mid CM_A(x)\geq\alpha_i\}$ is a characteristic subgroup of X, $\forall i=0,\dots,n$. Let $\theta\in Aut(X)$. As

$$CM_{A^{\theta}}(x) = CM_A(\theta(x)) = CM_{\theta^{-1}(A)}(x) = CM_A(x),$$

it follows that $Im(A^{\theta})=IM(A)$. Moreover, $\forall k=0,\ldots,n$, we get $(A^{\theta})_{[\alpha_k]} = A_{[\alpha_k]}$, since

$$x \in (A^{\theta})_{[\alpha_k]} \Leftrightarrow CM_{A^{\theta}}(x) \geq \alpha_k \Leftrightarrow CM_A(\theta(x)) \geq \alpha_k \Leftrightarrow \theta(x) \in A_{[\alpha_k]} \Leftrightarrow x \in \theta^{-1}\left(A_{[\alpha_k]}\right) \Leftrightarrow x \in A_{[\alpha_k]}$$

Thus $A^{\theta}=A$. The result follows.

In the next theorem, we show that Theorems 4.2 and 4.3 hold if the finiteness on the order of X is discarded.

Theorem 4.4 *For a fuzzy submultigroup A of $B\in FMG(X)$, the following statements are equivalent:*

1. A is a characteristics fuzzy submultigroup of B.
2. $A_{[\alpha]}$ for $\alpha\in[0,1]$ is a characteristic subgroup of X $\forall$ $\alpha\leq CM_A(e)$, where e is the identity element of X (also $A^{[\alpha]}$ $\forall$ $\alpha\geq CM_A(e)$).

Proof. (i)⇒ (ii). Let $\alpha\in Im(A)$, $\theta\in Aut(X)$ and $x\in A_{[\alpha]}$. Then $CM_A(\theta(x))=CM_A(x)\geq\alpha$, since A is a characteristic submultigroup of B. It follows that $\theta(x)\in A_{[\alpha]}$ and so, $\theta(A_{[\alpha]})\subseteq A_{[\alpha]}$. We prove that $A_{[\alpha]}\subseteq\theta(A_{[\alpha]})$ by symmetry. Let $x\in A_{[\alpha]}$ and let $y\in X$ such that $\theta(y)=x$. Then $CM_A(y) = CM_A(\theta(y)) = CM_A(x)\geq\alpha$, thus $y\in A_{[\alpha]}$, therefore $x\in\theta(A_{[\alpha]})$. Hence $A_{[\alpha]}\subseteq\theta(A_{[\alpha]})$. Thus $A_{[\alpha]}$ is a characteristic subgroup of X.

(ii)⇒ (i). Let $x\in X$, $\theta\in Aut(X)$ and $CM_A(x)=\alpha$. Then $x\in A_{[\alpha]}$ and $x\notin A_{[\beta]}$ $\forall\beta>\alpha$ by Proposition 2.1. By hypothesis, $\theta(A_{[\alpha]})= A_{[\alpha]}$, so $\theta(x)\in A_{[\alpha]}$ and hence $CM_A(x) = CM_A(\theta(x))\geq\alpha$.

Let $\beta=CM_A(\theta(x))$. Assume $\beta>\alpha$. Then $\theta(x)\in A_{[\beta]}=\theta(A_{[\beta]})$. Since θ is one-to-one, it follows that $x\in A_{[\beta]}$, which is a contradiction. Hence $CM_A(\theta(x))=\alpha=CM_A(x)$, implying that A is a characteristic fuzzy submultigroup of B.

Theorem 4.5 *Let θ: $X\rightarrow Y$ be a homomorphism. Let A and B be fuzzy multigroups of X and Y, respectively. Then $\theta(A)\subseteq B$ if and only if $CM_B(f(x))\geq CM_A(x)$ $\forall$ $x\in X$.*

Proof. Let $\theta(A)\subseteq B$ and $x\in X$. Then we have

$$\begin{aligned} CM_B(\theta(x)) &\geq CM_{\theta(A)}(\theta(x)) \\ &= \vee\{CM_A(z) \mid \theta(z)=\theta(x)\} \\ &= CM_A(x). \end{aligned}$$

Conversely, let $CM_B(\theta(x))\geq CM_A(x)$ for every $x\in X$. Then we have

$$\begin{aligned}CM_{\theta(A)}(y) &= \vee\{CM_A(x) \mid \theta(x)=y\}\\ &\leq \vee\{CM_B(\theta(x)) \mid \theta(x)=y\}\\ &= CM_B(y)\end{aligned}$$

for every $y\in Y$. Hence $\theta(A)\subseteq B$.

Proposition 4.5 *Let X be a cyclic group and B∈FMG(X). Then every fuzzy submultigroup of B is f-invariant.*

Proof. Let $X=\langle x\rangle$ and A be a fuzzy submultigroup of $B\in FMG(X)$. Then $CM_A(x^n)\geq CM_A(x)$ for every $m\in\mathbb{Z}^+$. This implies that $CM_A(\theta(x))\geq CM_A(x)$ for every $\theta\in Aut(X)$. By Theorem 4.5, it follows that A is f-invariant.

5. DIRECT PRODUCT OF FUZZY MULTIGROUPS

Suppose X and Y are two groups. Then the direct product, $X\times Y$ is the Cartesian product of ordered pair (x,y) such that $x\in X$ and $y\in Y$, and the group operation is component-wise, so $(x_1,y_1)\times(x_2,y_2)=(x_1x_2,y_1y_2)$. The resulting algebraic structure satisfies the axioms for a group. Since the ordered pair (x,y) such that $x\in X$ and $y\in Y$ is an element of $X\times Y$, we simply write $(x,y)\in X\times Y$. In this section, we discuss the notion of direct product of two fuzzy multigroups defined over X and Y, respectively.

Definition 5.1 Let X and Y be groups, $A\in FMG(X)$ and $B\in FMG(Y)$, respectively. The direct product of A and B depicted by $A\times B$ is a function $CM_{A\times B}$: $X\times Y\to Q$ defined by

$$CM_{A\times B}((x,y)) = CM_A(x)\wedge CM_B(y)\,\forall x\in X, \forall y\in Y,$$

where Q is the set of all multisets from the unit interval $I=[0,1]$.

Example 5.1 Let $X=\{1,x\}$ be a group, where $x^2=1$ and $Y=\{e,a,b,c\}$ be a Klein 4-group, where $a^2=b^2=c^2=e$. Suppose

$$A=\left\{\langle\frac{1,0.8}{1}\rangle,\langle\frac{0.8,0.5}{x}\rangle\right\} \text{ and } B=\left\{\langle\frac{1,0.9}{e}\rangle,\langle\frac{0.6,0.5}{a}\rangle,\langle\frac{0.7,0.6}{b}\rangle,\langle\frac{0.6,0.5}{c}\rangle\right\}$$

are fuzzy multigroups of X and Y by Definition 2.8. Now

$X\times Y=\{(1,e),(1,a),(1,b),(1,c),(x,e),(x,a),(x,b),(x,c)\}$

is a group from the classical sense. By Definition 5.1, we get

$$A\times B=\left\{\langle\frac{1,0.8}{(1,e)}\rangle,\langle\frac{0.6,0.5}{(1,a)}\rangle,\langle\frac{0.7,0.6}{(1,b)}\rangle,\langle\frac{0.6,0.5}{(1,c)}\rangle,\langle\frac{0.8,0.5}{(x,e)}\rangle,\langle\frac{0.6,0.5}{(x,a)}\rangle,\langle\frac{0.7,0.5}{(x,b)}\rangle,\langle\frac{0.6,0.5}{(x,c)}\rangle\right\}.$$

Certainly, $A\times B$ is a fuzzy multigroup of $X\times Y$ in accordance to Definition 2.8.

Next, we consider an example to investigate what happens of the direct product of a fuzzy multigroup of a group X and a fuzzy multiset of a group Y.

Example 5.2 Let X and Y be groups as in Example 5.1. Suppose we have a fuzzy multigroup of X given as

$$A=\left\{\langle\frac{1,0.5}{1}\rangle,\langle\frac{0.7,0.4}{x}\rangle\right\},$$

and a fuzzy multiset of Y as

$$B=\left\{\langle\frac{0.7,0.5}{e}\rangle,\langle\frac{0.6,0.4}{a}\rangle,\langle\frac{0.7,0.6}{b}\rangle,\langle\frac{0.6,0.4}{c}\rangle\right\}.$$

Synthesizing Definitions 2.8 and 5.1, we get

$$A\times B=\left\{\langle\frac{0.7,0.5}{(1,e)}\rangle,\langle\frac{0.6,0.4}{(1,a)}\rangle,\langle\frac{0.7,0.5}{(1,b)}\rangle,\langle\frac{0.6,0.4}{(1,c)}\rangle,\langle\frac{0.7,0.4}{(x,e)}\rangle,\langle\frac{0.6,0.4}{(x,a)}\rangle,\langle\frac{0.7,0.4}{(x,b)}\rangle,\langle\frac{0.6,0.4}{(x,c)}\rangle\right\},$$

and it follows that $A\times B$ is a fuzzy multigroup of $X\times Y$ although B is not a fuzzy multigroup of Y.

Theorem 5.1 *Let $A\in FMG(X)$ and $B\in FMG(Y)$, respectively. Then for all* $\alpha\in[0,1]$,

1. $(A\times B)_{[\alpha]}=A_{[\alpha]}\times B_{[\alpha]}$.
2. $(A\times B)^{[\alpha]}=A^{[\alpha]}\times B^{[\alpha]}$.

Proof. (i) Let $(x,y)\in(A\times B)_{[\alpha]}$. Using Definition 2.14, we have

$$CM_{A\times B}((x,y))=(CM_A(x)\wedge CM_B(y))\geq\alpha.$$

This implies that $CM_A(x)\geq\alpha$ and $CM_B(y)\geq\alpha$, then $x\in A_{[\alpha]}$ and $y\in B_{[\alpha]}$. Thus, $(x,y)\in A_{[\alpha]}\times B_{[\alpha]}$. Also, let $(x,y)\in A_{[\alpha]}\times B_{[\alpha]}$. Then $CM_A(x)\geq\alpha$ and $CM_B(y)\geq\alpha$. That is,

$$(CM_A(x)\wedge CM_B(y))\geq\alpha.$$

This yields us $(x,y)\in(A\times B)_{[\alpha]}$. Therefore, $(A\times B)_{[\alpha]}=A_{[\alpha]}\times B_{[\alpha]}$ $\forall\alpha\in[0,1]$.

(ii) Similar to (i).

Corollary 5.1 *Let $A \in FMG(X)$ and $B \in FMG(Y)$, respectively. Then*

1. $(A\times B)_* = A_* \times B_*$,
2. $(A\times B)^* = A^* \times B^*$.

Proof. Similar to Theorem 5.1.

Theorem 5.2 *Let A and B be fuzzy multigroups of X and Y, respectively. Then $A\times B$ is a fuzzy multigroup of $X\times Y$.*

Proof. Let $(x,y) \in X\times Y$ and let $x=(x_1,x_2)$ and $y=(y_1,y_2)$. We have

$$\begin{aligned} CM_{A\times B}(xy) &= CM_{A\times B}\left((x_1,x_2)(y_1,y_2)\right) \\ &= CM_{A\times B}\left((x_1y_1,x_2y_2)\right) \\ &= CM_A(x_1y_1) \wedge CM_B(x_2y_2) \\ &\geq \wedge\left(CM_A(x_1)\wedge CM_A(y_1), CM_B(x_2)\wedge CM_B(y_2)\right) \\ &= \wedge\left(CM_A(x_1)\wedge CM_B(x_2), CM_A(y_1)\wedge CM_B(y_2)\right) \\ &= CM_{A\times B}\left((x_1,x_2)\right)\wedge CM_{A\times B}\left((y_1,y_2)\right) \\ &= CM_{A\times B}(x)\wedge CM_{A\times B}(y). \end{aligned}$$

Also,

$$\begin{aligned} CM_{A\times B}\left(x^{-1}\right) &= CM_{A\times B}\left((x_1,x_2)^{-1}\right) = CM_{A\times B}\left(\left(x_1^{-1},x_2^{-1}\right)\right) \\ &= CM_A\left(x_1^{-1}\right)\wedge CM_B\left(x_2^{-1}\right) = CM_A(x_1)\wedge CM_B(x_2) \\ &= CM_{A\times B}\left((x_1,x_2)\right) = CM_{A\times B}(x). \end{aligned}$$

Hence $A\times B \in FMG(X\times Y)$.

Corollary 5.2 *Let $A_1, B_1 \in FMG(X_1)$ and $A_2, B_2 \in FMG(X_2)$, respectively such that $A_1 \subseteq B_1$ and $A_2 \subseteq B_2$. If A_1 and A_2 are normal fuzzy submultigroups of B_1 and B_2, then $A_1\times A_2$ is a normal fuzzy submultigroup of $B_1\times B_2$.*

Proof. By Theorem 5.2, $A_1\times A_2$ is a fuzzy multigroup of $X_1\times X_2$. Also, $B_1\times B_2$ is a fuzzy multigroup of $X_1\times X_2$. We show that $A_1\times A_2$ is a normal fuzzy submultigroup of $B_1\times B_2$. Let $(x,y) \in A_1\times A_2$ such that $x=(x_1,x_2)$ and $y=(y_1,y_2)$. Then we get

$$\begin{aligned} CM_{A_1\times A_2}(xy) &= CM_{A_1\times A_2}((x_1,x_2)(y_1,y_2)) \\ &= CM_{A_1\times A_2}((x_1y_1,x_2y_2)) \\ &= CM_{A_1}(x_1y_1)\wedge CM_{A_2}(x_2y_2) \\ &= CM_{A_1}(y_1x_1)\wedge CM_{A_2}(y_2x_2) \\ &= CM_{A_1\times A_2}((y_1x_1,y_2x_2)) \\ &= CM_{A_1\times A_2}((y_1,y_2)(x_1,x_2)) \\ &= CM_{A_1\times A_2}(yx). \end{aligned}$$

Hence $A_1 \times A_2$ is a normal fuzzy submultigroup of $B_1 \times B_2$ by Proposition 2.3.

Theorem 5.3 *Let A and B be fuzzy multigroups of X and Y, respectively. Then*

1. $(A\times B)_*$ is a subgroup of $X\times Y$,
2. $(A\times B)^*$ is a subgroup of $X\times Y$,
3. $(A\times B)_{[\alpha]}$ is a subgroup of $X\times Y$, $\forall \alpha \leq CM_{A\times B}(e,e')$ and $\alpha\in[0,1]$,
4. $(A\times B)^{[\alpha]}$ is a subgroup of $X\times Y$, $\forall \alpha \geq CM_{A\times B}(e,e')$ and $\alpha\in[0,1]$.

Proof. Combining Proposition 2.2, Theorems 2.2 and 5.2, the results follow.

Corollary 5.3 *Let $A,C\in FMG(X)$ such that $A\subseteq C$ and $B,D\in FMG(Y)$ such that $B\subseteq D$, respectively. If A and* B *are normal, then*

1. $(A\times B)_*$ is a normal subgroup of $(C\times D)_*$,
2. $(A\times B)^*$ is a normal subgroup of $(C\times D)^*$,
3. $(A\times B)_{[\alpha]}$ is a normal subgroup of $(C\times D)_{[\alpha]}$, $\forall \alpha \leq CM_{A\times B}(e,e')$ and $\alpha\in[0,1]$,
4. $(A\times B)^{[\alpha]}$ is a normal subgroup of $(C\times D)^{[\alpha]}$, $\forall \alpha \geq CM_{A\times B}(e,e')$ and $\alpha\in[0,1]$.

Proof. Combining Proposition 2.2, Theorems 2.2, 5.2 and Corollary 5.2, the results follow.

Proposition 5.1 *Let $A\in FMG(X)$ and $B\in FMG(Y)$, respectively. Then $\forall(x,y)\in X\times Y$, we have*

1. $CM_{A\times B}((x^{-1},y^{-1})) = CM_{A\times B}((x,y))$,
2. $CM_{A\times B}((e,e')) \geq CM_{A\times B}((x,y))$,
3. $CM_{A\times B}((x,y)^n) \geq CM_{A\times B}((x,y))$,

where e and e' are the identity elements of X and Y, respectively and $n\in\mathbb{N}$.

Proof. Let $x\in X$, $y\in Y$ and $(x,y)\in X\times Y$. By Theorem 5.2, it follows that $A\times B\in FMG(X\times Y)$. Now,

(i)
$$\begin{aligned} CM_{A\times B}\left(\left(x^{-1},y^{-1}\right)\right) &= CM_A\left(x^{-1}\right)\wedge CM_B\left(y^{-1}\right) \\ &= CM_A\left(x\right)\wedge CM_B\left(y\right) \\ &= CM_{A\times B}\left(\left(x,y\right)\right). \end{aligned}$$

Clearly, $CM_{A\times B}\left(\left(x^{-1},y^{-1}\right)\right)=CM_{A\times B}\left(\left(x,y\right)\right)\ \forall\left(x,y\right)\in X\times Y$.

(ii)
$$\begin{aligned} CM_{A\times B}\left(\left(e,e'\right)\right) &= CM_{A\times B}\left(\left(x,y\right)\left(x^{-1},y^{-1}\right)\right) \\ &\geq CM_{A\times B}\left(\left(x,y\right)\right)\wedge CM_{A\times B}\left(\left(x^{-1},y^{-1}\right)\right) \\ &= CM_{A\times B}\left(\left(x,y\right)\right)\wedge CM_{A\times B}\left(\left(x,y\right)\right) \\ &= CM_{A\times B}\left(\left(x,y\right)\right)\ \forall\left(x,y\right)\in X\times Y. \end{aligned}$$

Hence, $CM_{A\times B}\left(\left(e,e'\right)\right)\geq CM_{A\times B}\left(\left(x,y\right)\right)$.

(iii) (iii)
$$\begin{aligned} CM_{A\times B}\left(\left(x,y\right)^n\right) &= CM_{A\times B}\left(\left(x^n,y^n\right)\right) \\ &= CM_{A\times B}\left(\left(x^{n-1},y^{n-1}\right)\left(x,y\right)\right) \\ &\geq CM_{A\times B}\left(\left(x^{n-1},y^{n-1}\right)\right)\wedge CM_{A\times B}\left(\left(x,y\right)\right) \\ &\geq CM_{A\times B}\left(\left(x^{n-2},y^{n-2}\right)\right)\wedge CM_{A\times B}\left(\left(x,y\right)\right)\wedge CM_{A\times B}\left(\left(x,y\right)\right) \\ &\geq CM_{A\times B}\left(\left(x,y\right)\right)\wedge CM_{A\times B}\left(\left(x,y\right)\right)\wedge\ldots\wedge CM_{A\times B}\left(\left(x,y\right)\right) \\ &= CM_{A\times B}\left(\left(x,y\right)\right), \end{aligned}$$

$$\Rightarrow CM_{A\times B}\left(\left(x,y\right)^n\right)=CM_{A\times B}\left(\left(x^n,y^n\right)\right)\geq CM_{A\times B}\left(\left(x,y\right)\right)\ \forall\left(x,y\right)\in X\times Y.$$

Theorem 5.4 *Let A and B be fuzzy multisets of groups X and Y, respectively. Suppose that e and* e' *are the identity elements of X and Y, respectively. If A×B is a fuzzy multigroup of X×Y, then at least one of the following statements hold.*

1. $CM_B\left(e'\right)\geq CM_A\left(x\right)$ ∀x∈X,
2. $CM_A\left(e\right)\geq CM_B\left(y\right)$ ∀y∈Y.

Proof. Let *A×B∈FMG(X×Y)*. By contrapositive, suppose that none of the statements holds. Then suppose we can find *a* in *X* and *b* in *Y* such that

$CM_A\left(a\right)>CM_B\left(e'\right)$ and $CM_B\left(b\right)>CM_A\left(e\right)$.

From these we have

$$\begin{aligned} CM_{A\times B}((a,b)) &= CM_A(a) \wedge CM_B(b) \\ &> CM_A(e) \wedge CM_B(e') \\ &= CM_{A\times B}((e,e')). \end{aligned}$$

Thus, $A\times B$ is not a fuzzy multigroup of $X\times Y$ by Proposition 5.1. Hence, either

$$CM_B(e') \geq CM_A(x)\ \forall x \in X \ \ or \ \ CM_A(e) \geq CM_B(y)\ \forall y \in Y.$$

This completes the proof.

Theorem 5.5 *Let A and B be fuzzy multisets of groups X and Y, respectively, such that*

$$CM_A(x) \leq CM_B(e')\ \ \forall x \in X,\ e'$$

being the identity element of Y. If A×B is a fuzzy multigroup of X×Y, then A is a fuzzy multigroup of X.

Proof. Let $A\times B$ be a fuzzy multigroup of $X\times Y$ and $x,y \in X$. Then $(x,e'),(y,e') \in X\times Y$. Now, using the property $CM_A(x) \leq CM_B(e')\ \forall x \in X$, we get

$$\begin{aligned} CM_A(xy) &= CM_A(xy) \wedge CM_B(e'e') \\ &= CM_{A\times B}((xy,e'e')) \\ &= CM_{A\times B}((x,e')(y,e')) \\ &\geq CM_{A\times B}((x,e')) \wedge CM_{A\times B}((y,e')) \\ &= \wedge(CM_A(x) \wedge CM_B(e'), CM_A(y) \wedge CM_B(e')) \\ &= CM_A(x) \wedge CM_A(y). \end{aligned}$$

Also,

$$\begin{aligned} CM_A(x^{-1}) &= CM_A(x^{-1}) \wedge CM_B(e'^{-1}) = CM_{A\times B}((x^{-1},e'^{-1})) \\ &= CM_{A\times B}((x,e')^{-1}) = CM_{A\times B}((x,e')) \\ &= CM_A(x) \wedge CM_B(e') = CM_A(x). \end{aligned}$$

Hence, A is a fuzzy multigroup of X. This completes the proof.

Theorem 5.6 *Let A and B be fuzzy multisets of groups X and Y, respectively, such that* $CM_B(x) \leq CM_A(e)$ $\forall x \in Y$*, e being the identity element of X. If A×B is a fuzzy multigroup of X×Y, then B is a fuzzy multigroup of Y.*

Proof. Similar to Theorem 5.5.

Corollary 5.4 *Let A and B be fuzzy multisets of groups X and Y, respectively. If A×B is a fuzzy multigroup of X×Y, then either A is a fuzzy multigroup of X or B is a fuzzy multigroup of Y.*

Proof. Combining Theorems 5.4, 5.5 and 5.6, the result follows.

Theorem 5.7 *If A and C are conjugate fuzzy multigroups of a group X, and B and D are conjugate fuzzy multigroups of a group Y. Then A×B is a conjugate of C×D.*

Proof. Since *A* and *C* are conjugate, it implies that for $g_1 \in X$, we have

$$CM_A(x) = CM_C(g_1^{-1}xg_1) \quad \forall x \in X.$$

Also, since *B* and *D* are conjugate, for $g_2 \in Y$, we get

$$CM_B(y) = CM_D(g_2^{-1}yg_2) \quad \forall y \in Y.$$

Now,

$$\begin{aligned} CM_{A\times B}((x,y)) &= CM_A(x) \wedge CM_B(y) = CM_C(g_1^{-1}xg_1) \wedge CM_D(g_2^{-1}yg_2) \\ &= CM_{C\times D}((g_1^{-1}xg_1),(g_2^{-1}yg_2)) \\ &= CM_{C\times D}((g_1^{-1},g_2^{-1})(x,y)(g_1,g_2)) \\ &= CM_{C\times D}((g_1,g_2)^{-1}(x,y)(g_1,g_2)). \end{aligned}$$

This completes the proof.

Theorem 5.8 *Let A∈FMG(X) and B∈FMG(Y). Suppose C and D are two fuzzy submultisets of A and B, respectively. Then C×D is a fuzzy submultigroup of A×B if and only if both C and D are fuzzy submultigroups of A and B, respectively.*

Proof. Suppose *C* and *D* are two fuzzy submultigroups of *A* and *B*, respectively. Then *C*∈*FMG*(*X*) and *D*∈*FMG*(*Y*) by Remark 2.2. It then follows that *C*×*D* ∈ *FMG*(*X*×*Y*) by Theorem 5.2. Since *A*×*B* is a fuzzy multigroup of *X*×*Y* by the same reason, and *C*⊑*A* and *D*⊑*B*, thus, *C*×*D* is a fuzzy submultigroup of *A*×*B*.

Conversely, If *C*×*D* is a fuzzy submultigroup of *A*×*B*. Then, it follows that *C*⊑*A* and *D*⊑*B*. These complete the proof.

Corollary 5.5 *Let $A \in FMG(X)$ and $B \in FMG(Y)$. Suppose C and D are two fuzzy submultigroups of A and B, respectively. Then $C \times D$ is a normal fuzzy submultigroup of $A \times B$ if and only if both C and D are normal fuzzy submultigroups of A and B, respectively.*

Proof. Combining both Definition 2.17, Theorems 5.2 and 5.8, the proof follows.

Corollary 5.6 *Let $A \in FMG(X)$ and $B \in FMG(Y)$. Suppose C and D are two fuzzy submultigroups of A and B, respectively. Then $C \times D$ is a characteristic fuzzy submultigroup of $A \times B$ if and only if both C and D are characteristic fuzzy submultigroups of A and B, respectively.*

Proof. Combining both Theorems 5.2 and 5.8, the proof follows.

Remark 5.1 With the same hypothesis as in Corollary 5.6, it follows that $C \times D$ is a normal fuzzy submultigroup of $A \times B$ if both C and D are characteristic fuzzy submultigroups of A and B, respectively.

Corollary 5.7 *Let $A \in FMG(X)$ and C be a fuzzy submultiset of A. Then $C \times C$ is a fuzzy submultigroup of $A \times A$ if and only if C is a fuzzy submultigroup of A.*

Proof. The proof is straightforward from Theorem 5.8.

Remark 5.2 Let $A \in FMG(X)$ and C be a fuzzy submultigroup of A. Then

1. $C \times C$ is a normal fuzzy submultigroup of $A \times A$ if and only if C is a normal fuzzy submultigroup of A.
2. $C \times C$ is a characteristic fuzzy submultigroup of $A \times A$ if and only if C is a characteristic fuzzy submultigroup of A.
3. $C \times C$ is a normal fuzzy submultigroup of $A \times A$ if C is a characteristic fuzzy submultigroup of A.

Theorem 5.9 *Let A and B be fuzzy multigroups of groups X and Y, respectively. Then A and B are commutative if and only if $A \times B$ is a commutative fuzzy multigroup of $X \times Y$.*

Proof. Suppose A and B are commutative. We show that $A \times B$ is a commutative fuzzy multigroup of $X \times Y$. It is a known fact that $A \times B \in FMG(X \times Y)$ by Theorem 5.2. Let $(x,y) \in X_1 \times X_2$ such that $x=(x_1,x_2)$ and $y=(y_1,y_2)$. Then we get

$$\begin{aligned}
CM_{A\times B}(xy) &= CM_{A\times B}((x_1,x_2)(y_1,y_2)) \\
&= CM_{A\times B}(x_1y_1, x_2y_2) \\
&= CM_A(x_1y_1) \wedge CM_B(x_2y_2) \\
&= CM_A(y_1x_1) \wedge CM_B(y_2x_2) \\
&= CM_{A\times B}(y_1x_1, y_2x_2) \\
&= CM_{A\times B}((y_1,y_2)(x_1,x_2)) \\
&= CM_{A\times B}(yx).
\end{aligned}$$

Hence, $A \times B$ is a commutative fuzzy multigroup of $X \times Y$ by Definition 2.17.

Conversely, suppose $A \times B$ is a commutative fuzzy multigroup of $X \times Y$. Then, it is clear that both A and B are commutative fuzzy multigroups of groups X and Y, respectively.

CONCLUSION

Since the concept of fuzzy multigroups is relatively new, we have studied some notions like fuzzy co-multisets, characteristic fuzzy submultigroups and direct product of fuzzy multigroups, whose analogous concepts have been hitherto established in group theory. Some results on these notions were obtained and proved. For future studies, more notions in group theory could be established in fuzzy multigroup setting by redefining the notions in the light of fuzzy multigroups.

REFERENCES

Ajmal, N., & Prajapati, A. S. (1992). Fuzzy cosets and fuzzy normal subgroups. *Information Sciences*, *64*(1-2), 17–25. doi:10.1016/0020-0255(92)90107-J

Baby, A., Shinoj, T. K., & Sunil, J. J. (2015). On abelian fuzzy multigroups and order of fuzzy multi-groups. *J. New Theory*, *5*(2), 80–93.

Bhattacharya, P., & Mukherjee, N. P. (1987). Fuzzy groups: Some group theoretic analogs II. *Information Sciences*, *41*(1), 77–91. doi:10.1016/0020-0255(87)90006-5

Biswas, R. (1999). An application of Yager's bag theory in multicriteria based decision making problems. *International Journal of Intelligent Systems*, *14*(12), 1231–1238. doi:10.1002/(SICI)1098-111X(199912)14:12<1231::AID-INT4>3.0.CO;2-6

Ejegwa, P. A. (2014). Correspondence between fuzzy multisets and sequences. *Global J. Sci. Frontier Research: Math. Decision Sci.*, *14*(7), 61–66.

Ejegwa, P. A. (2018a). On fuzzy multigroups and fuzzy submultigroups. *J. Fuzzy Math.*, *26*(3), 641–654.

Ejegwa, P. A. (2018b). On abelian fuzzy multigroups. *J. Fuzzy Math.*, *26*(3), 655–668.

Ejegwa, P. A. (2018c). On normal fuzzy submultigroups of a fuzzy multigroup. *Theory Appl. Math. Comp. Sci.*, *8*(1), 64–80.

Ejegwa, P. A. (2018d). Homomorphism of fuzzy multigroups and some of its properties. *Applications and Applied Mathematics*, *13*(1), 114–129.

Kim, D. S. (2003). Characteristic fuzzy groups. *Commun. Korean Math. Soc.*, *18*(1), 21–29. doi:10.4134/CKMS.2003.18.1.021

Kumar, R. (1992). Fuzzy characteristic subgroups of a group. *Fuzzy Sets and Systems*, *45*(3), 397–398. doi:10.1016/0165-0114(92)90356-9

Miyamoto, S. (1996). Basic operations of fuzzy multisets. *J. Japan Soc. Fuzzy Theory Systems*, *8*(4), 639–645. doi:10.3156/jfuzzy.8.4_47

Miyamoto, S., & Mizutani, K. (2004). *Fuzzy multiset model and methods for nonlinear document clustering for information retrieval*. Springer-Verlag Berlin Heidelberg.

Mizutani, K., Inokuchi, R., & Miyamoto, S. (2008). Algorithms of nonlinear document clustering based on fuzzy multiset model. *International Journal of Intelligent Systems*, *23*(2), 176–198. doi:10.1002/int.20263

Mordeson, J. M., Bhutani, K. R., & Rosenfeld, A. (2005). *Fuzzy group theory*. Heidelberg, Germany: Springer-Verlag Berlin. doi:10.1007/b12359

Mukherjee, N. P., & Bhattacharya, P. (1984). Fuzzy normal subgroups and fuzzy cosets. *Information Sciences*, *34*(3), 255–239. doi:10.1016/0020-0255(84)90050-1

Mukherjee, N. P., & Bhattacharya, P. (1986). Fuzzy groups: Some group theoretic analogs. *Information Sciences*, *39*(3), 247–268. doi:10.1016/0020-0255(86)90039-3

Onasanya, B. O., & Ilori, S. A. (2013). On fuzzy subgroups and fuzzy cosets. *International Journal of Computers and Applications*, *81*(14), 20–22. doi:10.5120/14184-2353

Ray, A. K. (1999). On product of fuzzy subgroups. *Fuzzy Sets and Systems*, *105*(1), 181–183. doi:10.1016/S0165-0114(98)00411-4

Rosenfeld, A. (1971). Fuzzy subgroups. *Journal of Mathematical Analysis and Applications*, *35*(3), 512–517. doi:10.1016/0022-247X(71)90199-5

Seselja, B., & Tepavcevic, A. (1997). A note on fuzzy groups. *Yugoslav J. Operations Research*, *7*(1), 49–54.

Shinoj, T. K., Baby, A., & Sunil, J. J. (2015). On some algebraic structures of fuzzy multisets. *Ann. Fuzzy Math. Inform.*, *9*(1), 77–90.

Sidky, F. I., & Mishref, M. A. A. (1991). Fully invariant, characteristic and S-fuzzy subgroups. *Information Sciences*, *55*(1-3), 27–33. doi:10.1016/0020-0255(91)90003-D

Syropoulos, A. (2012). On generalized fuzzy multisets and their use in computation. *Iranian J. Fuzzy Syst.*, *9*(2), 113–125.

Yager, R. R. (1986). On the theory of bags. *International Journal of General Systems*, *13*(1), 23–37. doi:10.1080/03081078608934952

Zadeh, L. A. (1965). Fuzzy sets. *Information and Control*, *8*(3), 338–353. doi:10.1016/S0019-9958(65)90241-X

Chapter 4
Neighborhood-Based Classification of Imprecise Data

Sampath Sundaram
Kalasalingam Academy of Higher Education and Research, Srivilliputhur, India

Miriam Kalpana Simon
Madras Christian College, India

ABSTRACT

This chapter considers k-nn classification of objects that assume values of imprecise nature with respect to attributes being considered. In order to handle imprecise data values, two approaches of crisp conversion of fuzzy data sets are considered. The approaches considered in this chapter are borrowed from Credibility Theory of Liu. A comparative study on the choice of approach for crisp conversion of fuzzy data has been carried out with the help of certain multivariate simulated data sets. Conclusions drawn from the study are presented.

1. INTRODUCTION

Classification is one of the major branches of data mining. While clustering algorithms attempt to identify the presence of natural subgroups in a data set, classifiers aim at assigning membership for a test set object with one of the pre-identified classes. In the literature various types of approaches are available for classification. Among them neighborhood based methods, decision trees, naïve Bayesian classifiers and support vector machines are widely studied non parametric classifiers. Neighborhood based methods (Fix and Hodges (1951); Cover and Hart, 1967)) like *k* nearest neighborhood works in a manner that assigns membership to a test set object based on the memberships of the objects which are close to the *k* nearest neighbors. Here *k* denotes a pre-determined user specific positive integer. Decision trees which are developed by taking into account various measures of information/impurity leads to the classification of test objects based on the traversal along its branches. Naïve Bayesian classifier which uses class conditional independence producing posterior probabilities can be used for classification purpose. The naïve Bayesian classifier used Bayes theorem. Details related to these two classifiers can be found in Tan,

DOI: 10.4018/978-1-7998-0190-0.ch004

Steinbach and Kumar (2006). Support vector machines (Corinna and Vapnik, 1995) which are based on sound mathematical principles perform extremely well for various types of benchmark datasets. Apart from these classifiers which do not make use of statistical principles, several classifiers like Fisher's linear discriminant functions, quadratic discriminant functions based on statistical theory (Johnson and Wichern, 2008) are also available in literature.

The classifiers mentioned above are predominantly meant for datasets consisting of crisp values. To be precise, values assumed by the objects (both training set as well as test set) with respect to variables under consideration assume crisp values which are free from any kind of impreciseness (both subjective as well as objective). Even though majority of the classifiers assign a test object into one of the predefined classes based on the trained classifiers in a precise (crisp) manner, researchers have made attempts to design classifiers which give scope for assignment of membership in an imprecise manner. Such classifiers become relevant in real life situations particularly when clear cut assignment of membership with an identified class becomes a difficult task. This type of problem arises in several areas of scientific investigations. For example, based on clinical symptoms and diagnostic tests often a medical practitioner finds it to difficult to classify a patient into one of the possible diseases. Hence, assigning the membership with different groups of people associated with varying ailments using different membership values (which indicate the belongingness of the test patient) appears to be prudent.

Even though, several classifiers are available in literature for imprecise or fuzzy classification, the question of classifying data objects which assume imprecise values is not considered by many. The main difficulty in handling such data sets is the choice of metrics which can quantify the amount of dissimilarities/similarities between objects in the data set. It is to be noted that in the literature some fuzzy distance metrics are available. However, such fuzzy distances which are once again fuzzy quantities. This renders direct application of existing methods to classify the objects almost impossible. Recently, Sampath and Kalaivani (2010), Sampath and Senthil Kumar (2014a,2014b,2015a,2015b) have explored the tools available in Credibility theory due to Liu (2008) in clustering of imprecise data objects. In this chapter, the two different approaches (based on credibility critical values and credibility expectation) pursued by them are explored for the purpose of classification. This chapter contains five sections. The second section of this chapter is devoted for an account on k-nn classification and imprecise data sets. The third section is devoted for credibility theory. Important terminologies associated with credibility theory which find applications in this chapter are also discussed. Two different approaches based on credibility theory meant for crisp conversion of imprecise data sets are explained in the fourth section of this chapter. The fifth section gives a complete description on simulation of imprecise data sets and the results of experimental studies carried out to compare the two different approaches when classification is performed using k -nn classifier. Concluding remarks are provided in the sixth section of this chapter.

2. NEIGHBORHOOD BASED CLASSIFICATION

In data mining literature several classifiers are available. Different classifiers use different approaches. k nearest neighborhood method is one of the classifiers based on the notion of nearest neighbors. This algorithm uses the proximity of the test object to the most similar (as identified by the distance metrics) training set objects. This is one of the most popular classifiers that has received the attention of data mining researchers. k -nn classifier has found applications in different domains of scientific investigation. Yihua and Vemuri (2002) used k-nn in Intrusion Detection. Hall, Park, Samworth, (2008) made a detailed study on the choice of neighbor order in nearest-neighbor classification. Shouman, Turner and Stocker (2012) considered the application of k-nn in diagnosing heart disease patients. Imandoust and Bolandraftar (2013) made a study on the application of k-nn classifier for predicting the occurrence of certain economic events. Alkhatib, Najadat, Hmeidi and Shatnawi (2013) have utilized k-nn algorithm in stock price prediction. Tsigkritis, Groumas and Schneider (2018) described an algorithm developed by modifying the conventional k-nn classifer to help detect threatening behavior in a computer network or cloud. The works mentioned above are related to classification of crisp data sets. By crisp data sets, it is meant that the values assumed by various variables/attributes/features associated with the data objects are crisp values which are free from any kind of impreciseness. In addition to this, the assignment of membership of a test object with the identified classes is also done in a precise manner leaving no scope for vague or unclear assignments. Several forms of modifications of k -nn classification are being done to increase the accuracy of the classifier.

Details of k -nn classification is furnished below.

As in the case of other classifiers, it assumes the existence of a training set. That is, it assumes the availability of a collection of objects whose status with respect to different classes are completely known. A test object is assigned a membership by using the following steps.

1. Compute the distances between the test object and every object in the training data set.
2. Arrange the distances in ascending order.
3. Identify the objects corresponding to k smallest distances.
4. Form the class distribution of k objects identified in the previous step.
5. Assign the test object to the class corresponding to the maximum frequency.

From the steps explained above, it is clear that the application k-nn method requires the quantification of distances between objects in the test set and those in training set. The choice of metric to be used in such quantification depends on the nature of the underlying variables. For interval valued variables one can use metrics such as Euclidean distance, Canberra metric etc. On the other hand, for binary valued variables, metrics like Hemming's distance, Jaccard coefficient etc can be used. Metrics mentioned above can be applied in a straight forward manner for data sets involving crisp values. For example, consider a hypothetical data set given below which describes the scores obtained by five candidates who have applied for a position in a company.

Table 1. A crisp data set

S.No	Quantitative	Language	General Awareness
1	65	75	45
2	71	83	36
3	38	65	49
4	57	50	58
5	15	40	61
6	37	60	35

One can use an appropriate metric to measure the level of dissimilarity between the candidates. It is to be noted that scores obtained in a test given in just one instance may not reflect the true ability of candidates due to various factors like the mental status of the candidate at the time of taking the test, the quality (reliability) of questions framed in the assessment process and so on. Hence, a data of the above kind may not a perfect picture about the candidates' ability. A situation like this gives an opportunity for the entry of impreciseness. Instead of rating the students on the basis of just one assessment, one can think of eliciting the opinion of people who have monitored them in their academic pursuits (with due credit to the scores obtained in tests) to grade the candidates. These grades may be given in the form of belief levels (lying between 0 and 1). A specific score with belief level close to 1 indicates a strong agreement and one close to zero indicates a strong disagreement. A table like the one given below will throw a clear picture about this.

For example, $\begin{pmatrix} 60 & 65 & 70 \\ 0.8 & 1 & 0.4 \end{pmatrix}$ indicates that candidate with serial 1 is given three scores for quantitative possibly 60,65 and 70 with belief levels 0.8, 1 and 0.4 respectively. The most possible score is 65 and the score 60 is more possible when compared to 70.

Sensing the need for a theory to handle these kinds of imprecise quantities, Zadeh (1965) introduced fuzzy set theory. Following the popularity of Fuzzy set theory, modifications on *k*-nn classifier has been done to perform fuzzy classification. It is Keller, Gray, and Givens (1985) who introduced the idea of fuzzy neighborhood-based classification by modifying the traditional *k*-nn classifier. In fuzzy *k*-nn, the classifier receives as input crisp data and gives the membership values of test objects with respect to

Table 2. An imprecise data set

S.No	Quantitative	Language	General Awareness
1	$\begin{pmatrix} 60 & 65 & 70 \\ 0.8 & 1 & 0.4 \end{pmatrix}$	$\begin{pmatrix} 71 & 75 & 80 \\ 0.1 & 0.8 & 1 \end{pmatrix}$	$\begin{pmatrix} 40 & 45 & 50 \\ 0.5 & 1 & 0.8 \end{pmatrix}$
2	$\begin{pmatrix} 71 & 75 & 79 \\ 0.4 & 1 & 0.6 \end{pmatrix}$	$\begin{pmatrix} 80 & 83 & 86 \\ 0.2 & 0.8 & 1 \end{pmatrix}$	$\begin{pmatrix} 30 & 35 & 40 \\ 0.1 & 0.8 & 1 \end{pmatrix}$
3	$\begin{pmatrix} 32 & 35 & 40 \\ 0.1 & 0.8 & 1 \end{pmatrix}$	$\begin{pmatrix} 57 & 65 & 70 \\ 0.3 & 1 & 0.7 \end{pmatrix}$	$\begin{pmatrix} 45 & 50 & 55 \\ 0.4 & 0.8 & 1 \end{pmatrix}$
4	$\begin{pmatrix} 51 & 57 & 60 \\ 0.6 & 1 & 0.8 \end{pmatrix}$	$\begin{pmatrix} 46 & 51 & 53 \\ 0.7 & 0.8 & 1 \end{pmatrix}$	$\begin{pmatrix} 57 & 60 & 65 \\ 0.4 & 1 & 0.6 \end{pmatrix}$
5	$\begin{pmatrix} 11 & 15 & 20 \\ 0.3 & 1 & 0.7 \end{pmatrix}$	$\begin{pmatrix} 41 & 45 & 49 \\ 0.1 & 1 & 0.2 \end{pmatrix}$	$\begin{pmatrix} 61 & 65 & 69 \\ 0.1 & 0.8 & 1 \end{pmatrix}$
6	$\begin{pmatrix} 31 & 35 & 39 \\ 0.6 & 0.8 & 1 \end{pmatrix}$	$\begin{pmatrix} 63 & 65 & 69 \\ 0.3 & 0.8 & 1 \end{pmatrix}$	$\begin{pmatrix} 31 & 36 & 40 \\ 1 & 0.8 & 0.2 \end{pmatrix}$

different identified classes. Following this, several works have been done in this direction. It is to be highlighted that these classifiers receive as input, crisp data sets and create classifications in an imprecise manner. Thus, it is observed that there are two approaches while performing classification. The first one being creation of crisp classification using crisp data sets. The second direction of studies is fuzzy classification based on crisp data sets. However, in real life situations one may come across imprecise data sets and face the situation of creating crisp classification. This chapter attempts to consider the application of *k*-nn classifier when the training data set contains imprecise values and crisp classification of test objects needs to be performed. The main difficulty in the implementation is computing the distances between objects possessing imprecise values as explained in Table 2. It is pertinent to note that there is no dearth of measures for computing the distances between fuzzy sets. Unfortunately, almost all the measures available in the literature are once again fuzzy quantities. Examples of such measures can be found in Chakraborty (2001). However, the application of *k*-nn classifier requires a single crisp value representing the distance between the objects. In this chapter it is proposed to adopt the paths used for such data sets in clustering by Sampath and Kalaivani (2010) and Senthil Kumar (2014a). Their studies make use of Credibility theory due to Liu (2008).

Following the work of Zadeh(1965), several competing theories are being evolved to study mathematical and application aspects of imprecise data sets. Recently, Liu (2008) has introduced a theory called Credibility theory to handle fuzziness. It is to be noted that credibility theory is based on possibility measures proposed by Zadeh (1978). Ever since credibility theory was evolved, researchers have contributed significantly on both the theoretical aspects as well as applications of credibility theory in real life situations. To maintain the readability of the chapter a brief account on credibility theory is furnished below.

3. CREDIBILITY THEORY

Credibility Measure and Credibility Space: Let Θ be a non-empty set and *P* be the power set of Θ. Each element in $P^{`}$ is called an event. In order to present an axiomatic definition of credibility measure, it is necessary to assign a number *Cr*(*A*) to each event *A*, where *Cr*(*A*) indicates the credibility that *A* will occur. The following four axioms are satisfied by a credibility measure.

1. (Normality) $Cr(\Theta)=1$
2. (Monotonicity) $Cr(A)\leq Cr(B)$ whenever $A \subset B$
3. (Self Duality) $Cr(A)+Cr(A^c)=1$
4. (Maximality) $Cr\{\cup_i A_i\} = Sup_i Cr\{A_i\}$ for any events $\{A_i\}$ with $Sup_i Cr\{A_i\}<0.5$

Fuzzy Variable: A credibility measurable function from credibility space (Θ, *P*, *Cr*) to the set of real numbers is called a fuzzy variable. It is usually denoted by ξ.

Membership Function: Let ξ defined on the credibility space (Θ, *P*, *Cr*). Then its membership function μ is derived from the credibility measure by

$$\mu(x) = (2Cr\{\xi = x\}) \wedge 1, \forall x \in \Re$$

If x is an impossible point then the membership degree $\mu(x)=0$ and $\mu(x)=1$. If x is the most possible point that ξ takes.

Types of Fuzzy Variables and Credibility Distribution

The various types of fuzzy variables are determined by the nature of the membership functions. In this section, definitions of some important fuzzy variables that are most often used in literature namely, equiossible fuzzy variable, triangular fuzzy variable and trapezoidal fuzzy variable are given below.

Equipossible Fuzzy Variable: A fuzzy variable fully determined by the pair (a,b) of crisp numbers with $a<b$ and membership function f the form

$$\mu(x) = \begin{cases} 1 \text{ if } a < b \\ 0 \text{ otherwise} \end{cases} \tag{1}$$

is called an equipossible fuzzy variable.

Triangular Fuzzy Variable: A fuzzy variable fully determined by the triplet (a,b,c) of crisp numbers with $a<b<c$ and membership function of the form

$$\mu(x) = \begin{cases} \dfrac{x-a}{b-1} & \text{if } a \le x \le b \\ \dfrac{x-c}{b-c} & \text{if } b \le x \le c \\ 0 & \text{otherwise} \end{cases} \tag{2}$$

is called a triangular fuzzy variable. If b-a=c-b then the triangular fuzzy variable is said to be symmetric.

Trapezoidal Fuzzy Variable: A fuzzy variable fully determined by the quadruplet (a,b,c,d) of crisp numbers with $a<b<c<d$ and membership function of the form

$$\mu(x) = \begin{cases} \dfrac{x-a}{b-1} & \text{if } a \le x \le b \\ 1 & \text{if } b \le x \le c \\ \dfrac{x-d}{c-d} & \text{if } c \le x \le d \\ 0 & \text{otherwise} \end{cases} \tag{3}$$

is called a trapezoidal fuzzy variable. If b-a=d-c then the trapezoidal fuzzy variable is said to be symmetric.

Credibility Distribution: The credibility distribution $\Phi : \Re \to [0,1]$ of a fuzzy variable ξ is defined as

$$\Phi(x) = Cr\{\theta \in \Theta \mid \xi(\theta) \le x\}, \forall x \in \Re \tag{4}$$

That is, Φ is the credibility that the fuzzy variable ξ takes a value less than or equal to x. Generally, the credibility distribution is neither left-continuous nor right-continuous.

Critical Values: For every fuzzy variable ξ, Liu (2008) defines two crisp critical values, namely, Optimistic andPessimistic values as follows.

Let ξ be a fuzzy variable, and $\alpha \in (0,1]$. The α–optimistic value of ξ is defined as

$$\xi_{\sup}(\alpha) = \sup\{r \,|\, Cr\{\xi \geq r\} \geq \alpha\} \tag{5}$$

Let ξ be a fuzzy variable, and $\alpha \in (0,1]$. The α–pessimistic value of ξ is defined as

$$\xi_{\inf}(\alpha) = \inf\{r \,|\, Cr\{\xi \leq r\} \geq \alpha\} \tag{6}$$

It shows that theα–optimistic value $\xi_{\sup}(\alpha)$ is the supremum value that ξ achieves with credibility at leastα, and theα–pessimistic value $\xi_{\inf}(\alpha)$ is the infimum value that ξ achieves with credibility at least α.

Distance between Fuzzy Variables: Liu (2008) defines the distance between two fuzzy variables ξ and η as $E(|\xi\text{-}\eta|)$. If $\xi_1=(a,b,c)$ and $\xi_2=(d,e,f)$ are triangular fuzzy variables then the distance between ξ_1 and ξ_2 is given by

$$\frac{1}{4}(|\,a-f\,|+|\,b-e\,|+|\,c-d\,|).$$

This holds only if $(a,c) \cap (d,f) = \varnothing$ and is true. However, when this condition is not satisfied the computation of distance between the two fuzzy variables ξ_1 and ξ_2 is difficult. The distances between fuzzy variables could not be of much in the context of classification related imprecise data sets. Hence, alternative approaches used by Sampath and Kalaivani (2010) and Sampath and Senthil Kumar (2014a) are given in the following section.

4. DISTANCE BETWEEN DATA OBJECTS WITH IMPRECISE VALUES

As mentioned in the beginning of this chapter, the problem considered in this paper is classification of test objects using the proximity of test objects with neighboring training objects. Notations used in representing an imprecise training set are explained below. This explanation assumes the N training set objects are measured using p attributes.

Let $O_1, O_2, \ldots, O_N$ be the N training objects in the data set each having p attributes. The value assumed by the ith object with respect to jth attribute is considered as a fuzzy variable $\xi_{ij}(i=1,2,\ldots,N; j=1,2,\ldots,p)$ with corresponding membership function denoted by $\mu_{ij}(i=1,2,\ldots,N; j=1,2,\ldots,p)$. It is assumed that the underlying fuzzy variables have well defined credibility distributions. If each of these fuzzy variables takes m possible values, then its membership function assumes m values. In this work it is assumed that all the fuzzy variables same number of values. Taking into account all the p attributes the distance between the objects O_i and O_j is defined as

$$d(O_i, O_j) = d_{ij} = \sum_{k=1}^{p} E(|\, \xi_{ik} - \xi_{jk} \,|.$$

Obtaining compact general expressions which can be used for the distance between imprecise data objects with the help of this formula is difficult. Hence, Sampath and Kalaivani (2010) proposed an alternative way based on critical values of fuzzy variables for measuring the distance between two multidimensional objects having fuzzy attributes. The method pursued by them is given below.

Corresponding to every fuzzy variable one can compute optimistic and pessimistic values for a given α. These optimistic and pessimistic values of the fuzzy variable ξ_{ij} are denoted by $\overline{\xi}_{ij}$ and $\underline{\xi}_{ij}$ respectively. Further, on averaging the optimistic and pessimistic values one can develop another crisp value which is denoted as

$$\overset{*}{\xi}_{ij} = \frac{1}{2}\left[\overline{\xi}_{ij} + \underline{\xi}_{ij}\right].$$

Thus, on using these three values namely, optimistic, pessimistic and average of optimistic and pessimistic values one can define three p-component crisp vectors namely,

$$\overline{\xi}_i = \left(\overline{\xi}_{i1}, \overline{\xi}_{i2}, \ldots, \overline{\xi}_{ip}\right), \underline{\xi}_j = \left(\underline{\xi}_{j1}, \underline{\xi}_{j2}, \ldots, \underline{\xi}_{jp}\right)$$

and

$$\overset{*}{\xi}_i = \left(\overset{*}{\xi}_{i1}, \overset{*}{\xi}_{i2}, \ldots, \overset{*}{\xi}_{ip}\right)$$

for every object in the data set. Note that these vectors are of crisp nature. Once these values are identified, any appropriate distance metric can be used to measure the levels of dissimilarity between objects O_i and O_j. If Euclidean distance is used to measure the levels of dissimilarity between O_i and O_j, then we have

(i) distance based on optimistic value $\underline{d}_{ij} = \sqrt{\sum_{k=1}^{p}\left(\underline{\xi}_{ik} - \underline{\xi}_{kj}\right)^2}$

(ii) distance based on pessimistic value $\overline{d}_{ij} = \sqrt{\sum_{k=1}^{p}\left(\overline{\xi}_{ik} - \overline{\xi}_{kj}\right)^2}$ and

(iii) distance based on average of optimistic and pessimistic values $\overset{*}{d}_{ij} = \sqrt{\sum_{k=1}^{p}\left(\overset{*}{\xi}_{ik} - \overset{*}{\xi}_{kj}\right)^2}$

Crisp Conversion Using Credibility Expectation: The method explained above considers crisp conversion based on a suitably chosen critical value. The crisp values depend on the choice of the order of critical value. Sampath and Senthil Kumar (2014b) advocated a method of crisp conversion using the notion of credibility expectation due to Liu (2008). The main advantage of this approach is that unlike critical value approach value it does not require specifying the orders of critical values to be used.

In this approach, each element of the data matrix presented is treated as a discrete fuzzy variable. Here, values assumed by each discrete fuzzy variable are associated with certain membership values which lie between 0 and 1. The process of computing the expected value of the fuzzy variable ξ_{ij} which assumes m possible values namely $\xi_{ij}^{u}(u = 1,2,..,m)$, with membership values $\mu_{ij}^{u}(u = 1,2,..,m)$ is explained below.

For every possible value $\xi_{ij}^{u}(u = 1,2,..,m)$, a weight is computed based on the membership values associated with the possible values assumed by the fuzzy variable on using the formula

$$w_{ij}^{u} = \frac{1}{2}\left[\max_{1\le k\le m}\left\{\mu_{ij}^{k} \mid x_{ij}^{k} \le x_{ij}^{u}\right\} - \max_{1\le k\le m}\left\{\mu_{ij}^{k} \mid x_{ij}^{k} < x_{ij}^{u}\right\} + \max_{1\le k\le m}\left\{\mu_{ij}^{k} \mid x_{ij}^{k} \ge x_{ij}^{u}\right\} - \max_{1\le k\le m}\left\{\mu_{ij}^{k} \mid x_{ij}^{k} > x_{ij}^{u}\right\}\right]$$

$$i=1,2,\ldots,N;\ j=1,2,\ldots,p \tag{7}$$

The expected value of the discrete fuzzy variable ξ_{ij} is the sum of product of values assumed by the fuzzy variable and corresponding weights. That is,

$$E[\xi_{ij}] = \sum_{u=1}^{m} w_{ij}^{u}\xi_{ij}^{u} \tag{8}$$

This is based on the definition of expectation of fuzzy variable given by Liu (2008). Thus two approaches of crisp conversion have been discussed in this chapter. In the following section a comparative study on the choice of method to be used in crisp conversion while implementing k-nn classification is considered.

5. EXPERIMENTAL STUDY

While developing machine learning algorithms, researchers often make use of bench mark data sets available in popular repositories. However, bench mark data sets pertaining to imprecise data values are not available in such repositories. Hence, it is decided to make use of simulated imprecise data sets. The process of simulating imprecise data sets as carried out by Sampath and Senthil Kumar(2014a) is given below. objects are simulated from each of the three populations of the following data sets generated from multivariate normal populations as explained below. It is to be noted that there are several practical situations where practitioners come across such situations as one can find in Johnson and Wichern (2008). In addition to such practical situations, diagnostic tools for assessing multivariate normality are discussed in Chapter 4 of Johnson and Wichern (2008).

It is a well-known fact that the data sets considered in data mining studies are multivariate data sets. If there are N objects in the training data set with p attributes and a class label then the data matrix will be of order $N\times(p+1)$ whose first p columns represent the values of N with respect to the p variables / attributes /features and the last column gives the class labels of the training set objects.

In the above data matrix X_{ij} denotes the j-th value simulated from the p- variate distribution with the i-th parametric setting. This value will be treated as the value of the j-th attribute corresponding to the i-th object in the data set.

The process of simulating imprecise data sets begins with the creation of crisp data sets. Consider the situation where there are m classes and the number of p- variate instances from the ith class is N_i.

Step 1: m sets of parametric values which characterize the inherent characteristics of m classes are identified.

Step 2: Corresponding to each set of parametric value a total of N_i, p- component instances are simulated.

The total of $N = \sum_{i=1}^{N_i} N_i$ data instances obtained in Step 2 form the foundation of imprecise data set simulation. Thus, a total of Np scalar values are simulated.

Each one of these Np values will be used to generate a fuzzy variable which assumes a predetermined set of possible values (depending on the domain of the study) along with respective membership values. For example, if the number of possible values X_{ij} is t then we create a fuzzy variable ξ_{ij} which can take t values with given membership values. The credibility distribution corresponding to ξ_{ij} can be represented as

$$\xi_{ij} = \begin{pmatrix} \xi_{ij}^{(1)} & \xi_{ij}^{(2)} & .. & .. & .. & \xi_{ij}^{(t-1)} & \xi_{ij}^{(t)} \\ \mu_{ij}^{(1)} & \mu_{ij}^{(2)} & .. & .. & .. & \mu_{ij}^{(t-1)} & \mu_{ij}^{(t)} \end{pmatrix}, i = 1,2,..,N, j = 1,2,..,p\,.$$

Even though the number of possible values a fuzzy variable assume can be different, in this work they are assumed to be constant, namely t. The t possible values assumed by the fuzzy variable ξ_{ij} are expected to be around X_{ij} and their corresponding membership values are randomly generated from the interval [0,1].

In this work, it is proposed to take the value of t as 5 where the middlemost value will be the value X_{ij} and the other values are X_{ij} -2, X_{ij} -1, X_{ij} +1 and X_{ij} +2. For example, if $X_{ij} = 45$ then the remaining four values are 43, 44, 46 and 47. Each of these values are assigned randomly simulated membership values. It is pertinent to note that at least one of the possible values of the fuzzy variable must be assigned the membership value 1. According to Peng and Iwamura (2010), a credibility distribution is well defined if and only if the membership value assumes 1 at least for one possible value of the fuzzy variable.

Based on the above discussion, one should observe one crisp quantity X_{ij} leads to a fuzzy variable which assumes five different values with different levels of membership values. Such values are used in this experimental study. It is to be emphasized that the possible value the fuzzy variable assumes is domain dependent and the membership values are elicited from the expects who are familiar with the system. It is proposed to use two different types of data sets are to be used. The first type contains objects whose values over lap and the second type contains non over lapping values. The details and outcomes of the experimental study are explained below.

Over lapping data: The first group of data set is generated from three trivariate normal populations with identical mean vectors $\begin{bmatrix}\mu_1\\ \mu_2\\ \mu_3\end{bmatrix}$ with variance covariance matrices of the form $\Sigma = \begin{bmatrix}\sigma_1^2 & \rho\sigma_1\sigma_2 & \rho\sigma_1\sigma_3\\ \rho\sigma_2\sigma_1 & \sigma_2^2 & \rho\sigma_2\sigma_3\\ \rho\sigma_3\sigma_1 & \rho\sigma_3\sigma_2 & \sigma_3^2\end{bmatrix}$.

Three sets of data are obtained using the parametric values listed below.

(i) Data Set 1:

Class 1: $\mu_1 = 45, \mu_2 = 35, \mu_3 = 25, \sigma_1^2 = 3, \sigma_2^2 = 2, \sigma_3^2 = 1, \rho = 0$

Class 2: $\mu_1 = 45, \mu_2 = 35, \mu_3 = 25, \sigma_1^2 = 3, \sigma_2^2 = 2, \sigma_3^2 = 1, \rho = 0.25$

Class 3: $\mu_1 = 45, \mu_2 = 35, \mu_3 = 25, \sigma_1^2 = 3, \sigma_2^2 = 2, \sigma_3^2 = 1, \rho = 0.5$

(ii) Data Set 2:

Class 1: $\mu_1 = 45, \mu_2 = 35, \mu_3 = 25, \sigma_1^2 = 3, \sigma_2^2 = 2, \sigma_3^2 = 1, \rho = 0.2$

Class 2: $\mu_1 = 45, \mu_2 = 35, \mu_3 = 25, \sigma_1^2 = 3, \sigma_2^2 = 2, \sigma_3^2 = 1, \rho = 0.4$

Class 3: $\mu_1 = 45, \mu_2 = 35, \mu_3 = 25, \sigma_1^2 = 3, \sigma_2^2 = 2, \sigma_3^2 = 1, \rho = 0.6$

(iii) Data Set 3:

Class 1: $\mu_1 = 45, \mu_2 = 35, \mu_3 = 25, \sigma_1^2 = 3, \sigma_2^2 = 2, \sigma_3^2 = 1, \rho = 0.4$

Class 2: $\mu_1 = 45, \mu_2 = 35, \mu_3 = 25, \sigma_1^2 = 3, \sigma_2^2 = 2, \sigma_3^2 = 1, \rho = 0.6$

Class 3: $\mu_1 = 45, \mu_2 = 35, \mu_3 = 25, \sigma_1^2 = 3, \sigma_2^2 = 2, \sigma_3^2 = 1, \rho = 0.8$

It is to be noted that all the three sets have the same mean vector. This choice makes the datasets have low separation between the data values. The three classes are characterized by the structure of the variance covariance matrices used for their respective simulation. In Data Set 1, the variables are simulated from distributions with low correlations ranging from 0 to 0.5. The third data set simulates from a system where the variables have correlations on the higher side ranging from 0.4 to 0.8. The second data set ensures medium level correlations between the variables lying between 0.2 and 0.6. Each data set is made up of 150 objects belonging to three classes as characterized by the parameters. Equal number of objects (namely, 50) are simulated from each class.

The entries in the second row of Table 3, namely 0.5981, 0.5811 and 0.5515 give error rates of *k*-nn classification when the imprecise data sets generated using the parametric settings given in Data Set 1, Data Set 2 and Data Set 3 are converted into crisp data sets using credibility expectation. The entries given in the other rows correspond to error rates of *k*-nn classifier for different values of α ranging from 0.55 to 0.95 for the above explained three data sets when optimistic, pessimistic critical values and their average are used for crisp conversion. It is to be noted that the error rates given in the table are the averages of error rates based on 30 replications carried out for each parametric setting. For each replication, the error rate is based on the outcome of tuning done to the *k*-nn classifier using *tune.knn* function available in the R package *e1071*. Values ranging from 1 to 50 have been used in the tuning process for the number of neighbors to be used in classification and 10-fold cross validation has been used.

Table 3. Error rates corresponding to crisp conversion using expectation and critical values for overlapping data

	Data Set 1			Data Set 2			Data Set 3		
Exp	**0.5981**			**0.5811**			**0.5515**		
α	**Opt**	**Pes**	**Avg**	**Opt**	**Pes**	**Avg**	**Opt**	**Pes**	**Avg**
0.50	0.6021	***0.5968***	0.6024	0.6123	***0.6067***	0.6168	0.5859	***0.5835***	0.5843
0.55	0.6064	***0.5944***	0.5995	0.6221	***0.6115***	0.6120	***0.5864***	0.5928	0.5891
0.60	0.6048	0.6075	***0.6064***	0.6037	0.6053	***0.6027***	0.5912	0.5877	***0.5736***
0.65	0.6016	0.6045	***0.5933***	0.6053	***0.5976***	0.5984	0.5899	0.5787	***0.5757***
0.70	0.6093	0.5992	***0.5949***	0.6027	0.5997	***0.5995***	0.5821	0.5715	***0.5563***
0.75	0.6003	***0.5864***	0.6008	0.5971	0.5939	***0.5896***	0.5672	0.5635	***0.5619***
0.80	0.5944	***0.5864***	0.5909	0.5933	***0.5811***	0.5864	0.5731	0.5584	***0.5496***
0.85	0.5888	0.5941	***0.5741****	0.5867	0.5861	***0.5773***	0.5531	0.5515	***0.5411***
0.90	0.5960	0.5931	***0.5829***	0.5776	0.5787	***0.5709***	0.5485	0.5395	***0.5368***
0.95	0.5923	0.5877	***0.5853***	0.5712	***0.5699****	0.5824	0.5328	***0.5197****	0.5237

A close scrutiny of the contents of Table 3 clearly supports the use of crisp conversion using credibility expectation rather than the one based on critical values when lower level critical values are used. However, if the critical value of higher order are used then either pessimistic or average of pessimistic and optimistic work well. Optimistic conversion fails to work well at all levels of conversions. Overall, conversion using the average of pessimistic and average at a level close to 0.90 appears to work well in all the three data sets.

Non overlapping Data: The second group of data sets are generated from three trivariate normal populations with different mean vectors and variance covariance matrices as explained below.

(iv) Data Set 4:

Class 1: $\mu_1 = 45, \mu_2 = 35, \mu_3 = 25, \sigma_1^2 = 3, \sigma_2^2 = 2, \sigma_3^2 = 1, \rho = 0$

Class 2: $\mu_1 = 25, \mu_2 = 10, \mu_3 = 15, \sigma_1^2 = 3, \sigma_2^2 = 2, \sigma_3^2 = 1, \rho = 0.25$

Class 3: $\mu_1 = 30, \mu_2 = 25, \mu_3 = 20, \sigma_1^2 = 3, \sigma_2^2 = 2, \sigma_3^2 = 1, \rho = 0.5$

(v) Data Set 5:

Class 1: $\mu_1 = 45, \mu_2 = 35, \mu_3 = 25, \sigma_1^2 = 3, \sigma_2^2 = 2, \sigma_3^2 = 1, \rho = 0.2$

Class 2: $\mu_1 = 25, \mu_2 = 10, \mu_3 = 15, \sigma_1^2 = 3, \sigma_2^2 = 2, \sigma_3^2 = 1, \rho = 0.4$

Class 3: $\mu_1 = 30, \mu_2 = 25, \mu_3 = 20, \sigma_1^2 = 3, \sigma_2^2 = 2, \sigma_3^2 = 1, \rho = 0.6$

(vi) Data Set 6:

Class 1: $\mu_1 = 45, \mu_2 = 35, \mu_3 = 25, \sigma_1^2 = 3, \sigma_2^2 = 2, \sigma_3^2 = 1, \rho = 0.4$

Class 2: $\mu_1 = 25, \mu_2 = 10, \mu_3 = 15, \sigma_1^2 = 3, \sigma_2^2 = 2, \sigma_3^2 = 1, \rho = 0.6$

Class 3: $\mu_1 = 30, \mu_2 = 25, \mu_3 = 20, \sigma_1^2 = 3, \sigma_2^2 = 2, \sigma_3^2 = 1, \rho = 0.8$

In the above choices, the variance covariance matrices are identical to those used in the overlapping cases. However, the mean vectors are selected so that the ranges of values assumed by the variables are significantly different from class to class. The number of objects in the data sets (namely 150) and the size distribution (namely 50 from each class) of various classes are identical to those used in the overlapping case.

Entries in Table 4 clearly demonstrates that crisp conversion using credibility expectation gives better classification of imprecise data sets when compared to conversion using critical values of lower order (approximately less than 0.65). When critical values of higher order (exceeding 0.65) are used either pessimistic or the average of pessimistic give better results. If the data sets which represent the three classes are made up of variables with smaller correlations, then conversion using pessimistic values works well. On the other hand, average conversion works well. Based on the entries of Table 4, the conversion using average of pessimistic and optimistic with level around 0.95 can be recommended.

Another noticeable future one can observe is that pessimistic and averages of optimistic and pessimistic conversion give relatively better results when compared to optimistic conversion for higher level (roughly exceeding 0.65) critical values. On the other end for lower levels optimistic conversion seems to be better among the three conversion methods based on critical values. As the critical level increases the error rate decreases in almost all the cases. Also interesting note that as the critical level increase in all the three approaches, namely, optimistic, pessimistic and average of pessimistic and optimistic the error rates have tendency to decrease.

6. CONCLUSION

This chapter studies the conversion of imprecise data sets to crisp data by credibility critical values and credibility expectation. Conversion of these data sets facilitates the application of k-nn classifier which primarily depends on the distances which are free from any kind of impreciseness. In this work, two

Table 4. Error rates corresponding to crisp conversion using expectation and critical values for Non overlapping data

	Data Set 4			Data Set 5			Data Set 6		
Exp	0.5891			0.5912			0.5976		
α	Opt	Pes	Avg	Opt	Pes	Ave	Opt	Pes	Avg
0.5000	0.6032	0.6080	***0.6029***	0.5971	0.5912	***0.5928***	0.6149	***0.6131***	0.6152
0.5500	0.6064	***0.5939***	0.6107	***0.6019***	0.6032	0.6032	0.6136	0.6147	***0.6067***
0.6000	0.5955	***0.5901***	0.5971	***0.6032***	0.6149	0.6064	0.6179	0.6147	***0.6075***
0.6500	0.5869	0.5901	***0.5845***	***0.5960***	0.6072	0.6045	0.6104	0.6157	***0.5995***
0.7000	0.5997	*0.5829*	***0.5917***	0.6051	0.6064	***0.5979***	0.6101	0.6064	***0.5963***
0.7500	0.5883	0.5851	***0.5768***	0.5960	0.5949	***0.5949***	0.6131	***0.5992***	0.6072
0.8000	0.6016	***0.5757***	0.5867	0.5944	***0.5827***	0.5859	0.5963	***0.5925***	0.6003
0.8500	0.5800	***0.5795***	0.5805	0.5947	0.5795	***0.5816***	0.5957	0.5979	***0.5907***
0.9000	0.5909	***0.5840***	0.5859	0.6016	0.5763	***0.5819***	0.5888	0.5891	***0.5821****
0.9500	0.5920	***0.5795****	0.5867	0.5957	0.5845	***0.5805****	0.5832	0.5789	***0.5824***

directions of crisp conversion have been considered. Three approaches of obtaining crisp quantities make use of credibility optimistic, pessimistic and the average of optimistic and pessimistic values. The last approach uses credibility expectation. Extensive simulation studies using two types of imprecise data sets were simulated from three trivariate normal populations. One group was generated from populations with identical mean vectors but with different variance covariance matrices and the other from different mean vectors. The following are broad conclusions reached based on the study carried out in this chapter. The conclusions hold good for both the type of data sets considered in this work.

1. Crisp conversion using credibility expectation gives smaller error rates when compared to the conversions done using critical values for smaller values of α (approximately less than 0.6)
2. When conversion using critical values is carried out, error rates have the tendency to decrease as the critical level increases
3. Both pessimistic and average of pessimistic and optimistic conversions yield smaller error rates when compared to optimistic conversion in vast majority of the cases. There are few cases where optimistic conversion gives better results when compared to the other two critical values based conversions. But in such situations, credibility expectation conversion works well.
4. Among the four methods considered smallest error rates are in almost all the cases (barring very few cases) those based on the average of pessimistic and optimistic values corresponding to high values for α, say above 0.90

REFERENCES

Alkhatib, K., Najadat, H., Hmeidi, I., & Shatnawi, M. K. A. (2013). Stock Price Prediction Using K-Nearest Neighbor (kNN) Algorithm. *Int. J. Bus. Humanit. Technol.*, *3*(3), 32–44.

Chakraborty, D. (2001). Structural quantification of vagueness in linguistic expert opinion in an evaluation programme. *Fuzzy Sets and Systems*, *119*(1), 171–186. doi:10.1016/S0165-0114(99)00044-5

Corinna, C., & Vapnik, V. (1995). Support vector networks. *Machine Learning*, *20*(3), 273–297. doi:10.1007/BF00994018

Cover, T. M., & Hart, P. E. (1967). Nearest neighbor pattern classification. *IEEE Transactions on Information Theory*, *13*(1), 21–27. doi:10.1109/TIT.1967.1053964

Fix, E., & Hodges, J. L. (1951). *Discriminatory analysis, Nonparametric discrimination; consistency properties.* Tech. Rep. 4, USAF School of Aviation Medicine, Randolph Field, TX.

Hall, P., Park, B. U., & Samworth, R. J. (2008). Choice of neighbor order in nearest-neighbor classification. *Annals of Statistics*, *36*(5), 2135–2152. doi:10.1214/07-AOS537

Imandoust, S. B., & Bolandraftar, M. (2013). Application of K-Nearest Neighbor (KNN) Approach for Predicting Economic Events: Theoretical Background. *Int. J. Eng. Res. Appl.*, *3*(5), 605–610.

Johnson, R. A., & Wichern, D. W. (2008). *Applied Multivariate Statistical Analysis*. Pearson.

Keller, J. M., Gray, M. R., & Givens, J. A. (1985). A Fuzzy K-nearest Neighbor Algorithm. *IEEE Transactions on Systems, Man, and Cybernetics*, *SMC-15*(4), 580–585. doi:10.1109/TSMC.1985.6313426

Liu, B. (2008). *Uncertainty Theory* (3rd ed.). Retrieved from http://orsc.edu.cn/liu/ut.pdf

Peng, Z. X., & Iwamura, K. (2010). A Sufficient and Necessary Condition of Uncertainty Distribution. *Journal of Interdisciplinary Mathematics*, *13*(3), 277–285. doi:10.1080/09720502.2010.10700701

Sampath, S., & Kalaivani, R. (2010). Clustering of Fuzzy data using credibilistic critical values. *IEEE International Conference on Signal and Image Processing*, 227–232. 10.1109/ICSIP.2010.5697474

Sampath, S., & Senthil Kumar, R. (2014a). Fuzzy Clustering using Credibilistic Critical Values. *International Journal of Computational Intelligence and Informatics*, *3*, 213–221.

Sampath, S., & Senthil Kumar, R. (2014b). Clustering of Fuzzy Data using Credibilistic Expected and Critical Values. *IEEE Explore*, 176-181. 1 doi:0.1109/ICCCS.2014.7068189

Sampath, S., & Senthil Kumar, R. (2015a). Rough Clustering Using Credibilistic Critical Values. *International Journal of Data Mining and Emerging Technologies*, *5*(1), 24–30. doi:10.5958/2249-3220.2015.00005.1

Sampath, S., & Senthil Kumar, R. (2015b). Image Segmentation Using Credibilistic Critical Values. *Journal of Image Processing & Pattern Recognition Progress*, *2*, 46–55.

Shouman, M., Turner, T. L., & Stocker, R. C. (2012). Applying k-nearest neighbour in diagnosing heart disease patients. *International Journal of Information and Education Technology (IJIET)*, *2*, 220–223. doi:10.7763/IJIET.2012.V2.114

Tan, P. N., Steinbach, M., & Kumar, V. (2006). *Introduction to Data Mining*. Pearson.

Tsigkritis, T., Groumas, G., & Schneider, M. (2018). On the Use of k-NN in Anomaly Detection. *Journal of Information Security*, *9*(01), 70–84. doi:10.4236/jis.2018.91006

Yihua, L., & Vemuri, V. R. (2002). Use of K-NN for Intrusion Detection. *Computers & Society*, *21*(5), 439–448.

Zadeh, L. A. (1965). Fuzzy sets. *Information and Control*, *8*(3), 338–353. doi:10.1016/S0019-9958(65)90241-X

Zadeh, L. A. (1978). Fuzzy sets as a basis for a theory of possibility. *Fuzzy Sets and Systems*, *1*(1), 3–28. doi:10.1016/0165-0114(78)90029-5

Chapter 5
Multiset Approach to Algebraic Structures

Suma P.
National Institute of Technology, Calicut, India

Sunil Jacob John
https://orcid.org/0000-0002-6333-2884
National Institute of Technology, Calicut, India

ABSTRACT

This chapter explores the various algebraic structures that exist in the fuzzy set environment in a multiset context. The authors have also established analogs of groups, rings, and ideals of sets in multiset environment. The conditions under which a multiset that is derived from a group becomes a multiset group (shortly mset group) and that derived from a ring to become a multiset ring (shortly mset ring) have been mentioned. The distinguishing between multiset subring and multiset ideals has been discussed on in the chapter. The authors have also delved into properties such as intersection and union of mset groups, mset rings, and ideals.

1. INTRODUCTION

Fuzzy Sets have been along for quite a while now. The concept was first introduced by Zadeh in 1965 and since then, a wide variety of applications and developments have been explored. On the other hand, Multisets were introduced almost two decades after fuzzy sets by Knuth, yet the level of advancement is quite pale in comparison. A comprehensive account of fundamentals of multiset and its applications in various forms can be found in (Chris Brink, 1988), C.S.calude et al. (C. S. Calude, G. Paun, G. Rozenberg & A. Salomaa, 2001), D. Singh et al. (D. Singh, A. M. Ibrahim, T. Yohanna & J. N. Singh, 2007), (K. P. Girish & Sunil Jacob John, 2009), (K. P. Girish & S.J. John 2011), (K. P. Girish & S.J. John, 2012), (K. P. Girish & S.J. John, 2012). (N.J.Wildberger, 2003). and (Wayne D. Blizard, 1991). Applications of cardinality bounded multisets can be found in these papers.

DOI: 10.4018/978-1-7998-0190-0.ch005

Since Crisp Sets have a membership value of 0 and 1, Fuzzy Sets take on a value in [0,1]. But, from this, when we come on to the case of multisets, any integer value both positive and negative is possible. The results of Fuzzy Set Theory have a wide range of application, especially in Computer Science field. Multiset is also now emerging in various fields although it is largely unexplored. A prime example would be *GUAVA* a version of JAVA developed by Google. It is primarily based on Multisets and their properties rather than Fuzzy Sets.

Multisets are used more and more in computer science for quantitative analysis, information retrieval and models of resources. The concept of multiset was introduced in order to capture the idea of multiplicity of appearance, or resource. Multisets are defined by assuming that for a given set M, an element x occurs a finite number of times. For example, the prime factorization of a natural number 'n' is a multiset whose elements are primes. The invariants of a finite abelian group also form a multiset. Even processes in an operating system can be seen as a multiset, and the examples can continue.

Various algebraic structures like Groups, Rings etc. based on Fuzzy Sets are already being used in many fields such as Computer Science, Chemistry, Physics and other subjects. These structures on Fuzzy Sets have been deliberated on for quite a while now. Some of the basic concepts of Fuzzy Groups and Fuzzy Groupoids are explained by Azriel Rosenfeld (Azriel Rosenfeld, 1971). Fuzzy rings, Fuzzy left and right ideals, fuzzy prime and maximal ideals of a general ring R are discussed by Souriar Sebastian et al. (Souriar Sebastian, Mercy K. Jacob, V. M. Mary & Divya Mary Daise. S, 2012). Yuying Li et al. (Yuying Li, Xuzhu Wang & Liqiong Yang, 2013) have presented a special type of (λ, μ) Fuzzy subgroup and its basic properties. Even if these exist in multisets, their possibilities and applications have remained largely unexplored. A few like multi-groups and some of their results have been discussed by (A. M. Ibrahim & P.A. Ajegwa 2016), Binod Chandra Ttripathy et al. (Binod ChandraTtripathy, Shyamal Debnath & Debjani Rakshit, 2018), (Johnson Aderemi Awolola & Adeku Musa Ibrahim 2016), (P.A. Ejegwa, 2017), S.K..Nazmul et al. (S.K..Nazmul, P.Majumdar & S.K.Samanta, 2013) and (Y. Tella & S. Daniel, 2013). Different aspects and applications of Fuzzy Multisets are investigated by P. Rajarajeswari and N. Uma (P. Rajarajeswari & N. Uma, 2013) and Sabu Sebastian and T.V. Ramakrishnan (Sabu Sebastian & T.V. Ramakrishnan, 2011).

The authors attempt in this article is to extend the algebraic concepts that are utilized in the Fuzzy environment to a Multiset context. It was noted that while Arrays had an ordered structure, yet allowed for duplicity, and sets had uniqueness in its elements while being unordered, multisets combined the better aspects of the two while removing the drawbacks as the elements of a multiset are both ordered and unique. This observation led to the thought that the usage of multisets and structures derived from them could be used in place of Cantor sets in various places for an improved result, which is what resulted in this paper.

In section 2, some of the definitions of both fuzzy sets and of multisets and few basic concepts about multisets are mentioned. Section 3 deals with group structure, while section 4, ring structure. Meanwhile, the ideals under multiset context are discussed in section 5.

2. PRELIMINARIES

In this section, they cite the fundamental definitions of fuzzy set, Fuzzy group, Fuzzy ring and Fuzzy ideal, followed by a few basic results about Multisets.

2.1. Definition: (L. A. Zadeh, 1965) *Let* X be any non-empty set. A fuzzy subset 'A' of X is a function from X to [0,1]

2.2. Definition: (K. P. Girish & Sunil Jacob John, 2009) Let A be a fuzzy set on a group G. Then A is said to be a fuzzy subgroup of G if for all x, y ∈G,

(i) $A(xy) \geq \min\{A(x), A(y)\}$

(ii) $A(x^{-1}) = A(x)$.

2.3. Definition: (Souriar Sebastian, Mercy K. Jacob, V. M. Mary & Divya Mary Daise. S, 2012) Let R be a ring. A fuzzy set A of R is called a fuzzy ring of R if

(i) $A(x - y) \geq \min(A(x), A(y))$, for all x, y in R

(ii) $A(x\,y) \geq \min(A(x), A(y))$, for all x, y in R

2.4. Definition: (Souriar Sebastian, Mercy K. Jacob, V. M. Mary & Divya Mary Daise. S, 2012)R being a ring, a fuzzy set A of R is called a fuzzy ideal of R if

(i) $A(x - y) \geq \min(A(x), A(y))$, for all x, y in R

(ii) $A(x\,y) \geq \max(A(x), A(y))$, for all x, y in R

2.5. Definition:(K. P. Girish & Sunil Jacob John, 2009) A collection of elements containing duplicate is called multiset. If X is a set of elements, a multiset M drawn from X is represented by a function C_M defined as C_M: $X \rightarrow N$ where N is the set of non negative integers. For each $x \in X$, $C_M(x)$ is the characteristic value of x in M and indicates the number of occurrence of x in M.

The function C_M is known as the count function.

The word multiset is often shortened to mset.

2.6. Notation: Let M be an mset from $X=\{x_1,x_2,\ldots,x_n\}$ with x_1 appearing k_1 times, x_2 appearing k_2 times and so on, x_n appearing k_n times. Then M is written as $M=\{k_1|x_1,k_2|x_2,\ldots,k_n|x_n\}$.

2.7. Definition: (K. P. Girish & Sunil Jacob John, 2009) Let M_1 and M_2 be two msets drawn from a set X M_1 is a *submultiset* of

$$M_2, (M_1 \subseteq M_2) \text{ if } C_{M_1}(x) \leq C_{M_2}(x)$$

for all x in X.

2.8. Definition: (K. P. Girish & Sunil Jacob John, 2009) Two msets M_1 and M_2 are *equal* if

$$M_1 \subseteq M_2 \text{ and } M_2 \subseteq M_1.$$

2.9. Definition: (D. E. Knuth, 1981) For the mset $M=\{k_1|x_1,k_2|x_2,\ldots,k_n|x_n\}$, the set $S=\{x_1,x_2,\ldots,x_n\}$ is known as the *Root set* of M.

2.10. Definition: (K. P. Girish & Sunil Jacob John, 2009) The union of M_1 and M_2 is a multiset, denoted by $M = M_1 \cup M_2$, with the count function $C_M(x) = \max\{C_{M_1}(x), C_{M_2}(x)\}$, for every $x \in X$.

2.11. Definition: (K. P. Girish & Sunil Jacob John, 2009) The intersection of M_1 and M_2 is a multiset, denoted by $M = M_1 \cap M_2$, with the count function $C_M(x) = \min\{C_{M_1}(x), C_{M_2}(x)\}$, for every $x \in X$.

3. MULTISET GROUP

3.1. Definition: (N.J.Wildberger, 2003).Let (G, *) be a group and M be a multiset drawn from G. Then M is a **multiset group** or shortly **Mset group** if the characteristic function satisfies the following properties

(1) $C_M(x*y) \geq \text{Min}\{C_M(x), C_M(y): x,y \in G\}$.

(2) $C_M(x) = C_M(x^{-1})$ for all $x \in G$ where x^{-1} is the inverse of x in G.

3.2. Example: The multiset M = {3|0, 2|1, 3|2, 2|3}, drawn from the group $(Z_4, +_4)$ is a multiset group.

3.3. Theorem: Let (G, *) be a group and M be an mset group derived from G. Then the root set of M is a subgroup of G.

Proof: Let $x,y \in S$ the root set of M. Then $C_M(x)>0$ and $C_M(y)>0$. $C_M(x*y^{-1}) \geq \text{Min}\{C_M(x), C_M(y^{-1})\} = \text{Min}\{C_M(x), C_M(y)\} > 0$ means that $x*y^{-1} \in S$.

3.4. Proposition: If (G, *) is a group with identity element e and M is a multiset group derived from G then

(1) $C_M(e) \geq C_M(x)\ \forall x \in G$

(2) $C_M(x^n) \geq C_M(x)\ \forall x \in G$

and all natural number n. Here, x^n means $x*x*\cdots n$ times.

Proof: (1) Since $e = x*x^{-1}\ \forall x \in G$, $C_M(e) \geq \text{Min}\{C_M(x), C_M(x^{-1})\} = C_M(x)$. (2) The proof is by induction on n. For $n=1$, $C_M(x) = C_M(x)$, and hence the result is true.

Assume that the result is true for n-1 ie., $C_M(x^{n-1}) \geq C_M(x)$. Now,

$$\begin{aligned} C_M(x^n) &= C_M(x^{n-1}*x) \\ &\geq \text{Min}\{C_M(x^{n-1}), C_M(x)\} \\ &= C_M(x), \end{aligned}$$

by induction hypothesis.

3.5. Notations:

MS(G): Set of all multisets derived from a group G.

MG(G): Set of all mset groups derived from a group G.

3.6. Definition: Let (G, *) be a group. For $M_1, M_2 \in MS(G)$, define $M_1 \circledast M_2$ by,

$$C_{M_1 \circledast M_2}(x) = \text{Max}\{\text{Min}\left(C_{M_1}(y), C_{M_2}(z)\right), y, z \in G \text{ with } x = y*z\}.$$

3.7. Proposition:

1. $\circledast$ is associative in $MS(G)$.
2. $\circledast$ .is commutative in $MS(G)$, if $(G,*)$ is commutative.
3. $(M_1 \circledast M_2)^{-1} = M_2^{-1} \circledast M_1^{-1}$.

3.8. Definition: Let (G, *) be a group. For $M \in MS(G)$, define M^{-1} by, $C_{M^{-1}}(x) = C_M(x^{-1})$, where, x^{-1} is the inverse of x in (G, *) .

3.9. Proposition:

1. For $M \in MS(G)$, $(M^{-1})^{-1}=M$.
2. If $M_1,M_2 \in MS(G)$ and $M_1 \subseteq M_2$, then $M_1^{-1} \subseteq M_2^{-1}$.
3. $\cup M_i^{-1} = (\cup M_i)^{-1}$ for $i \in I$.
4. $\cap M_i^{-1} = (\cap M_i)^{-1}$ for $i \in I$.

3.10. Definition: Let $M \in MG(G)$. A subset N of M is an m set subgroup if N itself is an m set group on G.

3.11. Definition: Let G be a group with identity element e. The identity m setgroups E_n of G are defined as $C_{E_n}(x) = n$ for $x=e$ but $C_{E_n}(x) = 0$ for $x \neq e$.

3.12. Example: $M=\{3|1,1|1,2|i,2|-i\}$ is a multisetgroup derived from the group $G=\{1,-1,i,-i\}$ under usual multiplication.

3.13. Example: Consider $G=\{1,-1,i,-i\}$. Then $(G,\times)$ is a group with $e=1$. Let $M_1=\{4|1,1|-1,3|i\}$ and $M_2\{3|1,2|-i\}$. Then $M_1,M_2 \in MS(G)$.

$$M_1 \circledast M_2 = \{3|1,1|-1,3|i,2|-i\}\, M_1^{-1} = \{4|1,1|-1,3\,|-i\}.$$

3.14. Theorem: Let $M \in MG(G)$. Then $M=M^{-1}$.

Proof: $C_{M^{-1}}(x) = C_M(x^{-1}) = C_M(x)$. Since this is true for all x in G $M=M^{-1}$.

3.15. Theorem: Let $M \in MG(G)$. Then $M \circledast M = M$.

Proof: For $x \in M$ with $x=x*z$, $C_M(x) \geq \text{Min}\{C_M(y),C_M(z))\}$. $C_{M\circledast M}(x) = \text{Max}\{\text{Min}(C_M(y),C_M(z))$: $y*z=x \leq C_M(x)$. So, $M \circledast M \subseteq M$. We have $x*e=x\ \forall x \in M$ and $C_M(e) \geq C_M(x)$.

$$\therefore Min\{C_M(e),C_M(x)\} = C_M(x)\text{Max}\{\text{Min}(C_M(y),C_M(z)): y*z=x\} \geq C_M(x)$$

ie.

$$C_{M\circledast M}(x) \geq C_M(x), \forall x \in M \quad M \circledast M \supsetneq M$$

Hence Proved.

3.16. Theorem: Let $M \in MS(G)$. Then $M \in MG(G)$ iff $C_M(x*y^{-1}) \geq \text{Min}\{C_M(x),C_M(y)\}\ \forall x,y \in G$.

Proof: Suppose $M \in MG(G)$.. Then $C_M(x*y^{-1}) \geq \text{Min}\{C_M(x),C_M(y^{-1})\}=\{\text{Min}\{C_M(x),C_M(y)\}$. Conversely, assume that $C_M(x*y^{-1}) \geq \text{Min}\{C_M(x),C_M(y)\}\ \forall x,y \in G$. Now, $C_M(e)=C_M(x*x^{-1})\ \forall x \in G \geq \text{Min}\{C_M(x),(C_M(x)\}$, by assumption. So, $C_M(e) \geq C_M(x)\ \forall x \in G$. Now, $C_M(x^{-1})=C_M(e*x^{-1}) \geq \text{Min}\{C_M(e),C_M(x)\} \geq C_M(x)$. Similarly, $C_M(x)=C_M(e*x) \geq \text{Min}\{C_M(e),C_M(x^{-1})\} \geq C_M(x^{-1})$. Hence we get $C_M(x)=C_M(x^{-1})\ \forall x \in G$ which is one of the condition of mset group. Now to show the other condition, consider two arbitrary elements x and y in G. $C_M(x*y)=C_M(x*(y^{-1})^{-1}) \geq \text{Min}\{C_M(x),C_M(y^{-1})\}$, by assumption $=\text{Min}\{C_M(x),C_M(y)\}$ which completes the proof.

3.17. Theorem: Let $(G,*)$ be a group with identity e and M be an mset group derived from G. If $N=\{x \in G$: $C_M(x)=C_M(e)\}$, then N is a subgroup of G.

Proof: Take x and y from N. Then, $C_M(x)=C_M(y)=C_M(e)$. Now, $C_M(x*y^{-1})\geq \text{Min}\{C_M(x),C_M(y)\}$, by theorem 3.10. $=C_M(e)$, means $x*y^{-1}\in N$ and this implies N is a subgroup of G.

3.18. Theorem: Let $(G,*)$ be a group with identity e and M be an mset group derived from G. If $N_n=\{x: C_M(x)\geq n\}$. Then N_n are all subgroups of G, for every positive integer n.

Proof: Let $x,y\in N_n$. Then, $C_M(x)\geq n$ and $C_M(y)\geq n$. Now, $C_M(x*y^{-1})\geq \text{Min}\{C_M(x),C_M(y)\}\geq n$. So N_n are all subgroups of G for all positive integer n.

3.19. Theorem: Let $(G,*)$ be a group with identity element e and M be an mset group derived from G. If $C_M(x*y^{-1})=C_M(e)$, for some x and y in G then $C_M(x)=C_M(y)$.

Proof: $-C_M(x)=C_M(x*e)=C_M(x*(y^{-1}*y))=C_M((x*y^{-1})*y)\geq \text{Min}\{C_M(x*y^{-1}),C_M(y)\}=C_M(y)$. Similarly starting from $C_M(y)$, we can show that $C_M(y)\geq C_M(x)$.

4. MULTISET RING

As in the case of multiset group, now we define Multiset Ring or shortly Mset ring

4.1. Definition: Let $(R,+,.)$ be a ring and M be a multiset derived from R. Then M is a Multiset ring (shortly Mset ring) if the charectistic function satisfies the following properties

(1) $C_M(x+y)\geq \min\{C_M(x),C_M(y): x,y\in R\}$.

(2) $C_M(x,y)\geq \min\{C_M(x),C_M(y): x,y\in R\}$.

(3) $C_M(-x)=C_M(x),\ \forall x\in R$.

4.2. Example: The multiset M = {3|0, 2|2, 2|4} drawn from the ring $(Z_6,+6,\times 6)$ is a multiset ring

4.3. Theorem: If M is multiset ring derived from a ring $(R,+,.)$, then the root set S of M is a subring of R.

Proof: If $x,y\in S$ then $C_M(x)>)$ and $C_M(y)>0)$. Then, $C_M(x-y)=C_M(x+(-y))\geq \text{Min}\{C_M(x),C_M(y)\}>0$ which means $x\text{-}y\in S$. Similarly, $C_M(x)>0$ and $C_M(y)>0$ will imply $C_M(x,y)>0$ ie, $x,y\in S$. So s is a subring of R.

4.4. Example: Let $R=(Z_6,+6,\times 6)$ and $M=\{3|0,2|2,2|4\}$. Then M is a multiset ring.

4.5. Notations: MS[R] be the collection of all msets over a ring R MS[R] be the collection of all multiset rings over a ring R.

4.6. Proposition: If $M\in MR[R]$. Then,

(a) $C_M(0)\geq C_M(x)\ \forall x\in R$.

(b) If $a+a=b$ and $b+b=a$, for $a,b\in M$ then $C_M(a)=C_M(b)$.

(c) If $a.a=b$ and $b.b=a$ for $a,b\in M$ then $C_M(a)=C_M(b)$.

Proof: (a) Since $0=x+(-x)\ \forall x\in R$, $C_M(0)\geq \text{Min}\{C_M(x),C_M(-x)\}=C_M(x)\ \forall x\in R$. (b) $a+a=b$ implies $C_M(b)\geq C_M(a)$ and $b+b=a$ implies $C_M(a)\geq C_M(b)$. Combining these, we have the result. (c) $a.a=b$ implies $C_M(b)\geq C_M(a)$ and $b.b=a$ implies $C_M(a)\geq C_M(b)$.

Combining these, we have the result.

4.7. Definition: On MS[R], define two operations $\oplus$ and $\odot$ by

$$C_{M_1\oplus M_2}(x)=Max\left\{Min\left(C_{M_1}(y),C_{M_2}(z):\ y,z\in R,\ x=y+z\right)\right\}$$

$$C_{M_1 \odot M_2}(x) = Max\left\{Min\left(C_{M_1}(y), C_{M_2}(z):\ y,z \in R,\ x = y.z\right)\right\}$$

4.8. Definition: For an $M \in MS[R]$, define $-M$ by $C_{-M}(x) = C_{-M}(-x)$

4.9. Proposition: For msets in MS[R]

(a) $-(-M) = M$.

(b) If $M_1 \subseteq M_2$, then $-M_1 \subseteq -M_2$.

(c) $-(\cup M_i) = \cup(-M_i)$, for $i \in I$.

(d) $-(\cap M_i) = \cap(-M_i)$, for $i \in I$.

(e) $-(M_1 \oplus M_2) = (-M_1) \oplus (-M_2)$..

(f) $-(M_1 \odot M_2) = (-M_1) \odot (-M_2)$..

Proof: (a) $C_{-(-M)}(x) = C_{-M}(-x) = C_M(-(-x)) = C_M(x)\,\forall x \in R$.

(b) For $M_1 \subseteq M_2, C_{M_1}(x) \leq C_{M_2}(x)\,\forall x \in R$,

$$C_{-M_1}(x) = C_{M_1}(-x) \leq C_{M_2}(-x) = C_{-M_2}(x)\,\forall x \in R.$$

So, $-M_1 \subseteq M_2$.

(c)
$$\begin{aligned} C_{-(M_1 \cup M_2)}(x) &= C_{M_1 \cup M_2}(x) = Max\left\{C_{M_1}(-x), C_{M_2}(-x)\right\} \\ &= Max\left\{C_{-M_1}(x), C_{-M_2}(x)\right\} = C_{(-M_1)\cup(-M_2)}(x) \end{aligned}.$$

The result can be extended by using Mathematical Induction.

(d)
$$\begin{aligned} C_{-(M_1 \cap M_2)}(x) &= C_{M_1 \cap M_2}(x) = Min\left\{C_{M_1}(-x), C_{M_2}(-x)\right\} \\ &= Min\left\{C_{-M_1}(x), C_{-M_2}(x)\right\} = C_{(-M_1)\cap(-M_2)}(x) \end{aligned}$$

The result can be extended by using Mathematical Induction

(e)
$$\begin{aligned} C_{-(M_1 \oplus M_2)}(x) &= C_{(M_1 \oplus M_2)}(-x) = Max\left\{Min\left(C_{M_1}(y), C_{M_2}(z):\ y,z \in R,\ -x = y+z\right)\right\} \\ &= Max\left\{Min\left(C_{M_1}(-y), C_{M_2}(-z):\ y,z \in R,\ x = y+z\right)\right\} \\ &= Max\left\{Min\left(C_{-M_1}(y), C_{-M_2}(z):\ y,z \in R,\ x = y+z\right)\right\} = C_{(-M_1)\oplus(-M_2)}(x)\,\forall x \in R. \end{aligned}$$

(f)
$$\begin{aligned} C_{-(M_1 \odot M_2)}(x) &= C_{(M_1 \odot M_2)}(-x) = Max\left\{Min\left(C_{M_1}(y), C_{M_2}(z):\ y,z \in R,\ -x = y.z\right)\right\} \\ &= Max\left\{Min\left(C_{M_1}(-y), C_{M_2}(-z):\ y,z \in R,\ x = y.z\right)\right\} \\ &= Max\left\{Min\left(C_{-M_1}(y), C_{-M_2}(z):\ y,z \in R,\ x = y.z\right)\right\} = C_{(-M_1)\odot(-M_2)}(x)\,\forall x \in R. \end{aligned}$$

4.10. Theorem: If $M \in MR[R]$, then $-M \in MR[R]$.
Proof: Let $x, y \in R$.

$$\begin{aligned} C_{-M}(x+y) &= C_M(-(x+y)) \\ &= C_M((-x)+(y)) \\ &\geq \text{Min}\{C_M(-x), C_M(-y)\} \\ &= \text{Min}\{C_{-M}(x), C_{-M}(y)\}. \end{aligned}$$

$$\begin{aligned} C_{-M}(x.y) &= C_M(-(x.y)) \\ &= C_M((-x).(y)) \\ &\geq \text{Min}\{C_M(-x), C_M(y)\} \\ &= \text{Min}\{C_M(-x), C_M(-y)\}. \\ &= \text{Min}\{C_{-M}(x), C_{-M}(y)\} \end{aligned}$$

$$C_{-M}(-x) = C_M(x) = C_M(-x) = C_{-M}(x)$$

4.11. Definition.: Let R be a ring with additive identity 0. The identity mset rings O_n of R are defined as $C_{O_n}(x) = 0$ if $x \neq 0$ and $C_{O_n}(x) = n$ if $x=0$.

4.12. Proposition: If $M_1, M_2 \in MR[R]$

(a) $M_1 \oplus M_2 \in MR[R]$.

(b) $M_1 \odot M_2 \in MR[R]$ if R is a commutative ring.

(c) $M_1 \cap M_2 \in MR[R]$.

4.13. Note: $M_1, M_2 \in MR[R]$ need not imply that $M_1 \cup M_2 \in MR[R]$..

4.14. Example: Consider $R=Z_6$ with operations $+_6$ and X_6. Let $M_1=\{3|0,2|2,2|4\}$ and $M_2=\{2|0,1|3\}$. Then M_1 and M_2 belongs to $MR[R]$.

$$M_1 \oplus M_2 = \{2|0,1|1,2|2,1|3,2|4,1|5\},$$

$$M_1 \odot M_2 = \{2|0\} \quad M_1 \cap M_2 = \{2|0\}.$$

Here, $M_1 \oplus M_2$, $M_1 \odot M_2$ and $M_1 \cap M_2$ are elements of $MR[R]$. But

$$M_1 \cup M_2 = \{3|0,2|2,1|3,2|4\} \notin MR[R],$$

because $3 +_6 4 = 1$ but,

$$C_{M_1 \cup M_2}(1) = 0 < Min\{C_{M_1 \cup M_2}(3), C_{M_1 \cup M_2}(4)\}.$$

4.15. Proposition:(a) $\oplus$ is commutative and associative.

(b) $\odot$ is associative.

(c) $\odot$ is commutative if R is commutative

4.16. Definition: If R is a ring with unity 1, an mset ring M in $MR[R]$ is said to be a ring with unity, if $C_M(1)>0$.

4.17. Definition: If M_1 is an mset ring derived from a ring R and $M_2 \subseteq M_1$ with M_2 itself an mset ring, then M_2 is known as a sub mset ring of M_1.

4.18. Definition: Let $M \in MR[R]$. Then $M_n = \{r \in R: C_M(r) \geq n\}$.

Note: M_1 is the root set of M.

4.19. Proposition: M_n are all subrings of R, $\forall n$.

Proof: Let $x,y \in M_n$. Then $C_M(x) \geq n$ and $C_M(y) \geq n$. Now, $C_M(x\text{-}y) \geq \text{Min}\{C_M(x), C_M(-y)\} = \text{Min}\{C_M(x), C_M(y)\} \geq n \Rightarrow x\text{-}y \in M_n$. Similarly $C_M(x.y) \geq \text{Min}\{C_M(x), C_M(y)\} \geq n \Rightarrow x.y \in M_n$. So M_n is a Subring of R.

4.20. Definition: Let $M \in MR[R]$. Then $M_0 = \{r \in R: C_M(r) = C_M(0)\}$.

4.21. Proposition: M_0 is a subring of R.

Proof: Let $x,y \in M_0$. Then $\text{IC}_M(x) = C_M(0)$ and $C_M(y) = C_M(0)$. Now $C_M(x\text{-}y) \geq \text{Min}\{C_M(x), C_M(-y)\} = \text{Min}\{C_M(x), C_M(y)\} \geq C_M(0)$. But we have $C_M(0) \geq C_M(x)$ $\forall x \in R$.

$$\therefore C_M(0) \geq C_M(x-y).$$

So, $C_M(x\text{-}y) = C_M(0) \Rightarrow x\text{-}y \in M_0$. Similarly $C_M(x.y) \geq \text{Min}\{C_M(x), C_M(y)\} \geq C_M(0) \Rightarrow x.y \in M_0$. So M_0 is a Subring of R.

4.22. Definition: Let $M \in MR[R]$. Then M^1 is defined as $M^1 = \{r \in R: C_M(r) = C_M(1)\}$.

4.23. Proposition: M^1 is a subring of R.

Proof: Similar as in proof of Proposition 4.20

5. MULTISET IDEALS

5.1. Definition: Let $(R,+,.)$ be a ring and M an mset derived from R. Then M is called a left Multiset Ideal, if

1. $C_M(x\text{-}y) \geq \text{Min}\{C_M(x), C_M(y)\}$.
2. $C_M(r.x) \geq C_M(x)$ for $x \in M$, $r \in R$.

5.2. Definition: Let $(R,+,.)$ be a ring and M an mset derived from R. Then M is called a right Multiset Ideal, if

1. $C_M(x\text{-}y) \geq \text{Min}\{C_M(x), C_M(y)\}$.
2. $C_M(x,r) \geq C_M(x)$ for $x \in M$, $r \in R$.

5.3. Definition : An mset M derived from a ring R is a multiset ideal if it is both left mset ideal and right mset ideal.

Note: An mset M is an mset ideal if,

1. $C_M(x\text{-}y) \geq \text{Min}\{C_M(x), C_M(y)\}$.
2. $C_M(x.y) \geq \text{Max}\{C_M(x), C_M(y)\}$.

5.4. Proposition: Every mset ideal is an mset ring, but an mset ring need not be an mset ideal.

5.5. Example: {5|0,1|1,3|2,1|3,3|4,1|5} is an mset ideal of $(Z_6,+_6,\times_6)$.

5.6. Example: {5|0,3|1,1|2,1|3,1|4,3|5} is only an mset ring but not an mset ideal of $(Z_6,+_6,\times_6)$..

5.7. Theorem: The intersection of two mset ideals is again an mset ideal.

Proof: Let M_1 and M_2 be two mset ideals over a ring *R*. Let $x, y \in M_1 \cap M_2$. Since M_1 .is an mset ideal,

$$C_{M_1}(x-y) \geq \text{Min}\{C_{M_1}(x), C_{M_1}(y)\} \text{ and } C_{M_1}(x.y) \geq \text{Max}\{C_{M_1}(x), C_{M_1}(y)\}.$$

Similarly, since M_2 is an mset ideal,

$$C_{M_2}(x-y) \geq \text{Min}\{C_{M_2}(x), C_{M_2}(y)\} \text{ and } C_{M_2}(x.y) \geq \text{Max}\{C_{M_2}(x), C_{M_2}(y)\}$$

$$C_{M_1 \cap M_2}(x-y) \geq \text{Min}\{C_{M_1 \cap M_2}(x), C_{M_1 \cap M_2}(y)\},$$

since $M_1 \cap M_2$ is an mset ring. Now,

$$\begin{aligned} C_{M_1 \cap M_2}(x.y) &= \text{Min}\{C_{M_1}(x.y), C_{M_2}(x.y)\} \\ &\geq \text{Min}\{\text{Max}(C_{M_1}(x), C_{M_1}(y)),\ \text{Max}(C_{M_2}(x), C_{M_2}(y))\} \\ &\geq \text{Min}\{\text{Min}(C_{M_1}(x), C_{M_1}(y)),\ \text{Min}(C_{M_2}(x), C_{M_2}(y))\} \\ &= \text{Min}\{\text{Min}(C_{M_1}(x), C_{M_1}(y), C_{M_2}(x), C_{M_2}(y))\} \\ &= \text{Min}\{\text{Min}(C_{M_1}(x), C_{M_2}(x)), \text{Min}(C_{M_1}(y), C_{M_2}(y))\} = \text{Min}\{C_{M_1 \cap M_2}(x), C_{M_1 \cap M_2}(y)\} \end{aligned}$$

5.8. Definition: Let *M* be an mset ideal of a ring *R*. For *r*∈*R* the Multiset Coset M_r is defined as $C_{M_r}(x) = C_M(x-r)$..

5.9. Definition: Let *M* be an mset ideal of a ring *R*. Then the set of all cosets M_r determined by *M* is a ring under the operations

$$C_{M_r+M_s}(x) = C_{M_{r+s}}(x) \text{ and } C_{M_r.M_s}(x) = C_{M_{r.s}}(x).$$

This is called *Factor Ring* and denoted as *R*/*M*.

5.10. Proposition: If *M* is an mset ideal with unity, then $C_M(1) \leq C_M(x)$ ∀*x*∈*M*.

Proof: We have $x=x.1\ \forall x$. M being an ideal, $C_M(x)=C_M(x.1)\geq \text{Max}\{C_M(x),C_M(1)\}$, which completes the proof.

5.11. Notation: Denote R^k be the mset derived from a ring R having the multiplicity k for all elements.

5.12. Proposition: R^k is an ideal.

Proof: $C_{R^k}(x-y)=k=\text{Min}\{C_{R^k}(x),C_{R^k}(y)\}\ C_{R^k}(x.y)=k=\text{Max}\{C_{R^k}(x),C_{R^k}(y)\}$.

5.13. Definition: An mset ideal M is said to be an Mset prime ideal, if there exists two proper ideals M_1 and M_2 with $M\subseteq M_1\odot M_2$, then either $M\subseteq M_1$ or $M\subseteq M_2$.

5.14. Proposition: R^k is a prime ideal.

5.15. Proposition: Let $(R,+,.)$ be a ring and M be a subring of R. Then the mset M^n defined as $C_{M^n}(x)=n$ if $x\in M$ and 0 otherwise, is an mset prime ideal for every positive integer n.

5.16. Example: $M^2=\{2|0, 2|2, 2|4\}$ is a prime mset ideal as $M=\{0,2,4\}$ is a subring of $(Z_6,+_6,\times_6)$.

5.17. Theorem: Let M be an mset prime ideal over a ring R with unity 1. If $S=\{x\in R: C_M(x)=C_M(1)\}$ then S is a prime ideal of R.

Proof: Let $x,y\in S$. Then, $C_M(x)=C_M(y)=C_M(1)$.

$C_M(x\text{-}y)\geq \text{Min}\{C_M(x),C_M(y)\}=C_M(1)$.

But M being an mset ideal, $C_M(1)\leq C_M(x)\ \forall x\in R$. So, $C_M(x\text{-}y)=C_M(1)$ means that $x\text{-}y\in S$. Now for $r\in R$ and $x\in S$, $C_M(r.x)\geq \text{Max}\{C_M(r),C_M(x)\}=\text{Max}\{C_M(r),C_M(1)\}\geq C_M(1)$.

$\therefore C_M(r.x)=C_M(1)$.

Similarly we can show that $C_M(x.r)=C_M(1)$. So, S is an ideal of R. Let $a,b\in R$ and $a.b\in S$. Then, $C_M(a.b)=C_M(1)$. Since M is an mset ideal, $C_M(a.b)\geq \text{Max}\{C_M(a),C_M(b)\}$ i.e. $C_M(1)\geq \text{Max}\{C_M(a), C_M(b)\}$ $C_M(a)=C_M(1)$ or $C_M(b)=C_M(1)$.

So, S is a prime ideal of R

CONCLUSION AND FUTURE WORK

Rings, Ideals and such have a wide range of application. But their possibilities in multiset have not yet been explored. In this work, by standing in the fuzzy platform, the authors have tried to explain the algebraic structures like groups, rings and ideals in multiset context and established links between the said structures in both set and multiset environments. As stated previously, many of these structeres are used in various fields of science and even more in different technological aspects. Although the possibility of msets in place of these have been touched upon, there is still a huge window of possibilities to explore the mset context in place of fuzzy concepts.

Here, they have only barely scratched the surface of the wide range of applications of Algebraic Structure in msets. The authors' future work will consist of deeper exploration into these applications such as the different types of ideals, fields and so on.

REFERENCES

Awolola, J.A., & Ibrahim, A.M. (2016). Some results on multi-groups. *Quasigroups and Related Systems*, (24), 169-177.

Blizard, W.D. (1991). The Developement of Multiset theory. *Modern Logic*, (1), 319-352.

Brink, C. (1988). Multiset and the Algebra of Relevance Logic. *Non-Classical Logic*, (5), 75-95.

Calude, C. S., Paun, G., Rozenberg, G., & Salomaa, A. (2001). Mathematics of Multisets. Springer Verlag.

Ejegwa, P.A. (2017). Upper and Lower Cuts of Multigroups. *Prajna International Journal of Mathematical Sciences and Applications,* (1), 19-26.

Girish, K. P., & John, S. J. (2009). Relations and functions in multiset context. Information Sciences, (179), 758–768.

Girish, K. P., & John, S. J. (2011). Rough Multiset and its Multiset Topology. Lecture Notes in Computer Science, 6600, 62-80.

Girish, K. P., & John, S. J. (2012). Multiset Topologies induced by multiset relations. Information Sciences, (188), 298–313.

Girish, K.P., & John, S.J. (2012). On multiset Topologies. *Theory and Applications of Mathematics and Computer Science*, (2), 37-52.

Ibrahim, A.M. & Ajegwa, P.A. (2016). A Survey on the Concept of Multigroups. *Journal of the Nigerian Association of Mathematical Physics*, (38), 1-8.

Knuth, D.E. (1981). The Art of Computer Programming. *Semi Numerical Algorithms, 2.*

Li, Y., Wang, X., & Yang, L. (2013). A Study of Fuzzy Subgroups. Journal of Applied Mathematics. doi:10.1155/2013/485768

Nazmul, S.K., Majumdar, P., & Samanta, S.K. (2013). On Multisets and Multi-groups. *Annals of Fuzzy Mathematics and Informatics,* (3), 643-656.

Rajarajeswari, R., & Uma, N. (2013). On Distance and Similarity Measures of Intuitionistic Fuzzy Multi Set. *IOSR Journal of Mathematics*, (4), 19-23.

Rosenfeld, A. (1971). Fuzzy Groups. Journal of Mathematical Analysis and Applications, (35), 512–517.

Sebastian, S., & Ramakrishnan, T.V. (2011). Multi-fuzzy Sets: An Extension of Fuzzy Sets. *Fuzzy Information and Engineering,* (1), 35-43.

Sebastian, S., Jacob, M. K., Mary, V. M., & Daise, D. M. (2012). On Fuzzy Ideals and Rings. Journal of Computer and Mathematical Sciences, *3*(1), 115–120.

Singh, D., Ibrahim, A.M., Yohanna, T., & Singh, J.N. (2007). An overview of the application of multiset. *Novi Sad Journal of Math*ematics, (37), 73-92.

Tella, Y., & Daniel, S. (2013). A Study of Group Theory in the Context of Multiset Theory. International Journal of Science and Technology, *2*(8), 609–615.

Ttripathy, B.C., Debnath, S., & Rakshit, D. (2018). On multiset group. *Proyecciones Journal of Mathematics*, (33), 479-489.

Wildberger, N.J. (2003). *A New Look at Multisets*. School of Mathematics, UNSW.

Zadeh, L. A. (1965). Fuzzy Sets. Information and Control, (8), 338–353.

Chapter 6
Intuitionistic Fuzzy Soft Ideals

Shuker Khalil
https://orcid.org/0000-0002-7635-3553
University of Basrah, Iraq

ABSTRACT

The basic notions of soft sets theory are introduced by Molodtsov to deal with uncertainties when solving problems in practice as in engineering, social science, environment, and economics. This notion is convenient and easy to apply as it is free from the difficulties that appear when using other mathematical tools as theory of theory of fuzzy sets, rough sets, and theory of vague sets. The soft set theory has recently gaining significance for finding rational and logical solutions to various real-life problems, which involve uncertainty, impreciseness, and vagueness. The concepts of intuitionistic fuzzy soft left almost semigroups and the intuitionistic fuzzy soft ideal are introduced in this chapter, and some of their basic properties are studied.

1. INTRODUCTION

The basic notions of soft sets theory are introduced by (Molodtsov, 1999) to deal with uncertainties when solving problems in practice as in engineering, social science, environment, and economics. This notion is convenient and easy to apply as it is free from the difficulties that appear when using other mathematical tools as theory of theory of fuzzy sets, rough sets and theory of vague sets etc. The soft set theory has recently gaining significance for finding rational and logical solutions to various real life problems, which involve uncertainty, impreciseness, and vagueness. The concepts of intuitionistic fuzzy soft left almost semigroups and the intuitionistic fuzzy soft ideal are introduced in this chapter, and some of their basic properties are studied. The fuzzy set (FS) as proposed by (Zadeh, 1965) is a framework to encounter uncertainty, vagueness, and partial truth; an FS represents the degree of membership for each member of the universe of discourse to a subset of the said universe. By adding the degree of non-membership to a fuzzy set (Atanassov,1986) proposed the intuitionistic fuzzy set (IFS), which appears more accurate for uncertainty quantification and provides opportunities to precisely model a problem based on existing knowledge and observations. The concept of an LA-semigroup was first introduced by (Kazim and Naseerudin, 1972). A non-empty set S with binary operation $*$ is said to be a left almost

DOI: 10.4018/978-1-7998-0190-0.ch006

semigroup (LA-semigroup), if the identity $(x * y) * z = (z * y) * x$ for all $x,y,z \in S$ holds. Subsequently, Mushtaq and others have further investigated this structure and added several useful concepts to the theory of LA-semigroups. The soft set concept was proposed by (Molodtsov, 1999) as a completely new approach for modeling and uncertainty. A soft set is a parameterized family of subsets of the universal set. (Maji et al., 2001) presented the definition of fuzzy soft set. Next, many applications of soft and fuzzy sets are shown (Aygunoglu and Aygun, 2009) introduced the concept of a fuzzy soft group and discussed some of its properties. (Maji et al. 2001, 2004) introduced the intuitionistic fuzzy soft set theory, which is based on a combination of the intuitionistic fuzzy set and soft set models. (Shuker, 2014) generated the $\xi_{H(A)}$ (induced fuzzy soft set by (H,A) and $\eta_{H(A)}$ (induced intuitionistic fuzzy soft set by H_A), where (H,A) is a soft set over the universe of the given set with a fixed set of parameters, and H_A is a fuzzy soft set. The concept of a soft set was also studied by other mathematicians. (Yaqoob et al., 2012) applied soft set theory to Γ-hyperideals in left almost Γ-semihypergroups, as also discussed see (Yaqoob et al., 2012, 2013). Moreover, (Feng et al., 2008) defined soft semirings and several related notions to establish a connection between soft sets and semirings. (Ali and Shabir, 2009) defined soft ideals and generalized fuzzy ideals in semigroups. A special structure satisfying this intuitionistic fuzzy soft algebra is given, and the relation between this structure and the theory of fuzzy soft sets is studied.In the present work, the concepts of intuitionistic fuzzy soft LA-semigroups and the intuitionistic fuzzy soft ideal are introduced, and some of their basic properties are studied.

2. DEFINITIONS AND NOTATIONS

The following definitions have been used to obtain the results and properties developed in this work.

Definition 2.1 (Aktas and Cagman, 2007)

An intuitionistic fuzzy set A (IFS, in short) over the universe U can be defined as follows $A=\{(x,\mu_A(x),\nu_A(x)), x\in U\}$ where $\mu_A(x)$: $U\rightarrow [0; 1]$, $\nu_A(x)$: $U\rightarrow [0; 1]$ with the property $0 \leq \mu_A(x)+\nu_A(x) \leq 1, \forall x\in U$. The values $\mu_A(x)$ and $\nu_A(x)$ represent the degree of membership and non-membership of x to A respectively.

Definition 2.2 (Atanassov,1986)

The intuitionistic fuzzy sets $\bar{0}=\{(x,0,1): x\in U\}$ and $\bar{1}=\{(x,0,1): x\in U\}$ are the empty set and the whole set of U, respectively.

Definition 2.3 Basic Relations and Operations on Intuitionistic Fuzzy Sets (Ejegwa, 2014)

Let $A=\{(x,\mu_A(x),\nu_A(x), x\in U\}$ and $B=\{(x,\mu_B(x),\nu_B(x), x\in U\}$ be two intuitionistic fuzzy sets of U. Then we deduced the following relations:

(1) [inclusion] $A\subseteq B$ if and only if $\mu_A(x)\leq\mu_B(x)$ and $\nu_A(x)\geq\nu_B(x)$ for all $\forall x\in U$.

Example: Let $A=\{(x,\mu_A(x),v_A(x), x\in U\}$ and $B=\{(x,\mu_B(x),v_B(x), x\in U\}$ be two intuitionistic fuzzy sets of U. Where $U=\{a,b,c\}$, $A=\{(a,0.3,0.7), (b,0,1), (c,0.6,0.4)\}$ and $B=\{(a,0.8,0,2), (b,0.2,0,8), (c,0.9,0.1)$. Then we have $A\subseteq B$

(2) [equality] $A=B$ if and only if $A\subseteq B$ and $B\subseteq A$.
Example: Let $A=\{(x,\mu_A(x),v_A(x), x\in U\}$ and $B=\{(x,\mu_B(x),v_B(x), x\in U\}$ be two intuitionistic fuzzy sets of U. Where $U=\{a,b,c\}$, $A=\{(a,0.3,0.7), (b,0,1), (c,0.6,0.4)\}$ and $B=\{(a,0.3,0.7), (b,0,1), (c,0.6,0.4)\}$. Then we have $A=B$.

(3) [intersection] $A\tilde{\bigcap}B=\{(x, \min\{\mu_A(x),\mu_B(x)\}, \max\{v_A(x),v_B(x)\}): x\in U\}$.
Example: Let $A=\{(x,\mu_A(x),v_A(x), x\in U\}$ and $B=\{(x,\mu_B(x),v_B(x), x\in U\}$ be two intuitionistic fuzzy sets of U. Where $U=\{a,b,c\}$, $A=\{(a,0.3,0.7), (b,0,1), (c,0.6,0.4)\}$ and $B=\{(a,0.8,0.2), (b,0.2,0.8), (c,0.9,0.1)\}$. Then we have $A\tilde{\bigcap}B=\{(a,0.3,0.7), (b,0,1), (c,0.6,0.4)\}$.

(4) [union] $A\tilde{\bigcup}B=\{(x, \max\{\mu_A(x),\mu_B(x)\}, \min\{v_A(x),v_B(x)\}): x\in U\}$
Example: Let $A=\{(x,\mu_A(x),v_A(x), x\in U\}$ and $B=\{(x,\mu_B(x),v_B(x), x\in U\}$ be two intuitionistic fuzzy sets of U. Where $U=\{a,b,c\}$, $A=\{(a,0.3,0.7), (b,0,1), (c,0.6,0.4)\}$ and $B=\{(a,0.8,0.2), (b,0.2,0.8), (c,0.9,0.1)\}$. Then we have $A\tilde{\bigcup}B=\{(a,0.8,0.2), (b,0.2,0.8), (c,0.9,0.1)\}$.

(5) [complement]$A^c=\{(x,v_A(x)\mu_A(x)), x\in U\}$.
Example: Let $A=\{(x,\mu_A(x),v_A(x), x\in U\}$ be an intuitionistic fuzzy sets of U. Where $U=\{a,b,c\}$ and $A=\{(a,0.3,0.7), (b,0,1), (c,0.6,0.4)\}$. Then we have $A^c=\{(a,0.7,0.3\}, (b,1,0), (c,0.4,0.6)\}$.

(6) [difference] $A\text{-}B=\{(x, \min\{\mu_A(x),v_B(x)\}, \max\{v_A(x),\mu_B(x)\}): x\in U\}$,
Example: Let $A=\{(x,\mu_A(x),v_A(x), x\in U\}$ and $B=\{(x,\mu_B(x),v_B(x), x\in U\}$ be two intuitionistic fuzzy sets of U. Where $U=\{a,b,c\}$, $A=\{(a,0.3,0.7), (b,0,1), (c,0.6,0.4)\}$ and $B=\{(a,0.8,0.2), (b,0.2,0.8), (c,0.9,0.1)\}$. Then we have A-B=$\{(a,0.2,0.8), (b,0,1), (c,0.1,0.9)\}$.

(7) [Cartesian product] $A\times B=\{(x,\mu_A(x),\mu_B(x),v_A(x),v_B(x)), x\in U\}$,
Example: Let $A=\{(x,\mu_A(x),v_A(x), x\in U\}$ and $B=\{(x,\mu_B(x),v_B(x), x\in U\}$ be two intuitionistic fuzzy sets of U. Where $U=\{a,b,c\}$, $A=\{(a,0.3,0.7), (b,0,1), (c,0.6,0.4)\}$ and $B=\{(a,0.8,0.2), (b,0.2,0.8), (c,0.9,0.1)\}$. Then we have $A\times B=\{(a,0.24,0.14), (b,0,8), (c,0.54,0.04)\}$.

(8) $A\text{-}B=\varnothing$ iff $A\tilde{\bigcap}B=\varphi$.
Example: Let $A=\{(x,\mu_A(x),v_A(x), x\in U\}$ and $B=\{(x,\mu_B(x),v_B(x), x\in U\}$ be two intuitionistic fuzzy sets of U. Where $U=\{a,b,c\}$, $A=\{(a,0,1),(b,0,1),(c,0,1)\}$ and $B=\{(a,0.8,0.2),(b,0.2,0.8),(c,0.9,0.1)\}$. Then we have $A\text{-}B=\{(a,0,1),(b,0,1), (c,0,1)\}=\varnothing$. Also, $A\tilde{\bigcap}B=\{(a,0,1),(b,0,1), (c,0,1)\}=\varnothing$.

Definition 2.4 (Yang, 2008)

A pair (F,A) is called a soft set (over U) where F is a mapping $F: A\rightarrow P(U)$. The soft set is a parameterized family of subsets of the set U. Each set $F(e)$, $e\in E$, from this family may be considered as the set of e-elements of the soft set (F,A) or as the set of e-approximate elements of the soft set. Clearly, a soft set is not a set. For two soft sets (F,A) and (G,B) over the common universe U, we say that (F,A) is a soft subset of (G,B) if $A\subseteq B$ and for all $e\in A$, $F(e)$ and $G(e)$ are identical approximations. We write $(F,A)\subseteq(G,B)$. (F,A) is said to be a soft superset of (G,B), if (G,B) is a soft subset of (F,A). Two soft sets (F,A) and (G,B) over a common universe U are said to be soft equals if (F,A) is a soft subset of (G,B) and (G,B) is a soft subset of (F,A). A soft set (F,A) over U is called a null soft set, which is denoted by $\Phi=(\phi,\phi)$, if for each $F(e)=\varnothing$, $\forall e\in A$. Similarly, the universal soft set is denoted by (U,E), if for each $F(e)=U$, $\forall e\in A$.

Definition 2.5 (Maji et al., 2001)

Let U be an initial universe set and let E be a set of parameters. Let I^U denote the collection of all fuzzy subsets of U and $A \subseteq E$. Then, the mapping F_A: $A \to I^U$, which is defined by $F_A(e) = \mu_{F_A}^e$ (a fuzzy subset of U), is called a fuzzy soft set over (u,E), where $\mu_{F_A}^e = \bar{0}$ if $e \in E \setminus A$ and $\mu_{F_A}^e \neq \bar{0}$ if $e \in A$.

Definition 2.6(Maji et al., 2001)

The fuzzy soft set F_A is said to be a fuzzy soft subset of the fuzzy soft set G_B over a common universe U if $F_A(e) \subseteq G_B(e)$ for all $e \in E$, i.e., if $\mu_{F_A}^e(x) \leq \mu_{G_B}^e(x)$ for all $x \in U$ and for all $e \in E$.

Definition 2.7(Maji et al., 2001)

The union of two fuzzy soft sets F_A and G_B over the common universe U is the fuzzy soft set H_C, which is defined by $H_C(e) = \mu_{H_C}^e = \mu_{F_A}^e \cup \mu_{G_B}^e$ for all $e \in E$, where $C = A \cup B$. Thus, $H_C = F_A \amalg G_B$.

Definition 2.8 (Maji et al., 2001)

Let F_A and G_B be two fuzzy soft sets, then the intersection of F_A and G_B is the fuzzy soft set H_C, which is defined by $H_C(e) = \mu_{H_C}^e = \mu_{F_A}^e \cap \mu_{G_B}^e$ for all $e \in E$, where $C = A \cap B$. Thus, $H_C = F_A \sqcap G_B$.

Definition 2.9 (IF soft set) (Maji et al., 2001)

Let U be an initial universe set and E be the set of parameters. Let IF^U denote the collection of all IF subsets of U. Let $A \subseteq E$. A pair $(F, A)'$ is called an IF soft set over U, where F is a mapping given by F: $A \to IF^U$. In general, for every $e \in A$, $F(e)$ is an IF set of U and is called the IF value set of parameter e. Clearly, $F(e)$ can be written as an IF set, such that $F(e) = \{(x, \mu_{F(e)}(x), \nu_{F(e)}(x)), x \in U\}$. The set of all IF soft sets over U with parameters from E is called an IF soft class and is denoted by IFS(U_E).

Definition 2.10 (Maji et al., 2001)

The union of two IF soft sets $(F, A)'$ and $(G, B)'$ over the common universe U is the IF soft set.

Definition 2.11 (Kazim and Naseerudin,1972)

A non-empty set S with the binary operation $*$ is said to be an LA-semigroup, if the identity $(x * y) * z = (z * y) * x$ for all $x, y, z \in S$ holds.

Definition 2.12 (Maji et al., 2001)

An IF soft set $(F, A)'$ is said to be an IF soft subset of a IF soft set $(G, B)'$ over a common universe U if $A \subseteq B$ and $F(e) \subseteq G(e)$, for all $e \in A$.

Definition 2.13 (Kazim and Naseerudin,1972)

A non-empty set S with binary operation $*$ is said to be a left almost semigroup (LA-semigroup), if the identity $(x * y) * z = (z * y) * x$ for all $x, y, z \in S$ holds.

3. FUZZY SOFT Γ-IDEALS AND INTUITIONISTIC FUZZY SOFT SETS OF Γ-LA-SEMIGROUPS

In this section, we introduce some new concepts, such as the intuitionistic fuzzy soft LA-semigroup and the intuitionistic fuzzy soft ideal. Moreover, some basic properties are studied.

Definition 3.1

Let $A=(\mu_A,\nu_A)$ and $B=(\mu_B,\nu_B)$ be two IF sets of a semigroup S. Then the product $A \circ B$ is defined as follows:

$$A \circ B(x) = \begin{cases} (\underset{x=yz}{\vee}(\mu_A(y) \vee \mu_B(z)), \underset{x=yz}{\wedge}(\nu_A(y) \wedge \nu_B(z))) \; if \; for \; every \; y, z \in S, \ni x = yz \\ (0,1), \qquad otherwise \end{cases}$$

for all $x \in S$. The operation $\circ$ is associative.

Remark 3.2

We will refer to $\bar{\coprod}$ by $\bar{\coprod}_E$ and the union in definition 2.10 is called extended union and then the extended intersection of two IF soft sets are defined as flows:

Definition 3.3

The extended intersection of two IF soft sets $(F, A)'$ and $(G, B)'$ over the common universe U is the IF soft set $(H, C)' = (F, A)' \bar{\prod}_E (G, B)'$, where $C = A \cup B$ and for all $e \in C$,

$$H(e) = \begin{cases} F(e) & if \; e \in A \setminus B, \\ G(e) & if \; e \in B \setminus A, \\ F(e) \tilde{\cap} G(e) & if \; e \in A \cap B. \end{cases}$$

Definition 3.4

An IF set $A=(\mu_A,\nu_A)$ of a semigroup S is called an intuitionistic fuzzy right (left) ideal of S if $\mu_A(xy)\geq\mu_A(x)$ and $\nu_A(xy)\leq\nu_A(x)$ ($\mu_A(xy)\geq\mu_A(y)$ and $\nu_A(xy)\leq\nu_A(y)$) for all $x,y\in S$. An intuitionistic fuzzy set $A=(\mu_A,\nu_A)$ of a semigroup S is called an intuitionistic fuzzy ideal of S if it is both a fuzzy left and a fuzzy right ideal of S.

Definition 3.5

If $(F,A)'$ and $(G,B)'$ are two IF soft sets over a common universe U then

$(F,A)'$ AND $(G,B)'$

is an IF soft set denoted by $(F,A)'\bar{\Pi}'(G,B)'$ and defined as,

$$(F,A)'\bar{\Pi}'(G,B)'=(H,A\times B)',$$

where $H(a,b)=F(a)\tilde{\bigcap}G(b)$ for all $(a,b)\in A\times B$.

Definition 3.6

If $(F,A)'$ and $(G,B)'$ are two IF soft sets over a common universe U then

$(F,A)'$ OR $(G,B)'$

is an IF soft set denoted by $(F,A)'\bar{\amalg}'(G,B)'$ and defined as,

$$(F,A)'\bar{\amalg}'(G,B)'=(H,A\times B)',$$

where $H(a,b)=F(a)\tilde{\bigcup}G(b)$ for all $(a,b)\in A\times B$.

Definition 3.7

If $(F,A)'$ and $(G,B)'$ are two IF soft sets over a common universe U then their restricted intersection is an intuitionistic fuzzy soft set $(H,C)'$ denoted by

$$(F,A)'\bar{\Pi}_r(G,B)'=(H,C)'$$

where $C=A\cap B\neq\phi$ and H is a function from C to IF^U, defined as $H(c)=F(c)\tilde{\bigcap}G(c)$ for all $c\in C$.

Definition 3.8

If $(F,A)'$ and $(G,B)'$ are two IF soft sets over a common universe U then their restricted union is an intuitionistic fuzzy soft set $(H,C)'$ denoted by

$$(F,A)'\bar{\amalg}_r(G,B)'=(H,C)'$$

where $C=A\cap B\neq\phi$ and H is a function from C to IF^U, defined as $H(c)=F(c)\tilde{\bigcup}G(c)$ for all $c\in C$.

Definition 3.9

If $(F,A)'$ is an IF soft set of an LA-semigroup *S*. Then for any $a\in A$, $F(a)$ is called an intuitionistic fuzzy LA-semigroup if $F(a)\circ F(a)\leq F(a)$.

Definition 3.10

Let $(F,A)'$ and $(G,B)'$ be any two IF soft sets over an LA-semigroup *S*. Then, the restricted product of $(F,A)'$ and $(G,B)'$ is defined as the IF soft set

$$(H,C)'=(F,A)'\circ'(G,B)',$$

where $C=A\cap B$ and H is a set valued function from C to IF^U, defined as $H(c)=F(c)\circ G(c)$ for all $c\in C$.

Definition 3.11

Let *S* be an LA-semigroup and $(F,A)'$. An IF soft set over *S*, then $(F,A)'$ is an IF soft LA-semigroup over *S* if $(F,A)'\circ'(F,A)'\subseteq(F,A)'$.

Definition 3.12

Let *S* be an LA-semigroup and $(F,A)'$. An IF soft set over *S*, then $(F,A)'$ is said to be an IF soft left (right) ideal over *S* if $F(a)$ is a fuzzy left (right) ideal of *S* and for all $a\in A$. An IF soft set $(F,A)'$ over *S* an IF soft ideal if it is both an IF soft left and an IF soft right ideal over *S*.

Theorem 3.13

Let *S* be an LA-semigroup and $(F,A)'$. An IF soft set over *S*, then $(F,A)'$ is an IF soft LA-semigroup iff each $F(a)\neq\phi$ is an IF LA-semigroup of S for all $a\in A$.

Proof

Assume $(F,A)'$ is an IF soft LA-semigroup. By definition (3.11),

$$(F,A)'\circ'(F,A)'\subseteq(F,A)'\Rightarrow F(a)\circ F(a)\leq F(a)$$

for all $a\in A$. Therefore $F(a)$is an IF LA-semigroup of S. Conversely, assume $F(a)$is an IF LA-semigroup of S. By definition (3.9), $F(a)\circ F(a)\leq F(a)$ for all

$$a\in A\Rightarrow(F,A)'\circ'(F,A)'\subseteq(F,A)'.$$

Therefore $(F,A)'$ is an IF soft LA-semigroup.

Theorem 3.14

Let $(F,A)'$ and $(G,B)'$ be two IF soft sets LA-semigroups over S. Then, $(F,A)'\ \overline{\prod}'\ (G,B)'$ is also an IF soft LAsemigroup over S.

Proof

By the definition (3.5),

$$(F,A)'\overline{\Pi}'(G,B)'=(H,A\times B)',$$

where $H(a,b)=F(a)\tilde{\cap}G(b)$ for all $(a,b)\in A\times B$. Since $F(a)$ and $G(b)$ are fuzzy LA-semigroups over S. Therefore, either

$$F(a)\tilde{\cap}G(b)=\varphi\ \text{(or)}\ F(a)\tilde{\cap}G(b)$$

is an IF LA-semigroup of S. which implies, $H(a,b)$ is an IF LA-semigroup of S. Hence $(H,A\times B)'$ is an IF soft LA-semigroup over S. Hence $(F,A)'\overline{\Pi}'(G,B)'$ is an IF soft LA-semigroup over S.

Theorem 3.15

Let $(F,A)'$ and $(G,B)'$ be two IF soft LA-semigroups over S, then $(F,A)'\overline{\Pi}_E(G,B)'$ is also an IF soft LA-semigroup over S.

Proof

Assume $(F,A)'$ and $(G,B)'$ be two IF soft LA-semigroups over S. By definition (3.3),

$$(F,A)'\overline{\Pi}_E(G,B)'=(H,C)',$$

where $C = A\cup B$ and for all $e \in C$,

$$H(e)=\begin{cases} F(e) & if\ e\in A\setminus B, \\ G(e) & if\ e\in B\setminus A, \\ F(e)\tilde{\bigcap}G(e) & if\ e\in A\bigcap B. \end{cases}$$

Then for all $e \in C \Rightarrow$ either $e \in$ A-B *(or)* $e \in$ A-B *(or)* $e \in A \cap B$.

If $e \in$ A-B $\Rightarrow H(e) = F(e)$
If $e \in$ A-B $\Rightarrow H(e) = G(e)$
If $e \in$ A-B $\Rightarrow H(e) = F(e)\tilde{\bigcap}G(e)$

*F*rom the above, we have $F(e)\tilde{\bigcap}G(e)$ is an IF LA-semigroup of S $\Rightarrow$ *H* (*e*) is a*n* IF LA-semigroup of S. Therefore $(H,C)'$ i*s* an IF soft LA-semigroup over S. Hence $(F,A)'\bar{\Pi}_E(G,B)'$ is a*n* IF soft LA-semigroup over S.

Theorem 3.*16*

Let $(F,A)'$ and $(G,B)'$ be two IF soft LA-semigroups over S such that A∩B≠ϕ. Then $(F,A)'\bar{\Pi}_r(G,B)'$ is a IF soft LA-semigroup over S.

Proof

Assume $(F,A)'$ and $(G,B)'$ are IF soft LA-semigroups over S. By de*fi*nition (3.7), $(F,A)'\bar{\Pi}_r(G,B)'=(H,C)'$ where $C=A\bigcap B\neq\varphi$ and $H(c)=F(e)\tilde{\bigcap}G(e)$ for all e∈C. *I*n the above, *F*(*e*) and *G*(*e*) are IF LA-semigroups of S and a*l*so $F(e)\tilde{\bigcap}G(e)$ is an IF LA-semigroup of S is e*i*ther empty or an IF LA-semigroup of S. Hen*ce*, $(F,A)'\bar{\Pi}_r(G,B)'=(H,C)'$ is a IF soft LA-semigroup over S.

T*he*orem 3.17

Let $(F,A)'$ and $(G,B)'$ be two IF soft LA-semigroups over S *s*uch that A∩*B*≠*ϕ*. Then $(F,A)'\bar{\amalg}_E(G,B)'$ is also a IF soft LA-semigroup over S.

Proof

Assume $(F,A)'$ and $(G,B)'$ be two IF soft LA-semigroups. By definition 2.10 and remark 3.2 $(F,A)'\bar{\amalg}_E(G,B)'=(H,C)'$, where C = A∪B and for all e ∈ C,

$$H(e)=\begin{cases} F(e) & if\ e\in A\setminus B, \\ G(e) & if\ e\in B\setminus A, \\ F(e)\tilde{\bigcup}G(e) & if\ e\in A\bigcap B. \end{cases}$$

Then for all $e \in C$, either $e \in$ A-B *(or)* $\in$ A-B, *if* $e \in$ A-B $\Rightarrow H(e) = (e)$ and if $e \in$ A-B $\Rightarrow H(e) =$ (e). From both the cases $H(e)$ is an IF LA-semigroup of S. Therefore $(H,C)'$ is an IF soft LA-semigroup over S. Hence $(F,A)'\bar{\amalg}_E(G,B)'$ *is* an IF soft LA-semigroup over S.

Theorem 3.18

An IF soft set $(F,A)'$ over S is an *I*F soft ideal over S iff $F(a) \neq \varnothing$ is an IF ideal of S for all $a \in$A.
Proo*f*
It is obvious from the definition (3.12).

Theorem 3.19

Let $(F,A)'$ and $(G,B)'$ be two IF soft ideals over S. Th*en* $(F,A)'\bar{\Pi}'(G,B)'$ is also an IF soft ideal over S.

Proof

By the definition (3.5), $(F,A)'\bar{\Pi}'(G,B)' = (H, A\times B)'$, where $H(a,b) = F(a)\tilde{\bigcap}G(b)$ for all *(a,b)*∈*A*×*B*. Since $(F,A)'$ and $(G,B)'$ are IF soft ideals over *S.* Then *(a)* and *(b)* are IF ideals of S *for* every *a*∈A *and* b∈B $\Rightarrow F(a)\tilde{\bigcap}G(b)$ is an IF ideal of S. *T*herefore $(H, A\times B)'$ is an IF soft ideal over S. Hence $(F,A)'\bar{\Pi}'(G,B)'$ is an IF soft ideal over *S.*

Theorem 3.20

Let $(F,A)'$ and $(G,B)'$ be two IF soft ideals over *S.* Then $(F,A)'\bar{\amalg}'(G,B)'$ is also an IF soft ideal over *S.*

Proof

By the definition (3.6), $(F,A)'\bar{\amalg}'(G,B)' = (H, A\times B)'$, where $H(a,b) = F(a)\tilde{\bigcup}G(b)$ for all *(a,b)*∈*A*×*B*. Since $(F,A)'$ and $(G,B)'$ are IF soft ideals over *S.* Then *(a)* and *(b)* are IF ideals of S *for* every *a*∈A *and* *b*∈B $\Rightarrow F(a)\tilde{\bigcup}G(b)$ is an IF ideal of S. *Th*erefore $(H, A\times B)'$ is an IF soft ideal over S. *H*ence $(F,A)'\bar{\amalg}'(G,B)'$ is an IF soft ideal over S.

Theorem 3.22

Let $(F,A)'$ and $(G,B)'$ be two IF soft ideals over S, such that *A*∩*B*≠ ∅. Then $(F,A)'\bar{\Pi}_r(G,B)'$ is also an IF soft ideal over *S.*

Proof

By the definition (3.7), $(F,A)'\bar{\Pi}_r(G,B)' = (H,C)'$ where

$C = A \bigcap B \neq \varphi$ and $H(c) = F(c) \tilde{\bigcap} G(c)$

for all $c \in C$. Then $F(c) \tilde{\bigcap} G(c)$ is an IF ideal of $S \Rightarrow (e)$ is an IF ideal of S. Therefore $(H,C)'$ is an IF soft ideal over S. Hence $(F,A)' \overline{\Pi}_r (G,B)'$ is an IF soft ideal over S.

Theorem 3.23

Let $(F,A)'$ and $(G,B)'$ be two IF soft ideals over S. Then $(F,A)' \overline{\Pi}_E (G,B)'$ is also an IF soft ideal over S.

Proof

By the definition (3.3), $(F,A)' \overline{\Pi}_E (G,B)' = (H,C)'$, where $C = A \cup B$ and for all $e \in C$,

$$H(e) = \begin{cases} F(e) & if \ e \in A \setminus B, \\ G(e) & if \ e \in B \setminus A, \\ F(e) \tilde{\bigcap} G(e) & if \ e \in A \bigcap B. \end{cases}$$

Then, for all $e \in C \Rightarrow$ either $e \in$ A-B (or) $e \in$ A-B (or) $e \in A \cap B$.

If $e \in$ A-B $\Rightarrow H(e) = (e)$
If $e \in$ A-B $\Rightarrow H(e) = (e)$
If $e \in A \cap B \Rightarrow H(e) = F(e) \tilde{\bigcap} G(e)$

From the above, we have $F(e) \tilde{\bigcap} G(e)$ is an IF ideal of S $\Rightarrow H(e)$ is an IF ideal of S. Therefore $(H,C)'$ is an IF soft ideal over S. Hence $(F,A)' \overline{\Pi}_E (G,B)'$ is an IF soft ideal over S.

Theorem 3.24

Let $(F,A)'$ and $(G,B)'$ be two IF soft ideals over S. Then $(F,A)' \overline{\coprod}_E (G,B)'$ is also an IF soft ideal over S.

Pro*of*

By definition (2.10) and remark (3.2), $(F,A)' \overline{\coprod}_E (G,B)' = (H,C)'$, where $C = A \cup B$ *a*nd for all $e \in C$,

$$H(e) = \begin{cases} F(e) & if \ e \in A \setminus B, \\ G(e) & if \ e \in B \setminus A, \\ F(e) \tilde{\bigcup} G(e) & if \ e \in A \bigcap B. \end{cases}$$

Then, for all $e \in C$, either $e \in$ A-*B* (or) $e \in$ A-B (or) $e \in A \cap B$.

If $e \in$ A-B $\Rightarrow H(e) = (e)$

If $e \in$ A-B $\Rightarrow H(e) = (e)$
If e∈A∩B⇒ $H(e) = F(e)\tilde{\bigcup}G(e)$

From the above, we have $F(e)\tilde{\bigcup}G(e)$ is an IF ideal of S $\Rightarrow H(e)$ is an IF ideal of S. Therefore $(H,C)'$ is an IF soft ideal over S. Hence $(F,A)'\bar{\coprod}_E(G,B)'$ *is* an IF soft ideal over S.

Theorem 3.25

Let $(F,A)'$ and $(G,B)'$ be two IF soft ideals over S. Then $(F,A)'\bar{\coprod}_r(G,B)'$ is also an IF soft ideal over S.

Proof

Using the definition (3.8), $(F,A)'\bar{\coprod}_r(G,B)' = (H,C)'$ where

$$C = A \bigcap B \neq \varphi \text{ and } H(e) = F(e)\tilde{\bigcup}G(e)$$

for all *e*∈*C*. Then $F(e)\tilde{\bigcup}G(e)$ is an IF ideal of $S \Rightarrow H(e)$ is an IF ideal of *S*. Therefore $(H,C)'$ is an IF soft ideal over *S*. Hence $(F,A)'\bar{\coprod}_r(G,B)'$ is an IF soft ideal over *S*.

4. CONCLUSION

In this chapter, the concepts of the intuitionistic fuzzy soft LA-semigroups and the intuitionistic fuzzy soft ideals are introduced. Also, their some basic properties are studied. In this chapter some of the binary operations like

$$\bar{\coprod}, \bar{\coprod}_E, \bar{\sqcap}_E, \tilde{\bigcap}, \bar{\sqcap}', \circ, \tilde{\bigcup}, \circ', \bar{\coprod}_r,$$

and others are defined on intuitionistic fuzzy soft sets to consider some results about the intuitionistic fuzzy soft LA-semigroups and the intuitionistic fuzzy soft ideals. Therefore, if we define new binary operations on intuitionistic fuzzy soft sets, then we will consider new interesting results about the intuitionistic fuzzy soft LA-semigroups and the intuitionistic fuzzy soft ideals.

REFERENCES

Adnan Abdul-Ghani, Mahmood Khalil, Abd Ulrazaq, & Firas Muhammad Jawad Al-Musawi. (2018). New Branch of Intuitionistic Fuzzification in Algebras with Their Applications. *International Journal of Mathematics and Mathematical Sciences.*

Aktas, H., & Cagman, N. (2007). Soft sets and soft groups. *Information Science*, *177*, 2726–2735.

Ali, M. I., & Shabir, M. (2009). Soft ideals and generalized fuzzy ideals in semigroups. *New Mathematics and Natural Computation*, *5*, 599–615.

Ali Abbas, N.M., Khalil, S.M. & Sadiq, H.J. (2019). A Note on the New Closed Classes and GP-Regular Classes in Soft. *Journal of Southwest Jiaotong University, 54*(1). doi:10.3969/j.issn.0258-2724.2019.155

Atanassov, K. T. (1986). Intuitionistic fuzzy sets. *Fuzzy Sets and Systems*, *20*, 87–96.

Aygunoglu, A., & Aygun, H. (2009). Introduction to fuzzy soft groups. *Computers & Mathematics with Applications (Oxford, England)*, *58*, 1279–1286.

Chen, B. (2013). Soft semi-open sets and related properties in soft topological spaces. *Applied Mathematics & Information Sciences*, *7*, 287–294.

Ejegwa, P. J., Akubo, A. J., & Joshua, O. M. (2014). Intuitionistic Fuzzy Sets and its Application in Career Determination Via Normalized Euclidean Distance Method. *European Scientific Journal*, *10*, 529–536.

Fenga, F., Jun, Y. B., & Zhao, X. (2008). Soft semirings. *Computers & Mathematics with Applications (Oxford, England)*, *56*, 2621–2628.

Kazim, M. A., & Naseerudin, M. (1972). On almost semigroups. *Alig. Bull. Math.*, *2*, 1–7.

Mahmood, S. (2014). Soft Regular Generalized b-Closed Sets in Soft Topological Spaces. *Journal of Linear and Topological Algebra*, *3*, 195–204.

Mahmood, S. (2015). On intuitionistic fuzzy soft b - closed sets in intuitionistic fuzzy soft topological spaces. *Annals of Fuzzy Mathematics and Informatics*, *10*, 221–233.

Mahmood, S. (2016). Dissimilarity Fuzzy Soft Points and their Applications. *Fuzzy Information and Engineering*, *8*, 281–294. doi:10.1016/j.fiae.2016.11.003

Mahmood, S. (2017). Soft Sequentially Absolutely Closed Spaces. *International Journal of Applications of Fuzzy Sets and Artificial Intelligence*, *7*, 73–92.

Mahmood, S. (2017). Characteristics of Soft Tychonoff Spaces with New Soft Separation Axioms. *Journal of The College of Education*, *2*, 1423–1446.

Mahmood, S. (2017). Tychonoff Spaces in Soft Setting and their Basic Properties. *International Journal of Applications of Fuzzy Sets and Artificial Intelligence*, *7*, 93–112.

Mahmood, S. (2018). S. and Abd Ulrazaq, M., Soft *BCH*-Algebras of the Power Sets. *American Journal of Mathematics and Statistics*, *8*(1), 1–7.

Mahmood, S., & Abd Ulrazaq, M. (2018). New Category of Soft Topological Spaces. *American Journal of Computational and Applied Mathematics*, *8*(1), 20–25.

Mahmood, S., & Alradha, M. A. (2017). Soft Edge $\rho-$Algebras of the power sets. *International Journal of Applications of Fuzzy Sets and Artificial Intelligence*, *7*, 231–243.

Mahmood, S., & Hameed, F. (2018). An algorithm for generating permutations in symmetric groups using soft spaces with general study and basic properties of permutations spaces. *J Theor Appl Inform Technol*, *96*(9), 2445–2457.

Mahmood, S., & Hameed, F. (2018). An algorithm for generating permutation algebras using soft spaces. *Journal of Taibah University for Science, 12*(3), 299–308.

Mahmood, S., & Muhamad, A. (2017). Soft BCL-Algebras of the Power Sets. *International Journal of Algebra, 11*(7), 329–341.

Mahmood Khalil, S., Ulrazaq, M., & Abdul-Ghani, S. (2018). Abu Firas Al-Musawi, $\sigma-$Algebra and $\sigma-$Baire in Fuzzy Soft Setting. Advances in Fuzzy Systems.

Maji, P. K., Biswas, R., & Roy, A. R. (2001). Intuitionistic fuzzy soft sets. *Journal of Fuzzy Mathematics, 9*, 677–692.

Maji, P. K., & Roy, A. R. (2004). On intuitionistic fuzzy soft sets. *Journal of Fuzzy Mathematics, 12*, 669–683.

Maji, P. K., Roy, A. R., & Biswas, R. (2001). Fuzzy soft sets. *Journal of Fuzzy Mathematics, 9*, 589–602.

Molodtsov, D. (1999). Soft set theory–First results. *Computers & Mathematics with Applications (Oxford, England), 37*, 19–31.

Mushtaq, Q., & Kamran, M. S. (1989). On LA-semigroup with weak associative law. *Scientic Khyber, 1*, 69–71.

Mushtaq, Q., & Khan, M. (2008). Direct product of Abel Grassmann's Groupoids. *J. Interdiscip. Math., 11*, 461–467.

Mushtaq, Q., & Khan, M. (2009). M-System in LA-semigroups. *Southeast Asian Bulletin of Mathematics, 33*, 321–327.

Mushtaq, Q., & Khan, M. (2003). Ideals in Left Almost semigroups. *5th Proc. Int. Conf. Pur. Math.*, 23-29.

Shabir, M. & Naz, M. (2011). On Soft topological spaces. *Comp. and Math. With Applications, 61*, 1786-1799.

Yang, C. (2008). A not on soft set theory. *Computers & Mathematics with Applications (Oxford, England), 56*, 1899–1900.

Yaqoob, N., Aslam, M., & Ansari, M. A. (2012). Structures of N-G-hyperideals in left almost G-semihypergroups. *World Applied Sciences Journal, 17*, 1611–1617.

Yaqoob, N., Aslam, M. & Hila, K. (2012). Rough Fuzzy Hyperideals in Ternary Semihypergroups. *Advances in Fuzzy Systems.*

Yaqoob, N., Aslam, M., & Yousafzai, F. (2012). On soft Γ-hyperideals over left almost Γ-semihypergroups. *Journal of Advanced Research in Dynamical and Control Systems, 4*, 1–2.

Yaqoob, N., Mostafa, S. M., & Ansari, M. A. (2013). On Cubic KU-Ideals of KU-Algebras. ISRN Algebra.

Yuksel, S., Tozlu, N., & Guzel, Z. (2013). On Soft Generalized Closed Sets in Soft Topological Spaces. *Journal of Theoretical and Appl. Inform. Technology, 55*, 17–21.

Zadeh, L. A. (1965). Fuzzy sets. *Information and Control, 8*, 338–353.

Chapter 7
Characterizations of Fuzzy Sublattices Based on Fuzzy Point

Chiranjibe Jana
https://orcid.org/0000-0002-4541-5336
Vidyasagar University, India

Faruk Karaaslan
Cankiri Karatekin University, Turkey

ABSTRACT

In a lattice $\mathfrak{L}$, the authors used the concept of belongingness and quasi-coincidence of fuzzy point to a fuzzy set, and by this notion, $(\in, \in\vee q)$-fuzzy sublattice, $(\in, \in\vee q)$-fuzzy ideal, cartesian product of $(\in, \in\vee q)$-fuzzy sublattice, $(\in, \in\vee q)$-fuzzy complemented sublattice, and cartesian product of $(\in, \in\vee q)$-fuzzy complemented sublattice are introduced, and their properties are briefly studied. The relationship between fuzzy sublattice and $(\in, \in\vee q)$-fuzzy sublattice, fuzzy ideal and $(\in, \in\vee q)$-fuzzy ideal of L are established. The authors prove that the cartesian product of two $(\in, \in\vee q)$-fuzzy ideals of a lattice is not necessarily a fuzzy ideal of a lattice. The theory of image and inverse image of an $(\in, \in\vee q)$-fuzzy sublattice and $(\in, \in\vee q)$-fuzzy ideal, an $(\in, \in\vee q)$-fuzzy complemented sublattice, and $(\in, \in\vee q)$-fuzzy complemented ideal of $\mathfrak{L}$ on the basis of homomorphism of lattices are also significantly established.

1 INTRODUCTION

The concept of fuzzy sets proposed by Zadeh in 1965 (Zadeh, 1965). After the initiation of fuzzy sets by Zadeh, there is a huge number of generalization of fuzzy sets in many areas, fuzzification of algebraic structure is one of them. Fuzzy algebraic structures play an important role in mathematics and have a huge scientific applications in theoretical physics, computer sciences, control engineering, information science, coding theory, topological spaces and so on. Then researchers are strongly motivated to review their concepts and results in various areas of abstract algebras in broader frame of fuzzy setting.

DOI: 10.4018/978-1-7998-0190-0.ch007

Rosenfeld introduced the notion of fuzzy subgroup of a group, base on this fundamental concept many researchers generalizes this idea in many structures (Rosenfeld, 1971), Mordeson and Malik study on fuzzy subring and fuzzy ideal in ring theory (Mordeson, & Malik., 1998). Jun, (Jun, 2005), Jun et al. (Jun, Öztürk, &Muhiuddin., 2016), Meng and Guo (Meng, & Guo., 2005), Ma et al.(Ma, Zhan, & Davvaz., 2008), (Ma, Zhan, & Jun., 2009), (Ma, Zhan, & Jun., 2011) applied fuzzy set theory based on fuzzy points in *BCK*/*BCI*-subalgebras and ideals of *BCK*/*BCI* -algebras. He et al.(He, Xin, & Zhan., 2014), (He, Xin, & Zhan., 2016) and Zhang and Zhan (Zhang, & Zhan., 2017) have been done lot of work on lattices, Khan et al. (Khan, Davvaz, Yaqoob, Gulistan, & Khalaf., 2015) studied briefly on $(\in,\in\vee q_k)$ -fuzzy *KU*-ideals and $(\in,\in\vee q_k)$-fuzzy ideals on subtraction algebras, Muhiuddin and Al-roqi (Muhiuddin, & Al-roqi., 2015) focused to the study of subalgebras of *BCK*/*BCI* -algebras based on (α,β) -type fuzzy sets, Jana et al. (Jana et al., 2015, 2015a, 2016, 2016a, 2017, 2017a), Bej and Pal (Bej, & Pal., 2015) have done detailed investigation in *BCK*/*BCI* -algebras and *BG*/*G*-algebras. Bhakat and Das analyzed a new kind of fuzzy subgroup by using the notion of belongings to $(\in)$and quasi-coincident of a fuzzy point x_a to a fuzzy set A in group theory (Bhakat, & Das., 1996a), (Bhakat, & Das., 1996b). Now, we review one of the structures that are most extensively used and discussed in mathematics and its applications are lattices done by Birkhoof one hand which is considered as a relational, ordered structure, on the other hand considered as an algebras (Birkhoff, 2009). Thereafter many researchers generalized lattices by using the fuzzy set theory and developed different fuzzy lattice structures such as Bo and Wangming introduced the concept of fuzzy ideal theory in distributive lattices (Bo, & Wangming., 1990). The definition of $\mathfrak{L}$ -fuzzy sets introduced by Tepavcevic and Trajkovski (Tepavcevic, & Trajkovski., 2001), Seselja et al. introduced the notion of representation of fuzzy lattice and codes by using the $\mathfrak{L}$ -fuzzy lattices (Seselja, & Tepavcevic., 1994), (Seselja, Tepavcevic, & Vojvodic., 1993), Ali and Ray studied product on fuzzy sublattices (Ali, & Ray., 2007), Feng et al. introduced the concept of (λ,γ) -fuzzy sublattice and (λ,γ)-fuzzy subhyperlattice, and described the Cartesian product of (λ,γ)-fuzzy sublattices, and product of subhyperlattice and interval-valued subhyperlattice (Feng, & Zou., 2010), (Feng, Duan, & Zeng., 2010), (Feng, Zeng, Zou, & Zhang., 2010). Recently, Kazanci and Davvaz studied the generalized fuzzy sublattices by means of fuzzy points (Kazancı, & Davvaz., 2010). Ali et al. studied $(\in,\in\vee q)$-fuzzy equivalence relation by using indistinguishability operator (Ali, Feng, & Shabir., 2012). Motivated by the above works and best of our knowledge there is no work available on $(\in,\in\vee q)$-fuzzy sublattice and $(\in,\in\vee q)$-fuzzy ideal of a complemented lattices, of their homomorphism and their Cartesian product of $(\in,\in\vee q)$-fuzzy sublattice of a complemented lattice. For this reason we developed a theoretical study on characterization of fuzzy sublattice and ideals.

Inspired by the previous works, a natural problem is weather of a lattices can be characterized by the corresponding properties of fuzzy sets. In this paper, we combine the theories of fuzzy sets with fuzzy sublattice and fuzzy ideal of a lattice to broaden application fields of theory of fuzzy sets and provide more ways to study fuzzy algebras. We study the notion of $(\in,\in\vee q)$-fuzzy sublattice and $(\in,\in\vee q)$-fuzzy ideal, and Cartesian product of $(\in,\in\vee q)$-fuzzy sublattice. Then $(\in,\in\vee q)$-fuzzy sublattice and $(\in,\in\vee q)$-fuzzy ideal of a complemented lattice, and Cartesian product of $(\in,\in\vee q)$-fuzzy sublattice of complemented lattices are studied. Also, we have shown that the image and inverse image on a lattice should be an $(\in,\in\vee q)$-fuzzy sublattice (complemented) and $(\in,\in\vee q)$-fuzzy ideal (complemented) on the basis of homomorphism of a lattice.

2 PRELIMINARIES

In this section, we recall some elementary aspects that are necessary for this paper are included. In this paper, we denote a lattice ($\mathfrak{L}$,∨,∧,0,1) by $\mathfrak{L}$, where the join and met operations are respectively denoted by ∨ and ∧ in $\mathfrak{L}$. The greatest element of $\mathfrak{L}$ is denoted by 1 and least element is denoted by 0.

Definition 2.1 (Kazancı, & Davvaz., 2010). Let $\mathfrak{L}$ *be a lattice and* $\mathfrak{L}' \neq \varnothing$ *be a subset of* $\mathfrak{L}$. Then $\mathfrak{L}'$ *is called a sublattice of* $\mathfrak{L}$ if *for every pair of elements* $x, y \in \mathfrak{L}'$ *implies* $x \vee y \in \mathfrak{L}'$ *and* $x \wedge y \in \mathfrak{L}'$.

Definition 2.2 (Kazancı, & Davvaz., 2010). *Let I be a non-empty subset of* $\mathfrak{L}$. T*hen I is called an ideal of* $\mathfrak{L}$ if *for all x,y*∈$\mathfrak{L}$ impl*y*

1. $x \in \mathfrak{L}$ and $y \in I \Rightarrow x \wedge y \in I$
2. $x \in \mathfrak{L}$, $y \in \mathfrak{L}$ and $y \leq x \Rightarrow y \in I$.

Definition 2.3 (Ali, Feng, & Shabir., 2012). Let $\mathcal{F} : \mathfrak{L} \to \mathfrak{M}$ *be a function from the lattice* $\mathfrak{L}$ to *the lattice* $\mathfrak{M}$ and *is called lattice homomorphism if it satisfies the conditions:*

$$\mathcal{F}(x \wedge y) = \mathcal{F}(x) \wedge \mathcal{F}(y),\ \mathcal{F}(x \vee y) = \mathcal{F}(x) \vee \mathcal{F}(y)$$

for all *x,y*∈$\mathfrak{L}$.

Definition 2.4 *Let X be a non-empty set. A fuzzy set* $\hat{\gamma}$ *of X is defined as a mapping* $\hat{\gamma} : X \to [0,1]$ *where* [0,1] *is the usual interval of real numbers. We take* $\mathbb{F}(X)$ *as the set of all fuzzy subsets of X.*

In this paper, {[0,1],∨,∧) is denoted for a complete lattice, where [0,1] is the set of real numbers between 0 and 1, and *x*∨*y*=max{*x,y*} and *x*∧*y*=min{*x,y*}.

Let $\mathfrak{L}$ be a non-empty set. A fuzzy subset $\hat{\gamma}$ of $\mathfrak{L}$ is a function $\hat{\gamma} : \mathfrak{L} \to [0,1]$. Then $\hat{\gamma}_\alpha = \{x \in \mathfrak{L} \mid \hat{\gamma}(x) \geq \alpha\}$ is called a level subset of $\hat{\gamma}$.

Definition 2.5 (Bo, & Wangmin., 1990). *A fuzzy set* $\hat{\gamma}$ *is called a fuzzy sublattice of* $\mathfrak{L}$ *if* $\hat{\gamma}(x) \wedge \hat{\gamma}(y) \leq \hat{\gamma}(x \vee y) \wedge \hat{\gamma}(x \wedge y)$ *for all x,y*∈$\mathfrak{L}$.

Definition 2.6 (Bhakat, & Das., 1996a). *A fuzzy set* $\hat{\gamma}$ *in a lattice* $\mathfrak{L}$ i*s of the form*

$$\hat{\gamma}(y) = \begin{cases} \alpha \in (0,1] & \text{if } y = x \\ 0 & \text{if } y \neq x. \end{cases}$$

is said to be a fuzzy point with support *x* and value α and is denoted by x_α.

Definition 2.7 (Bhakat, & Das., 1996a). *Let* $\hat{\gamma}$ *be a fuzzy subsets of* $\mathfrak{L}$ and *x*α be *a fuzzy point.*

1. If $\hat{\gamma}(x) \geq \alpha$, then x_α belong to $\hat{\gamma}$, denoted $x_\alpha \in \hat{\gamma}$
2. If $\hat{\gamma}(x) + \alpha \geq 1$, then x_α is quasi-coincident with $\hat{\gamma}$, and denoted by $x_\alpha q \hat{\gamma}$
3. $x_\alpha \in \wedge q \hat{\gamma} \Leftrightarrow x_\alpha \in \hat{\gamma}$ and $x_\alpha q \hat{\gamma}$
4. $x_\alpha \in \vee q \hat{\gamma} \Leftrightarrow x_\alpha \in \hat{\gamma}$ or $x_\alpha q \hat{\gamma}$.

Definition 2.8 (Feng, Duan, & Zeng., 2010). *A fuzzy set* $\hat{\gamma}$ *is called a fuzzy sublattice of* $\mathfrak{L}$ *if for all x,y∈*$\mathfrak{L}$ *satisfies the following conditions:*

1. $\hat{\gamma}(x \wedge y) \geq \hat{\gamma}(x) \wedge \hat{\gamma}(y)$
2. $\hat{\gamma}(x \vee y) \geq \hat{\gamma}(x) \wedge \hat{\gamma}(y)$.

Definition 2.9 (Feng, Duan, & Zeng., 2010). *A fuzzy set* $\hat{\gamma}$ *of lattice* $\mathfrak{L}$ *is called fuzzy ideal of* $\mathfrak{L}$ *if for all x,y∈*$\mathfrak{L}$

1. $\hat{\gamma}(x \wedge y) \geq \hat{\gamma}(x) \vee \hat{\gamma}(y)$
2. $\hat{\gamma}(x \vee y) \geq \hat{\gamma}(x) \wedge \hat{\gamma}(y)$.

3 (∈,∈∨*q*)-FUZZY SUBLATTICE AND (∈,∈∨*q*)-FUZZY IDEALS

In this section, we have to study (∈,∈∨*q*)-fuzzy sublattice and (∈,∈∨*q*)-fuzzy ideals of a lattice and homomorphism of these notion on the basis of (∈,∈∨)-fuzzy sublattice and (∈,∈∨*q*)-fuzzy ideal.

Definition 3.1 *A fuzzy set* $\hat{\gamma}$ *is called an* (∈,∈∨*q*)*-fuzzy sublattice of* $\mathfrak{L}$ *if for all x,y∈*$\mathfrak{L}$ *satisfies the following conditions:*

1. $\hat{\gamma}(x \wedge y) \geq \hat{\gamma}(x) \wedge \hat{\gamma}(y) \wedge 0.5$
2. $\hat{\gamma}(x \vee y) \geq \hat{\gamma}(x) \wedge \hat{\gamma}(y) \wedge 0.5$.

Example 3.2 *For a lattice* $\mathfrak{L}$=$\{0,a,b,c,d,1\}$ *as shown in Hasse diagram of Figure 1 and the Caley tables for the operations* ∧ *and* ∨ *of* $\mathfrak{L}$ *are respectively shown below:*

Let us define the fuzzy set $\hat{\gamma}$ on $\mathfrak{L}$ as follows:

$$\hat{\gamma}(x) = \begin{cases} 0.3 & \text{if} \quad x = 0 \\ 0.8 & \text{if} \quad x = a \\ 0.6 & \text{if} \quad x = b \\ 0.4 & \text{if} \quad x = c, d \\ 0 & \text{if} \quad x = 1 \end{cases}$$

Figure 1. The lattice in example 3.2

∧	0	a	b	c	d	1
0	0	0	0	0	0	0
a	0	a	a	a	a	a
b	0	a	b	a	b	b
c	0	a	a	c	c	c
d	0	a	b	c	d	d
1	0	a	b	c	d	1

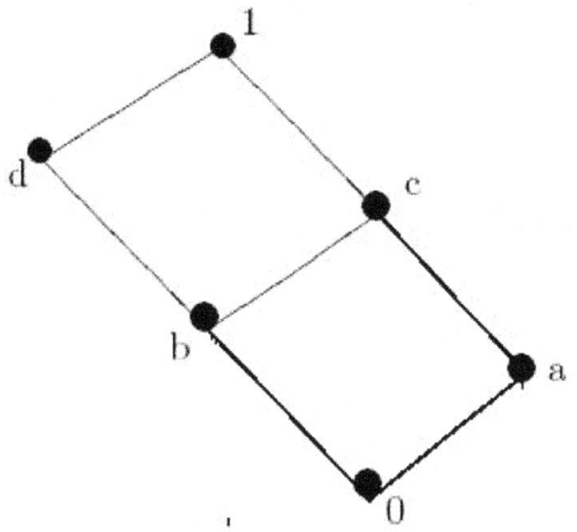

Table 1.

∨	0	a	b	c	d	1
0	0	a	b	c	d	1
a	a	a	b	c	d	1
b	b	b	b	d	d	1
c	c	c	d	c	d	1
d	d	d	d	d	d	1
1	1	1	1	1	1	1

Then it is verified that $\hat{\gamma}$ is an $(\in,\in\vee q)$-fuzzy sublattice of $\mathfrak{L}$.

Theorem 3.3 *Let* $\hat{\gamma}$ *be an* $(\in,\in\vee q)$*-fuzzy sublattice of* $\mathfrak{L}$. *Then followings are equivalent:*

1. $\hat{\gamma}$ is an $(\in,\in\vee q)$-fuzzy sublattice of $\mathfrak{L}$.
2. $\hat{\gamma}_t$ is a sublattice of $\mathfrak{L}$ for any $\alpha\in(0,0.5]$, where $\hat{\gamma}_\alpha \neq 0$.

Proof: (1)⇒(2). Let $\hat{\gamma}$ be an (∈,∈∨*q*)-fuzzy sublattice of 𝔏. Then for any α∈(0,0.5] such that $\hat{\gamma}_\alpha \neq 0$, we have to show that

$$x \wedge y \in \hat{\gamma}_\alpha \text{ and } x \vee y \in \hat{\gamma}_\alpha$$

for all *x,y*∈𝔏. From the definition of $\hat{\gamma}_\alpha$, we have for $x \in \hat{\gamma}_\alpha$ implies $\hat{\gamma}(x) \geq \alpha$, and so $\hat{\gamma}(y) \geq \alpha$. Hence,

$$\hat{\gamma}(x \wedge y) \geq \hat{\gamma}(x) \wedge \hat{\gamma}(y) \wedge 0.5 = \alpha \wedge 0.5 = \alpha,$$

and hence $x \wedge y \in \hat{\gamma}_\alpha$. Similarly, we can proved $x \vee y \in \hat{\gamma}_\alpha$.

(2)⇒(1). Let $\hat{\gamma}_\alpha$ be a sublattice of 𝔏. Suppose that there exists x_0, y_0∈𝔏 such that

$$\hat{\gamma}(x \wedge y) < \alpha = \hat{\gamma}(x) \wedge \hat{\gamma}(y) \wedge 0.5,$$

where α∈(0,0.5], then $\hat{\gamma}(x) \wedge \hat{\gamma}(y) \geq \alpha$ imply that $x_0 \in \hat{\gamma}_\alpha$ and $y_0 \in \hat{\gamma}_\alpha$, but $\hat{\gamma}(x_0 \wedge y_0) < \alpha$, i.e. $x_0 \wedge y_0 \notin \hat{\gamma}_\alpha$, which contradict that $\hat{\gamma}_\alpha$ is sublattice of 𝔏 for α∈(0,0.5]. Thus,

$$\hat{\gamma}(x \wedge y) \geq \hat{\gamma}(x) \wedge \hat{\gamma}(y) \wedge 0.5$$

for all *x,y*∈𝔏. Similarly, we can prove that $\hat{\gamma}(x \vee y) \geq \hat{\gamma}(x) \wedge \hat{\gamma}(y) \wedge 0.5$. Therefore, $\hat{\gamma}$ is an (∈,∈∨*q*)-fuzzy sublattice of 𝔏. ■

Theorem 3.4 *Let* $\hat{\gamma}$ *be an* (∈,∈∨*q*)*-fuzzy sublattice of* 𝔏. *Then followings are equivalent:*

1. $\hat{\gamma}$ is an (∈,∈∨*q*)-fuzzy sublattice of 𝔏.
2. $\hat{\gamma}_t$ is a sublattice of 𝔏 for any α∈(0,0.5], where $\hat{\gamma}_\alpha \neq 0$.

Proof: Straight forward. ■

Theorem 3.5 *Every fuzzy sublattice of a lattice* 𝔏 *is an* (∈,∈∨q)*-fuzzy sublattice of* 𝔏.

The converse of the Theorem 3.5 is not true in general justified in the following example.

Example 3.6 *Consider a lattice* 𝔏={0,a,*b,c,d,1*} *with the Caley tables for the operations* ∧ and ∨ *respectively given in the Example 3.2, then defined fuzzy set* $\hat{\gamma}$ on 𝔏 *such as follows:*

$$\hat{\gamma}(x)=\begin{cases}0.3 & \text{if} \quad x=0\\ 0.8 & \text{if} \quad x=a\\ 0.7 & \text{if} \quad x=b,c\\ 0.6 & \text{if} \quad x=d\\ 0 & \text{if} \quad x=1\end{cases}$$

It is easy to verify that $\hat{\gamma}$ is an $(\in,\in\vee q)$-fuzzy sublattice of $\mathfrak{L}$ but not a fuzzy sublattice of $\mathfrak{L}$, because

$$\hat{\gamma}(b\vee c)=\hat{\gamma}(d)=0.6\Leftrightarrow 0.7=\hat{\gamma}(b)\wedge\hat{\gamma}(c).$$

Definition 3.7 *A fuzzy set* $\hat{\gamma}$ *of lattice* $\mathfrak{L}$ *is called an* $(\in,\in\vee q)$*-fuzzy ideal of* $\mathfrak{L}$ if *it satisfies the following conditions*

(a) $\hat{\gamma}(x\wedge y)\geq\left(\hat{\gamma}(x)\vee\hat{\gamma}(y)\right)\wedge 0.5$

(b) $\hat{\gamma}(x\vee y)\geq\left(\hat{\gamma}(x)\wedge\hat{\gamma}(y)\right)\wedge 0.5$ for all $x,y\in\mathfrak{L}$.

Example 3.8 Consider a lattice $\mathfrak{L}=\{0,a,b,c,d,1\}$ with the Caley tables for the operations $\wedge$ and $\vee$ respectively given in Example 3.2. Now, we defined fuzzy set $\hat{\gamma}$ on $\mathfrak{L}$ as follows:

$$\hat{\gamma}(x)=\begin{cases}0.9 & \text{if } x=0\\ 0.8 & \text{if } x=a\\ 0.6 & \text{if } x=b\\ 0.5 & \text{if } x=c\\ 0.7 & \text{if } x=d\\ 0 & \text{if } x=1.\end{cases}$$

It is verified that $\hat{\gamma}$ is an $(\in,\in\vee q)$-fuzzy ideal of $\mathfrak{L}$.

Proposition 3.9 *Let* $\hat{\gamma}$ *be a fuzzy subset of a lattice* $\mathfrak{L}$, *then following conditions are equivalent for all* $x,y\in\mathfrak{L}$,

1. $\hat{\gamma}(x\wedge y)\geq\left(\hat{\gamma}(x)\vee\hat{\gamma}(y)\right)\wedge 0.5$
2. $x\leq y\Rightarrow\hat{\gamma}(x)\geq\hat{\gamma}(y)\wedge 0.5$.

Proof: (1)⇒(2). If $x\leq y$, then $x\wedge y=x$, and so

$$\hat{\gamma}(x)=\hat{\gamma}(x\wedge y)\geq\left(\hat{\gamma}(x)\vee(y)\right)\wedge 0.5\geq\hat{\gamma}(y)\wedge 0.5.$$

(2) ⇒ (1). For $x\wedge y\leq x$, then $\hat{\gamma}(x\wedge y)\geq\hat{\gamma}(x)\wedge 0.5$ and from $x\wedge y\leq y$ implies $\hat{\gamma}(x\wedge y)\geq\hat{\gamma}(y)\wedge 0.5$.

Thus, $\hat{\gamma}(x\wedge y)\geq\left(\hat{\gamma}(x)\wedge 0.5\right)\vee\left(\hat{\gamma}(y)\wedge 0.5\right)\geq\left(\hat{\gamma}(x)\vee\hat{\gamma}(y)\right)\wedge 0.5$. ■

Theorem 3.10 *Let $\hat{\gamma}$ be an (∈,∈∨q)-fuzzy sublattice of $\mathfrak{L}$. Then following are equivalent:*

1. $\hat{\gamma}$ is an (∈,∈∨q)-fuzzy ideal of $\mathfrak{L}$
2. $\hat{\gamma}_t$ is an ideal of $\mathfrak{L}$ for any α∈(0,0.5], with $\hat{\gamma}_\alpha\neq 0$.

Proof: (1)⇒(2). Let $\hat{\gamma}$ be an (∈,∈∨q)-fuzzy ideal of $\mathfrak{L}$. For any α∈(0,0.5] such that $\hat{\gamma}_\alpha\neq 0$. We have to prove $x\wedge y\in\hat{\gamma}_\alpha$ for all $x\in\hat{\gamma}_\alpha$ and for $y\in L$. From the definition of $\hat{\gamma}_\alpha$, we get $\hat{\gamma}(x)\geq\alpha$, and so

$$\alpha\leq\alpha\wedge 0.5\leq 0.5\wedge\left(\hat{\gamma}(y)\vee\hat{\gamma}(x)\right)\leq\hat{\gamma}(x\wedge y).$$

Hence, $\hat{\gamma}(x\wedge y)\geq\alpha$ which imply $x\wedge y\in\hat{\gamma}_\alpha$.

(2) ⇒ (1). Let $\hat{\gamma}_\alpha$ is an ideal of $\mathfrak{L}$, for all α∈(0,0.5], where $\hat{\gamma}_\alpha\neq 0$. If there exists $x_0,y_0\in\mathfrak{L}$ such that

$$\hat{\gamma}(x_0\wedge y_0)<\alpha=\left(\hat{\gamma}(x_0)\vee\hat{\gamma}(y_0)\right)\wedge 0.5$$

which imply that

$$\hat{\gamma}(x_0)\vee\hat{\gamma}(y_0)\geq\alpha \text{ and } \hat{\gamma}(x_0\wedge y_0)<\alpha$$

from which we get $x_0\in\hat{\gamma}_\alpha$ and $y_0\in\hat{\gamma}_\alpha$ but $x_0\wedge y_0\notin\hat{\gamma}_\alpha$ which contradicts that $\hat{\gamma}_\alpha$ is an ideal of $\mathfrak{L}$. Thus, $\hat{\gamma}(x\wedge y)\geq\left(\hat{\gamma}(x)\vee\hat{\gamma}(y)\right)\wedge 0.5$ for all $x,y\in\mathfrak{L}$.

Hence the theorem. ■

Theorem 3.11 *Let $\hat{\gamma}$ be an (∈,∈∨q)-fuzzy sublattice of $\mathfrak{L}$. Then following are equivalent:*

1. $\hat{\gamma}$ is an (∈,∈∨q)-fuzzy ideal of $\mathfrak{L}$.
2. $\hat{\gamma}_\alpha$ is an ideal of $\mathfrak{L}$ for any α∈(0,0.5], where $\hat{\gamma}_\alpha\neq 0$.

Proof: Straight forward. ■

Theorem 3.12 *Every fuzzy ideal of* $\mathfrak{L}$ *is an* (∈,∈∨q)*-fuzzy ideal of* $\mathfrak{L}$.

The converse of the Theorem 3.12 is not true in general as seen in the following example.

Example 3.13 *Consider a lattice* $\mathfrak{L}=\{0,a,b,c,d,1\}$ *with the Caley tables for the operations* ∧ and ∨ *respectively given in the Example 3.2, then we defined fuzzy set* $\hat{\gamma}$ on $\mathfrak{L}$ *such as follows:*

$$\hat{\gamma}(x)=\begin{cases}0.7 & \text{if } x=0\\ 0.6 & \text{if } x=a,c\\ 0.9 & \text{if } x=b\\ 0.5 & \text{if } x=d\\ 0.8 & \text{if } x=1.\end{cases}$$

Then it is verified that $\hat{\gamma}$ is an (∈,∈∨*q*)-fuzzy ideal of $\mathfrak{L}$ but not a fuzzy ideal of $\mathfrak{L}$.
Let $\mathfrak{L}_1$, $\mathfrak{L}_2$ be two lattices in which sup(∅)=0, where ∅ is the empty-set.

Theorem 3.14 Le*t* $\mathcal{F}:\mathfrak{L}_1\rightarrow\mathfrak{L}_2$ *be a homomorphism and let* $\hat{\gamma}$ *be an* (∈,∈∨*q*)*-fuzzy sublattice of* $\mathfrak{L}_1$. *Then* $\mathcal{F}(\hat{\gamma})$ *is an* (∈,∈∨*q*)*-fuzzy sublattice of* $\mathfrak{L}_2$, *where*

$$\mathcal{F}(\hat{\gamma})(y)=\sup_{x\in\mathfrak{L}_1}\{\hat{\gamma}(x)\mid\mathcal{F}(x)=y,\forall y\in\mathfrak{L}_2\}.$$

Proof: Let $\mathcal{F}^{-1}(y_1)=\varnothing$ or $\mathcal{F}^{-1}(y_2)=\varnothing$ for any $y_1,y_2\in\mathfrak{L}_2$, then

$$\mathcal{F}(y_1\vee y_2)\geq 0=\left(\mathcal{F}(\hat{\gamma})(y_1)\wedge\mathcal{F}(\hat{\gamma})(y_2)\right)\wedge 0.5.$$

Assume that $\mathcal{F}^{-1}(y_1)\neq\varnothing$, $\mathcal{F}^{-1}(y_2)\neq\varnothing$ for any $y_1,y_2\in\mathfrak{L}_2$ and so

$$\begin{aligned}\mathcal{F}(\hat{\gamma})(y_1\vee y_2)&=\sup_{m\in\mathfrak{L}_1}\{\hat{\gamma}(m)\mid\mathcal{F}(m)=y_1\vee y_2\}\\ &\geq\sup_{x_1,x_2\in\mathfrak{L}_1}\{\hat{\gamma}(x_1\vee x_2)\mid\mathcal{F}(x_1)=y_1,\mathcal{F}(x_2)=y_2\}\\ &\geq\sup_{x_1,x_2\in\mathfrak{L}_1}\{\left(\hat{\gamma}(x_1)\wedge\hat{\gamma}(x_2)\right)\wedge 0.5\mid\mathcal{F}(x_1)=y_1,\mathcal{F}(x_2)=y_2\}\\ &=(\sup_{x_1\in\mathfrak{L}_1}\{\hat{\gamma}(x_1)\mid\mathcal{F}(x_1)=y_1\}\wedge\sup_{x_2\in\mathfrak{L}_1}\{\hat{\gamma}(x_2)\mid\mathcal{F}(x_2)=y_2\})\wedge 0.5\\ &=\left(\mathcal{F}(\hat{\gamma})(y_1)\wedge\mathcal{F}(\hat{\gamma})(y_2)\right)\wedge 0.5.\end{aligned}$$

In similar way, we have to show that $\mathcal{F}(y_1 \wedge y_2) \geq \left(\mathcal{F}\left(\hat{\gamma}\right)(y_1) \wedge \mathcal{F}\left(\hat{\gamma}\right)(y_2)\right) \wedge 0.5$.

Hence, $\mathcal{F}\left(\hat{\gamma}\right)$ is an $(\in, \in \vee q)$-fuzzy sublattice of $\mathfrak{L}_2$. ■

Theorem 3.15 *Let $\mathcal{F}: \mathfrak{L}_1 \rightarrow \mathfrak{L}_2$ be a surjective homomorphism and let $\hat{\gamma}$ be an $(\in, \in \vee q)$-fuzzy ideal of $\mathfrak{L}_1$. Then $\mathcal{F}\left(\hat{\gamma}\right)$ is an $(\in, \in \vee q)$-fuzzy ideal of $\mathfrak{L}_2$.*

Proof: Let for any $y_1, y_2 \in \mathfrak{L}_2$, $\mathcal{F}^{-1}(y_1)$ and $\mathcal{F}^{-1}(y_2)$ are not empty. Then we have

$$\begin{aligned}
\mathcal{F}\left(\hat{\gamma}\right)(y_1 \wedge y_2) &= \sup_{m \in \mathfrak{L}} \{\hat{\gamma}(m) \mid \mathcal{F}(m) = y_1 \wedge y_2\} \\
&\geq \sup_{x_1, x_2 \in \mathfrak{L}_1} \{\hat{\gamma}(x_1 \wedge x_2) \mid \mathcal{F}(x_1) = y_1, \mathcal{F}(x_2) = y_2\} \\
&\geq \sup_{x_1, x_2 \in \mathfrak{L}_1} \{\left(\hat{\gamma}(x_1) \vee \hat{\gamma}(x_2)\right) \wedge 0.5 \mid \mathcal{F}(x_1) = y_1, \mathcal{F}(x_2) = y_2\} \\
&= (\sup_{x_1 \in \mathfrak{L}_1} \{\hat{\gamma}(x_1) \mid \mathcal{F}(x_1) = y_1\} \vee \sup_{x_2 \in \mathfrak{L}_1} \{\hat{\gamma}(x_2) \mid \mathcal{F}(x_2) = y_2\}) \wedge 0.5 \\
&= \left(\mathcal{F}\left(\hat{\gamma}\right)(y_1) \vee \mathcal{F}\left(\hat{\gamma}\right)(y_2)\right) \wedge 0.5.
\end{aligned}$$

Hence, $\mathcal{F}\left(\hat{\gamma}\right)$ is an $(\in, \in \vee q)$-fuzzy ideal of $\mathfrak{L}_2$. ■

Theorem 3.16 *Let $\mathcal{F}: \mathfrak{L}_1 \rightarrow \mathfrak{L}_2$ be a homomorphism and $\hat{\gamma}$ be an $(\in, \in \vee q)$-fuzzy sublattice of $\mathfrak{L}_2$. Then $\mathcal{F}^{-1}\left(\hat{\gamma}\right)$ is an $(\in, \in \vee q)$-fuzzy sublattice of $\mathfrak{L}_2$,* where

$$\mathcal{F}^{-1}\left(\hat{\gamma}\right)(x) = \hat{\gamma}\left(\mathcal{F}(x)\right)$$

for all $x \in \mathfrak{L}_1$.

Proof: Let for any $x_1, x_2 \in \mathfrak{L}_1$, we have

$$\begin{aligned}
\mathcal{F}^{-1}\left(\hat{\gamma}\right)(x_1 \vee x_2) &= \hat{\gamma}\left(\mathcal{F}(x_1 \vee x_2)\right) \\
&= \hat{\gamma}(\mathcal{F}(x_1) \vee \mathcal{F}(x_2)) \\
&\geq \left(\hat{\gamma}\left(\mathcal{F}(x_1)\right) \wedge \hat{\gamma}\left(\mathcal{F}(x_2)\right)\right) \wedge 0.5 \\
&= \left(\mathcal{F}^{-1}\left(\hat{\gamma}\right)(x_1) \wedge \mathcal{F}^{-1}\left(\hat{\gamma}\right)(x_2)\right) \wedge 0.5.
\end{aligned}$$

In this same way can prove

$$\mathcal{F}^{-1}\left(\hat{\gamma}\right)(x_1 \wedge x_2) \geq \left(\mathcal{F}^{-1}\left(\hat{\gamma}\right)(x_1) \wedge \mathcal{F}^{-1}\left(\hat{\gamma}\right)(x_2)\right) \wedge 0.5.$$

Thus, $\mathcal{F}^{-1}\left(\hat{\gamma}\right)$ is an $(\in,\in\vee q)$ -fuzzy sublattice of $\mathfrak{L}_1$. ■

Theorem 3.17 *Let* $\mathcal{F}: \mathfrak{L}_1 \rightarrow \mathfrak{L}_2$ *be a homomorphism and* $\hat{\gamma}$ *be an* $(\in,\in\vee q)$ *-fuzzy ideal of* $\mathfrak{L}_2$. Then $\mathcal{F}^{-1}\left(\hat{\gamma}\right)$ *is an* $(\in,\in\vee q)$ *-fuzzy ideal of* $\mathfrak{L}_1$.

Proof: Let for any $x_1,x_2\in\mathfrak{L}_1$, we have

$$\begin{aligned}\mathcal{F}^{-1}\left(\hat{\gamma}\right)(x_1 \wedge x_2) &= \hat{\gamma}\left(\mathcal{F}(x_1 \wedge x_2)\right)\\ &= \hat{\gamma}\left(\mathcal{F}(x_1) \wedge \mathcal{F}(x_2)\right)\\ &\geq \left(\hat{\gamma}\left(\mathcal{F}(x_1)\right) \vee \hat{\gamma}\left(\mathcal{F}(x_2)\right)\right) \wedge 0.5\\ &= \left(\mathcal{F}^{-1}\left(\hat{\gamma}\right)(x_1) \vee \mathcal{F}^{-1}\left(\hat{\gamma}\right)(x_2)\right) \wedge 0.5.\end{aligned}$$

Therefore, $\mathcal{F}^{-1}\left(\hat{\gamma}\right)$ is an $(\in,\in\vee q)$ -fuzzy ideal of $\mathfrak{L}_1$. ■

4 CARTESIAN PRODUCT OF (∈,∈∨*q*) -FUZZY SUBLATTICES

In this section, we studied product of $(\in,\in\vee q)$ -fuzzy sublattice and ideals of a lattice $\mathfrak{L}$.

Definition 4.1 *Let* $\mathfrak{L}_1$ and $\mathfrak{L}_2$ be *two lattices. Then Cartesian product of two lattices* $\mathfrak{L}_1$ and $\mathfrak{L}_2$, is d*enoted by* $\mathfrak{L}_1\times\mathfrak{L}_2$ and is *defined by*

$$\mathfrak{L}_1 \times \mathfrak{L}_2 = \{(x,y) \mid x \in \mathfrak{L}_1, y \in \mathfrak{L}_2\}$$

und*er the partially ordered of* $\mathfrak{L}_1\times\mathfrak{L}_2$ by the *relation* $(x_1,y_1)\leq(x_2,y_2)$ *if* and *only if* $x_1\leq x_2$ and $y_1\leq y_2$ for *all*

$$(x_1,y_1),(x_2,y_2) \in \mathfrak{L}_1 \times \mathfrak{L}_2$$

is *a lattice, where* $(x_1,y_1)\wedge(x_2,y_2) = (x_1\wedge x_2,y_1\wedge y_2)$ and $(x_1,y_1)\vee(x_2,y_2) = (x_1\vee x_2,y_1\vee y_2)$ for all

$$(x_1,y_1),(x_2,y_2) \in \mathfrak{L}_1 \times \mathfrak{L}_2.$$

Theorem 4.2 *Let* $\hat{\gamma}_1$ and $\hat{\gamma}_2$ *be two* $(\in,\in\vee q)$ *-fuzzy sublattices of* $\mathfrak{L}_1$ and $\mathfrak{L}_2$ *respectively. Then Cartesian product of* $\hat{\gamma}_1$ and $\hat{\gamma}_2$ *is* $\hat{\gamma}_1 \times \hat{\gamma}_2$, *and is an* $(\in,\in\vee q)$ *-fuzzy sublattice of* $\mathfrak{L}_1\times\mathfrak{L}_2$, and *defined by*

$$\left(\hat{\gamma}_1 \times \hat{\gamma}_2\right)(x,y) = \hat{\gamma}_1(x) \wedge \hat{\gamma}_2(y)$$

for all $(x,y) \in \mathfrak{L}_1 \times \mathfrak{L}_2$.

Proof: Let $(x_1,x_2),(y_1,y_2) \in \mathfrak{L}_1 \times \mathfrak{L}_2$. Then $(x_1,y_1) \wedge (x_2,y_2) = (x_1 \wedge x_2, y_1 \wedge y_2)$ and $(x_1,y_1) \vee (x_2,y_2) = (x_1 \vee x_2, y_1 \vee y_2)$. Now, we have

$$\begin{aligned}\left(\hat{\gamma}_1 \times \hat{\gamma}_2\right)\left((x_1,y_1) \wedge (x_2,y_2)\right) &= \left(\hat{\gamma}_1 \times \hat{\gamma}_2\right)(x_1 \wedge x_2, y_1 \wedge y_2) \\ &= \left\{\hat{\gamma}_1(x_1 \wedge x_2) \wedge \hat{\gamma}_2(y_1 \wedge y_2)\right\} \\ &\geq \left\{\left(\hat{\gamma}_1(x_1) \wedge \hat{\gamma}_1(x_2)\right) \wedge 0.5 \wedge \left(\hat{\gamma}_2(y_1) \wedge \hat{\gamma}_2(y_2)\right) \wedge 0.5\right\} \\ &= \left\{\hat{\gamma}_1(x_1) \wedge \hat{\gamma}_1(x_2) \wedge \hat{\gamma}_2(y_1) \wedge \hat{\gamma}_2(y_2)\right\} \wedge 0.5 \\ &= \{(\hat{\gamma}_1(x_1) \wedge \hat{\gamma}_2(y_1) \wedge \hat{\gamma}_1(x_2) \wedge \hat{\gamma}_2(y_2)\} \wedge 0.5 \\ &= \left(\left(\hat{\gamma}_1 \times \hat{\gamma}_2\right)(x_1,y_1) \wedge \left(\hat{\gamma}_1 \times \hat{\gamma}_2\right)(x_2,y_2)\right) \wedge 0.5.\end{aligned}$$

and

$$\begin{aligned}\left(\hat{\gamma}_1 \times \hat{\gamma}_2\right)\left((x_1,y_1) \vee (x_2,y_2)\right) &= \left(\hat{\gamma}_1 \times \hat{\gamma}_2\right)(x_1 \vee x_2, y_1 \vee y_2) \\ &= \left\{\hat{\gamma}_1(x_1 \vee x_2) \wedge \hat{\gamma}_2(y_1 \vee y_2)\right\} \\ &\geq \left\{\left(\hat{\gamma}_1(x_1) \wedge \hat{\gamma}_1(x_2)\right) \wedge 0.5 \wedge \left(\hat{\gamma}_2(y_1) \wedge \hat{\gamma}_2(y_2)\right) \wedge 0.5\right\} \\ &= \left\{\hat{\gamma}_1(x_1) \wedge \hat{\gamma}_1(x_2) \wedge \hat{\gamma}_2(y_1) \wedge \hat{\gamma}_2(y_2)\right\} \wedge 0.5 \\ &= \{(\hat{\gamma}_1(x_1) \wedge \hat{\gamma}_2(y_1) \wedge \hat{\gamma}_1(x_2) \wedge \hat{\gamma}_2(y_2)\} \wedge 0.5 \\ &= \left(\left(\hat{\gamma}_1 \times \hat{\gamma}_2\right)(x_1,y_1) \wedge \left(\hat{\gamma}_1 \times \hat{\gamma}_2\right)(x_2,y_2)\right) \wedge 0.5.\end{aligned}$$

Therefore, $\hat{\gamma}_1 \times \hat{\gamma}_2$ is an $(\in,\in\vee q)$ -fuzzy sublattice of $\mathfrak{L}_1 \times \mathfrak{L}_2$. ■

The following example shows that the Cartesian product of two $(\in,\in\vee q)$ -fuzzy ideals of a lattice $\mathfrak{L}$ is not necessarily a fuzzy ideal of a lattice $\mathfrak{L}$.

Example 4.3 Co*nsider a lattice* $\mathfrak{L}=\{0,a,b,c,d,1\}$ with *the Caley tables for the operations* $\wedge$ and $\vee$ given *in the Example 3.2, the fuzzy sets* $\hat{\gamma}_1$ and $\hat{\gamma}_2$ *of a lattice* $\mathfrak{L}$ which *are defined respectively by the following tables:*

Table 2. Fuzzy set $\hat{\gamma}_1$ of $\mathfrak{L}$

$\mathfrak{L}$	0	a	b	c	d	1
$\hat{\gamma}_1$	0.6	0.7	0.6	0.9	0.9	0.8

Table 3. Fuzzy set $\hat{\gamma}_2$ of $\mathfrak{L}$

$\mathfrak{L}$	0	a	b	c	d	1
$\hat{\gamma}_2$	0.7	0.6	0.8	0.6	0.7	0.9

Here, $\hat{\gamma}_1$ and $\hat{\gamma}_2$ are $(\in,\in\vee q)$ -fuzzy ideals of $\mathfrak{L}$. Now, for $(b,c),(c,d)\in \mathfrak{L}\times\mathfrak{L}$, we have

$$\left(\hat{\gamma}_1\times\hat{\gamma}_2\right)\left((b,c)\wedge(c,d)\right)=\left(\hat{\gamma}_1\times\hat{\gamma}_2\right)(b\wedge c,c\wedge d)$$
$$=\left(\hat{\gamma}_1\times\hat{\gamma}_2\right)(c,a)=\hat{\gamma}_1(c)\wedge\hat{\gamma}_2(a)=0.6$$

$$\left(\hat{\gamma}_1\times\hat{\gamma}_2\right)\left((b,c)\vee(c,d)\right)=\left(\hat{\gamma}_1\times\hat{\gamma}_2\right)(b\vee c,c\vee d)$$
$$=\left(\hat{\gamma}_1\times\hat{\gamma}_2\right)(d,d)=\hat{\gamma}_1(d)\wedge\hat{\gamma}_2(d)=0.7$$

and

$$\left(\hat{\gamma}_1\times\hat{\gamma}_2\right)(b,c)=\hat{\gamma}_1(b)\wedge\hat{\gamma}_2(c)=0.6$$

$$\left(\hat{\gamma}_1\times\hat{\gamma}_2\right)(c,d)=\hat{\gamma}_1(c)\wedge\hat{\gamma}_2(d)=0.7$$

Thus,

$$\left(\hat{\gamma}_1\times\hat{\gamma}_2\right)\left((b,c)\wedge(c,d)\right)\geq\left(\hat{\gamma}_1\times\hat{\gamma}_2\right)(b,c)\vee\left(\hat{\gamma}_1\times\hat{\gamma}_2\right)(c,d)\wedge 0.5$$

$$\left(\hat{\gamma}_1\times\hat{\gamma}_2\right)\left((b,c)\vee(c,d)\right)\geq\left(\hat{\gamma}_1\times\hat{\gamma}_2\right)(b,c)\wedge\left(\hat{\gamma}_1\times\hat{\gamma}_2\right)(c,d)\wedge 0.5$$

holds. But,

$$\left(\hat{\gamma}_1\times\hat{\gamma}_2\right)\left((b,c)\wedge(c,d)\right)\Leftrightarrow\left(\hat{\gamma}_1\times\hat{\gamma}_2\right)(b,c)\vee\left(\hat{\gamma}_1\times\hat{\gamma}_2\right)(c,d)$$

$$\left(\hat{\gamma}_1\times\hat{\gamma}_2\right)\left((b,c)\vee(c,d)\right)\Leftrightarrow\left(\hat{\gamma}_1\times\hat{\gamma}_2\right)(b,c)\wedge\left(\hat{\gamma}_1\times\hat{\gamma}_2\right)(c,d)$$

do not hold. Therefore, $\hat{\gamma}_1 \times \hat{\gamma}_2$ is not a fuzzy ideal of a lattice $\mathfrak{L}\times\mathfrak{L}$.

5 $(\in,\in\vee q)$ -FUZZY SUBLATTICES OF COMPLEMENTED LATTICES

In this section, we studied the notion of $(\in,\in\vee q)$ -fuzzy sublattice of complemented lattice and homomorphism of complemented lattice by means of an $(\in,\in\vee q)$ -fuzzy sublattice and $(\in,\in\vee q)$ -fuzzy ideal of a lattice.

Let 0 and 1 be the least and greatest element of the lattice $\mathfrak{L}$. Let x∈L, then by complement of an element x in $\mathfrak{L}$ is an element y *in* $\mathfrak{L}$ such that x∧y=0 and x∨y=1.

Definition 5.1 (Ali, & Ray., 2007). A *lattice* $\mathfrak{L}$ wit*h greatest element 1* and *least element 0* is *called complemented if all its elements have unique complements.*

Definition 5.2 A *fuzzy set* $\hat{\gamma}$ *is called an* $(\in,\in\vee q)$ *-fuzzy sublattice of a complemented lattice* $\mathfrak{L}$ if *for all x,y,z*∈$\mathfrak{L}$ sat*isfies the following conditions:*

1. $\hat{\gamma}(x\wedge y)\geq\left(\hat{\gamma}(x)\wedge\hat{\gamma}(y)\right)\wedge 0.5$
2. $\hat{\gamma}(x\vee y)\geq\left(\hat{\gamma}(x)\wedge\hat{\gamma}(y)\right)\wedge 0.5$
3. $\hat{\gamma}(z')\geq\hat{\gamma}(z)\wedge 0.5$, where z' is the complement of z.

Proposition 5.3 *If* $\hat{\gamma}$ *is an* $(\in,\in\vee q)$ *-fuzzy sublattice of a complemented lattice* $\mathfrak{L}$, *then*

1. $\hat{\gamma}(0)\geq\hat{\gamma}(x)\wedge 0.5$.
2. $\hat{\gamma}(1)\geq\hat{\gamma}(x)\wedge 0.5$ for all $x\in\mathfrak{L}$.

Proof: Let x' be the complement of the element $x\in\mathfrak{L}$. Then

1. $$\begin{aligned}\hat{\gamma}(0)&=\hat{\gamma}(x\wedge x')\geq\left(\hat{\gamma}(x)\wedge\hat{\gamma}(x')\right)\wedge 0.5\\&=\left(\hat{\gamma}(x)\wedge 0.5\right)\wedge\hat{\gamma}(x')\geq\left(\hat{\gamma}(x)\wedge 0.5\right)\wedge\left(\hat{\gamma}(x)\wedge 0.5\right)=\hat{\gamma}(x)\wedge 0.5\end{aligned}.$$
2. $$\begin{aligned}\hat{\gamma}(1)&=\hat{\gamma}(x\vee x')\geq\left(\hat{\gamma}(x)\wedge\hat{\gamma}(x')\right)\wedge 0.5\\&=\left(\hat{\gamma}(x)\wedge 0.5\right)\wedge\hat{\gamma}(x')\geq\left(\hat{\gamma}(x)\wedge 0.5\right)\wedge\left(\hat{\gamma}(x)\wedge 0.5\right)=\hat{\gamma}(x)\wedge 0.5\end{aligned}.$$ ■

Proposition 5.4 *Let* $\hat{\gamma}$ *be an* $(\in,\in\vee q)$ *-fuzzy sublattice of a complemented lattice* $\mathfrak{L}$. *If* $\hat{\gamma}(0)\neq\hat{\gamma}(1)$, *then* $\hat{\gamma}(0)\wedge\hat{\gamma}(1)\geq 0.5$ *holds.*

Proof: Let us suppose that $\hat{\gamma}(0)<\hat{\gamma}(1)$ and if $\hat{\gamma}(0)\wedge\hat{\gamma}(1)<0.5$. Then two possible cases aries:

1. If $\hat{\gamma}(0) < 0.5$ and $\hat{\gamma}(0) < \hat{\gamma}(1)$, then $\hat{\gamma}(0) < \hat{\gamma}(1) \wedge 0.5$. This is a contradiction of the Proposition 5.3.
2. If $\hat{\gamma}(1) < 0.5$ and $\hat{\gamma}(0) < \hat{\gamma}(1)$, then $\hat{\gamma}(0) < \hat{\gamma}(1) \wedge 0.5$. This is a contradiction of the Proposition 5.3. If $\hat{\gamma}(1) < \hat{\gamma}(0)$, then we prove the above results dually. ■

Theorem 5.5 *Let* $\hat{\gamma}$ *be an* $(\in,\in\vee q)$ *-fuzzy sublattice of a complemented lattice* $\mathfrak{L}$. *Then followings are equivalent:*

1. $\hat{\gamma}$ is an $(\in,\in\vee q)$ -fuzzy sublattice of $\mathfrak{L}$.
2. $\hat{\gamma}_t$ is a sublattice of $\mathfrak{L}$ for any α∈(0,0.5], where $\hat{\gamma}_\alpha \neq 0$.

Proof: Straight forward. ■

Theorem 5.6 *Let* $\hat{\gamma}$ *be an* $(\in,\in\vee q)$ *-fuzzy sublattice of a complemented lattice* $\mathfrak{L}$. *Then followings are equivalent:*

1. $\hat{\gamma}$ is an $(\in,\in\vee q)$ -fuzzy sublattice of $\mathfrak{L}$.
2. $\hat{\gamma}_t$ is a sublattice of $\mathfrak{L}$ for any α∈(0,0.5], where $\hat{\gamma}_\alpha \neq 0$.

Proof: Straight forward. ■

Definition 5.7 *A fuzzy set* $\hat{\gamma}$ *of a complemented lattice* $\mathfrak{L}$ *is called an* $(\in,\in\vee q)$ *-fuzzy ideal if for all x,y,z*∈L

1. $\hat{\gamma}(x \wedge y) \geq (\hat{\gamma}(x) \vee \hat{\gamma}(y)) \wedge 0.5$
2. $\hat{\gamma}(x \vee y) \geq (\hat{\gamma}(x) \wedge \hat{\gamma}(y)) \wedge 0.5$
3. $\hat{\gamma}(z') \geq \hat{\gamma}(z) \wedge 0.5$. where z' is the complement of z.

Theorem 5.8 *Let* $\hat{\gamma}$ *be an* $(\in,\in\vee q)$ *-fuzzy sublattice of a complemented lattice* $\mathfrak{L}$. *Then followings are equivalent:*

1. $\hat{\gamma}$ is an $(\in,\in\vee q)$ -fuzzy ideal of $\mathfrak{L}$.
2. $\hat{\gamma}_t$ is an ideal of $\mathfrak{L}$ for any α∈(0,0.5], where $\hat{\gamma}_\alpha \neq 0$.

Proof: The proof of the theorem is obvious. ■

Theorem 5.9 *Let* $\hat{\gamma}$ *be an* $(\in,\in\vee q)$ *-fuzzy sublattice of a complemented lattice* $\mathfrak{L}$. *Then followings are equivalent:*

1. $\hat{\gamma}$ is an $(\in,\in\vee q)$ -fuzzy ideal of $\mathfrak{L}$.
2. $\hat{\gamma}_t$ is an ideal of $\mathfrak{L}$ for any α∈(0,0.5], where $\hat{\gamma}_\alpha \neq 0$.

Proof: The proof of the theorem is obvious. ■

Theorem 5.10 *Let* $\mathcal{F}:\mathfrak{L}_1 \to \mathfrak{L}_2$ *be a homomorphism, where* 𝔏1, 𝔏2 be *two complemented lattices and let* $\hat{\gamma}$ *be an* $(\in,\in\vee q)$ *-fuzzy sublattice of* $\mathfrak{L}_1$. *Then* $\mathcal{F}\left(\hat{\gamma}\right)$ *is an* $(\in,\in\vee q)$ *-fuzzy sublattice of* $\mathfrak{L}_2$.

Proof: Let y' be the complement of an any element $y \in \mathfrak{L}_2$. If $\mathcal{F}^{-1}(y)=\varnothing$, then

$$\mathcal{F}\left(\hat{\gamma}\right)(y') \geq \mathcal{F}(A)(y) \wedge 0.5.$$

If $\mathcal{F}^{-1}(y)\neq\varnothing$, then

$$\begin{aligned}\mathcal{F}\left(\hat{\gamma}\right)(y') &= \sup_{x'\in\mathfrak{L}_1}\{\hat{\gamma}(x') \mid \mathcal{F}(x')=y'\}\\ &\geq \sup_{x\in\mathfrak{L}_1}\{\hat{\gamma}(x)\wedge 0.5 \mid \mathcal{F}(x')=y'\}\\ &= \sup_{x\in\mathfrak{L}_1}\{\hat{\gamma}(x)\wedge 0.5 \mid \mathcal{F}(x)=y\}\\ &= \sup_{x\in\mathfrak{L}_1}\{\hat{\gamma}(x) \mid \mathcal{F}(x)=y\}\wedge 0.5\\ &= \mathcal{F}\left(\hat{\gamma}\right)(y)\wedge 0.5.\end{aligned}$$

Therefore, $\mathcal{F}\left(\hat{\gamma}\right)$ is an $(\in,\in\vee q)$ -fuzzy sublattice of a complemented lattice $\mathfrak{L}_2$. ■

Theorem 5.11 *Let* $\mathcal{F}:\mathfrak{L}_1 \to \mathfrak{L}_2$ *be a surjective homomorphism, where* $\mathfrak{L}_1$, $\mathfrak{L}_2$ are *two complemented lattices and let* $\hat{\gamma}$ *be an* $(\in,\in\vee q)$ *-fuzzy ideal of* $\mathfrak{L}_1$. *Then* $\mathcal{F}\left(\hat{\gamma}\right)$ *is an* $(\in,\in\vee q)$ *-fuzzy ideal of* $\mathfrak{L}_2$.

Proof: The proof of the theorem is obvious. ■

Theorem 5.12 *Let* $\mathcal{F}:\mathfrak{L}_1 \to \mathfrak{L}_2$ *be a homomorphism, where* $\mathfrak{L}_1$ and $\mathfrak{L}_2$ *are two complemented lattices and let* $\hat{\gamma}$ *be an* $(\in,\in\vee q)$ *-fuzzy sublattice of* $\mathfrak{L}_2$. *Then* $\mathcal{F}^{-1}(A)$ *is an* $(\in,\in\vee q)$ *-fuzzy sublattice of* $\mathfrak{L}_1$.

Proof: For any x∈$\mathfrak{L}_1$, we have

$$\begin{aligned}\mathcal{F}^{-1}\left(\hat{\gamma}\right)(x') &= \hat{\gamma}\left(\mathcal{F}(x')\right) \\ &= \hat{\gamma}(\mathcal{F}(x))' \\ &\geq \hat{\gamma}\left(\mathcal{F}(x)\right) \wedge 0.5 \\ &= \mathcal{F}^{-1}\left(\hat{\gamma}\right)(x) \wedge .5.\end{aligned}$$

Hence, $\mathcal{F}^{-1}\left(\hat{\gamma}\right)$ is an $(\in,\in\vee q)$ -fuzzy sublattice of $\mathfrak{L}1$ ■

Theorem 5.13 *Let* $\mathcal{F} : \mathfrak{L}_1 \rightarrow \mathfrak{L}_2$ *be a homomorphism, where* $\mathfrak{L}_1$ and $\mathfrak{L}_2$ *are two complemented lattices and let* $\hat{\gamma}$ *be an* $(\in,\in\vee q)$ *-fuzzy ideal of* $\mathfrak{L}_2$. Then $\mathcal{F}^{-1}\left(\hat{\gamma}\right)$ *is an* $(\in,\in\vee q)$ *-fuzzy ideal of* $\mathfrak{L}_1$.

Proof: The proof of the theorem is obvious. ■

6 CARTESIAN PRODUCT OF $(\in,\in\vee q)$ -FUZZY SUBLATTICES OF COMPLEMENTED LATTICES

With this section, we studied Cartesian product of $(\in,\in\vee q)$ -fuzzy sublattice of complemented lattice.

Let 0_1, I_1 be respectively least and greatest element of a complemented lattice $\mathfrak{L}1$ and, let 02 I2 are respectively least and greatest element of a complemented lattice $\mathfrak{L}2$, then $\mathfrak{L}1\times\mathfrak{L}_2$ is a complemented lattice with the least element (01,02) and greatest element (I1,I2). Moreover, the complement of any element

$(x,a)\in\mathfrak{L}_1\times\mathfrak{L}_2$ is $(y,b)\in\mathfrak{L}_1\times\mathfrak{L}_2$

if and only if y is the complement of x in $\mathfrak{L}1$ and b *is* the complement of a *in* $\mathfrak{L}2$, (Ali, & Ray., 2007).

Theorem 6.1 Le*t* $\hat{\gamma}_1$ and $\hat{\gamma}_2$ be *two* $(\in,\in\vee q)$ *-fuzzy sublattices of complemented lattices* $\mathfrak{L}_1$ and $\mathfrak{L}_2$ *respectively. Then Cartesian product of* $\hat{\gamma}_1$ and $\hat{\gamma}_2$, de*noted by* $\hat{\gamma}_1\times\hat{\gamma}_2$ *is an* $(\in,\in\vee q)$ *-fuzzy sublattice of a complemented lattice* $\mathfrak{L}_1\times\mathfrak{L}_2$.

Proof: Let (x',a') be the complement of an element $(x,a)\in\mathfrak{L}_1\times\mathfrak{L}_2$. Then x' is the complement of $x\in\mathfrak{L}_1$ and a' is the complement of $a\in\mathfrak{L}_2$. Therefore,

$$\hat{\gamma}_1(x') \geq \hat{\gamma}_1(x)\wedge 0.5 \text{ and } \hat{\gamma}_2(a') \geq \hat{\gamma}_2(a)\wedge 0.5.$$

Thus,

$$\begin{aligned}\left(\hat{\gamma}_1\times\hat{\gamma}_2\right)(x',a') &= \hat{\gamma}_1(x')\wedge\hat{\gamma}_2(a')\\ &\geq \left(\hat{\gamma}_1(x)\wedge 0.5\right)\wedge\left(\hat{\gamma}_2(a)\wedge 0.5\right)\\ &= \left(\hat{\gamma}_1(x)\wedge\hat{\gamma}_2(a)\right)\wedge 0.5\\ &= \left(\hat{\gamma}_1\times\hat{\gamma}_2\right)(x,a)\wedge 0.5.\end{aligned}$$

Hence, $\left(\hat{\gamma}_1\times\hat{\gamma}_2\right)$ is an $(\in,\in\vee q)$ -fuzzy sublattice of $\mathfrak{L}_1\times\mathfrak{L}_2$. ■

Theorem 6.2 *Let* $\hat{\gamma}_1$ and $\hat{\gamma}_2$ *be two fuzzy subsets of a complemented lattices* $\mathfrak{L}_1$ and $\mathfrak{L}_2$ re*spectively. If* $\hat{\gamma}_1\times\hat{\gamma}_2$ *is an* $(\in,\in\vee\mathrm{q})$ *-fuzzy sublattices of* $\mathfrak{L}_1\times\mathfrak{L}_2$, *then one of the following statements holds:*

$$\hat{\gamma}_1(I_1)\geq\hat{\gamma}_2(a)\wedge 0.5 \text{ for all } a\in\mathfrak{L}_2$$

and

$$\hat{\gamma}_2(I_2)\geq\hat{\gamma}_1(x)\wedge 0.5 \text{ for all } x\in\mathfrak{L}_1.$$

Proof: Let $\hat{\gamma}_1\times\hat{\gamma}_2$ be an $(\in,\in\vee q)$ -fuzzy sublattice of a complemented lattice $\mathfrak{L}1_\times\mathfrak{L}2$. Assume that none of the above statements hold, then find $x\in\mathfrak{L}_1$ and $a\in\mathfrak{L}_2$ such that

$$\hat{\gamma}_1(x)\wedge 0.5>\hat{\gamma}_2(I_2) \text{ and } \hat{\gamma}_2(a)\wedge 0.5>\hat{\gamma}_1(I_1).$$

Then

$$\begin{aligned}\left(\hat{\gamma}_1\times\hat{\gamma}_2\right)(x,a)\wedge 0.5 &= \hat{\gamma}_1(x)\wedge\hat{\gamma}_2(a)\wedge 0.5\\ &= \left(\hat{\gamma}_1(x)\wedge 0.5\right)\wedge\left(\hat{\gamma}_2(a)\wedge 0.5\right)\\ &> \hat{\gamma}_1(I_1)\wedge\hat{\gamma}_2(I_2)\\ &= \left(\hat{\gamma}_1\times\hat{\gamma}_2\right)(I_1,I_2).\end{aligned}$$

Therefore, $\hat{\gamma}_1\times\hat{\gamma}_2$ is an $(\in,\in\vee q)$ -fuzzy sublattice of a complemented lattice $\mathfrak{L}_1\times\mathfrak{L}_2$ satisfied by

$$\left(\hat{\gamma}_1\times\hat{\gamma}_2\right)(x,a)>\left(\hat{\gamma}_1\times\hat{\gamma}_2\right)(I_1,I_2)\wedge 0.5,$$

which is a contradiction that (I_1,I_2) is the greatest element of $\mathfrak{L}_1\times\mathfrak{L}_2$. ■

Theorem 6.3 *Let* $\hat{\gamma}_1$ and $\hat{\gamma}_2$ *be two fuzzy subsets of a complemented lattices* $\mathfrak{L}_1$ and $\mathfrak{L}_2$ *respectively. If* $\hat{\gamma}_1 \times \hat{\gamma}_2$ *is an* $(\in,\in\vee q)$ *-fuzzy sublattices of* $\mathfrak{L}_1\times\mathfrak{L}_2$, then *one of the following statements holds:*

$$\hat{\gamma}_1(0_1) \geq \hat{\gamma}_2(a) \wedge 0.5 \text{ for all } a\in\mathfrak{L}_2$$

and

$$\hat{\gamma}_2(0_2) \geq \hat{\gamma}_1(x) \wedge 0.5 \text{ for all } x\in\mathfrak{L}_1.$$

Proof: The proof of the theorem is similar as that of Theorem 6.2

Theorem 6.4 *Let* $\hat{\gamma}_1$ and $\hat{\gamma}_2$ *be two fuzzy subsets of the complemented lattices* $\mathfrak{L}_1$ and $\mathfrak{L}_2$ *respectively, such that*

$$\hat{\gamma}_2(I_2) \geq \hat{\gamma}_1(x) \wedge 0.5 \text{ or } \hat{\gamma}_2(0_2) \geq \hat{\gamma}_1(x) \wedge 0.5$$

for all $x\in\mathfrak{L}_1$. *If* $\hat{\gamma}_1 \times \hat{\gamma}_2$ *is an* $(\in,\in\vee q)$ *-fuzzy sublattices of* $\mathfrak{L}_1\times\mathfrak{L}_2$, then $\hat{\gamma}_1$ *is an* $(\in,\in\vee q)$ *-fuzzy sublattice of* $\mathfrak{L}_1$.

Let $(x,I_2),(y,I_2)\in \mathfrak{L}_1\times\mathfrak{L}_2$. If $\hat{\gamma}_2(I_2) \geq \hat{\gamma}_1(x) \wedge 0.5$ for all $x\in\mathfrak{L}_1$, then

$$\begin{aligned}
\hat{\gamma}_1(x \wedge y) &\geq \hat{\gamma}_1(x \wedge y) \wedge \hat{\gamma}_2(I_2 \vee I_2) \\
&= (\hat{\gamma}_1 \times \hat{\gamma}_2)(x \wedge y, I_2 \wedge I_2) \\
&= (\hat{\gamma}_1 \times \hat{\gamma}_2)((x, I_2) \wedge (y, I_2)) \\
&\geq ((\hat{\gamma}_1 \times \hat{\gamma}_2)(x, I_2) \wedge (\hat{\gamma}_1 \times \hat{\gamma}_2)(y, I_2)) \wedge 0.5 \\
&= \hat{\gamma}_1(x) \wedge \hat{\gamma}_2(I_2) \wedge \hat{\gamma}_1(y) \wedge \hat{\gamma}_2(I_2) \wedge 0.5 \\
&= (\hat{\gamma}_1(x) \wedge \hat{\gamma}_1(y)) \wedge 0.5
\end{aligned}$$

and

$$\begin{aligned}\hat{\gamma}_1(x\vee y) &\geq \hat{\gamma}_1(x\vee y)\wedge\hat{\gamma}_2(I_2\vee I_2)\\ &=(\hat{\gamma}_1\times\hat{\gamma}_2)(x\vee y, I_2\vee I_2)\\ &=(\hat{\gamma}_1\times\hat{\gamma}_2)((x,I_2)\vee(y,I_2))\\ &\geq((\hat{\gamma}_1\times\hat{\gamma}_2)(x,I_2)\wedge(\hat{\gamma}_1\times\hat{\gamma}_2)(y,I_2))\wedge 0.5\\ &=\hat{\gamma}_1(x)\wedge\hat{\gamma}_2(I_2)\wedge\hat{\gamma}_1(y)\wedge\hat{\gamma}_2(I_2)\wedge 0.5\\ &=\hat{\gamma}_1(x)\wedge\hat{\gamma}_1(y)\wedge 0.5.\end{aligned}$$

Let us take x' be the complement of x in $\mathfrak{L}_1$, then $(x',0_2)$ is the complement of (x,I_2). Then,

$$\begin{aligned}\hat{\gamma}_1(x') &\geq \hat{\gamma}_1(x')\wedge\hat{\gamma}_2(0_2)=(\hat{\gamma}_1\times\hat{\gamma}_2)(x',0_2)\\ &\geq(\hat{\gamma}_1\times\hat{\gamma}_2)(x,I_2)\wedge 0.5=\hat{\gamma}_1(x)\wedge\hat{\gamma}_2(I_2)\wedge 0.5=\hat{\gamma}_1(x)\wedge 0.5\end{aligned}$$

Hence, $\hat{\gamma}_1$ is an $(\in,\in\vee q)$ -fuzzy sublattice of $\mathfrak{L}_1$. ■

Theorem 6.5 *Let* $\hat{\gamma}_1$ and $\hat{\gamma}_2$ *be two fuzzy subsets of the complemented lattices* $\mathfrak{L}_1$ and $\mathfrak{L}_2$ *respectively, so that*

$$\hat{\gamma}_1(I_1)\geq\hat{\gamma}_2(a)\wedge 0.5 \text{ or } \hat{\gamma}_1(0_1)\geq\hat{\gamma}_2(a)\wedge 0.5$$

for all $a\in\mathfrak{L}_2$. *If* $\hat{\gamma}_1\times\hat{\gamma}_2$ *is an* $(\in,\in\vee q)$ *-fuzzy sublattice of* $\mathfrak{L}_1\times\mathfrak{L}_2$, then $\hat{\gamma}_2$ *is an* $(\in,\in\vee q)$ *-fuzzy sublattice of* $\mathfrak{L}2$.

Proof: The proof of the theorem is similar as that of the Theorem 6.3. ■

From the previous theorem, we find the following corollary.

Corollary 6.6 *Let* $\hat{\gamma}_1$ and $\hat{\gamma}_2$ *be two fuzzy subsets of a complemented lattices* $\mathfrak{L}_1$ and $\mathfrak{L}_2$ *respectively. If* $\hat{\gamma}_1\times\hat{\gamma}_2$ *is an* $(\in,\in\vee q)$ *-fuzzy sublattice of* $\mathfrak{L}_1\times\mathfrak{L}_2$, then *either* $\hat{\gamma}_1$ *is an* $(\in,\in\vee q)$ *-fuzzy sublattice of* $\mathfrak{L}_1$ or $\hat{\gamma}_2$ *is an* $(\in,\in\vee q)$ *-fuzzy sublattice of* $\mathfrak{L}_2$.

CONCLUSION

The special properties of sublattice, ideals and cartesian product plays an important role in the study of the structure of an algebraic system. In this paper, we apply fuzzy set and concept of belongingness and quasi- coincidence to the fuzzy sublattice in a lattice. Although, some results in this paper have already constituted a platform for further discussion concerning other algebraic structures (such as *BCK-*

algebras, *BCI*-algebras, *MTL*-algebras etc.). For consider these results have some possible applications in computer sciences and information systems in the future. The purpose of this paper is to introduce the idea of $(\in,\in\vee q)$ -fuzzy sublattice, $(\in,\in\vee q)$ -fuzzy ideals, $(\in,\in\vee q)$ -fuzzy sublattice of complemented lattices and their homomorphism. Their Cartesian product of $(\in,\in\vee q)$ -fuzzy sublattice is provided and investigated some properties and relations.

REFERENCES

Ali, M. I., Feng, F., & Shabir, M. A. (2012). Note on $(\in,\in\vee \mathrm{q})$ -fuzzy equivalence relations and indistinguishability operators. *Hacettepe Journal of Mathematics and Statistics*, *40*, 383–400.

Ali, T., & Ray, A. K. (2007). On product of fuzzy sublattices. *Journal of Fuzzy Mathematics*, *15*, 375–381.

Bej, T., & Pal, M. (2015). Doubt Atanassovs intuitionistic fuzzy Sub-implicative ideals in *BCI*. *International Journal of Computational Intelligent Systems*, *8*(2), 240–249. doi:10.1080/18756891.2015.1001948

Bhakat, S. K., & Das, P. (1996a). -fuzzy subgroup. *Fuzzy Sets and Systems*, *80*(3), 359–368. $(\in,\in\vee \mathrm{q})$ doi:10.1016/0165-0114(95)00157-3

Bhakat, S. K., & Das, P. (1996b). Fuzzy subrings and ideals redefined. *Fuzzy Sets and Systems*, *81*(3), 383–393. doi:10.1016/0165-0114(95)00202-2

Birkhoff, G. (1984). *Lattice theory*. American Mathematical Society Providence.

Birkhoof, G. (2009). *Lattice Theory*. New York: American Mathematical Society.

Bo, Y., & Wangming, W. (1990). Fuzzy ideals on a distributive lattice. *Fuzzy Sets and Systems*, *35*(2), 231–240. doi:10.1016/0165-0114(90)90196-D

Davey, B. A., & Priestley, H. A. (1990). *Introduction to Lattices and Order*. Cambridge, UK: Cambridge University Press.

Feng, Y., Duan, H., & Zeng, Q. (2010). $\left(\lambda,\hat{\gamma}\right)$-fuzzy sublattices and $\left(\lambda,\hat{\gamma}\right)$-fuzzy subhyperlattices. *Fuzzy Information and Engineering, 78*, 17-26.

Feng, Y., Zeng, Q., Zou, L., & Zhang, B. (2010). On Cartesian product of(», $\hat{\gamma}$). *Fuzzy Information and Engineering*, *4*(4), 413–422. doi:10.100712543-010-0059-0

Feng, Y., & Zou, L. (2010). On product of fuzzy subhyperlattice and interval-valued fuzzy subhyperlattice. *Discrete Mathematics, Algorithms, and Applications*, *2*(02), 239–246. doi:10.1142/S1793830910000619

He, P., Xin, X., & Zhan, J. (2014). (Fuzzy) hyperlattices and fuzzy preordered lattices. *Journal of Intelligent & Fuzzy Systems*, *26*, 2369–2381.

He, P., Xin, X., & Zhan, J. (2016). On derivations and their fixed point sets in residuated lattices. *Fuzzy Sets and Systems*, *303*, 97–113. doi:10.1016/j.fss.2016.01.006

Jana, C., & Pal, M. (2016a). Applications of (α,β) *BCK/BCI. International Journal of Intelligent Systems Technologies and Applications*, *16*, 269–288. doi:10.1504/IJISTA.2017.085362 PMID:27160752

Jana, C., & Pal, M. (2017). On(∈α,∈α∨qβ) BCI. *Missouri Journal of Mathematical Sciences*, *29*(2), 197–215. doi:10.35834/mjms/1513306831

Jana, C., Pal, M., & Saied, A. B. (2017a). -bipolar fuzzy BCK/BCI-algebras. *Missouri Journal of Mathematical Sciences*, *29*(2), 139–160. (∈,∈∨q) doi:10.35834/mjms/1513306827

Jana, C., Pal, M., Senapati, T., & Bhowmik, M. (2015a). Atanassov's intutionistic 𝔏-fuzzy G-subalgebras of G-algebras. *Journal of Fuzzy Mathematics, 23,* 195–209.

Jana, C., Senapati, T., Bhowmik, M., & Pal, M. (2015). On intuitionistic fuzzyG. *Fuzzy Information and Engineering*, *7*(2), 195–209. doi:10.1016/j.fiae.2015.05.005

Jana, C., Senapati, T., & Pal, M. (2016). (∈,∈∨q)-intuitionistic fuzzy BCI. *Journal of Intelligent & Fuzzy Systems*, *31*(1), 613–621. doi:10.3233/IFS-162175

Jun, Y. B. (2005). On (α,β) *BCK/BCI. Bulletin of the Korean Mathematical Society*, *42*(4), 703–711. doi:10.4134/BKMS.2005.42.4.703

Jun, Y. B., Öztürk, M. A., & Muhiuddin, G. (2016). A generalization of(∈,∈∨q). *International Journal of Algebra and Statistics*, *5*(1), 7–18. doi:10.20454/ijas.2016.1041

Kazanci, O., & Davvaz, B. (2010). More on fuzzy lattice. *Computers & Mathematics with Applications (Oxford, England)*, *64*(9), 2917–2925. doi:10.1016/j.camwa.2012.05.006

Khan, M., Davvaz, B., Yaqoob, N., Gulistan, M., & Khalaf, M. M. (2015). On (∈,∈∨qk) -intuitionistic (fuzzy ideals, fuzzy soft ideals) of subtraction algebras. *Songklanakarin Journal of Science and Technology*, *37*, 465–475.

Luo, F., & Zhan, J. (2013). Fuzzy parameterized fuzzy soft lattice implication algebras. *Ars Combinatoria*, *112*, 353–360.

Ma, X., Zhan, & J., Jun, Y.B. (2009). Some types of (∈,∈∨q) -interval-valued fuzzy ideals of BCI algebras. *Iranian Journal of Fuzzy Systems*, *6*, 53–63.

Ma, X., Zhan, J., Davvaz, B., & Jun, Y. B. (2008). Some kinds of(∈,∈∨q) BCI. *Information Sciences*, *178*(19), 3738–3754. doi:10.1016/j.ins.2008.06.006

Ma, X., Zhan, J., & Jun, Y. B. (2011). Some kinds of(∈γ,∈γ∨qδ) BCI. C*omputers & Mathematics with Applications (Oxford, England), 61(*4), 1005–1015. doi:10.1016/j.camwa.2010.12.049 PMID:21542996

Meng, J., & Guo, X. (2005). On fuzzy ideals in *BCK/BCI. Fuzzy Sets and Systems*, *149*(3), 509–525. doi:10.1016/j.fss.2003.11.014

Mordeson, J. N., & Malik, D. S. (1998). Fuzzy commutative algebra: World Scientific.

Muhiuddin, G., & Al-roqi, A. (2015). Subalgebras of *BCK/BCI*-algebras based on (α,β) -type fuzzy sets. *Journal of Computational Analysis and Applications*, *18*, 1057–1064.

Pu, P., & Liu, Y. (1980). Fuzzy topology I. Neighbourhood structures of a fuzzy point and Moore-Smith convergence. *Journal of Mathematical Analysis and Applications*, *76*(2), 571–579. doi:10.1016/0022-247X(80)90048-7

Rosenfeld, A. (1971). Fuzzy groups. *Journal of Mathematical Analysis and Applications*, *35*(3), 512–517. doi:10.1016/0022-247X(71)90199-5

Seselja, B., Tepavcevic, A., & Vojvodic, G. (1993). 𝔏-fuzzy sets and codes. F*uzzy Sets and Systems, 53,* 217-222.

Seselja, B., & Tepavcevic, A. (1994). Representation of lattices by fuzzy sets. *Information Sciences*, *79*(3-4), 171–180. doi:10.1016/0020-0255(94)90117-1

Tepavcevic, A., & Trajkovski, G. (2001). -fuzzy lattices: An introduction. *Fuzzy Sets and Systems*, *123*(2), 209–216. L doi:10.1016/S0165-0114(00)00065-8

Zadeh, L. A. (1965). Fuzzy sets. *Information and Control*, *8*(3), 338–353. doi:10.1016/S0019-9958(65)90241-X

Zhang, H., & Zhan, J. (2017). Rough soft lattice implication algebras and corresponding decision making methods. *International Journal of Machine Learning and Cybernetics*, *8*(4), 1301–1308. doi:10.100713042-016-0502-6

Chapter 8
Uni-Hesitant Fuzzy Set Approach to the Ideal Theory of BCK-Algebras

G. Muhiuddin
https://orcid.org/0000-0002-5596-5841
University of Tabuk, Saudi Arabia

ABSTRACT

In this chapter, the author studies the uni-hesitant fuzzy set-theoretical approach to the ideals of BCK-algebras. For a hesitant fuzzy set H on S and a subset of [0,1], the set $L(H;ʎ):=\{x \in S \mid xH \subseteq ʎ\}$*, is called the uni-hesitant level set of H. Moreover, the author discusses the relations between uni-hesitant fuzzy commutative ideals and uni-hesitant fuzzy ideals. Further, he considered the characterizations of uni-hesitant fuzzy commutative ideals in BCK-algebras. Finally, he proved some conditions for a uni-hesitant fuzzy ideal to be a uni-hesitant fuzzy commutative ideal.*

1 INTRODUCTION

To deal with hesitant and incongruous problems, Torra and Narukawa (Torra & Narukawa, 2009) and Torra (Torra, 2010) proposed the concept of hesitant fuzzy sets, a new generalization of fuzzy sets, which allows the membership of an element of a set to be represented by several possible values. They also discussed relationships among hesitant fuzzy sets and other generalizations of fuzzy sets such as intuitionistic fuzzy sets, type-2 fuzzy sets, and fuzzy multisets. Some set theoretic operations such as union, intersection and complement on hesitant fuzzy sets have also been proposed by Torra (Torra, 2010). Hesitant fuzzy sets can be used as an efficient mathematical tool for modeling peopleâ€™s hesitancy in daily life than the other classical extensions of fuzzy sets. The motivation for introducing these sets is that it is sometimes difficult to determine the membership of an element into a set and in some circumstances this difficulty is caused by a doubt between a few different values. For example, two experts discuss the membership of x into A, and one wants to assign 0.3 and the other 0.4. So, the uncertainty on the possible values is somehow limited.

DOI: 10.4018/978-1-7998-0190-0.ch008

The hesitant fuzzy set provides a more accurate representation of peoples hesitancy in stating their preferences over objects than the fuzzy set or its classical extensions. Hesitant fuzzy set theory has been applied to several practical problems, primarily in the area of decision making (see (Liao & Xu, 2017; Rodriguez, Martinez & Herrera, 2012; Torra & Narukawa, 2009; Wei, 2012; Xia & Xu, 2011; Xia, Xu & Chen, 2013; Xu & Xia, 2011; Zhu, Xu & Xia, 2012)). Recently, Jun et al. applied the notion of hesitant fuzzy sets to the different fields (see for e.g., (Jun & Song, 2014; Jun, Song & Muhiuddin, 2016; Jun, Ahn, Muhiuddin, 2014)). Also, Muhiuddin et al. studied the notion of hesitant fuzzy sets on the various aspects (see for e.g., (. Muhiuddin & Aldhafeeri, 2018; Muhiuddin, 2016; Muhiuddin, Roh, Ahn & Jun, 2017; Muhiuddin, Kim, Song & Jun, 2017; Muhiuddin & Al-roqi, 2018)).

In this chapter, we study hesitant fuzzy commutative ideals in *BCK*-algebras, and investigate related properties. We discuss relations between uni-hesitant fuzzy commutative ideals and uni-hesitant fuzzy ideals. We consider characterizations of uni-hesitant fuzzy commutative ideals, and provide conditions for a uni-hesitant fuzzy ideal to be a uni-hesitant fuzzy commutative ideal.

2 PRELIMINARIES

A an extensively investigated by several researchers.

An algebra $(X;*,0)$ of type (2,0) is called a *BCI-algebra* if it satisfies the following conditions:

$(\forall x,y,z\in X)\ (((x*y)*(x*z))*(z*y)=0)$,
$(\forall x,y\in X)\ ((x*(x*y))*y=0)$,
$(\forall x\in X)\ (x*x=0)$,
$(\forall x,y\in X)\ (x*y=0,\ y*x=0 \Rightarrow x=y)$.

If a *BCI*-algebra X satisfies the following identity:

$(\forall x\in X)\ (0*x=0)$,

then X is called a *BCK-algebra*. Any *BCK*/*BCI*-algebra X satisfies the following axioms:

$(\forall x\in X)\ (x*0=x)$,
$(\forall x,y,z\in X)\ (x\leq y \Rightarrow x*z\leq y*z,\ z*y\leq z*x)$,
$(\forall x,y,z\in X)\ ((x*y)*z=(x*z)*y)$,
$(\forall x,y,z\in X)\ ((x*z)*(y*z)\leq x*y)$

where $x\leq y$ if and only if $x*y=0$ A *BCK*-algebra X is s said to be *commutative* if $x\wedge y=y\wedge x$ for all $x,y\in X$ where $x\wedge y=y*(y*x)$. A nonempty subset S of a *BCK*/*BCI*-algebra X is called a *subalgebra* of X if $x*y\in S$ for all $x,y\in S$. A subset A of a *BCK*/*BCI*-algebra X is called an *ideal* of X if it satisfies:

$0\in E, (\forall x\in X)\ (\forall y\in A)\ (x*y\in A \Rightarrow x\in A)$.

An ideal A of a *BCI*-algebra S is said to be *closed* if $0*x\in A$ for all $x\in A$. A subset A of a *BCK*-algebra X is called a *commutative ideal* if it satisfies and

$$(\forall x, y \in X)(\forall z \in A)\big((x*y)*z \in A \Rightarrow x*(y*(y*x)) \in A\big). \tag{1}$$

Observe that every commutative ideal is an ideal, but the converse is not true (Meng & Jun, 2014).

Theorem 2.1 An ideal A of a BCK-algebra X is commutative if and only if the following implication holds:

$$(\forall x, y \in X)\big(x*y \in A \Rightarrow x*(y*(y*x)) \in A\big). \tag{2}$$

We refer the reader to the books (Huang, 2006; Meng & Jun, 2014) for further information regarding *BCK*/*BCI*-algebras.

Torra (Torra, 2010) defined hesitant fuzzy sets in terms of a function that returns a set of membership values for each element in the domain. We display the basic notions on hesitant fuzzy sets. For more details, we refer to references (Liao & Xu, 2017; Torra, 2010; Xu, 2014).

Let *S* be a reference set. Then we define a hesitant fuzzy set on *S* in terms of a function $\mathcal{H}$ that when applied to *S* returns a subset of [0,1], and the image of $x \in S$ under $\mathcal{H}$ is denoted by $x\mathcal{H}$.

For a hesitant fuzzy set $\mathcal{H}$ on *S* and a subset of [0,1], the set

$$L(\mathcal{H};\lambda) := \{x \in S \mid x\mathcal{H} \subseteq \lambda\}, \tag{3}$$

is called the uni-hesitant level set of $\mathcal{H}$.

Let $\mathcal{H}$ and $\mathcal{K}$ be two hesitant fuzzy sets on *S*. We define

1. $\mathcal{H} \sqsubseteq \mathcal{K}$ if $\mathcal{H}_x \subseteq \mathcal{K}_x$ for all $x \in S$.

3 UNI-HESITANT FUZZY COMMUTATIVE IDEALS

In what follows, let *S* denote a *BCK*-algebra unless otherwise specified.

Definition 3.1 () A hesitant fuzzy set $\mathcal{H}$ on S is called a uni-hesitant fuzzy ideal on S if it satisfies $0\mathcal{H} \subseteq x\mathcal{H}$ for all x∈S and

$$(\forall x, y \in S)\big(x\mathcal{H} \subseteq (x*y)\mathcal{H} \cup y\mathcal{H}\big). \tag{4}$$

Definition 3.2 A hesitant fuzzy set $\mathcal{H}$ on S is called a uni-hesitant fuzzy commutative ideal on S if $0\mathcal{H} \subseteq x\mathcal{H}$ for all x∈S and

$$(\forall x, y, z \in S)\Big(\big(x*(y*(y*x))\big)\mathcal{H} \subseteq \big((x*y)*z\big)\mathcal{H} \cup z\mathcal{H}\Big). \tag{5}$$

*Table 1. Cayley table for the *-operation*

*	0	*a*	*b*	*c*
0	0	0	0	0
a	*a*	0	0	0
b	*b*	*c*	0	*b*
c	*c*	*a*	*c*	0

Example 3.3 *Let S={0,a,b,c} be a BCK-algebra with the Cayley table which is appeared in Table 1.*

Let $\mathcal{K}$ be a hesitant fuzzy set on *S* defined as follows:

$$\mathcal{K}: S \to \mathcal{P}([0,1]), x \mapsto \begin{cases} (0,3,0.4) \cup \{0.6\} & \text{if } x = 0, \\ [0.2,0.4] \cup [0.5,0.6] & \text{if } x \in \{a,b\}, \\ (0.2,0.4] \cup (0.5,0.6] & \text{if } x = c. \end{cases} \tag{6}$$

It is routine to verify that $\mathcal{K}$ is a uni-hesitant fuzzy commutative ideal on *S*.

Example 3.4 *Let S={0,a,b,c,d} be a BCK-algebra with the Cayley table which is appeared in Table 2.*

Let $\mathcal{H}$ be a hesitant fuzzy set on *S* defined as follows:

$$\mathcal{H}: S \to \mathcal{P}([0,1]), x \mapsto \begin{cases} (0,4,0.5) \cup \{0.8\} & \text{if } x = 0, \\ [0.3,0.5] \cup (0.7,0.8] & \text{if } x \in \{a,b\}, \\ [0.4,0.5] \cup \{0.7,0.8\} & \text{if } x = c, \\ [0.4,0.6) \cup \{0.7,0.8,0.9\} & \text{if } x = d. \end{cases} \tag{7}$$

Routine calculations show that $\mathcal{H}$ is a uni-hesitant fuzzy commutative ideal on *S*.

Theorem 3.5 Every uni-hesitant fuzzy commutative ideal on S is a uni-hesitant fuzzy ideal on S.

*Table 2. Cayley table for the *-operation*

*	0	*a*	*b*	*c*	*d*
0	0	0	0	0	0
a	*a*	0	0	*a*	*a*
b	*b*	*a*	0	*b*	*b*
c	*c*	*c*	*a*	0	*c*
d	*d*	*d*	*d*	*d*	0

Proof. Let $\mathcal{H}$ be a uni-hesitant fuzzy commutative ideal on S. If we take y=0 in ?? and use (a1) and (V), then

$$x\mathcal{H} = \left(x*\left(0*\left(0*x\right)\right)\right)\mathcal{H} \subseteq \left(\left(x*0\right)*z\right)\mathcal{H} \cup z\mathcal{H} = \left(x*z\right)\mathcal{H} \cup z\mathcal{H}$$

for all $x.z \in S$. Hence $\mathcal{H}$ is a uni-hesitant fuzzy ideal on *S*.

The following example shows that the converse of Theorem 3.5 is not true.

Example 3.6 *Let S={0,a,b,c,d} be a BCK-algebra with the Cayley table which is appeared in Table 3.*

Let $\mathcal{K}$ be a hesitant fuzzy set on *S* defined as follows:

$$\mathcal{K}: S \to \mathcal{P}\left(\left[0,1\right]\right), x \mapsto \begin{cases} \{0.1,0.3,0.5,0.7\} & \text{if } x = 0, \\ \{0.1,0.3,0.5,0.7,0.9\} & \text{if } x = a, \\ \{0.1,0.2,0.3,0.4,0.5,0.6,0.7,0.9\} & \text{if } x \in \{b,c,d\}. \end{cases} \tag{8}$$

Then $\mathcal{K}$ is a uni-hesitant fuzzy ideal on *S*. But it is not a uni-hesitant fuzzy commutative ideal on *S* since

$$\begin{aligned} \left(b*\left(c*\left(c*b\right)\right)\right)\mathcal{K} &= \{0.1,0.2,0.3,0.4,0.5,0.6,0.7,0.9\} \\ &\text{Ú}\{0.1,0.3,0.5,0.7\} \\ &= \left(\left(b*c\right)*0\right)\mathcal{K} \cup 0\mathcal{K}. \end{aligned} \tag{9}$$

We provide conditions for a uni-hesitant fuzzy ideal to be a uni-hesitant fuzzy commutative ideal.

Theorem 3.7 *For a hesitant fuzzy set* $\mathcal{H}$ *on S, the following are equivalent.*

1. $\mathcal{H}$ is a uni-hesitant fuzzy commutative ideal on *S*.
2. 2. $\mathcal{H}$ is a uni-hesitant fuzzy ideal on *S* such that

*Table 3. Cayley table for the *-operation*

*	0	a	b	c	d
0	0	0	0	0	0
a	a	0	c	0	0
b	b	b	0	0	0
c	c	c	a	0	0
d	d	d	d	a	0

$$(\forall x, y \in S)\left(\left(x*\left(y*\left(y*x\right)\right)\right)\mathcal{H} \subseteq \left(x*y\right)\mathcal{H}\right). \tag{10}$$

Proof. Assume that $\mathcal{H}$ is a uni-hesitant fuzzy commutative ideal on *S*. Then $\mathcal{H}$ is a uni-hesitant fuzzy ideal on *S* by Theorem 3.5. Taking z=0 in (3.2) and using (a1), we have $\left(x*\left(y*\left(y*x\right)\right)\right)\mathcal{H} \subseteq \left(x*y\right)\mathcal{H}$ for all $x,y \in S$.

Conversely, let $\mathcal{H}$ be a uni-hesitant fuzzy ideal on *S* satisfying the condition (3.3). Using (3.1), we have

$$(\forall x, y, z \in S)\left(\left(x*y\right)\mathcal{H} \subseteq \left(\left(x*y\right)*z\right)\mathcal{H} \cup z\mathcal{H}\right). \tag{11}$$

Combining (3.3) and (3.4) induces (3.2). Hence $\mathcal{H}$ is a uni-hesitant fuzzy commutative ideal on *S*.

Corollary 3.8 *Let $\mathcal{H}$ be a hesitant fuzzy set on S. Then $\mathcal{H}$ is a uni-hesitant fuzzy commutative ideal on S if and only if $\mathcal{H}$ is a uni-hesitant fuzzy ideal on S satisfying the following condition:*

$$(\forall x, y \in S)\left(\left(x*\left(y*\left(y*x\right)\right)\right)\mathcal{H} \subseteq \left(x*y\right)\mathcal{H}\right). \tag{12}$$

We consider conditions for a uni-hesitant fuzzy ideal to be a uni-hesitant fuzzy commutative ideal.

Lemma 3.9 (7) *Every uni-hesitant fuzzy ideal $\mathcal{H}$ on S satisfies the following condition:*

$$(\forall x, y, z \in S)\left(\left(x*y\right)*z = 0 \Rightarrow x\mathcal{H} \subseteq y\mathcal{H} \cup z\mathcal{H}\right). \tag{13}$$

Theorem 3.10 *In a commutative BCK-algebra S, every uni-hesitant fuzzy ideal on S is a uni-hesitant fuzzy commutative ideal on S.*

Proof. Let $\mathcal{H}$ be a uni-hesitant fuzzy ideal on *S*. Note that

$$\begin{aligned}\left(\left(x*\left(y*\left(y*x\right)\right)\right)*\left(\left(x*y\right)*z\right)\right)*z &= \left(\left(x*\left(y*\left(y*x\right)\right)\right)*z\right)*\left(\left(x*y\right)*z\right)\\ &\leq \left(x*\left(y*\left(y*x\right)\right)\right)*\left(x*y\right)\\ &= \left(x*\left(x*y\right)\right)*\left(y*\left(y*x\right)\right) = 0,\end{aligned} \tag{14}$$

that is, $\left(\left(x*\left(y*\left(y*x\right)\right)\right)*\left(\left(x*y\right)*z\right)\right)*z = 0$ for all $x,y,z \in S$. It follows from Lemma 3.9 that

$$\left(x*\left(y*\left(y*x\right)\right)\right)\mathcal{H} \subseteq \left(\left(\left(x*y\right)*z\right)\right)\mathcal{H} \cup z\mathcal{H}. \tag{15}$$

Therefore $\mathcal{H}$ is a uni-hesitant fuzzy commutative ideal on *S*.

Lemma 3.11 () *Every uni-hesitant fuzzy ideal* $\mathcal{H}$ *on S satisfies the following condition.*

$$(\forall x, y \in S)(x^*y = 0 \Rightarrow x\mathcal{H} \subseteq y\mathcal{H}). \tag{16}$$

Theorem 3.12 *If S satisfy the following condition:*

$$(\forall x, y \in S)\left(\left(x^*\left(x^*y\right)\right)^*\left(y^*\left(y^*x\right)\right) = 0\right), \tag{17}$$

then every uni-hesitant fuzzy ideal is a uni-hesitant fuzzy commutative ideal.

Proof. Let $\mathcal{H}$ be a uni-hesitant fuzzy ideal on *S*. For any *x,y*∈*S,* we have

$$\left(x^*\left(y^*\left(y^*x\right)\right)\right)^*\left(x^*y\right) = \left(x^*\left(x^*y\right)\right)^*\left(y^*\left(y^*x\right)\right) = 0 \tag{18}$$

by (a3) and (3.7), which implies from Lemma 3.11 that $\left(x^*\left(y^*\left(y^*x\right)\right)\right)\mathcal{H} \subseteq \left(x^*y\right)\mathcal{H}$. Using Theorem 3.7, we conclude that $\mathcal{H}$ is a uni-hesitant fuzzy commutative ideal on *S*.

Using the notion of uni-hesitant level sets, we consider a characterization of a uni-hesitant fuzzy commutative ideal.

Theorem 3.13 *For a hesitant fuzzy set* $\mathcal{H}$ *on S, the following are equivalent.*

1. $\mathcal{H}$ is a uni-hesitant fuzzy commutative ideal on *S*.
2. The nonempty uni-hesitant level set of $\mathcal{H}$ is a commutative ideal of *S* for any $\in \mathcal{P}([0,1])$.

Proof. Assume that $\mathcal{H}$ is a uni-hesitant fuzzy commutative ideal on *S*. Let $\in \mathcal{P}([0,1])$ be such that $L(\mathcal{H};) \neq \varnothing$. Then $x\mathcal{H} \subseteq$ for some *x*∈*S* which implies that $0\mathcal{H} \subseteq x\mathcal{H} \subseteq$. Hence $0 \in L(\mathcal{H};)$. Let *x,y,z*∈*S* be such that

$$\left(x^*y\right)^*z \in L(\mathcal{H};) \text{ and } z \in L(\mathcal{H};).$$

Then $\left(\left(x^*y\right)^*z\right)\mathcal{H} \subseteq$ and $z\mathcal{H} \subseteq$. It follows from (3.2) that

$$\left(x^*\left(y^*\left(y^*x\right)\right)\right)\mathcal{H} \subseteq \left(\left(x^*y\right)^*z\right)\mathcal{H} \cup z\mathcal{H} \subseteq \tag{19}$$

so that $x^*\left(y^*\left(y^*x\right)\right) \in L(\mathcal{H};)$. Therefore $L(\mathcal{H};)$ is a commutative ideal of *S* for any $\in \mathcal{P}([0,1])$ with $L(\mathcal{H};) \neq \varnothing$.

Conversely, suppose that the nonempty uni-hesitant level set of $\mathcal{H}$ is a commutative ideal of *S* for any $\in \mathcal{P}([0,1])$. Then $L(\mathcal{H};)$ is a subalgebra of *S*. Let *x,y*∈*S* be such that $x\mathcal{H} =_x$ and $y\mathcal{H} =_y$. Taking $=_x \cup_y$ implies that $x, y \in L(\mathcal{H};)$, and so $x*y \in L(\mathcal{H};)$. Thus

$$(x*y)\mathcal{H} \subseteq =_x \cup_y = x\mathcal{H} \cup y\mathcal{H}, \tag{20}$$

which implies that $0\mathcal{H} = (x*x)\mathcal{H} \subseteq x\mathcal{H}$ for all *x*∈*S*. Let *x,y,z*∈*S* be such that $((x*y)*z)\mathcal{H} =_1$ and $z\mathcal{H} =_2$. Let us take $=_1 \cup_2$. Then $(x*y)*z \in L(\mathcal{H};)$ and $z \in L(\mathcal{H};)$. Since $L(\mathcal{H};)$ is a commutative ideal, it follows from ?? that $x*(y*(y*x)) \in L(\mathcal{H};)$. Hence

$$(x*(y*(y*x)))\mathcal{H} \subseteq =_1 \cup_2 = ((x*y)*z)\mathcal{H} \cup z\mathcal{H}. \tag{21}$$

Therefore $\mathcal{H}$ is a uni-hesitant fuzzy commutative ideal on *S*.

The commutative ideals $L(\mathcal{H};)$ in Theorem 3.13 are called *uni-hesitant level commutative ideals* of $\mathcal{H}$.

Theorem 3.14 *Let $\mathcal{H}$ be a hesitant fuzzy set on S. For any $\in \mathcal{P}([0,1]$, define a hesitant fuzzy set $\mathcal{H}^*$ on S by*

$$\mathcal{H}^* : S \to \mathcal{P}([0,1]), x \mapsto \begin{cases} x\mathcal{H} & \text{if } x \in L(H;), \\ [0,1] & \text{otherwise.} \end{cases} \tag{22}$$

If $\mathcal{H}$ is a uni-hesitant fuzzy commutative ideal on *S*, then so is $\mathcal{H}^*$.

Proof. If $\mathcal{H}$ is a uni-hesitant fuzzy commutative ideal on *S*, then $L(\mathcal{H};)$ is a commutative ideal of *S* for any $\in \mathcal{P}([0,1])$ with $L(\mathcal{H};) \neq \varnothing$. Hence $0 \in L(\mathcal{H};)$, and so

$$0\mathcal{H}^* = 0\mathcal{H} \subseteq x\mathcal{H} \subseteq x\mathcal{H}^*$$

for all *x*∈*S*. Let *x,y,z*∈*S*. If

$$(x*y)*z \in L(\mathcal{H};) \text{ and } z \in L(\mathcal{H};),$$

then $x*(y*(y*x)) \in L(\mathcal{H};)$ and so

$$\begin{aligned}(x*(y*(y*x)))\mathcal{H}^* &= (x*(y*(y*x)))\mathcal{H} \\ &\subseteq ((x*y)*z)\mathcal{H} \cup z\mathcal{H} = ((x*y)*z)\mathcal{H}^* \cup z\mathcal{H}^*.\end{aligned} \tag{23}$$

If

$$(x*y)*z \notin L(\mathcal{H};) \text{ or } z \notin L(\mathcal{H};),$$

then

$$((x*y)*z)\mathcal{H}^* = [0,1] \text{ or } z\mathcal{H}^* = (0,1).$$

Hence

$$(x*(y*(y*x)))\mathcal{H}^* \subseteq [0,1] = ((x*y)*z)\mathcal{H}^* \cup z\mathcal{H}^*. \tag{24}$$

This shows that $\mathcal{H}^*$ is a uni-hesitant fuzzy commutative ideal on *S*.

Theorem 3.15 *Every commutative ideal of S can be realized as a uni-hesitant level commutative ideal of some uni-hesitant fuzzy commutative ideal on S.*

Proof. Let *A* be a commutative ideal of *S*. For any $\in \mathcal{P}([0,1])$ with ≠[0,1], let $\mathcal{H}$ be a hesitant fuzzy set on *S* defined by

$$\mathcal{H}: S \to \mathcal{P}([0,1]), x \mapsto \begin{cases} & \text{if } x \in A, \\ [0,1] & \text{if } x \notin A. \end{cases} \tag{25}$$

Obviously, $0\mathcal{H} \subseteq x\mathcal{H}$ for all *x*∈*S* For any *x*,*y*,*z*∈*S* if (*x***y*)**z*∈*A* and *z*∈*A* then *x**(*y**(*y***x*))∈*A*. Hence

$$((x*y)*z)\mathcal{H} \cup z\mathcal{H} == (x*(y*(y*x)))\mathcal{H}. \tag{26}$$

If (*x***y*)**z*∉*A* or *z*∉*A* then $((x*y)*z)\mathcal{H} = [0,1]$ or $z\mathcal{H} = [0,1]$. It follows that

$$(x*(y*(y*x)))\mathcal{H} \subseteq [0,1] = ((x*y)*z)\mathcal{H} \cup z\mathcal{H}. \tag{27}$$

Therefore $\mathcal{H}$ is a uni-hesitant fuzzy commutative ideal on *S*, and clearly $L(\mathcal{H};) = A$. This completes the proof.

Lemma 3.16 () *For a hesitant fuzzy set $\mathcal{H}$ on S, the following are equivalent.*

1. $\mathcal{H}$ is a uni-hesitant fuzzy ideal on *S*.
2. The nonempty hesitant level set $L(\mathcal{H};)$ of $\mathcal{H}$ is an ideal of *S* for all $\in \mathcal{P}([0,1])$.

Theorem 3.17 *Let* $\mathcal{H}$ and $\mathcal{K}$ *be hesitant fuzzy sets on S such that*

1. $\mathcal{H} \sqsubseteq \mathcal{K}$,
2. $\mathcal{K}$ is a uni-hesitant fuzzy ideal on *S*.

If $\mathcal{H}$ is a uni-hesitant fuzzy commutative ideal on *S*, then so is $\mathcal{K}$.

Proof. Let $\in \mathcal{P}([0,1])$ be such that $L(\mathcal{K};) \neq \varnothing$. From the condition (ii) and Lemma 3.16, we know that $L(\mathcal{K};)$ is an ideal of *S*. Assume that $\mathcal{H}$ is a uni-hesitant fuzzy commutative ideal on *S*. Then $L(\mathcal{H};)$ is a commutative ideal of *S* for every $\in \mathcal{P}([0,1])$ with $L(\mathcal{H};) \neq \varnothing$. Let $x,y \in S$ and $\in \mathcal{P}([0,1])$ be such that $x*y \in L(\mathcal{K};)$. Since

$$(x*(x*y))*y = (x*y)*(x*y) = 0 \in L(\mathcal{H};), \tag{28}$$

it follows from (a3), Theorem 2.1 and (i) that

$$(x*(y*(y*(x*(x*y)))))*(x*y) = (x*(x*y))*(y*(y*(x*(x*y)))) \in L(\mathcal{H};) \subseteq L(\mathcal{K};) \tag{29}$$

so that

$$x*(y*(y*(x*(x*y)))) \in L(\mathcal{K};) \tag{30}$$

since $L(\mathcal{K};)$ is an ideal and $x*y \in L(\mathcal{K};)$. Note that $x*(x*y) \leq x$, and so

$$y*(y*(x*(x*y))) \leq y*(y*x) \tag{31}$$

by (a2). Hence

$$x*(y*(y*x)) \leq x*(y*(y*(x*(x*y)))). \tag{32}$$

Using (3.8) and (3.9), we get $x*(y*(y*x)) \in L(\mathcal{K};)$. Thus $L(\mathcal{K};)$ is a commutative ideal by Theorem 2.1, and so $\mathcal{K}$ is a uni-hesitant fuzzy commutative ideal on *S* by Theorem 3.13.

REFERENCES

Huang, Y. (2006). BCI-algebra. Science Press.

Jun, Y. B., Ahn, S. S., & Muhiuddin, G. (2014). Hesitant fuzzy soft subalgebras and ideals in BCK/BCI-algebras. The Scientific World Journal.

Jun, Y. B., & Song, S. Z. (2014). Hesitant fuzzy set theory applied to filters in MTL-algebras. *Honam Math. J.*, *36*(4), 813–830. doi:10.5831/HMJ.2014.36.4.813

Jun, Y. B., & Song, S. Z. (2015). Hesitant fuzzy prefilters and filters of EQ-algebras. *Applied Mathematical Sciences*, *9*, 515–532. doi:10.12988/ams.2015.411962

Jun, Y. B., Song, S. Z., & Muhiuddin, G. (2016). Hesitant fuzzy semigroups with a frontier. *Journal of Intelligent & Fuzzy Systems*, *30*(3), 1613–1618. doi:10.3233/IFS-151869

Liao, H., & Xu, Z. (2017). *Hesitant Fuzzy Decision Making Methodologies and Applications*. Springer. doi:10.1007/978-981-10-3265-3

Meng, J., & Jun, Y. B. (1994). BCK-algebras. Kyungmoon Sa Co.

Muhiuddin, G. (2016). Hesitant fuzzy filters and hesitant fuzzy G-filters in residuated lattices. *Journal of Computational Analysis and Applications*, *21*(2), 394–404.

Muhiuddin, G., & Al-roqi, M., A. (2018). Regular hesitant fuzzy filters and MV -hesitant fuzzy filters of residuated lattices. *Journal of Computational Analysis and Applications*, *24*(6), 1133–1144.

Muhiuddin, G., & Aldhafeeri, S. (2018). Subalgebras and ideals in *BCK/BCI. Eur. J. Pure Appl. Math*, *11*(2), 417–430. doi:10.29020/nybg.ejpam.v11i2.3246

Muhiuddin, G., Kim, H. S., Song, S. Z., & Jun, Y. B. (2017). Hesitant fuzzy translations and extensions of subalgebras and ideals in BCK/BCI-algebras. *Journal of Intelligent & Fuzzy Systems*, *32*(1), 43–48. doi:10.3233/JIFS-151031

Muhiuddin, G., Roh, E. H., Ahn, S. S., & Jun, Y. B. (2017). Hesitant fuzzy filters in lattice implication algebras. *Journal of Computational Analysis and Applications*, *22*(6), 1105–1113.

Rodriguez, M., Martinez, L., & Herrera, F. (2012). Hesitant fuzzy linguistic term sets for decision making. *IEEE Transactions on Fuzzy Systems*, *20*(1), 109–119. doi:10.1109/TFUZZ.2011.2170076

Torra, V. (2010). Hesitant fuzzy sets. *International Journal of Intelligent Systems*, *25*, 529–539.

Torra, V., & Narukawa, Y. (2009). On hesitant fuzzy sets and decision. *The 18th IEEE International Conference on Fuzzy Systems*, 1378–1382.

Wei, G. (2012). Hesitant fuzzy prioritized operators and their application to multiple attribute decision making. *Knowledge-Based Systems*, *31*, 176–182. doi:10.1016/j.knosys.2012.03.011

Xia, X., & Xu, Z. S. (2011). Hesitant fuzzy information aggregation in decision making. *International Journal of Approximate Reasoning*, *52*(3), 395–407. doi:10.1016/j.ijar.2010.09.002

Xia, X., Xu, Z. S., & Chen, N. (2013). Some hesitant fuzzy aggregation operators with their application in group decision making. *Group Decision and Negotiation*, *22*(2), 259–279. doi:10.100710726-011-9261-7

Xu, Z. S. (2014). *Hesitant Fuzzy Sets Theory*. Springer. doi:10.1007/978-3-319-04711-9

Xu, Z. S., & Xia, X. (2011). Distance and similarity measures for hesitant fuzzy sets. *Inform. Sci.*, *181*(11), 2128–2138. doi:10.1016/j.ins.2011.01.028

Zhu, B., Xu, Z. S., & Xia, X. (2012). Hesitant fuzzy geometric Bonferroni means. *Inform. Sci.*, *205*, 72–85. doi:10.1016/j.ins.2012.01.048

Chapter 9
Fuzzy Bi-Interior Ideals of Semirings

Venkateswarlu Bollineni
GITAM University, Bengaluru Rural, India

Sri Lakshmi T.
GITAM University, Visakhapatnam, India

Adi Narayana Y.
NRI IT, Vijayawada, India

ABSTRACT

In this chapter, the authors introduce the notion of a fuzzy bi-interior ideal of semiring and they characterize the regular semiring in terms of fuzzy bi-interior ideal of semiring. And also they proved Let I be a non-empty subset of a semiring M and χ_I be the characteristic function of I. Then I is a bi interior ideal of semiring if and only if χ_I is a fuzzy bi-interior ideal of semiring M.

1 INTRODUCTION

Semiring is an algebraic structure which is a common generalization of rings and distributive lattices, was first introduced by Vandiver (1934) but non-trivial examples of semirings had appeared in the studies on the theory of commutative ideals of rings by Dedekind in 19th century. Semiring is a universal algebra with two binary operations called addition and multiplication, where one of them distributive over the other. Bounded distributive lattices are commutative semirings which are both additively idempotent and multiplicatively idempotent. A natural example of semiring is the set of all natural numbers under usual addition and multiplication of numbers. In particular, if I is the unit interval on the real line, then (I, max, min) is a semiring in which 0 is the additive identity and 1 is the multiplicative identity. The theory of rings and the theory of semigroups have considerable impact on the development of the theory of semirings. In structure, semirings lie between semigroups and rings. Additive and multiplicative structures of a semiring play an important role in determining the structure of a semiring. Semiring as

DOI: 10.4018/978-1-7998-0190-0.ch009

the basic algebraic structure was used in the areas of theoretical computer science as well as in the solutions of graph theory, optimization theory and in particular for studying automata, coding theory and formal languages. Semiring theory has many applications in other branches. The notion of ideals was introduced by Dedekind for the theory of algebraic numbers, was generalized by Noether for associative rings. The one and two sided ideals introduced by her, are still central concepts in ring theory. We know that the notion of an one sided ideal of any algebraic structure is a generalization of notion of an ideal. The quasi ideals are generalization of left and right ideals where as the bi-ideals are a generalization of quasi ideals. The notion of bi-ideals in semigroups were introduced by Lajos. Iseki(1958, 1956, 1958) introduced the concept of quasi ideal for a semiring. Quasi ideals in Γ-semirings studied by Jagtap & Pawar (2009). Henriksen(1958) studied ideals in semirings. As a further generalization of ideals, Steinfeld (1956) first introduced the notion of quasi ideals for semigroups and then for rings. We know that the notion of the bi-ideal in semirings is a special case of (m,n)- ideal introduced by Lajos (1969). The concept of bi-ideals was first introduced by Good and Hughes (1952) for a semigroup. Lajos and Szasz (1970) introduced the concept of bi-ideals for rings.

The fuzzy set theory was developed by Zadeh (1965). Many papers on fuzzy sets appeared showing the importance of the concept and its applications to logic, set theory, group theory, ring theory, real analysis, topology, measure theory, etc. The falsification of algebraic structure was introduced by Rosenfeld (1971) and he introduced the notion of fuzzy subgroups. Swamy and Swamy (1988) studied fuzzy prime ideals in rings. Liu(1982) defined and studied fuzzy subrings as well as fuzzy ideals in rings. Mandal (2014) studied fuzzy ideals and fuzzy interior ideals in an ordered semiring. Kuroki (1981) studied fuzzy interior ideals in semigroups. Murali Krishna Rao(2018, 2017, 2018, 2017, 2018, 2018) introduced bi-quasi-ideals, bi-interior ideals in semirings, bi-quasi-ideals and fuzzy bi-quasi - ideals in Γ-semigroups and T-fuzzy ideals in ordered Γ-semirings. In this paper, we introduce the notion of a fuzzy bi-interior ideal of semiring and we characterize the regular semiring in terms of fuzzy bi-interior ideal of semiring.

2 PRELIMINARIES

In this section we will recall some of the fundamental concepts and definitions, which are necessary for this paper.

Definition 2.1 *A set S together with two associative binary operations called addition and multiplication (denoted by + and • respectively) will be called a semiring provided*

- addition is a commutative operation.
- multiplication distributes over addition both from the left and from the right.
- there exists $0 \in S$ such that $x+0=x$ and $x \cdot 0 = 0 \cdot x = 0$, for each $x \in S$.

A semiring M is said to be commutative semiring if $xy=yx$, for all $x,y \in M$. An element $a \in M$ is said to be regular element of M if there exist $x \in M$ such that $a=axa$. If every element of semiring M is a regular, then M is said to be regular semiring.

Definition 2.2 *A non-empty subset A of semiring M is called*

- a subsemiring of M if $(A,+)$ is a subsemigroup of $(M,+)$ and $AA\subseteq A$.
- a quasi ideal of M if A is a subsemiring of M and $AM \cap M \subseteq A$.
- a bi-ideal of M if A is a subsemiring of M and $AMA\subseteq A$.
- an interior ideal of M if A is a subsemiring of M and $MAM\subseteq A$.
- a left (right) ideal of M if A is a subsemiring of M and $MA\subseteq A(AM\subseteq A)$.
- an ideal if A is a subsemiring of $M, A\subseteq A$ and $MA\subseteq A$.
- a k-ideal if A is a subsemiring of $M, A\subseteq A$, $MA\subseteq A$ and $x\in M$, $x+y\in A$, $y\in A$ then $x\in A$.
- a left bi-quasi ideal (right bi-quasi ideal) of M if A is a subsemigroup of $(M,+)$ and $MA \cap LMA \subseteq A\left(AM \cap AMA \subseteq A\right)$.
- a bi-quasi ideal of M if it is both a left bi-quasi and a right bi-quasi ideal of M.
- a bi-interior ideal of M if A is a subsemiring of M and $MAM \cap AMA \subseteq A$.

Definition 2.3 *A semiring M is called*

- a right bi-quasi simple semiring if M has no right bi-quasi ideal other than M itself.
- a left (right) simple semiring if M has no proper left (right) ideal of M.
- a bi-quasi simple semiring if M has no proper bi-quasi ideal of M.
- simple semring if M has no proper ideals .

Theorem 2.4 *Let M be a semiring. Then the following are hold*

- Every left ideal is a bi-interior ideal of M.
- Every right ideal is a bi-interior ideal of M.
- Every quasi ideal is bi-interior ideal of M.
- Arbitrary intersection of bi-interior of M is also bi-interior ideal of M.
- Every ideal is a bi-interior ideal of M.
- If B is a bi-interior ideal of M then BM and MB are bi-interior ideals of M.

Definition 2.5 *Let M be a non-empty set. A mapping* $\mu: M\rightarrow[0,1]$ *is called a fuzzy subset of M.*

Definition 2.6 *If* μ *is a fuzzy subset of M, for* $t\in[0,1]$ *then the set* $\mu_t= \{x\in M \mid \mu(x)\geq t\}$ *is called a level subset of M with respect to a fuzzy subset* μ.

Definition 2.7 *A fuzzy subset* $\mu: M\rightarrow[0,1]$ *is a non-empty fuzzy subset if* μ *is not a constant function.*

Definition 2.8 *For any two fuzzy subsets* λ *and* μ *of M,* $\lambda\subseteq\mu$ *means* $\lambda(x)\leq\mu(x)$, *for all* $x\in M$.

Definition 2.9 *Let A be a non-empty subset of M. The characteristic function of A is a fuzzy subset of M is defined by*

$$\chi_A\left(x\right)=\begin{cases}1, & \text{if } x\in A;\\ 0, & \text{if } x\notin A.\end{cases}$$

Definition 2.10 *Let f and g be fuzzy subsets of S. Then* $f \cup g, f \cap g$ *are fuzzy subsets of S defined by*

$$f \cup g(x) = max\{f(x), g(x)\}, f \cap g(x) = min\{f(x), g(x)\},$$

for all x∈S. And f∘g is defined by

$$f \circ g(z) = \begin{cases} \sup\limits_{z=xy,x,y\in S} \{\min\{f(x), g(y)\}\} \\ 0 \quad \text{otherwise} \end{cases}, \quad \text{for all} \quad z \in S.$$

Definition 2.11 *A fuzzy subset* μ *of semiring M is called*

- a fuzzy subsemiring of M if it satisfies the following conditions
  $\mu(x+y)\geq\min\{\mu(x),\mu(y)\}$
  $\mu(xy)\geq\min\{\mu(x),\mu(y)\}$ for all $x,y\in M$.
- a fuzzy left (right) ideal of M if it satisfies the following conditions
  $\mu(x+y)\geq\min\{\mu(x),\mu(y)\}$
  $\mu(xy)\geq\mu(y),(\mu(y))$ for all $x,y\in M$.
- a fuzzy ideal of M if it satisfies the following conditions
  $\mu(x+y)\geq\min\{\mu(x),\mu(y)\}$
  $\mu(xy)\geq\max\{\mu(x),\mu(y)\}$ for all $x,y\in M$.
- a fuzzy bi-ideal of M if it satisfies the following conditions
  (a) $\mu(x+y)\geq\min\{\mu(x),\mu(y)\}$ for all $x,y\in M$.
  (b) $\mu\circ\chi_M\circ\mu\subseteq\mu$.
- a fuzzy quasi ideal of M if it satisfies the following conditions
  (a) $\mu(x+y)\geq\min\{\mu(x),\mu(y)\}$ for all $x,y\in M$.
  (b) $\mu \circ \chi_M \cap \chi_M \circ \mu \subseteq \mu$.
- a fuzzy left (right) bi-quasi ideal if it satisfies the following conditions
  (a) $\mu(x+y)\geq\min\{\mu(x),\mu(y)\}$ for all $x,y\in M$.
  (b) $\chi_M \circ \mu \cap \mu \circ \chi_M \circ \mu \subseteq \mu(\mu \circ \chi_M \cap \mu \circ \chi_M \circ \mu \subseteq \mu)$.

A fuzzy subset μ of semiring M is called a fuzzy bi-quasi ideal if it is both a fuzzy left and a fuzzy right bi-quasi ideal of M.

Theorem 2.12 *If* μ *is a fuzzy quasi-ideal of a regular semiring M then* μ *is a fuzzy ideal of M.*

Theorem 2.13 *M is a regular semiring if and only* $AB = A \cap B$, *for any right ideal A and left ideal B of semiring M.*

Theorem 2.14 *A semiring M is a regular if and only if* $\lambda \circ \mu = \lambda \cap \mu$, *for any fuzzy right ideal* λ *and fuzzy left ideal* μ *of M.*

Theorem 2.15 *Let M be a regular semiring. Then μ is a fuzzy left (right) bi-quasi ideal of M if and only if μ is a fuzzy quasi ideal of M.*

Theorem 2.16 *Let M be a semiring. Then M is a regular semiring if and only if* $B = MBM \cap BMB$ *for every bi-interior ideal of M.*

3 FUZZY BI-INTERIOR IDEALS OF SEMIRING

In this section we introduce the notion of a fuzzy bi-interior ideal of semiring and we characterize the regular semiring in terms of fuzzy bi-interior ideal of semiring.

Definition 3.1 *A fuzzy subset μ of a semiring M is called a fuzzy bi-interior ideal if*

- $\mu(x+y) \geq \min\{\mu(x), \mu(y)\}$ for all $x, y \in M$.
- $\chi_M \circ \mu \circ \chi_M \cap \mu \circ \chi_M \circ \mu \subseteq \mu$

Example 3.2 *Let Q be the set of all rational numbers,*

$$M = \left\{ \begin{pmatrix} a & b \\ 0 & c \end{pmatrix} \mid a, b, c \in Q \right\} \text{ and } A = \left\{ \begin{pmatrix} a & 0 \\ 0 & b \end{pmatrix} \mid a, 0 \neq b \in Q \right\}.$$

Then M is a semiring with respect to usual addition of matrices and ternary operation is defined as the usual matrix multiplication and A is a bi-interior ideal but not a bi-ideal of semiring M.

Define μ: $M \rightarrow [0,1]$ such that $\mu(x) = \begin{cases} 1 & \text{if } x \in A, \\ 0, & \text{otherwise.} \end{cases}$

Then μ is a fuzzy bi-interior ideal of M.

Theorem 3.3 *Every fuzzy left ideal of semiring M is a fuzzy bi-interior ideal of M.*

Proof. Let μ be a fuzzy left ideal of semiring M and $x \in M$. Then we have $\mu(xy) \geq \mu(y)$. It is enough to prove that $\chi_M \circ \mu \circ \chi_M \cap \mu \circ \chi_M \circ \mu \subseteq \mu$. Now

$$\begin{aligned} \mu \circ \chi_M(x) &= \sup_{x=ab} \min\{\chi_M(a), \mu(b)\} \\ &= \sup_{x=ab} \min\{1, \mu(b)\} \\ &\leq \sup_{x=ab} \mu(ab) = \mu(x). \end{aligned}$$

Therefore,

$\mu \circ \chi_M(x) \leq \mu(x)$

Now

$$\mu \circ \chi_M \circ \mu(x) = \sup_{x=uvs} \min\{\mu(u), \mu \circ \chi_M(vs)\}$$
$$\leq \sup_{x=uvs} \min\{\mu(u), \mu(vs)\} = \mu(x).$$

Now

$$\chi_M \circ \mu \circ \chi_M \cap \mu \circ \chi_M \circ \mu(x) = \min\{\chi_M \circ \mu \circ \chi_M(x), \mu \circ \chi_M \circ \mu(x)\}$$
$$\leq \min\{\chi_M \circ \mu \circ \chi_M(x), \mu(x)\} \leq \mu(x).$$

Hence μ is a fuzzy bi-interior ideal of semiring *M.*

Corollary 3.4 *Every fuzzy right ideal of semiring M is a fuzzy bi-interior ideal of M.*

Corollary 3.5 *Every fuzzy ideal of semiring M is a fuzzy bi-interior ideal of M.*

Theorem 3.6 *Let M be a semiring and* μ *be a non-empty fuzzy subset of M. A fuzzy subset* μ *is a fuzzy bi-interior ideal of semiring if and only if the level subset* μ_t *of* μ *is a left bi-interior ideal of semiring M for every* $t \in [0,1]$, *where* $\mu_t \neq \emptyset$.

Proof. Let *M* be a semiring and μ be a non-empty fuzzy subset of *M*. Suppose μ is a fuzzy bi-interior ideal of semiring, $\mu_t \neq \emptyset$, $t \in [0,1]$ and $a, b \in \mu_t$. Then

$$\mu(a) \geq t, \mu(b) \geq t$$
$$\Rightarrow \mu(a+b) \geq \min\{\mu(a), \mu(b)\} \geq t$$
$$\Rightarrow a+b \in \mu_t.$$

Let $x \in M\mu_t M \cap \mu_t M \mu_t$. Then $x=bau=cde$, where $b,d,u \in M$, $a,c,e \in \mu_t$. Then

$$\chi_M \circ \mu \circ \chi_M(x) \geq t \text{and} \mu \circ \chi_M \circ \mu(x) \geq t$$
$$\Rightarrow \mu(x) \geq (\chi_M \circ \mu \circ \chi_M \cap \mu \circ \chi_M \circ \mu)(x) \geq t$$

Therefore, $x \in \mu_t$.
Hence μ_t is a bi-interior ideal of *M*.
Conversely suppose that μ_t is a bi-interior ideal of semiring *M*, for all $t \in Im(\mu)$.
Let $x,y \in M$, $\mu(x)=t_1$, $\mu(y)=t_2$ and $t_1 \geq t_2$. Then $x, y \in \mu_{t_2}$.

$$\Rightarrow x+y \in \mu_{t_2} \text{ and } xy \in \mu_{t_2}$$
$$\Rightarrow \mu(x+y) \geq t_2 = \min\{t_1,t_2\} = \min\{\mu(x),\mu(y)\}.$$

Therefore $\mu(x+y)\geq\min\{\mu(x),\mu(y)\}$.
We have $M\mu_l M \cap \mu_l M \mu_l \subseteq \mu_l$, for all $l \in Im(\mu)$.
Suppose $t=\min\{Im(\mu)\}$. Then $M\mu_t M \cap \mu_t M \mu_t \subseteq \mu_t$.
Therefore $\chi_M \circ \mu \circ \chi_M \cap \mu \circ \chi_M \circ \mu \subseteq \mu$.
Hence μ is a fuzzy bi-interior ideal of semiring M.

Theorem 3.7 *Let I be a non-empty subset of a semiring M and χ_I be the characteristic function of I. Then I is a bi-interior ideal of semiring M if and only if χ_I is a fuzzy bi-interior ideal of semiring M.*

Proof. Let I be a non-empty subset of a semiring M and χ_I be the characteristic function of I. Suppose I is a bi-interior ideal of semiring M. Obviously χ_I is a fuzzy subsemiring of M. We have $MIM \cap IMI \subseteq I$. Then

$$\chi_M \circ \chi_I \circ \chi_M \cap \chi_I \circ \chi_M \circ \chi_I = \chi_{MIM} \cap \chi_{IMI}$$
$$= \chi_{MIM \cap IMI} \subseteq \chi_I.$$

Therefore χ_I is a fuzzy bi-interior ideal of semiring M.
Conversely suppose that χ_I is a fuzzy bi-interior ideal of semiring M.
Then I is a subsemiring of M. We have

$$\chi_M \circ \chi_I \circ \chi_M \cap \chi_I \circ \chi_M \circ \chi_I \subseteq \chi_I$$
$$\Rightarrow \chi_{MIM} \cap \chi_{IMI} \subseteq \chi_I$$
$$\Rightarrow \chi_{MIM \cap IMI} \subseteq \chi_I.$$

Therefore $MIM \cap IMI \subseteq I$.
Hence I is a bi-interior ideal of semiring M.

Theorem 3.8 *If μ and λ are fuzzy bi-interior ideals of semiring M then $\mu \cap \lambda$ is a fuzzy bi-interior ideal of semiring M.*

Proof. Let μ and λ be fuzzy bi-interior ideals of semiring M. Then

$$\begin{aligned}\mu\cap\lambda(x+y) &= \min\{\mu(x+y),\lambda(x+y)\}\\ &\geq \min\{\min\{\mu(x),\mu(y)\},\min\{\lambda(x),\lambda(y)\}\}\\ &= \min\{\min\{\mu(x),\lambda(x)\},\min\{\mu(y),\lambda(y)\}\}\\ &= \min\{\mu\cap\lambda(x),\mu\cap\lambda(y)\}\end{aligned}$$

$$\begin{aligned}\chi_M \circ \mu \cap \lambda(x) &= \sup_{x=ab} \min\{\chi_M(a), \mu \cap \lambda(b)\} \\ &= \sup_{x=ab} \min\{\chi_M(a), \min\{\mu(b), \lambda(b)\}\} \\ &= \sup_{x=ab} \min\{\min\{\chi_M(a), \mu(b)\}, \min\{\chi_M(a), \lambda(b)\}\} \\ &= \min\left\{\sup_{x=ab} \min\{\chi_M(a), \mu(b)\}, \sup_{x=ab} \min\{\chi_M(a), \lambda(b)\}\right\} \\ &= \min\{\chi_M \circ \mu(x). \chi_M \circ \lambda(x)\} \\ &= \chi_M \circ \mu \cap \chi_M \circ \lambda(x).\end{aligned}$$

Therefore $\chi_M \circ \mu \cap \lambda = \chi_M \circ \mu \cap \chi_M \circ \lambda$.

$$\begin{aligned}\mu \cap \lambda \circ \chi_M \circ \mu \cap \lambda(x) &= \sup_{x=abc} \min\{\mu \cap \lambda(a), \chi_M \circ \mu \cap \lambda(bc)\} \\ &= \sup_{x=abc} \min\{\mu \cap \lambda(a), \chi_M \circ \mu \cap \chi_M \circ \lambda(bc)\} \\ &= \sup_{x=abc} \min\{\min\{\mu(a), \lambda(a)\}, \min\{\chi_M \circ \mu(bc), \chi_M \circ \lambda(bc)\}\} \\ &= \sup_{x=abc} \min\{\min\{\mu(a), \chi_M \circ \mu(bc)\}, \min\{\lambda(a), \chi_M \circ \lambda(bc)\}\} \\ &= \min\left\{\sup_{x=abc} \min\{\mu(a), \chi_M \circ \mu(bc)\}, \sup_{x=abc} \min\{\lambda(a), \chi_M \circ \lambda(bc)\}\right\} \\ &= \min\{\mu \circ \chi_M \circ \mu(x), \lambda \circ \chi_M \circ \lambda(x)\} \\ &= \mu \circ \chi_M \circ \mu \cap \lambda \circ \chi_M \circ \lambda(x).\end{aligned}$$

Therefore $\mu \cap \lambda \circ \chi_M \circ \mu \cap \lambda = \mu \circ \chi_M \circ \mu \cap \lambda \circ \chi_M \circ \lambda$.
Similarly $\chi_M \circ \mu \cap \lambda \circ \chi_M = \chi_M \circ \mu \circ \chi_M \cap \chi_M \circ \lambda \circ \chi_M$. Hence

$$\begin{aligned}&\chi_M \circ \mu \cap \lambda \circ \chi_M \cap \mu \cap \lambda \circ \chi_M \circ \mu \cap \lambda \\ &= (\chi_M \circ \mu \circ \chi_M) \cap (\mu \circ \chi_M \circ \mu) \cap (\chi_M \circ \lambda \circ \chi_M) \cap (\lambda \circ \chi_M \circ \lambda) \subseteq \mu \cap \lambda.\end{aligned}$$

Hence $\mu \cap \lambda$ is a fuzzy bi-interior ideal of *M*. Hence the theorem.

Corollary 3.9 *Let μ and λ be fuzzy right ideal and fuzzy left ideal of semiring M respectively. Then* $\mu \cap \lambda$ *is a fuzzy bi-interior ideal of semiring M.*

Theorem 3.10 *Let M be a semiring. Then M is a regular if and only if* $\mu = \chi_M \circ \mu \circ \chi_M \cap \mu \circ \chi_M \circ \mu$, *for any fuzzy bi-interior ideal μ of semiring M.*

Proof. Let μ be a fuzzy bi-interior ideal of regular semiring *M*. Then $\chi_M \circ \mu \circ \chi_M \cap \mu \circ \chi_M \circ \mu \subseteq \mu$.

$$\mu \circ \chi_M \circ \mu(x) = \sup_{x=xyx}\{\min\{\mu(x), \chi_M \circ \mu(yx)\}\}$$
$$= \sup_{x=xyx}\left\{\min\left\{\mu(x), \sup_{yx=rs}\{\min\{\chi_M(r), \mu(s)\}\}\right\}\right\}$$
$$= \sup_{x=xyx}\{\min\left\{\mu(x), \sup_{yx=rs}\{\min\{1, \mu(s)\}\}\right\}$$
$$\geq \sup_{x=xyx}\{\min\{\mu(x), \mu(x)\}\} = \mu(x).$$

Similarly $\chi_M \circ \mu \circ \chi_M(x) \geq \mu(x)$. Hence $\chi_M \circ \mu \cap \mu \circ \chi_M \circ \mu = \mu$.

Conversely suppose that $\mu = \chi_M \circ \mu \cap \mu \circ \chi_M \circ \mu$, for any fuzzy bi-interior ideal μ of semiring *M*. Let *B* be a bi-interior ideal of semiring *M*.

Then by Theorem 3.7, χ_B be a fuzzy bi-interior ideal of semiring *M*. Therefore

$$\chi_B = \chi_M \circ \chi_B \circ \chi_M \cap \chi_B \circ \chi_M \circ \chi_B$$
$$= \chi_{MBM} \cap \chi_{BMB}$$

$$B = MBM \cap BMB.$$

By Theorem 2.16, *M* is a regular semiring.

Theorem 3.11 *Let M be a regular semiring. Then μ is a fuzzy bi-interior ideal of M if and only if μ is a fuzzy quasi ideal of M.*

Proof. Let μ be a fuzzy bi-interior ideal of semiring *M* and *x*∈*M*. Then $\chi_M \circ \mu \circ \chi_M \cap \mu \circ \chi_M \circ \mu \subseteq \mu$.

Suppose $\chi_M \circ \mu(x) > \mu(x)$. Since *M* is a regular, there exists *y*∈*M* such that *x*=*xyx*. Then

$$\mu \circ \chi_M \circ \mu(x) = \sup_{x=xyx}\{\min\{\mu(x), \chi_M \circ \mu(yx)\}$$
$$> \sup_{x=xyx}\{\min\{\mu(x), \mu(x)\}\} = \mu(x).$$

Which is a contradiction. Therefore $\chi_M \circ \mu \subseteq \mu$. Similarly $\mu \circ \chi_M \subseteq \mu$. Thus $\mu \circ \chi_M \cap \chi_M \circ \mu \subseteq \mu$.

Hence μ is fuzzy quasi ideal of *M*.

By Theorem 2.12 and Corollary 3.5, converse is true.

Theorem 3.12 *Let M be a semiring. Then M is a regular if and only if*

$$\mu \cap \gamma \subseteq \gamma \circ \mu \circ \gamma \cap \mu \circ \gamma \circ \mu,$$

for every fuzzy bi-interior ideal μ and every fuzzy ideal γ of semiring M.

Proof. Let M be a regular semiring and $x \in M$. Then there exists $y \in M$ such that $x=xyx$.

$$\begin{aligned}\mu \circ \gamma \circ \mu(x) &= \sup_{x=xyx}\left\{\min\left\{\mu \circ \gamma(xy), \mu(x)\right\}\right\} \\ &= \min\left\{\sup_{xy=xyxy}\left\{\min\left\{\mu(x), \gamma(yxy)\right\}, \mu(x)\right\}\right\} \\ &\geq \min\{\min\left\{\mu(x), \gamma(x)\right\}, \mu(x\} \\ &= \min\left\{\mu(x), \gamma(x)\right\} = \mu \cap \gamma(x).\end{aligned}$$

$$\begin{aligned}\gamma \circ \mu \circ \gamma(x) &= \sup_{x=xyx}\left\{\min\left\{\gamma \circ \mu(xy), \mu(x)\right\}\right\} \\ &\geq \min\left\{\gamma(x), \mu(x)\right\} = \mu \cap \gamma(x).\end{aligned}$$

Therefore $\mu \cap \gamma \subseteq \mu \circ \gamma \circ \mu$ and $\mu \cap \gamma \subseteq \gamma \circ \mu \circ \gamma$.

Hence $\mu \cap \gamma \subseteq \gamma \circ \mu \circ \gamma \cap \mu \circ \gamma \circ \mu$.

Conversely suppose that the condition holds. Let μ be a fuzzy bi-interior ideal of semiring M. Then $\mu \cap \chi_M \subseteq \chi_M \circ \mu \circ \chi_M \cap \mu \circ \chi_M \circ \mu$.

Therefore $\mu \subseteq \chi_M \circ \mu \circ \chi_M \cap \mu \circ \chi_M \circ \mu$. By Theorem 2.16, M is a regular semiring.

REFERENCES

Allen, P. J. (1969). A fundamental theorem of homomorphism for semirings. *Proceedings of the American Mathematical Society*, *21*(2), 412–416. doi:10.1090/S0002-9939-1969-0237575-4

Good, R. A., & Hughes, D. R. (1952). Associated groups for a semigroup. *Bulletin of the American Mathematical Society*, *58*, 624–625.

Henriksen, M. (1958). Ideals in semirings with commutative addition. *Amer. Math. Soc. Notices*, *5*, 321.

Iseki, K. (1956). Ideal theory of semiring. *Proceedings of the Japan Academy*, *32*(8), 554–559. doi:10.3792/pja/1195525272

Iseki, K. (1958). Quasi-ideals in semirings without zero. *Proceedings of the Japan Academy*, *34*(2), 79–84. doi:10.3792/pja/1195524783

Iseki, K. (1958). Ideal in semirings. *Proceedings of the Japan Academy*, *34*(1), 29–31. doi:10.3792/pja/1195524845

Jagatap, R. D., & Pawar, Y. S. (2009). Quasi-ideals and minimal quasi-ideals in Γ-semirings. *Novi Sad J. Math.*, *39*(2), 79–87.

Kuroki, N. (1991). On fuzzy semigroups. *Information Sciences*, *53*(3), 203–236. doi:10.1016/0020-0255(91)90037-U

Lajos, S. (1969). On the bi-ideals in semigroups. *Proceedings of the Japan Academy*, *45*(8), 710–712. doi:10.3792/pja/1195520625

Lajos, S., & Szasz, F. A. (1970). On the bi-ideals in associative ring. *Proceedings of the Japan Academy*, *46*(6), 505–507. doi:10.3792/pja/1195520265

Liu, F. A. (1982). Fuzzy invariant subgroups and fuzzy ideals. *Fuzzy Sets and Systems*, *8*(2), 133–139. doi:10.1016/0165-0114(82)90003-3

Mandal, D. (2014). Fuzzy ideals and fuzzy interior ideals in ordered semirings. *Fuzzy Info. and Engg.*, *6*, 101-114.

Murali Krishna Rao, M. (2017). fuzzy ideals in ordered *T*-Γ-. *Anl. of Fuzzy Math. and Info.*, *13*(2), 253–276. doi:10.30948/afmi.2017.13.2.253

Murali Krishna Rao, M. (2017). Bi-quasi-ideals and fuzzy bi-quasi-ideals of Γ- semigroups. *Bull. Int. Math. Virtual Inst*, *7*(2), 231–242.

Murali Krishna Rao, M. (2018). Bi –Ideals of semigroups. *Disc. Math. Gen. Alg. and Appl.*, *38*, 69–78.

Murali Krishna Rao, M. (2018). Left bi-quasi ideals of semirings. *Bull. Int. Math. Virtual Inst.*, *8*, 45–53.

Murali Krishna Rao, M. (2018). Fuzzy left and right bi-quasi ideals of semirings. *Bull. Int. Math. Virtual Inst.*, *8*(3), 449–460.

Murali Krishna Rao, M. (2018). Fuzzy Bi-Interior Ideals of Semigroups, *Asia. Pacific Journal of Mathematics*, *5*, 208–218.

Murali Krishna Rao, M., Venkateswarlu, B., & Adinaryana, Y. (1971). (commnicated). Bi-Interior Ideals of Semirings, Rosenfeld, A.(1971). Fuzzy groups. *Journal of Mathematical Analysis and Applications*, *35*, 512–517.

Steinfeld, O. (1956). Uher die quasi ideals, *Von halbgruppend. Publicationes Mathematicae (Debrecen)*, *4*, 262–275.

Swamy, U. M., & Swamy, K. L. N. (1988). Fuzzy prime ideals of rings. *Journal of Mathematical Analysis and Applications*, *134*(1), 94–103. doi:10.1016/0022-247X(88)90009-1

Vandiver, H. S. (1934). Note on a simple type of algebra in which cancellation law of addition does not hold. *Bull. Amer. Math. Soc.(N.S.)*, *40*, 914--920.

Zadeh, L. A. (1965). Fuzzy sets. *Information and Control*, *8*(3), 338–353. doi:10.1016/S0019-9958(65)90241-X

Chapter 10
On More Generalized Fuzzy Interior Ideals in Semigroup

Muhammad Sajjadali Ali Khan
Hazara University Mansehra, Pakistan

ABSTRACT

In this chapter, the authors introduce the concepts of $(\in,\in\vee q_k^{\delta})$-fuzzy interior ideal in semigroup, which is a generalisation of $(\in,\in\vee qk)$-fuzzy interior ideal and $(\in,\in\vee q)$-fuzzy interior ideal in semigroup, and investigate some of its properties. They give example to show that $(\in,\in\vee q_k^{\delta})$-fuzzy interior ideal in semigroup is not an $(\in,\in\vee qk)$-fuzzy interior ideal and $(\in,\in\vee q)$-fuzzy interior ideal in semigroup. They show that every $(\in,\in\vee q_k^{\delta})$-fuzzy ideal is an $(\in,\in\vee q_k^{\delta})$-fuzzy interior ideal, but the converses are not true in general.

1 INTRODUCTION

In 1965, the concept of a fuzzy set was first introduced by, Zadeh in his important paper (Zadeh, 1965). This inspirational paper has opened up new awareness and application in a wide range of scientific fields. After the Zadeh's fuzzy sets many researches convoyed the researches on the generalizations of the of fuzzy sets ideas, with huge applications in computer science, artificial intelligence, control engineering, expert, robotics, automat theory, finite state machine, graph theory logics and many branches of pure and applied mathematics. Investigations of fuzzy groups and fuzzy semigroups have a long convention, starting with the opening of fuzzy period. Wang (Wang, 2014) extended the concept of fuzzy numbers to fuzzy arithmetics. Wang (Wang, 2015) studied the theory of fuzzy Probability for Fuzzy Inference and Computational Intelligence. Rosenfeld was the first who used the notion of a fuzzy subset and introduced the concept of fuzzy groups (Rosenfeld, 1971). Kuroki was the first who initiated the theory of fuzzy semigroups (Kuroki, 1979). Kuroki (Kuroki, 1981) initiated the notion fuzzy ideals, bi-ideals, quasi-ideals of semigroups. The book by Mordeson (Mordeson, Malik & Kuroki, 2003), deals with the theory of fuzzy semigroups and their use in fuzzy coding, fuzzy finite state machines and fuzzy languages. Mordeson and Malik (Mordeson & Malik, 2002) deal with the application of the fuzzy approach to the automata and formal languages. Pu and Lia (Pu & Lia, 1980), introduced the notion "belongs to"

DOI: 10.4018/978-1-7998-0190-0.ch010

relation (∈). Murali (Murali, 2004) initiated the notion of belongingness (q) of a fuzzy point to a fuzzy subset under an expected equality on a fuzzy subset. These two notions played a vital role in generating some different types of fuzzy subgroups. Bhakat (Bhakat, 1999) and Bhakat and Das (Bhakat & Das, 1996) provided the notion of (α,β) -fuzzy subgroups, where $\alpha,\beta \in \{\in, q, \in\vee q, \in\wedge q\}$ and $\alpha \neq \in\wedge q$. The concept of (α,β)-fuzzy subgroups are further studied in ((Bhakat, 1992; 2000). Davvaz (Davvaz, 2006) introduced the notion of $(\in,\in\vee q)$ -fuzzy sub near-rings and ideals of a near ring. Jun and Song (Jun & Song, 2006) initiated the study of (α,β)-fuzzy interior ideals of a semigroup. By using the concept of $(\in,\in\vee q)$-fuzzy ideals, Shabir, Jun and Nawaz (Shabir, Jun & Nawaz, 2010a) characterized regular semigroup. Khan, Sarmin, Khan and Khan (Khan, Sarmin, Khan & Khan, 2015) introduced the concepts of (α,β)-fuzzy bi-ideals and (α,β)-fuzzy bi-ideals of ordered semigroups and investigated some of its related properties. Kazanci and Yamak (Kazanci and Yamak, 2008) considered $(\in,\in\vee q)$-fuzzy bi-ideals of a semigroup and prove some motivated results. Generalizing the idea of the quasi-coincident of a fuzzy point with a fuzzy subset, Jun (Jun, 2009) introduced the concept of $(\in,\in\vee q_k)$-fuzzy subalgebras in **BCK/BCI**-algebras, respectively. Shabir, Jun and Nawaz (Shabir, Jun & Nawaz, 2010b) generalized the concept of an$(\in,\in\vee q)$-fuzzy bi-ideal, an $(\in,\in\vee q)$-fuzzy interior ideal, an $(\in,\in\vee q)$-fuzzy quasi-ideal of a semigroup and introduced the notion of an $(\in,\in\vee q_k)$-fuzzy bi-ideal, an $(\in,\in\vee q_k)$-fuzzy interior ideal, an $(\in,\in\vee q_k)$-fuzzy quasi-ideal in a semigroup. Shabir, Jun and Nawaz (Shabir, Jun & Nawaz, 2010c) characterized regular semigroups by the properties of $(\in,\in\vee q_k)$-fuzzy ideals. The concept of $\left(\in,\in\vee q_k^\delta\right)$-fuzzy subsemigroup was initiated by Kang (Kang, 2016). Khan, Abdullah, Jun and Rahman (Khan, Abdullah, Jun & Rahman, 2017) and Khan, Rahman, Fahmi and Shakeel (Khan, Rahman, Fahmi & Shakeel, 2018) introduced the concepts of $\left(\in,\in\vee q_k^\delta\right)$-fuzzy ideal, $\left(\in,\in\vee q_k^\delta\right)$-fuzzy quasi-ideals in semigroups. Khan, Abdullah, Ali, Amin and Rahman (Khan, Abdullah, Ali, Amin & Rahman, 2017) initiated the concept of $\left(\in,\in\vee q_k^\delta\right)$-fuzzy quasi-ideals in ordered semigroups. Khan, Abdullah, Rahman and Abbass (Khan, Abdullah, Rahman & Abbass, 2017) characetrized regular semigroup by the properties of $(\in, \in\vee(k^*, q_k))$-fuzzy ideal.

Motivated by the above discussion in we introduce the notion of an $\left(\in,\in\vee q_k^\delta\right)$-fuzzy interior ideal of a semigroup. The concept of an $\left(\in,\in\vee q_k^\delta\right)$-fuzzy interior ideals of a semigroup is the generalization of $(\in,\in\vee q_k)$-fuzzy interior ideals and $(\in,\in\vee q)$-fuzzy interior ideals. If we take δ=1, then we get, an $(\in,\in\vee q_k)$-fuzzy interior ideal of a semigroup. If we take δ=1 and k=0, then we get, an $(\in,\in\vee q)$-fuzzy interior ideal,of a semigroup. Which means that these fuzzy ideals become a special case of an $\left(\in,\in\vee q_k^\delta\right)$-fuzzy interior ideal, of a semigroup.

By means of the notion of $\left(\in,\in\vee q_k^\delta\right)$-fuzzy ideals of semigroups in this paper, we generalized the concept of $(\in,\in\vee q_k)$-fuzzy interior ideals and define $\left(\in,\in\vee q_k^\delta\right)$-fuzzy interior ideals in a semigroup. We show that every $\left(\in,\in\vee q_k^\delta\right)$-fuzzy ideal in $\left(\in,\in\vee q_k^\delta\right)$-fuzzy interior ideal of S, but the converse is not true.

2 PRELIMINARIES

Throughout this chapter, S will denote a semigroup unless otherwise stated. A non-empty subset A of S is said to be a subsemigroup of S if $A^2 \subseteq A$. A non-empty subset I of a S is said to be a left(right) ideal of S if $SI \subseteq I(IS \subseteq I)$. A non-empty subset I of S is said to be an ideal if it is both left and right ideal of S. A subsemigroup I of S is said to be an interior ideal of S if $SIS \subseteq I$.

A fuzzy subset λ of a universe X is a function from X into the unit closed interval [0,1], that is λ: $X \rightarrow 0,1]$.

Let λ be a fuzzy set of a semigroup S and $t \in 0,1]$, the set $U(\lambda;t)=\{x \in S | \lambda(x) \geq t\}$ is called a level subset of the fuzzy set λ.

Let A be a non-empty subset of S. We denote by λ_A, the characteristic function of A, that is the mapping of S into [0,1] defined by

$$\lambda_A(x)=\begin{cases}1, \text{if} x \in A \\ 0, \text{if} x \notin A\end{cases}$$

Obviously λ_A is a fuzzy subset of S.

A fuzzy subset in a universe X of the form

$$\lambda(y):=\begin{cases}t \in (0,1] & \text{if } y=x, \\ 0 & \text{if } y \neq x,\end{cases}$$

is called a fuzzy point with support x and value t and is denoted by (x,t).

For a fuzzy subset λ in S, a fuzzy point (x,t) said to

contained in λ denoted $(x,t) \in \lambda$, if $\lambda(x) \geq t$.
be quasi-coincident with λ, denoted by $(x,t)q\lambda$, if $\lambda(x)+t>1$.

For a fuzzy subset λ and fuzzy point (x,t) in a set S, we say that

$(x,t) \in \wedge q\lambda$ if $(x,t) \in \lambda$ or $(x,t)q\lambda$.

Generalizing the concept of $(x,t)q\lambda$, Jun (Jun, 2009), defined $(x,t)q_k\lambda$, where $k \in 0,1)$ as $(x,t)q_k\lambda$ if $\lambda(x)+t+k>1$ and $(x,t) \in \vee q_k\lambda$ if $(x,t) \in \lambda$ or$(x,t)q_k\lambda$. Kang (Kang, 2016), considered the general form of the symbol $(x,t)q_k\lambda$ and $(x,t) \in \vee q_k\lambda$ as follows: For a fuzzy point (x,t) and fuzzy subset λ in a set X, we say that

1. $(x,t)q^{\delta}\lambda$ if $\lambda(x)+t+k>\delta$,
2. $(x,t)q_k^{\delta}\lambda$ if $\lambda(x)+t>\delta$,
3. $(x,t) \in \vee q_k^{\delta}\lambda$ if $(x,t) \in \lambda$ or $(x,t)q_k^{\delta}\lambda$;
4. $(x,t)\overline{\alpha}\lambda$ if $(x,t)\alpha\lambda$ does not hold. For $\alpha \in \{\in, q, \in \vee q, \in \vee q_k, q_k^{\delta}, \in \vee q_k^{\delta}\}$.

where $k\in 0,1)$ and $k<\delta$ in $[0,1]$. Obviously, $(x,t)q^{\delta}\lambda$ implies $(x,t)q_k^{\delta}\lambda$.

Definition 2.1 (Kang, 2016) *A fuzzy subset λ of S is called an* $(\in,\in\vee q_k^{\delta})$*-fuzzy subsemigroup of S if for all* $t_1t_2\in(0,1]$ *and* $x,y\in S$,

$$(x,t_1)\in\lambda,(y,t_2)\in\lambda\Rightarrow(xy,t_1\wedge t_2)\in\vee q_k^{\delta}\lambda$$

Lemma 2.2 (Kang, 2016) *Let A be a subsemigroup of S and λ a fuzzy subset of S defined by*

$$\lambda(x)=\begin{cases}\varepsilon & \text{if } x\in A,\\ 0 & \textit{otherwise},\end{cases}$$

where $\varepsilon\geq\dfrac{1-k}{2}$. Then,

1. λ is $(q^{\delta},\in\vee q_k^{\delta})$-fuzzy subsemigroup of *S*.
2. λ is an $(\in,\in\vee q_k^{\delta})$-fuzzy subsemigroup of *S*.

Lemma 2.3 (Kang, 2016) *Let λ be a fuzzy subset of S. Then λ is an* $(\in,\in\vee q_k^{\delta})$*-fuzzy subsemigroup of S, if and only if* $\lambda(xy)\geq\lambda(x)\wedge\lambda(y)\wedge\dfrac{\delta-k}{2}$ *for all* $x,y\in S$ *and* $k\in 0,1)$.

Lemma 2.4 (Kang, 2016) *Let λ be a fuzzy subset of S. Then λ is an* $(\in,\in\vee q_k)$*-fuzzy subsemigroup of S if and only if* $U(\lambda;t)(\neq\varnothing)$ *is a subsemigroup of S for all* $t\in\left(0,\dfrac{\delta-k}{2}\right]$.

Definition 2.5 (Shabir et al., 2010b) *A fuzzy subset λ of S is said to be an* $(\in,\in\vee q_k)$*-fuzzy left (right) ideal of S if the following condition holds:*

$$(y,t)\in\lambda\Rightarrow(xy,t)\in\vee q_k\lambda\left((y,t)\in\lambda\Rightarrow(yx,t)\in\vee q_k\lambda\right)$$

Definition 2.6 (Shabir et al., 2010b) *Let λ be a fuzzy subset of S, then λ is said to be a* $(\in,\in\vee q_k)$*-fuzzy ideal of S if it is both left and right ideal of S.*

Lemma 2.7 (Shabir et al., 2010b) *Let A be a left (right) of S and λ a fuzzy subset of S defined by*

$$\lambda(x)=\begin{cases}\varepsilon & \text{if } x\in A,\\ 0 & \textit{otherwise},\end{cases}$$

where $\varepsilon\geq\dfrac{1-k}{2}$. Then,

1. λ is $(q,\in\vee q_k)$ -fuzzy left (right) of S.
2. λ is an $(\in,\in\vee q_k)$-fuzzy left (right) of S.

Lemma 2.8 (Shabir et al., 2010b) *Let λ be a fuzzy subset of S. Then λ is an $(\in,\in\vee q_k)$-fuzzy left (right) of S, if and only if*

$$\lambda(xy)\geq\lambda(x)\wedge\frac{1-k}{2}\left(\lambda(xy)\geq\lambda(y)\wedge\frac{1-k}{2}\right)$$

for all $x,y\in S$ and $k\in 0,1)$.

Lemma 2.9 (Shabir et al., 2010b) *Let λ be a fuzzy subset of S. Then λ is an $(\in,\in\vee q_k)$-fuzzy left (right) of S if and only if $U(\lambda;r)(\neq\varnothing)$ is a left (right) of S for all*

$$t\in\left(0,\frac{1-k}{2}\right].$$

3 $(\in,\in\vee q_k^{'})$-FUZZY INTERIOR IDEAL

In this section we define $(\in,\in\vee q_k^{\delta})$-fuzzy interior ideal of a semigroup of S and study some of its properties.

Definition 3.1 *Let λ be a fuzzy subset of S. Then λ is said to be an $(\in,\in\vee q_k^{\delta})$-fuzzy interior ideal of S if it satisfies the following conditions.*

1. $(x,t_1)\in\lambda$, $(y,t_2)\in\lambda\Rightarrow(xy,t_1\wedge t_2)\in\vee q_k^{\delta}\lambda$
2. $(a,t)\in\lambda,\Rightarrow(xay,t)\in\vee q_k^{\delta}\lambda$

for all $a,x,y\in S$ and $t,t_1,t_2\in(0,1]$, $k\in 0,1)$ and $k<\delta$ in $[0,1]$.

Example 3.2 *Let $S=\{a,b,c,d\}$ and "•"be a binary operation defined on S in Table 1.*

Table 1.

.	a	b	c	d
a	b	b	b	b
b	b	b	b	b
c	b	c	c	c
d	b	d	d	d

Then $(S,\bullet)$ is a semigroup. We define a fuzzy subset $\lambda: S\to[0,1]$ by $\lambda(a)=0.8$, $\lambda(b)=0.2$, $\lambda(c)=0.6$ and $\lambda(d)=0.4$. If we take $k=0.2$ and $\delta=.04$, then λ is an $\left(\in,\in\vee q_{0.2}^{0.4}\right)$-fuzzy interior ideal of S. Clearly λ is not a fuzzy ideal, not an $(\in,\in\vee q)$-fuzzy ideal and not an $(\in,\in\vee q_k)$-fuzzy interior ideal of S. Because,

(1): λ is not an $(\in,\in\vee q_{0.2})$ -fuzzy interior ideal of S. Because, $(a,t)=(a,0.2)\in\lambda$ but $\left(a\cdot a\cdot a,t\right)=\left(b,0.2\right)\overline{\in\vee q_{0.2}}\lambda$

(2): λ is not an $(\in,\in\vee q)$-fuzzy interior ideal of S. Because, $(a,t)=(a,0.2)\in\lambda$ but $\left(a\cdot a\cdot a,t\right)=\left(b,0.2\right)\overline{\in\vee q}\lambda$.

Theorem 3.3 *Let I be a interior ideal of S and λ a fuzzy subset of S defined by*

$$\lambda(x)=\begin{cases}\varepsilon & \text{if } x\in I,\\ 0 & \textit{otherwise},\end{cases}$$

where $\varepsilon\geq\frac{\delta-k}{2}$. Then, λ is $\left(\alpha,\in\vee q_k^{\delta}\right)$-fuzzy interior ideal of S for $\alpha\in\{\in,q^{\delta}\}$.

Proof. Let $x,y\in S$ and $t_x,t_y\in\{0,\delta]$ be such that $(x,t_x)q^{\delta}\lambda$ and $(y,t_y)q^{\delta}\lambda$. Then $\lambda(x)+t_x>\delta$ and $\lambda(y)+t_y>\delta$. If $x\notin I$ or $y\notin I$, then $\lambda(x)=0$ or $\lambda(y)=0$. Hence $t_x>\delta$ or $t_y>\delta$ which is a contradiction. Thus $x,y\in I$. Since, I is an interior ideal of S, we have $xy\in I$ and so $\lambda(xy)=\varepsilon\geq\frac{\delta-k}{2}$. If $t_x\wedge t_y\leq\frac{\delta-k}{2}$, then $\lambda(xy)\geq t_x\wedge t_y$ and so $(xy,t_x\wedge t_y)\in\lambda$. If $t_x\wedge t_y>\frac{\delta-k}{2}$, then

$$\lambda(xy)+t_x\wedge t_y+k>\frac{\delta-k}{2}+\frac{\delta-k}{2}+k=\delta.$$

Hence, $\left(xy,t_x\wedge t_y\right)q_k^{\delta}\lambda$. Therefore, $\left(xy,t_x\wedge t_y\right)\in q_k^{\delta}\lambda$. Now, let $a,x,y\in S$ and $t_a\in(0,\delta]$ be such that $(a,t_a)q^{\delta}\lambda$. Then $\lambda(a)+t_a>\delta$. If $a\notin I$, then $\lambda(a)=0$ Hence $t_a>\delta$ which is a contradiction. Thus, $a\in I$. Since, I is an interior ideal of S, we have $xay\in I$ and so $\lambda(a)=\varepsilon\geq\frac{\delta-k}{2}$. If $t_a\wedge t_a=t_a\leq\frac{\delta-k}{2}$, then $\lambda(xay)\geq t_a$ and so $(xay,t_a)\in\lambda$. If $t_a\wedge t_a=t_a>\frac{\delta-k}{2}$, then

$$\lambda(xay)+t_a+k>\frac{\delta-k}{2}+\frac{\delta-k}{2}+k=\delta.$$

Hence, $\left(xay,t_a\right)q_k^{\delta}\lambda$. Therefore, $\left(xay,t_a\right)\in q_k^{\delta}\lambda$. This shows that λ is $\left(q^{\delta},\in\vee q_k^{\delta}\right)$-fuzzy interior ideal of S.

Let $x,y\in S$ and $t_1,t_2\in(0,\delta]$ be such that $(x,t_1)\in\lambda$ and $(y,t_2)\in\lambda$. Then $\lambda(x)\geq t_1>0$ and $\lambda(y)+t_2>0$. Thus

$$\lambda(x)=\varepsilon \geq \frac{\delta-k}{2} \text{ and } \lambda(y)=\varepsilon \geq \frac{\delta-k}{2},$$

this implies that $x,y \in I$. Since I is an interior ideal of S, we have $xy \in I$. Hence $\lambda(xy)=\varepsilon \geq \frac{\delta-k}{2}$. If $t_1 \wedge t_2 \leq \frac{\delta-k}{2}$, then $\lambda(xy) \geq t_1 \wedge t_2$ and hence $(xy, t_1 \wedge t_2) \in \lambda$. If $t_1 \wedge t_2 > \frac{\delta-k}{2}$, then

$$\lambda(xy)+t_1 \wedge t_2+k > \frac{\delta-k}{2}+\frac{\delta-k}{2}+k=\delta.$$

Hence, $(xy, t_1 \wedge t_2) q_k^{\delta} \lambda$. Therefore, $(xy, t_1 \wedge t_2) \in q_k^{\delta} \lambda$. Now, let $a,x,y \in S$ and $t \in (0,\delta]$ be such that $(a,t) \in \lambda$. Then $\lambda(a) \geq t > 0$. If $a \notin I$, then $\lambda(a)=0$. Hence $t > \delta$ which is a contradiction. Thus $a \in I$. Since, I is an interior ideal of S, we have $xay \in I$ and so $\lambda(a)=\varepsilon \geq \frac{\delta-k}{2}$. If $t \wedge t = t \leq \frac{\delta-k}{2}$, then $\lambda(xay) \geq t$ and so $(xay,t) \in \lambda$. If $t \wedge t = t > \frac{\delta-k}{2}$, then

$$\lambda(xay)+t+k > \frac{\delta-k}{2}+\frac{\delta-k}{2}+k=\delta.$$

Hence, $(xay,t) q_k^{\delta} \lambda$. Therefore, $(xay,t) \in q_k^{\delta} \lambda$. This shows that λ is $(\in, \in \vee q_k^{\delta})$-fuzzy interior ideal of S.

If we put $\delta=1$ in Theorem 3.3 then we have the following two corollary.

Corollary 3.4 (Shabir et al., 2010b) *Let I be a interior ideal of S and λ a fuzzy subset of S defined by*

$$\lambda(x)=\begin{cases} \varepsilon & \text{if } x \in I, \\ 0 & \textit{otherwise}, \end{cases}$$

where $\varepsilon \geq \frac{1-k}{2}$. Then, λ is $(\alpha, \in \vee q_k)$ -fuzzy interior ideal of S for $\alpha \in \{\in, q\}$.

If we put $\delta=1$ and $k=0$ in Theorem 3.3 then we have the following two corollary.

Corollary 3.5 (Jun & Song, 2006) *Let I be a interior ideal of S and λ a fuzzy subset of S defined by*

$$\lambda(x)=\begin{cases} \varepsilon & \text{if } x \in I, \\ 0 & \textit{otherwise}, \end{cases}$$

where $\varepsilon \geq \frac{1}{2}$. Then, λ is $(\alpha, \in \vee q_k)$-fuzzy interior ideal of S for $\alpha \in \{\in, q\}$.

Theorem 3.6 *A fuzzy subset λ of S is an* $\left(\in,\in\vee q_k^{\delta}\right)$*-fuzzy interior ideal of S if and only if*

$$\lambda(xy)\geq\lambda(x)\wedge\lambda(y)\wedge\frac{\delta-k}{2} \text{ and } \lambda(xay)\geq\lambda(a)\wedge\frac{\delta-k}{2}$$

for all a,x,y∈S and k<δ in [0,1].

Proof. Let λ be an $\left(\in,\in\vee q_k^{\delta}\right)$-fuzzy interior ideal of *S*. Assume that there exist elements *x,y*∈*S* such that $\lambda(xy)<\lambda(x)\wedge\lambda(y)\wedge\frac{\delta-k}{2}$. Then we have the following two cases:

1. If $\lambda(x)\wedge\lambda(y)<\frac{\delta-k}{2}$, then we choose *t*∈(0,δ) such that

$$\lambda(xy)<t\leq\lambda(x)\wedge\lambda(y)\wedge\frac{\delta-k}{2}.$$

It follows that (*x,t*)∈λ and (*y,t*)∈λ, but $(xy,t)\overline{\in}\lambda$ and λ(*xy*)+*t*<2*t*<δ-*k* so $(xy,t)\overline{q_k^{\delta}}\lambda$, which is contradiction. Hence, $(xy,t)\overline{\in\vee q_k^{\delta}}\lambda$

2. If $\lambda(x)\wedge\lambda(y)\geq\frac{\delta-k}{2}$, then $\lambda(x)\geq\frac{\delta-k}{2}$, $\lambda(y)\geq\frac{\delta-k}{2}$ and $\lambda(xy)<\frac{\delta-k}{2}$. Thus

$$\left(x,\frac{\delta-k}{2}\right)\in\lambda \text{ and } \left(y,\frac{\delta-k}{2}\right)\in\lambda,$$

but $\left(xy,\frac{\delta-k}{2}\right)\overline{\in}\lambda$. Also,

$$\lambda(xy)+\frac{\delta-k}{2}<\frac{\delta-k}{2}+\frac{\delta-k}{2}=\delta-k,$$

i.e., $\left(xy,\frac{\delta-k}{2}\right)\overline{q_k^{\delta}}\lambda$. Hence, $\left(xy,\frac{\delta-k}{2}\right)\overline{\in\vee q_k^{\delta}}\lambda$, again a contradiction Therefore, $\lambda(xy)\geq\lambda(x)\wedge\lambda(y)\wedge\frac{\delta-k}{2}$.

Now, assume that there exist *a,x,y*∈*S* such that $\lambda(xay)<\lambda(a)\wedge\frac{\delta-k}{2}$. Then we have the following two cases:

1. If $\lambda(a)<\frac{\delta-k}{2}$, then we choose $t\in(0,\delta)$ such that $\lambda(xay)<t\leq\lambda(a)\wedge\frac{\delta-k}{2}$. It follows that $(a,t)\in\lambda$, but $(xay,t)\overline{\in}\lambda$ and $\lambda(xay)+t<2t<\delta-k$ so $(xay,t)\overline{q_k^{\delta}}\lambda$, which is contradiction. Hence, $(xay,t)\overline{\in\vee q_k^{\delta}}\lambda$.
2. If $\lambda(a)\geq\frac{\delta-k}{2}$, then we choose $t\in(0,\delta)$ such that $\lambda(xay)<t\leq\lambda(a)\wedge\frac{\delta-k}{2}$. Hence, $(a,t)\in\lambda$ and $(xay,t)\overline{\in\vee q_k^{\delta}}\lambda$. Therefore, $\lambda(xay)\geq\lambda(a)\wedge\frac{\delta-k}{2}$.

Conversely, Suppose that

$$\lambda(xy)\geq\lambda(x)\wedge\lambda(y)\wedge\frac{\delta-k}{2} \text{ and } \lambda(xay)\geq\lambda(a)\wedge\frac{\delta-k}{2}$$

for all $a,x,y\in S$. Let $(x,t_1)\in\lambda$ and $(y,t_2)\in\lambda$ for $t_1,t_2\in(0,\delta]$. Then $\lambda(x)\geq t_1$ and $\lambda(y)\geq t_2$. Now,

$$\lambda(xy)\geq\lambda(x)\wedge\lambda(y)\wedge\frac{\delta-k}{2}\geq t_1\wedge t_2\wedge\frac{\delta-k}{2}.$$

If $t_1\wedge t_2\leq\frac{\delta-k}{2}$, then $\lambda(xy)\geq t_1\wedge t_2$. So, $(xy,t_1\wedge t_2)\in\lambda$. If $t_1\wedge t_2>\frac{\delta-k}{2}$, then $\lambda(xy)\geq\frac{\delta-k}{2}$. So

$$\lambda(xy)+t_1\wedge t_2>\frac{\delta-k}{2}+\frac{\delta-k}{2}=\delta-k,$$

which implies that $(xy,t_1\wedge t_2)q_k^{\delta}\lambda$. Thus $(xy,t_1\wedge t_2)\in\vee q_k^{\delta}\lambda$. Now, let $t\in(0,\delta]$ and $(a,t)\in\lambda$. Then $\lambda(a)\geq t$. Now,

$$\lambda(xay)\geq\lambda(a)\wedge\frac{\delta-k}{2}\geq t\wedge\frac{\delta-k}{2}.$$

If $t\leq\frac{\delta-k}{2}$, then $\lambda(xay)\geq t$. So, $(xay,t)\in\lambda$.. If $t>\frac{\delta-k}{2}$, then $\lambda(xay)\geq\frac{\delta-k}{2}$. So

$$\lambda(xay)+t>\frac{\delta-k}{2}+\frac{\delta-k}{2}=\delta-k,$$

which implies that $(xay,t)q_k^{\delta}\lambda$. Thus $(xay,t)\in\vee q_k^{\delta}\lambda$. Therefore, λ is an $(\in,\in\vee q_k^{\delta})$-fuzzy interior ideal of S.

If we take $\delta=1$ in Theorem 3.6 then we have the following two corollary.

Corollary 3.7 (Shabir et al., 2010b) *A fuzzy subset λ of S is an* $(\in,\in\vee q_k)$*-fuzzy interior ideal of S if and only if*

$$\lambda(xy)\geq\lambda(x)\wedge\lambda(y)\wedge\frac{1-k}{2} \text{ and } \lambda(xay)\geq\lambda(a)\wedge\frac{1-k}{2}$$

for all a,x,y∈S.

If we take δ=1 and *k*=0 in Theorem 3.6 then we have the following two corollary.

Corollary 3.8 (Jun & Song 2006) *A fuzzy subset λ of S is an* $(\in,\in\vee q_k)$*-fuzzy interior ideal of S if and only if*

$$\lambda(xy)\geq\lambda(x)\wedge\lambda(y)\wedge\frac{1}{2} \text{ and } \lambda(xay)\geq\lambda(a)\wedge\frac{1}{2}$$

for all a,x,y∈S.

Theorem 3.9 *Let λ be a fuzzy subset of S. Then λ is an* $\left(\in,\in\vee q_k^\delta\right)$*-fuzzy interior ideal of S if and only if U(λ;t)(≠∅) is an interior ideal of S for all* $t\in\left(0,\frac{\delta-k}{2}\right]$.

Proof. Assume that λ is an $\left(\in,\in\vee q_k^\delta\right)$-fuzzy interior ideal of *S*. Let *x,y*∈*U*(λ;*t*) for $t\in\left(0,\frac{\delta-k}{2}\right]$. Then λ(*x*)≥*t* and λ(*y*)≥*t*. It follows from Theorem 3.6 that,

$$\lambda(xy)\geq\lambda(x)\wedge\lambda(y)\wedge\frac{\delta-k}{2}\geq t\wedge\frac{\delta-k}{2}=t.$$

So that *x,y*∈*U*(λ;*t*). Now, let *x,y*∈*S* and *a*∈*U*(λ;*t*). Then λ(*a*)≥*t* and so by Theorem 3.6, it implies that $\lambda(xay)\geq\lambda(a)\wedge\frac{\delta-k}{2}\geq t\wedge\frac{\delta-k}{2}=t$. Thus *xay*∈*U*(λ;*t*). Therefore, *U*(λ;*t*) is an interior ideal of *S*.

Conversely, let λ be a fuzzy subset of *S* such that *U*(λ;*t*) is an interior ideal of *S* for all $t\in\left(0,\frac{\delta-k}{2}\right]$. If there exist *x,y*∈*S* such that $\lambda(xy)<\lambda(x)\wedge\lambda(y)\wedge\frac{\delta-k}{2}$, then we can take *t*∈(0,δ) such that $\lambda(xy)<t<\lambda(x)\wedge\lambda(y)\wedge\frac{\delta-k}{2}$. Thus *x,y*∈*U*(λ;*t*) and $t<\frac{\delta-k}{2}$, and so *xy*∈*U*(λ;*t*), i.e., λ(*xy*)∈*t*. This is a contradiction. Therefore $\lambda(xy)\geq\lambda(x)\wedge\lambda(y)\wedge\frac{\delta-k}{2}$ for all *x,y*∈*S*. Now if there exist *a,x,y*∈*S*suchthat $\lambda(xay)<\lambda(a)\wedge\frac{\delta-k}{2}$ .thenwecanchoose*t*∈(0,δ)suchthat $\lambda(xay)<t<\lambda(a)\wedge\frac{\delta-k}{2}$. It follows that *a*∈*U*(λ;*t*) and $t<\frac{\delta-k}{2}$ so that *xay*∈*U*(λ;*t*), i.e., λ(*xay*)≤*t*. This is impossible. Hence

$\lambda(xay) \leq \lambda(a) \wedge \frac{\delta - k}{2}$ for all $a,x,y \in S$. Therefore by using Theorem 3.6, we have that λ is an $(\in, \in \vee q_k^\delta)$-fuzzy interior ideal of S.

If we take δ=1 in Theorem 3.9 then we have the following two corollary.

Corollary 3.10 (Shabir et al., 2010b) *Let λ be a fuzzy subset of S. Then λ is an* $(\in, \in \vee q_k)$*-fuzzy interior ideal of S if and only if* $U(\lambda;t)(\neq \varnothing)$ *is an interior ideal of S for all* $t \in \left(0, \frac{1-k}{2}\right]$.

If we take δ=1 and k=0 in Theorem 3.9 then we have the following two corollary.

Corollary 3.11 (Jun & Song, 2006) *Let λ be a fuzzy subset of S. Then λ is an* $(\in, \in \vee q)$*-fuzzy interior ideal of S if and only if* $U(\lambda;t)(\neq \varnothing)$ *is an interior ideal of S for all* $t \in \left(0, \frac{1}{2}\right]$.

Theorem 3.12 *Every* $(\in \vee q_k^\delta, \in \vee q_k^\delta)$*-fuzzy interior ideal is an* $(\in, \in \vee q_k^\delta)$*-fuzzy interior ideal.*

Proof. Let λ be an $(\in \vee q_k^\delta, \in \vee q_k^\delta)$-fuzzy interior ideal of S. Let $x,y \in S$ and $t_1,t_2 \in (0,\delta]$ be such that $(x,t_1) \in \lambda$ and $(y,t_2) \in I$. Then $(x,t_1) \in \vee q_k^\delta \lambda$ and $(y,t_2) \in \vee q_k^\delta \lambda$. Since λ is an $(\in \vee q_k^\delta, \in \vee q_k^\delta)$-fuzzy interior ideal, it follows that $(xy, t_1 \wedge t_2)$. Now let $a,x,y \in S$ and $t \in (0,\delta]$ be such that $(a,t) \in \lambda$. Then $(a,t) \in \vee q_k^\delta \lambda$, and so $\lambda(xay) \geq \lambda(a) \geq t$ or $\lambda(xay)+t+k \geq \lambda(a)+t+k > \delta$. This means that $(xay,t) \in \vee q_k^\delta \lambda$. Consequently, λ is an $(\in, \in \vee q_k^\delta)$-fuzzy interior ideal of S.

Lemma 3.13 *Every* $(\in,\in)$*-fuzzy interior ideal of S is a* $(\in, \in \vee q_k^\delta)$*-fuzzy interior ideal of S.*

Theorem 3.14 *Suppose that every fuzzy point has the value* $t \in \{0,1\}$. *Then every* $(\in, \in \vee q_k^\delta)$*-fuzzy interior ideal of S is a* $(q^\delta, \in \vee q_k^\delta)$*-fuzzy interior ideal of S.*

Proof. Let λ be an $(\in, \in \vee q_k^\delta)$-fuzzy interior ideal of S and $x,y \in S$, and $t_1, t_2 \in \left(0, \frac{\delta - k}{2}\right]$ be such that $(x,t_1) q^\delta \lambda$ and $(x,t_2) q^\delta \lambda$. Then, $\lambda(x)+t_1 > \delta$ and $\lambda(y)+t_2 > \delta$ it follows that $\lambda(x) > \delta - t_1 \geq t_1$, and $\lambda(y) > \delta - t_2 \geq t_2$ i.e. $(x,t_1) \in \lambda$ and $(y,t_2) \in \lambda$. By assumption it follows that $(xy, t_1 \wedge t_2) \in \vee q_k^\delta \lambda$. Now, let, $a,x,y \in S$ and $t \in \left(0, \frac{\delta - k}{2}\right]$ be such that $(a,t_1) q^\delta \lambda$. Then $\lambda(a)+t > \delta$ it follows that $\lambda(a) > \delta - t \geq t$ i.e, $(a,t) \in \lambda$. By assumption it follows that $(xay,t) \in \vee q_k^\delta \lambda$. Hence, λ is a $(q^\delta, \in \vee q_k^\delta)$-fuzzy interior ideal of S.

Theorem 3.15 *Let λ be an* $(\in, \in \vee q_k^\delta)$*-fuzzy interior ideal of S, then the fuzzy subset* $\lambda_k^\delta : S \to [0,1], x \mapsto \lambda(x) \wedge \frac{\delta - k}{2}$ *is a fuzzy interior ideal of S, which is contained in λ.*

Proof. For any $x,y \in S$, we have

$$\begin{aligned}\lambda_k^\delta(xy) &= \lambda(xy)\wedge\frac{\delta-k}{2}\\ &\geq\left\{\lambda(x)\wedge\lambda(y)\wedge\frac{\delta-k}{2}\right\}\wedge\frac{\delta-k}{2}\\ &=\left\{\lambda(x)\wedge\frac{\delta-k}{2}\right\}\wedge\left\{\lambda(y)\wedge\frac{\delta-k}{2}\right\}\\ &=\lambda_k^\delta(x)\wedge\lambda_k^\delta(y)\end{aligned}$$

Now, let $a,x,y\in S$, we have

$$\begin{aligned}\lambda_k^\delta(xay) &= \lambda(xay)\wedge\frac{\delta-k}{2}\\ &\geq\lambda(a)\wedge\frac{\delta-k}{2}=\lambda_k^\delta(a).\end{aligned}$$

Therefore, λ_k^δ is a fuzzy interior ideal of *S*. It is cleat that $\lambda\subseteq\lambda_k^\delta$.

Lemma 3.16 *Let* λ *be an* (∈,∈),

$$\left(\in,q_k^\delta\right),\ \left(\in,\in\vee q_k^\delta\right),\ \left(\in,\in\wedge q_k^\delta\right),\ \left(\in\vee q_k^\delta,\in\right),\ \left(\in\vee q_k^\delta,q^\delta\right),\ \left(\in\vee q_k^\delta,\in\wedge q_k^\delta\right),\ \left(\in\vee q_k^\delta,\in\vee q_k^\delta\right)$$

-fuzzy interior ideal of S. If every fuzzy point has the value $t\in\left(0,\frac{\delta-k}{2}\right]$, *then* λ *is an* $\left(q^\delta,\in\vee q_k^\delta\right)$*-fuzzy interior ideal of S.*

Theorem 3.17 *For any subset H of S,* χ_H *is an* $\left(\in,\in\vee q_k^\delta\right)$*-fuzzy interior ideal of S if and only if H is an interior ideal of S.*

Proof. Let χ_H be an $\left(\in,\in\vee q_k^\delta\right)$-fuzzy interior ideal of S. Let $x,y\in H$. Then $(x;1)\in\chi_H$ and $(y;1)\in\chi_H$ which imply that $(xy,1)=(xy,1\wedge1)\in\vee q_k^\delta\chi_H$. Hence $\chi_H(xy)>0$, and so $xy\in H$. Now let $x,y\in S$ and $a\in H$. Then $\chi_H(a)=1$, and thus $(a;1)\in\chi_H$. Since χ_H is an $\left(\in,\in\vee q_k^\delta\right)$-fuzzy interior ideal, it follows that $(xay,1)\in\chi_H$ so that $\chi_H(xay)=1$. Hence $xay\in H$, and *H* is an interior ideal of S. Conversely, if *H* is an interior ideal of *S*, then χ_H is an (∈,∈)-fuzzy interior ideal of *S* by Lemma 4.4 (June & Song 2016), and therefore χ_H is an $\left(\in,\in\vee q_k^\delta\right)$-fuzzy interior ideal of *S* by Lemma 3.13.

For a fuzzy set λ in *S* and $t\in(0,\delta]$, consider the Q_k^δ-set and $\left[\lambda_k^\delta\right]_t$-set with respect to *t* as follows:

$$Q_k^\delta(\lambda;t):=\left\{x\in X\mid(x,t)q_k^\delta\lambda\right\}\text{ and }\left[\lambda_k^\delta\right]_t:=\left\{x\in X\mid(x,t)\in\vee q_k^\delta\lambda\right\}.$$

Remark 3.18 *For any $t,r \in (0,\delta]$, if $t \geq r$, then*

$$Q_k^{\delta}(\lambda;r) \subseteq Q_k^{\delta}(\lambda;t) \text{ and } U(\lambda;r) \cup Q_k^{\delta}(\lambda;t) \subseteq U(\lambda;t) \cup Q_k^{\delta}(\lambda;t).$$

Obviously, $\left[\lambda_k^{\delta}\right]_t = U(\lambda;t) \cup Q_k^{\delta}(\lambda;t)$.

Theorem 3.19 *If λ is an $\left(\in, \in \vee q_k^{\delta}\right)$-fuzzy interior ideal of S, then*

$$(\forall t \in (0,\delta])\left(Q_k^{\delta}(\lambda;t) \neq \varnothing \Rightarrow Q_k^{\delta}(\lambda;t),\right.$$

is an interior ideal of S).

Proof. Let $x, y \in Q_k^{\delta}(\lambda;t)$ for $a,x,y \in S$. Then $\lambda(x)+t+k>\delta$ and $\lambda(y)+t+k>\delta$. It follows from Theorem 3.6 that,

$$\lambda(xy) \geq \lambda(x) \wedge \lambda(y) \wedge \frac{\delta-k}{2} \geq (\delta-t-k) \wedge \frac{\delta-k}{2} = \delta-t-k$$

and that $xy \in Q_k^{\delta}(\lambda;t)$, and $\lambda(a)+t+k>\delta$ It follows from Theorem 3.6 that,

$$\lambda(xay) \geq \lambda(a) \wedge \frac{\delta-k}{2} \geq (\delta-t-k) \wedge \frac{\delta-k}{2} = \delta-t-k$$

and that $xay \in Q_k^{\delta}(\lambda;t)$. Hence $Q_k^{\delta}(\lambda;t)$ is an interior ideal of S.

If we take $\delta=1$ in Theorem 3.19, then we have the following corollary.

Corollary 3.20 *If λ is an $(\in, \in \vee q_k)$-fuzzy interior ideal of S, then*

$$(\forall t \in (0,1])\left(Q_k(\lambda;t) \neq \varnothing \Rightarrow Q_k(\lambda;t) \text{is an interior ideal of } S\right)$$

Theorem 3.21 *A fuzzy set λ in S is an $\left(\in, \in \vee q_k^{\delta}\right)$-fuzzy interior ideal of S if and only if $\left[\lambda_k^{\delta}\right]_t$ is an interior ideal of S for all $t \in (0,1]$. We call $\left[\lambda_k^{\delta}\right]_t$ an $\left(\in \vee q_k^{\delta}\right)$-level interior ideal of λ.*

Proof. Let λ be an $\left(\in, \in \vee q_k^{\delta}\right)$-fuzzy interior ideal of S and let $x, y \in \left[\lambda_k^{\delta}\right]_t$ for $t \in (0,\delta]$. Then

$$(x;t) \in \vee q_k^{\delta} \lambda \text{ and } (y;t) \in \vee q_k^{\delta} \lambda,$$

that is, λ(*x*)≤*t* or λ(*x*)+*t*+*k*>δ, and λ(*y*)≤*t* or λ(*y*)+*t*+*k*>δ. Since $\lambda(xy)\geq\lambda(x)\wedge\lambda(y)\wedge\frac{\delta-k}{2}$ by Theorem 3.6, we have $\lambda(xy)\geq t\wedge\frac{\delta-k}{2}$. For, if not then $(x;t)\overline{q_k^\delta}\lambda$ or $(y;t)\overline{q_k^\delta}\lambda$, a contradiction. If $t\leq\frac{\delta-k}{2}$, then $\lambda(xy)\geq t\wedge\frac{\delta-k}{2}=t$ and so $xy\in U(\lambda;t)\subseteq\left[\lambda_k^\delta\right]_t$. If $t>\frac{\delta-k}{2}$, then

$$\lambda(xy)\geq t\wedge\frac{\delta-k}{2}=\frac{\delta-k}{2}$$

and thus $\lambda(xy)+t+k>\frac{\delta-k}{2}+\frac{\delta-k}{2}=\delta$. Hence $(xy;t)q_k^\delta\lambda$, and so $xy\in Q_k^\delta(\lambda;t)\subseteq\left[\lambda_k^\delta\right]_t$. Therefore $\left[\lambda_k^\delta\right]_t$ is a subsemigroup of S. Now let *x*,*y*∈*S* and $a\in\left[\lambda_k^\delta\right]_t$. Then $(a;t)\in\vee q_k^\delta\lambda$, i.e., λ(*a*)≥*t* or λ(*a*)+*t*+*k*>δ. Since $\lambda(xay)\geq\lambda(a)\wedge\frac{\delta-k}{2}$ by Theorem 3.6, it follows that $\lambda(xay)\geq t\wedge\frac{\delta-k}{2}$. Otherwise, we get $\lambda(xay)<t\wedge\frac{\delta-k}{2}$ and so $(a;t)\overline{\in\vee q_k^\delta}\lambda$; a contradiction. If $t\leq\frac{\delta-k}{2}$, then

$$\lambda(xay)\geq t\wedge\frac{\delta-k}{2}=t$$

and thus $xay\in U(\lambda;t)\subseteq\left[\lambda_k^\delta\right]_t$. If $t>\frac{\delta-k}{2}$, then $\lambda(xay)\geq t\wedge\frac{\delta-k}{2}=\frac{\delta-k}{2}$ and so

$$\lambda(xay)+t+k>\frac{\delta-k}{2}+\frac{\delta-k}{2}=\delta,$$

i.e., $(xay;t)q_k^\delta\lambda$. Hence $xay\in Q_k^\delta(\lambda;t)\subseteq\left[\lambda_k^\delta\right]_t$. Therefore, $\left[\lambda_k^\delta\right]_t$ is an interior ideal of *S*.

Conversely, let λ be a fuzzy set in *S* and *t*∈(0,δ] be such that $\left[\lambda_k^\delta\right]_t$ is an interior ideal of *S*. If possible, let

$$\lambda(xy)<t<\lambda(x)\wedge\lambda(y)\wedge\frac{\delta-k}{2}$$

for some $t\in\left(0,\frac{\delta-k}{2}\right)$. Then $x,y\in U(\lambda;t)\subseteq\left[\lambda_k^\delta\right]_t$, which implies that $xy\in\left[\lambda_k^\delta\right]_t$. Hence λ(*xy*)≥*t* or λ(*xy*)+*t*+*k*>δ, a contradiction. Therefore $\lambda(xy)\geq\lambda(x)\wedge\lambda(y)\wedge\frac{\delta-k}{2}$ for all *x*,*y*∈*S*. Now assume that $\lambda(xay)<\lambda(a)\wedge\frac{\delta-k}{2}$ for some *a*,*x*,*y*∈*S*. Then we can take $t\in\left(0,\frac{\delta-k}{2}\right)$ such that $\lambda(xay)<t<\lambda(a)\wedge\frac{\delta-k}{2}$. It follows that $a\in U(\lambda;t)\subseteq\left[\lambda_k^\delta\right]_t$ so that $xay\in\left[\lambda_k^\delta\right]_t$. This implies that $\lambda(xay)\geq t$ or $\lambda(xay)+t+k>\delta$, which is impossible. Hence $\lambda(xay)\geq\lambda(a)\wedge\frac{\delta-k}{2}$ for all $a,x,y\in S$. Therefore by using Theorem 3.6, we conclude that λ is an $(\in,\in\vee q_k^{\delta})$-fuzzy interior ideal of S.

Theorem 3.22 *If $\{\lambda_i\}_{i\in\Lambda}$ is a family of $(\in,\in\vee q_k^{\delta})$-fuzzy interior ideal of S, then $\bigcap_{i\in\Lambda}\lambda_i$ is an $(\in,\in\vee q_k^{\delta})$-fuzzy interior ideal of S.*

Proof. Let $x,y\in S$ and $t_1t_2\in(0,\delta]$ be such that $(x;t_1)\in\lambda$ and $(y;t_2)\in\lambda$. Assume that $(xy;t_1\wedge t_2)\overline{\in\vee q_k^{\delta}}\lambda$. Then $\lambda(xy)<t_1\wedge t_2$ and $\lambda(xy)+t_1\wedge t_2+k\leq\delta$. Thus, $\lambda(xy)<\frac{\delta-k}{2}$

Let $\Omega_1:=\{i\in\Lambda|\lambda_i(xy)\geq t_1\wedge t_2\}$, $\Omega_2:=\{i\in\Lambda|\lambda_i(xy)+t_1\wedge t_2+k>\delta\}$ and $\lambda_i(xy)<t_1\wedge t_2$. Then $\Lambda=\Omega_1\cup\Omega_2$ and $\Omega_1\cap\Omega_2=\varnothing$. If $\Omega_2=\varnothing$, then $(xy;t_1\wedge t_2)\in\lambda_i$ for all $i\in\Lambda$, i.e $\lambda_i(xy)\geq t_1\wedge t_2$ for all $i\in\Lambda$, and so $\lambda_i(xy)\geq t_1\wedge t_2$ which is a contradiction. Thus $\Omega_2=\varnothing$;, and hence for every $i\in\Omega_2$ we have $\lambda_i(xy)+t_1\wedge t_2>\delta-k$ and $\lambda_i(xy)<t_1\wedge t_2$. It follows that $t_1\wedge t_2>\frac{\delta-k}{2}$ so that $\lambda_i(x)\geq\lambda(x)\geq t_1\geq t_1\wedge t_2>\frac{\delta-k}{2}$ for all $i\in\Lambda$. Similarly, we get $\lambda_i(y)>\frac{\delta-k}{2}$ for all $i\in\Lambda$. Now suppose that $t:=\lambda i(xy)<\frac{\delta-k}{2}$ for some $i\in\Lambda$. Let $t_0\in\left(0,\frac{\delta-k}{2}\right)$ be such that $t<t_0$. Then

$$\lambda_i(x)>\frac{\delta-k}{2}>t_0 \text{ and } \lambda_i(y)>\frac{\delta-k}{2}>t_0,$$

that is, $(x;t_0)\in\lambda_i$ and $(y;t_0)\in\lambda_i$. But $\lambda_i(xy)=t<t_0$ and $\lambda_i(xy)+t_0+k<\delta$, that is, $(xy;t_0)\overline{\in\vee q_k^{\delta}}\lambda_i$. This contradicts that λ_i is an $(\in,\in\vee q_k^{\delta})$-fuzzy interior ideal of S. Hence $\lambda_i(xy)\geq\frac{\delta-k}{2}$ for all $i\in\Lambda$, and thus $\lambda(xy)\geq\frac{\delta-k}{2}$. This contradicts that, $\lambda(xy)<\frac{\delta-k}{2}$. Therefore $(xy;t_1\wedge t_2)\in\vee q_k^{\delta}\lambda_i$. Now, let $a,x,y\in S$ and $t\in(0,\delta]$ be such that $(a;t)\in\lambda$. If $(xay;t)\overline{\in\vee q_k^{\delta}}\lambda$; then $\lambda(xay)<t$ and $\lambda(xay)+t+k<\delta$. Thus, $\lambda(xay)<\frac{\delta-k}{2}$.

Let $\Omega_3:=\{i\in\Lambda|\lambda_i(xay)\geq t\}$, $\Omega_4:=\{i\in\Lambda|\lambda_i(xay)+t+k>\delta\}$ and $\lambda_i(xay)<t$. Then $\Lambda=\Omega_3\cup\Omega_4$ and $\Omega_3\cap\Omega_4=\varnothing$. If $\Omega_2=\varnothing$, then $\lambda_i(xay)\geq t$ for all $i\in\Lambda$, and thus $\lambda(xay)\geq t$, a contradiction. Hence $\Omega_4\neq\varnothing$. For every $i\in\Omega_4$ we have $\lambda_i(xay)+t+k>\delta$ and $\lambda_i(xay)<t$. It follows that $t>\frac{\delta-k}{2}$. Since $(a;t)\in\lambda$, we have $\lambda_i(a)\geq\lambda(a)\geq t>\frac{\delta-k}{2}$ for all $i\in\Lambda$. Suppose that $r:=\lambda_i(xay)<\frac{\delta-k}{2}$ for some $i\in\Lambda$. Taking $r_0\in(0,\delta]$ such that $r<r_0<\frac{\delta-k}{2}$, we get $\lambda_i(a)>\frac{\delta-k}{2}>r_0$, that is, $(a;r_0)\in\lambda_i$. But $(xay;r_0)\overline{\in\vee q_k^{\delta}}\lambda_i$ since $\lambda_i(xay)=r<r_0$ and $\lambda_i(xay)+r_0+k<\delta$. This contradicts that λ_i is an $(\in,\in\vee q_k^{\delta})$-fuzzy interior ideal of S. Hence $\lambda_i(xay) \geq \frac{\delta - k}{2}$ for all $i \in \Lambda$, and therefore $\lambda(xay) \geq \frac{\delta - k}{2}$ which contradicts that $\lambda(xay) < \frac{\delta - k}{2}$. Thus $(xay;t) \in \vee q_k^\delta \lambda$ and $\bigcap_{i \in \Lambda} \lambda_i$ is an $(\in, \in \vee q_k^\delta)$-fuzzy interior ideal of S.

We have an open question as follows.

Let $\{\lambda_i \mid i \in \Lambda\}$ be a family of $(\in, \in \vee q_k^\delta)$-fuzzy interior ideals of S. Is $A := \bigcup_{i \in \Lambda} \lambda_i$ an $(\in, \in \vee q_k^\delta)$-fuzzy interior ideal of S? For this we have the following example.

Example 3.23 *Consider a semigroup* $S=\{1,2,3,4\}$ *with Table 2.*

Let λ and μ be two fuzzy subset in S, defined by λ(1)=0.5, λ(2)=0.8, λ(3)=λ(4)=0.2 and μ(1)=0.3, μ(2)=μ(4)-0.2, μ(3)=0.4. If we take k=0.2 and δ=0.7, then $\frac{\delta - k}{2} = 0.3$, Hence for $t \in (0,0.2]$, we have

$$U(\lambda,t) = \begin{cases} S \text{ if } t \in (0,0.19] \\ \{0,1\} \text{ if } t \in (0.19,0.2] \end{cases}$$

and

$$U(\mu,t) = \begin{cases} S \text{ if } t \in (0,0.19] \\ \{0,2\} \text{ if } t \in (0.19,0.2] \end{cases}$$

Thus λ and μ are $(\in, \in \vee q_{0.2}^{0.7})$-fuzzy interior ideals of S. By Theorem 3.9. The union $\lambda \cup \mu$ of λ and μ is given by

$$(\lambda \cup \mu)(1) = 0.5, \ (\lambda \cup \mu)(2) = 0.8, \ (\lambda \cup \mu)(3) = 0.4, \ \text{and} \ (\lambda \cup \mu)(4) = 0.2,$$

Hence

$$U(\lambda \cup \mu,t) = \begin{cases} S \text{ if } t \in (0,0.19] \\ \{1,2,3\} \text{ if } t \in (0.19,0.2] \end{cases}$$

Table 2.

.	1	2	3	4
1	1	1	1	1
2	1	1	4	1
3	1	1	1	1
4	1	1	1	1

Since {1,2,3} is not a subsemigroup of *S* and so is not an interior ideal of *S*. Therefore, by Theorem 3.9, $\lambda \cup \mu$ is not an $\left(\in,\in \vee q_k^\delta\right)$-fuzzy interior ideals of *S*.

Theorem 3.24 *Let f: S→S_0 be a homomorphism of semigroups and let λ and μ be* $\left(\in,\in \vee q_k^\delta\right)$*-fuzzy interior ideals of S and S_0, respectively. Then*

1. $f^{-1}(\mu)$ is an $\left(\in,\in \vee q_k^\delta\right)$-fuzzy interior ideal of S.
2. If λ satisfies the sup property, i.e., for any subset *A* of *S* there exists $x_0 \in A$ such that $\lambda(x)=\vee\{\lambda(x) | x\in A\}$; then $f(\lambda)$ is an $\left(\in,\in \vee q_k^\delta\right)$-fuzzy interior ideal of S_0 when *f* is onto.

Proof. (*i*): Let $x,y\in S$ and $t_1,t_2\in(0,\delta]$ be such that $(x;t_1)\in f^{-1}(\mu)$ and $(y,t_2)\in f^{-1}(\mu)$. Then $(f(x),t_1)\in\mu$ and $(f(x),t_2)\in\mu$. Since μ is an $\left(\in,\in \vee q_k^\delta\right)$-fuzzy interior ideal of S_0, it follows that

$$\left(f(xy),t_1\wedge t_2\right)=\left(f(x)f(y),t_1\wedge t_2\right)\in\vee q_k^\delta\mu$$

so that $\left(xy,t_1\wedge t_2\right)\in\vee q_k^\delta f^{-1}(\mu)$. Now let $a,x,y\in S$ and $t\in(0,\delta]$ be such that $(a,t)\in f^{-1}(\mu)$. Then $(f(a),t)\in\mu$, which implies that

$$\left(f(xay),t\right)=\left(f(x)f(a)f(y),t\right)\in\vee q_k^\delta\mu,$$

since μ is an $\left(\in,\in \vee q_k^\delta\right)$-fuzzy interior ideal of S_0. Hence $\left(xay,t\right)\in\vee q_k^\delta f^{-1}(\mu)$. Therefore $f^{-1}(\mu)$ is an $\left(\in,\in \vee q_k^\delta\right)$-fuzzy interior ideal of *S*.

(ii): Let $x,y\in S_0$ and $t_1,t_2\in(0,\delta]$ be such that $(x,t_1)\in f(\lambda)$ and $(y,t_2)\in f(\lambda)$. Then $(f(\lambda))(a)\geq t_1$ and $(f(\lambda))(b)\geq t_2$. Since λ has the sup property, there exists $a\in f^{-1}(x)$ and $b\in f^{-1}(y)$ such that $\lambda(a)=\vee\{\lambda(z) | z\in f^{-1}(x)\}$ and $\lambda(b)=\vee\{\lambda(w) | w\in f^{-1}(y)\}$. Then $(a,t_1)\in\lambda$ and $(y,t_2)\in\lambda$. Since λ is an $\left(\in,\in \vee q_k^\delta\right)$-fuzzy interior ideal of *S*, we have $\left(xy,t_1\wedge t_2\right)\in\vee q_k\mu$. Now $ab\in f^{-1}(xy)$ and so $(f(\lambda))(xy)\geq\lambda(ab)$. Thus $(f(\lambda))(xy)\geq t_1\wedge t_2$ or $f(\lambda)(xy)+t_1\wedge t_2+k>\delta$ which means that $\left(xy,t_1\wedge t_2\right)\in\vee q_k^\delta f(\lambda)$.

Finally let $a,p,q\in S_0$ and $t\in(0,\delta]$ be such that $(a,t)\in f(\lambda)$. Then $(f(\lambda))(a)\geq t$. Since λ has the sup property, $\lambda(z)=\vee\{\lambda(w) | w\in f^{-1}(a)\}$ for some $z\in f^{-1}(a)$. Then $(z,t)\in\lambda$ and hence $\left(xzy,t\right)\in\vee q_k^\delta f(\lambda)$ for every $x\in f^{-1}(p)$ and $y\in f^{-1}(q)$. Since $xzy\in f^{-1}(paq)$, we get $(f(\lambda))(paq)\geq\lambda(xzy)$. It follows that $(f(\lambda))(paq)\geq t$ or $(f(\lambda))(paq)+t+k>\delta$ so that $\left(paq,t\right)\in\vee q_k^\delta f(\lambda)$. Therefore, $f(\lambda)$ is an $\left(\in,\in \vee q_k^\delta\right)$-fuzzy interior ideal of S_0.

Theorem 3.25 *Let $A_1\subseteq A_2\subseteq\ldots\subseteq A_r=S$ be a chain of interior ideals of S. Then there exists an* $\left(\in,\in \vee q_k^\delta\right)$*-fuzzy interior ideal whose* $\left(\in \vee q_k^\delta\right)$*-level interior ideal are precisely the members of the chain.*

Proof. Let $\left\{t_i\in\left(0,\frac{\delta-k}{2}\right] | i=1,2,\ldots,r\right\}$ be such that $t_1>t_2>\ldots>t_r$. Define a fuzzy subset λ in *S* by

$$\lambda(x)=\begin{cases} t_1 & \text{if } x\in A_1, \\ t_2 & \text{if } x\in A_1 \setminus A_2, \\ \dots & \\ t_r & \text{if } x\in A_1 \setminus A_r, \end{cases}$$

Then,

$$U(\lambda;t)=\begin{cases} \varnothing & \text{if } t\in\left(t_1,\dfrac{\delta-k}{2}\right], \\ A_1 & \text{if } t\in(t_2,t_1], \\ \dots & \\ A_r(=S) & \text{if } t\in(0,t_r], \end{cases}$$

and so λ is an $(\in,\in\vee q_k^{\delta}$-fuzzy subsemigroups of *S*. Obviously, $\left(\in\vee q_k^{\delta}\right)$-level subsemigroups of λ are precisely the member of the chain.

Given a fuzzy subset λ in *S*, an $\left(\in,\in\vee q_k^{\delta}\right)$-fuzzy interior ideal μ of *S* is said to be the $\left(\in,\in\vee q_k^{\delta}\right)$-fuzzy interior ideals generated by λ in *S* if λ⊆μ and for any $\left(\in,\in\vee q_k^{\delta}\right)$-fuzzy interior ideal η of *S* with λ⊆η it must be μ⊆η.

Theorem 3.26 *Let λ be a fuzzy subset in S with finite image. Define interior ideals A_i of S as follows:*

$$A_0=\left\{x\in S\mid \lambda(x)\geq\frac{\delta-k}{2}\right\},$$

$$A_i=A_{i-1}\cup\left\{x\in S\mid \lambda(x)=\sup\left\{\lambda(y)\mid y\in S\setminus A_{i-1}\right\}\right\}$$

for *i*=1,2,…,*k* where *k*≤#*Im*(λ), and A_k=*S*. Let λ* be a fuzzy subset in *S* defined by

$$\lambda^*(x)=\begin{cases} \lambda(x) & \text{if } x\in A_0 \\ \sup\left\{\lambda(y)\mid y\in S\setminus A_{i-1}\right\} & \text{if } x\in A_i\setminus A_{i-1} \end{cases}$$

Then, λ* is the $\left(\in,\in\vee q_k^{\delta}\right)$-fuzzy interior ideal of *S* generated by λ in *S*.

Proof. The construction of λ* shows that λ⊆λ*. Note that the $A_i^{'}s$ form a chain

$A_0\subseteq A_1\subseteq \dots \subseteq A_n=S$

We first show that λ^* is an $\left(\in,\in\vee q_k^\delta\right)$-fuzzy interior ideal of S. Let $x,y\in S$. If $x,y\in A_0$, then $xy\in A_0$ and $xay\in A_0$ for $a\in S$. so,

$$\lambda^*(xy)=\lambda(xy)\geq\lambda(x)\wedge\lambda(y)\wedge\frac{\delta-k}{2}=\lambda^*(x)\wedge\lambda^*(y)\wedge\frac{\delta-k}{2}.$$

and

$$\lambda^*(xay)=\lambda(xay)\geq\lambda(a)\wedge\frac{\delta-k}{2}=\lambda^*(a)\wedge\frac{\delta-k}{2}.$$

If $x\in A_i\backslash A_{i-1}$ and $y\in A_i\backslash A_{i-1}$, then we may assume that $i<j$ without loss of generality. It follows that $xy\in A_j$ and $xay\in A_j$ for $a\in S$ so that,

$$\begin{aligned}\lambda^*(xy)&=\sup\left\{\lambda(y)\mid y\in S\setminus A_{i-1}\right\}\\&\geq\sup\left\{\lambda(x)\mid x\in S\setminus A_{i-1}\right\}\wedge\sup\left\{\lambda(y)\mid y\in S\setminus A_{j-1}\right\}\wedge\frac{\delta-k}{2}\\&=\lambda^*(x)\wedge\lambda^*(y)\wedge\frac{\delta-k}{2}.\end{aligned}$$

and

$$\begin{aligned}\lambda^*(xay)&=\sup\left\{\lambda(a)\mid a\in S\setminus A_{i-1}\right\}\\&\geq\sup\left\{\lambda(a)\mid a\in S\setminus A_{i-1}\right\}\wedge\sup\left\{\lambda(a)\mid a\in S\setminus A_{j-1}\right\}\wedge\frac{\delta-k}{2}\\&=\lambda^*(a)\wedge\frac{\delta-k}{2}.\end{aligned}$$

Hence, λ^* is an $\left(\in,\in\vee q_k^\delta\right)$-fuzzy interior ideal of S whose $\left(\in\vee q_k^\delta\right)$-level interior ideal are precisely the members of the chain $A_0\subseteq A_1\subseteq\ldots\subseteq A_n=S$ by Theorem 3.25. Now let μ be any $\left(\in,\in\vee q_k^\delta\right)$-fuzzy interior ideal of S which contain λ. If $x\in A_0$, then $\lambda^*(x)=\lambda(x)\leq\mu(x)$. Let $\left\{B_{t_i}\right\}$ be the class of $\left(\in\vee q_k^\delta\right)$-level interior ideal of μ in S. Let $x\in A_1\backslash A_0$. Then,

$$\lambda^*(x)=\sup\left\{\lambda(y)\mid y\in S\setminus A_0\right\}$$

and $A_1=N_1$ where,

$$N_1 = A_0 \cup \{x \in S \mid \lambda(x) = \sup\{\lambda(y) \mid y \in S \setminus A_0\}\}.$$

Let $x \in N_1 \setminus A_0$. Then,

$$\lambda(x) = \sup\{\lambda(y) \mid y \in S \setminus A_0\}$$
$$\leq \inf\{\mu(x) \mid x \in N_1 \setminus A_0\} \leq \mu(x)$$

since λ is contained in μ. Putting $t_{i1}=\inf\{\mu(x) \mid x \in N_1 \setminus A_0\}$ induces $x \in B_{t_{i1}}$ and so $N_1 \setminus A_0 \subseteq B_{t_{i1}}$. Since, $A_0 \subseteq B_{t_{i1}}$, we have $A_1 = N_1 \subseteq B_{t_{i1}}$, and so $\mu(x) \geq t_{i1}$ for all $x \in A_1$. Therefore, $\lambda^*(x) = \sup\{\lambda(y) \mid y \in S \setminus A_0\} \leq t_{i1} \leq \mu(x)$ for all $x \in A_1 \setminus A_0$. Similarly, we can prove that $\lambda^*(x) \leq \mu(x)$ for all $\in A_i \setminus A_{i-1}$ for $2 \leq i \leq k$. Therefore, λ^* is the $(\in, \in \vee q_k^\delta)$-fuzzy interior ideal of *S* generated by λ in *S*.

Definition 3.27 (Khan et al., 2017) *A fuzzy subset λ of S is called an* $(\in, \in \vee q_k^\delta)$*-fuzzy left (resp. right) ideal of S if for all t∈*(0,1] *and x,y∈S,*

$$(y,t) \in \lambda \Rightarrow (xy,t) \in \vee q_k^\delta \lambda \left(\text{resp.} (y,t) \in \lambda \Rightarrow (yx,t) \in \vee q_k^\delta \lambda\right)$$

Definition 3.28 (Khan et al., 2017) *A fuzzy subset λ of S is called an* $(\in, \in \vee q_k^\delta)$*-fuzzy ideal of S if it is both an* $(\in, \in \vee q_k^\delta)$*-fuzzy left ideal and an* $(\in, \in \vee q_k^\delta)$*-fuzzy right ideal in S.*

Theorem 3.29 *Every* $(\in, \in \vee q_k^\delta)$*-fuzzy ideal λ of S is an* $(\in, \in \vee q_k^\delta)$*-fuzzy interior ideal of S.*

Proof. Let λ be an $(\in, \in \vee q_k^\delta)$-fuzzy ideal of *S* and *x,y*∈*S*. Now,

$$\lambda(xy) \geq \lambda(y) \wedge \frac{\delta-k}{2} \geq \lambda(x) \wedge \lambda(y) \wedge \frac{\delta-k}{2}.$$

So, $\lambda(xy) \geq \lambda(x) \wedge \lambda(y) \wedge \frac{\delta-k}{2}$ Also, for all *x,a,y*∈*S* we have

$$\lambda(x(ay)) \geq \lambda(ay) \wedge \frac{\delta-k}{2} \geq \lambda(a) \wedge \frac{\delta-k}{2}$$

So,

$$\lambda(x(ay)) \geq \lambda(a) \wedge \frac{\delta-k}{2}.$$

Hence, λ is an $(\in, \in \vee q_k^\delta)$-fuzzy interior ideal of *S*.

Table 3.

.	a	b	c	d
a	b	b	b	b
b	b	b	b	b
c	b	c	c	c
d	b	d	c	c

Remark 3.30 The converse of the above Lemma is not true in general. That is an $\left(\in,\in\vee q_k^\delta\right)$-fuzzy interior ideal of S need not be an $\left(\in,\in\vee q_k^\delta\right)$-fuzzy ideal of S. For this we have the following example. Let S={a,b,c,d} and "•"a binary operation define on S in *Table 3.*

Then, $(S,\bullet)$ is a semigroup. If we take, k=0.2, δ=0.8 and define $\lambda: S\rightarrow 0,1]$ by $\lambda(a)=\lambda(b)=0.2$, $\lambda(c)=0.3$, $\lambda(d)=0.7$, then λ is an $\left(\in,\in\vee q_k^\delta\right)$-fuzzy interior ideal of *S*. But not an $\left(\in,\in\vee q_k^\delta\right)$-fuzzy ideal of *S*, because $\lambda(ac)=\lambda(b)=0.2\geq\lambda(a)\wedge 0.3$ and $\lambda(ac)=\lambda(b)=0.2\ngeq\lambda(c)\wedge 0.3$. Which shows that λ is not a $\left(\in,\in\vee q_k^\delta\right)$-fuzzy left ideal of *S* and hence not an $\left(\in,\in\vee q_{0.2}^{0.8}\right)$-fuzzy ideal of *S*.

4 CONCLUSION AND DISCUSSION

In this chapter the concept of an $\left(\in,\in\vee q_k^\delta\right)$-fuzzy interior ideal of a semigroup is introduced which is a generalization of $(\in,\in\vee q_k)$-fuzzy interior ideals and $(\in,\in\vee q)$-fuzzy interior ideals. If we take δ=1, then we get, an $(\in,\in\vee q_k)$-fuzzy interior ideal, an of a semigroup. If we take δ=1 and k=0, then we get, an $(\in,\in\vee q)$-fuzzy interior ideal of a semigroup. Which means that these fuzzy ideals become a special case of an $\left(\in,\in\vee q_k^\delta\right)$-fuzzy interior ideal of a semigroup. We presented example which shows that an $\left(\in,\in\vee q_k^\delta\right)$-fuzzy interior ideal of a semigroup is not an $(\in,\in\vee q_k)$-fuzzy interior ideal and not an $(\in,\in\vee q)$-fuzzy interior ideal. Further we discussed some properties of the $\left(\in,\in\vee q_k^\delta\right)$-fuzzy interior ideal of a semigroup. We also showed that every $\left(\in,\in\vee q_k^\delta\right)$-fuzzy ideal of a semigroup is an $\left(\in,\in\vee q_k^\delta\right)$-fuzzy interior ideal of a semigroup but the converse is not true. In furthur we can apply the present concept to other algebraic structures, i.e Ring, Hemiring, Nearring etc. We can define $\left(\in,\in\vee q_k^\delta\right)$-fuzzy soft interior ideals.

REFERENCES

Bhakat, S. K. (1999). -level subset. *Fuzzy Sets and Systems*, *103*(3), 529–533. (∈∨q) doi:10.1016/S0165-0114(97)00158-9

Bhakat, S. K. (2000). -fuzzy normal, quasinormal and maximal subgroups. *Fuzzy Sets and Systems*, *112*(2), 299–312. (∈,∈∨q) doi:10.1016/S0165-0114(98)00029-3

Bhakat, S. K., & Das, P. (1992). On the definition of a fuzzy subgroup. *Fuzzy Sets and Systems*, *51*(2), 235–241. doi:10.1016/0165-0114(92)90196-B

Bhakat, S. K., & Das, P. (1996). Fuzzy subrings and ideals redefined. *Fuzzy Sets and Systems*, *81*(3), 383–393. doi:10.1016/0165-0114(95)00202-2

Davvaz, B. (2006). -fuzzy subnearrings and ideals. *Soft Computing*, *10*(3), 206–211. (∈,∈∨q) doi:10.100700500-005-0472-1

Jun, Y. B. (2009). Generalizations of(∈,∈∨q). *Computers & Mathematics with Applications (Oxford, England)*, *58*(7), 1383–1390. doi:10.1016/j.camwa.2009.07.043

Jun, Y. B., & Song, S. Z. (2006). Generalized fuzzy interior ideals in semigroups. *Information Science*, *176*(20), 3079–3093. doi:10.1016/j.ins.2005.09.002

Kang, J. G. (2016). General types of(∈,∈qk). *Honam Mathematical J.*, *38*, 795–807. doi:10.5831/HMJ.2016.38.4.795

Kazanci, O., & Yamak, S. (2008). Generalized fuzzy bi-ideals of semigroup. *Soft Computing*, *12*(11), 1119–1124. doi:10.100700500-008-0280-5

Khan, F. M., Sarmin, N. H., Khan, A., & Khan, H. U. (2015). Some innovative types of fuzzy bi-ideals in ordered semigroups. *Journal of Advanced Mathematics and Applications*, *4*(1), 24–36. doi:10.1166/jama.2015.1070

Khan, M.S.A., Abdullah, S., Ali, A., Amin, F., & Rahman, K. (2017). On generalized (∈, $\in\vee q_k$)-fuzzy quasi ideals in ordered semigroups. *Turkish Journal of Fuzzy Systems, 8*, 33-51.

Khan, M. S. A., Abdullah, S., Jun, Y. B., & Rahman, K. (2017). More generalized fuzzy subsemigroups/ideals in semigroups. *Honam Mathematical J.*, *39*, 527–559.

Khan, M. S. A., Abdullah, S. K., Rahman, K., & Abbass, S. Z. (2017). Characterization of regular semigroups by (∈, $\in\vee(k^*q_k)$. *Annals of Fuzzy Mathematics and Informatics*, *13*(3), 403–419. doi:10.30948/afmi.2017.13.3.403

Khan, M. S. A., Rahman, K. A., Fahmi, A., & Shakeel, M. (2018). Generalized (∈, $\in\vee q_k$) –fuzzy quasi-ideals in semigroups. *Punjab University Journal of Mathematics*, *50*, 35–53.

Kuroki, N. (1979). Fuzzy bi-ideals in semigroups. *Comment. Math. Univ. St. Pauli*, *28*, 17–21.

Kuroki, N. (1981). On fuzzy ideals and fuzzy bi-ideals in semigroups. *Fuzzy Sets and Systems*, 203–215.

Mordeson, J. N., & Malik, D. S. (2002). *Fuzzy automata and languages: Theory and applications*. Chapman and Hall/CRC. doi:10.1016/0165-0114(81)90018-X

Mordeson, J. N., Malik, D. S., & Kuroki, N. N. (2003). Fuzzy semigroups. In Studies in Fuzziness and Soft Computing (p. 131). Springer-Verlag, Berlin.

Murali, V. (2004). Fuzzy points of equivalent fuzzy subsets. *Information Science*, *158*, 277–288. doi:10.1016/j.ins.2003.07.008

Pu, P. M., & Liu, Y. M. (1980). Fuzzy topology I, neighborhood structure of a fuzzy point and Moore-Smith convergence. *Journal of Mathematical Analysis and Applications*, *76*(2), 571–599. doi:10.1016/0022-247X(80)90048-7

Rosenfeld, A. (1971). Fuzzy groups. *Journal of Mathematical Analysis and Applications*, *35*(3), 512–517. doi:10.1016/0022-247X(71)90199-5

Shabir, M., Jun, Y. B., & Nawaz, Y. (2010). Characterizations of regular semigroups by (α,β). *Computers & Mathematics with Applications (Oxford, England)*, *59*(1), 161–175. doi:10.1016/j.camwa.2009.07.062

Shabir, M., Jun, Y. B., & Nawaz, Y. (2010). Semigroups characterized by(∈,∈∨q). *Computers & Mathematics with Applications (Oxford, England)*, *60*(5), 1473–1493. doi:10.1016/j.camwa.2010.06.030

Shabir, M., Jun, Y. B., & Nawaz, Y. (2010). Characterizations of regular semigroups by(∈,∈∨qk). *Computers & Mathematics with Applications (Oxford, England)*, *59*, 539–549. doi:10.1016/j.camwa.2009.06.014

Wang, Y. (2014). The theory of fuzzy arithmetic in the extended domain of fuzzy numbers. *Journal of Advanced Mathematics and Applications*, *3*(2), 165–175. doi:10.1166/jama.2014.1063

Wang, Y. (2015). Fuzzy probability algebra (FPA): A theory of fuzzy probability for fuzzy inference and computational intelligence. *Journal of Advanced Mathematics and Applications*, *4*(1), 37–54. doi:10.1166/jama.2015.1071

Zadeh, L. A. (1965). Fuzzy sets. *Information and Control*, *8*(3), 338–353. doi:10.1016/S0019-9958(65)90241-X

Chapter 11
Bipolar Fuzzy Structure of H-Ideals in BCI-Algebras

Tapan Senapati
https://orcid.org/0000-0003-0399-7486
Southwest University, China

Guiyun Chen
Southwest University, China

ABSTRACT

In this chapter, the concepts of bipolar fuzzy H-ideals of BCI-algebras are introduced and their natures are investigated. Relations between bipolar fuzzy subalgebras, bipolar fuzzy ideals, and bipolar fuzzy H-ideals are discussed. Conditions for a bipolar fuzzy ideal to be a bipolar fuzzy H-ideal are provided. Some characterization theorems of bipolar fuzzy H-ideals are established. A bipolar fuzzy H-ideal is established by using a finite collection of H-ideals. The authors have shown that if every bipolar fuzzy H-ideal has the finite image, then every descending chain of H-ideals terminates at finite step.

1. INTRODUCTION

Zhang (1994) initiated the concept of bipolar fuzzy sets (BFSs) as a generalization of fuzzy sets (Zadeh, 1965). In fuzzy sets membership degree range is [0,1]. In a BFS membership degree range is increased from the interval [0,1] to the interval [-1,1]. The membership degree 0 means that elements are irrelevant to the corresponding property, the membership degrees on (0,1] indicate that elements somewhat satisfy the property and the membership degrees on [-1,0] indicate that elements somewhat satisfy the implicit counter-property. This domain has recently motivated new research in several directions (Akram, 2011, Dubois, 2008, Zadrozny & Kacprzyk, 2012, Zhang & Zhang, 2004).

BCK-algebras and *BCI*-algebras are two important classes of logical algebras introduced by Imai & Iseki (1966). It is known that the class of *BCK* -algebra is a proper subclass of the class of *BCI* -algebras. Xi (1991) applied the concept of fuzzy sets to BCK-algebras. Jun (1993) applied it to *BCI* -algebras. Khalid and Ahmad (1999) introduced fuzzy *H*-ideals in *BCI* -algebras. Senapati et al. (2013, 2015)

DOI: 10.4018/978-1-7998-0190-0.ch011

introduced (intuitionistic) fuzzy translations of (intuitionistic) fuzzy *H*-ideals in *BCK/BCI*-algebras and investigated their properties in details. The author (together with colleagues) have done lot of works on *B/BG/G*-algebras (Bhowmik et al, 2014, Senapati et al. 2012, 2014, 2015a, 2015b, 2015c, 2016) which is related to *BCK/BCI* -algebras.

Lee (2009) introduced the notion of bipolar fuzzy subalgebras and ideals in *BCK/BCI* -algebras. The concept of bipolar valued fuzzy translation and bipolar valued fuzzy *S*-extension of a bipolar valued fuzzy subalgebra in *BCK/BCI* -algebra was introduced by Jun et al.(2009). Lee et al. (2011) also extended this study to *a*-ideals of *BCI* -algebras.

Motivated by this, in this chapter, the notions of bipolar fuzzy *H*-ideals (BFHIs) of *BCI*-algebras is introduced and their properties are investigated. Relations among bipolar fuzzy subalgebras, bipolar fuzzy ideals and BFHIs are discussed. Conditions for a bipolar fuzzy ideal to be a BFHI are provided. Some characterization theorems of BFHIs are established. A BFHI is established by using a finite collection of *H*-ideals.

2. PRELIMINARIES

In this section, some elementary aspects that are necessary for this paper are included.

By a *BCI*-algebra we mean an algebra $(X,*,0)$ of type (2,0) satisfying the following axioms for all $x,y,z,\in X$:

1. $((x*y)*(x*z))*(z*y)=0$
2. $(x*(x*y))*y=0$
3. $x*x=0$
4. $x*y=0$ and $y*x=0$ imply $x=y$.

If a *BCI*-algebra X satisfies $0*x=0$ for all $x\in X$, then we say that X is a *BCK*-algebra.

We can define a partial ordering "$\leq$" by $x\leq y$ if and only if $x*y=0$.

A non-empty subset S of X is called a subalgebra (Jun, 1993) of X if $x*y\in S$ for any $x,y\in S$.

A non-empty subset I of X is called an ideal (Jun, 1993) of X if it satisfies (I_1) $0\in I$ and (I_2) $x*y\in I$ and $y\in I$ imply $x\in I$.

A non-empty subset I of X is said to be a *H*-ideal (Khalid & Ahmed, 1999) of X if it satisfies (I_1) and (I_3) $x*(y*z)\in I$ and $y\in I$ imply $x*x\in I$ for all $x,y,z\in X$.

A *BCI*-algebra is said to be associative (Hu & Iseki, 1980) if $(x*y)*z=x*(y*z)$ for all $x,y,z\in X$.

Definition 2.1 (Zadeh, 1965) Let X be the collection of objects denoted generally by x, then a fuzzy set A in X is defined as $A=\{\langle x,\lambda_A(x)\rangle : x\in X\}$, where $\lambda_A(x)$ is called the membership value of x in A and $0\leq\lambda_A(x)\leq 1$.

Definition 2.2 (Jun, 1993) A fuzzy set $A=\{\langle x,\lambda_A(x)\rangle : x\in X\}$ in X is called a fuzzy subalgebra of X if it satisfies the inequality $\lambda_A(x*y)\geq\min\{\lambda_A(x), \lambda_A(y)\}$ for all $x,y\in X$.

Definition 2.3 (Jun, 1993) A fuzzy set $A=\{\langle x,\lambda_A(x)\rangle : x\in X\}$ in X is called a fuzzy ideal of X if it satisfies (T_1) $\lambda_A(0)\geq\lambda_A(x)$ and (T_2) $\lambda_A(x)\geq\min\{\lambda_A(x*y),\lambda_A(y)\}$ for all $x,y\in X$.

Definition 2.4 (Khalid and Ahmad, 1999) A fuzzy set $A=\{\langle x,\lambda_A(x)\rangle : x\in X\}$ in X is called a fuzzy H-ideal of X if it satisfies (T_1) and (T_3) $\lambda_A(x*z)\geq\min\{\lambda_A(x(y*z)),\lambda_A(y)\}$ for all $x,y,z\in X$.

Definition 2.5 (Zhang & Zhang, 2004) Let X be a nonempty set. A BFS Ψ in X is an object having the form $\Psi=\{\langle x,\lambda_\psi^+(x),\lambda_\psi^-(x)\rangle : x\in X\}$ where $\lambda_\psi^+ : X\to[0,1]$ and $\lambda_\psi^- : X\to[-1,0]$ are mappings.

We use the positive membership degree $\lambda_\psi^+(x)$ to denote the satisfaction degree of an element x to the property corresponding to a BFS Ψ, and the negative membership degree $\lambda_\psi^-(x)$ to denote the satisfaction degree of an element x to some implicit counter-property corresponding to a BFS Ψ. If

$$\lambda_\psi^+(x)\neq 0 \text{ and } \lambda_\psi^-(x)=0,$$

it is the situation that x is regarded as having only positive satisfaction for Ψ. If

$$\lambda_\psi^+(x)=0 \text{ and } \lambda_\psi^-(x)\neq 0,$$

it is the situation that x does not satisfy the property of Ψ but somewhat satisfies the counter property of Ψ. It is possible for an element x to be such that

$$\lambda_\psi^+(x)\neq 0 \text{ and } \lambda_\psi^-(x)\neq 0$$

when the membership function of the property overlaps that of its counter property over some portion of X.

3. MAIN RESULTS

For the sake of simplicity, we shall use the symbol $\Psi=(\lambda_\psi^+,\lambda_\psi^-)$ for the BFS

$$\Psi=\{\langle x,\lambda_\psi^+(x),\lambda_\psi^-(x)\rangle : x\in X\}.$$

Throughout this paper, X always means a *BCI*-algebra without any specification.

Definition 3.1 (Lee, 2009) A BFS Ψ in X is called a bipolar fuzzy subalgebra of X if it satisfies:

(F1) $\lambda_\psi^+(x*y)\geq\min\{\lambda_\psi^+(x),\lambda_\psi^+(y)\}$
(F2) $\lambda_\psi^-(x*y)\leq\max\{\lambda_\psi^-(x),\lambda_\psi^-(y)\}$

for all $x,y \in X$.

Proposition 3.2 (Lee, 2009) If Ψ in X is a bipolar fuzzy subalgebra of X, then

$$\lambda_\psi^+(0) \geq \lambda_\psi^+(x) \text{ and } \lambda_\psi^-(0) \leq \lambda_\psi^-(x)$$

for all $x \in X$.

Definition 3.3 (Lee, 2009) A BFS Ψ in X is called a bipolar fuzzy ideal (BFI) of X if it satisfies:

(F3) $\lambda_\psi^+(0) \geq \lambda_\psi^+(x)$ and $\lambda_\psi^-(0) \leq \lambda_\psi^-(x)$
(F4) $\lambda_\psi^+(x) \geq \min\{\lambda_\psi^+(x*y), \lambda_\psi^+(y)\}$
(F5) $\lambda_\psi^-(x) \leq \max\{\lambda_\psi^-(x*y), \lambda_\psi^-(y)\}$

for all $x,y \in X$.

Definition 3.4 *A BFS Ψ in X is called a BFHI of X if it satisfies* (F3) *and*

(F6) $\lambda_\psi^+(x*z) \geq \min\{\lambda_\psi^+(x*(y*z)), \lambda_\psi^+(y)\}$
(F7) $\lambda_\psi^-(x*z) \leq \max\{\lambda_\psi^-(x*(y*z)), \lambda_\psi^-(y)\}$

for all $x,y,z \in X$.

Example 3.5 *Let $X=\{0,1,2,3,4\}$ be a BCK-algebra with the Cayley table (Table 1).*

Let $\Psi = (\lambda_\psi^+, \lambda_\psi^-)$ be a BFS of X defined by Table 2.

Then $\Psi = (\lambda_\psi^+, \lambda_\psi^-)$ is a BFHI of X.

Example 3.6 *Let $X=\{0,1,2,3,4,5\}$ be a BCI-algebra with the Cayley table (Table 3).*

Let $\Psi = (\lambda_\psi^+, \lambda_\psi^-)$ be a BFS of X defined by Table 4.
Then $\Psi = (\lambda_\psi^+, \lambda_\psi^-)$ is a BFHI of X.

Table 1.

*	0	1	2	3	4
0	0	0	0	0	0
1	1	0	1	0	0
2	2	2	0	0	0
3	3	3	3	0	0
4	4	3	4	1	0

Table 2.

X	0	1	2	3	4
λ_ψ^+	0.7	0.6	0.5	0.4	0.4
λ_ψ^-	-0.6	-0.5	-0.5	-0.3	-0.3

Table 3.

*	0	1	2	3	4	5
0	0	0	0	3	3	3
1	1	0	1	3	3	3
2	2	2	0	3	3	3
3	3	3	3	0	0	0
4	4	3	4	1	0	0
5	5	3	5	1	1	0

Table 4.

X	0	1	2	3	4	5
λ_ψ^+	0.8	0.7	0.6	0.5	0.5	0.5
λ_ψ^-	-0.7	-0.6	-0.5	-0.4	-0.4	-0.4

Table 5.

*	0	1	2	3	4
0	0	0	4	3	2
1	1	0	4	3	2
2	2	2	0	4	3
3	3	3	2	0	4
4	4	4	3	2	0

Table 6.

X	0	1	2	3	4
λ_ψ^+	0.8	0.6	0.5	0.5	0.5
λ_ψ^-	-0.7	-0.7	-0.4	-0.5	-0.4

Note that every BFHI of a *BCI*-algebra *X* is a fuzzy ideal of *X* by taking *z*=0 in Definition 3.4. But, the converse is not true as seen in the following example.

Example 3.7 *Let X*={0,1,2,3,4} *be a BCI-algebra with the Cayley table (Table 5).*

Let $\Psi = \left(\lambda_\psi^+, \lambda_\psi^-\right)$ be a BFS of *X* defined by Table 6.

Then $\Psi = \left(\lambda_\psi^+, \lambda_\psi^-\right)$ is a BFI of *X*. But, $\Psi = \left(\lambda_\psi^+, \lambda_\psi^-\right)$ is not a BFHI of *X* since

$$\lambda_\psi^+(4*2) = 0.5 \Leftrightarrow 0.8 = \min\left\{\lambda_\psi^+(4*(0*2)), \lambda_\psi^+(0)\right\}$$

and

$$\lambda_\psi^-(4*2) = -0.5 \Leftrightarrow -0.7 = \max\left\{\lambda_\psi^-(4*(0*2)), \lambda_\psi^-(0)\right\}.$$

We now discuss the relation between bipolar fuzzy subalgebras and BFHIs.

Theorem 3.8 *Every BFHI of X is a bipolar fuzzy subalgebra of X.*

Proof: Let $\Psi = \left(\lambda_\psi^+, \lambda_\psi^-\right)$ be a BFHI of *X*. Then we have

$$\lambda_\psi^+(x*z) \geq \min\left\{\lambda_\psi^+(x*(y*z)), \lambda_\psi^+(y)\right\} \text{ and } \lambda_\psi^-(x*z) \leq \max\left\{\lambda_\psi^-(x*(y*z)), \lambda_\psi^-(y)\right\}$$

for all $x,y,z \in X$. Substituting y for z we get

$$\lambda_\psi^+(x*y) \geq \min\{\lambda_\psi^+(x*(y*y)), \lambda_\psi^+(y)\} = \min\{\lambda_\psi^+(x*0), \lambda_\psi^+(y)\} = \min\{\lambda_\psi^+(x), \lambda_\psi^+(y)\}$$

and

$$\lambda_\psi^-(x*y) \leq \max\{\lambda_\psi^-(x*(y*y)), \lambda_\psi^-(y)\} = \max\{\lambda_\psi^-(x*0), \lambda_\psi^-(y)\} = \max\{\lambda_\psi^-(x), \lambda_\psi^-(y)\}.$$

Therefore, Ψ is a bipolar fuzzy subalgebra of X.

The converse of the Theorem 3.8 Is not true in general as seen in the following example.

Example 3.9 *Let X={0,1,2,3,4} be a BCI-algebra with the Cayley table (Table 7).*

Let $\Psi = (\lambda_\psi^+, \lambda_\psi^-)$ be a BFS of X defined by Table 8.

Then $\Psi = (\lambda_\psi^+, \lambda_\psi^-)$ is a bipolar fuzzy subalgebra of X. But, $\Psi = (\lambda_\psi^+, \lambda_\psi^-)$ is not a BFHI of X since

$$\lambda_\psi^+(4*2) = \lambda_\psi^+(4) = 0.4 \Leftrightarrow 0.6 = \min\{\lambda_\psi^+(4*(1*2)), \lambda_\psi^+(1)\}$$

and

$$\lambda_\psi^-(4*2) = \lambda_\psi^-(4) = -0.3 \Leftrightarrow -0.5 = \max\{\lambda_\psi^-(4*(1*2)), \lambda_\psi^-(1)\}.$$

Proposition 3.10 *Let* $\Psi = (\lambda_\psi^+, \lambda_\psi^-)$ *be a BFHI of X. If (x*a)*b=0 for all* $a,b,x \in X$, *then*

$$\lambda_\psi^+(x) \geq \min\{\lambda_\psi^+(a), \lambda_\psi^+(b)\} \text{ and } \lambda_\psi^-(x) \leq \max\{\lambda_\psi^-(a), \lambda_\psi^-(b)\}.$$

Proof: Let $a,b,x \in X$ be such that $(x*a)*b=0$. Then

Table 7.

*	0	1	2	3	4
0	0	0	0	0	0
1	1	0	1	0	0
2	2	2	0	0	2
3	3	2	1	0	2
4	4	1	4	1	0

Table 8.

X	0	1	2	3	4
λ_ψ^+	0.7	0.6	0.6	0.4	0.4
λ_ψ^-	-0.6	-0.5	-0.5	-0.3	-0.3

$$\lambda_\psi^+(x) \geq \min\{\lambda_\psi^+(x*a), \lambda_\psi^+(a)\} \geq \min\{\min\{\lambda_\psi^+((x*a)*b), \lambda_\psi^+(b)\}, \lambda_\psi^+(a)\}$$
$$= \min\{\lambda_\psi^+(a), \lambda_\psi^+(b)\} = \min\{\min\{\lambda_\psi^+(0), \lambda_\psi^+(b)\}, \lambda_\psi^+(a)\} = \min\{\lambda_\psi^+(b), \lambda_\psi^+(a)\}$$

and

$$\lambda_\psi^-(x) \leq \max\{\lambda_\psi^-(x*a), \lambda_\psi^-(a)\} \leq \max\{\max\{\lambda_\psi^-((x*a)*b), \lambda_\psi^-(b)\}, \lambda_\psi^-(a)\}$$
$$= \max\{\max\{\lambda_\psi^-(0), \lambda_\psi^-(b)\}, \lambda_\psi^-(a)\} = \max\{\lambda_\psi^-(b), \lambda_\psi^-(a)\} = \max\{\lambda_\psi^-(a), \lambda_\psi^-(b)\}.$$

The proof is complete.

The following can easily be proved by induction.

Corollary 3.11 *Let* $\Psi = (\lambda_\psi^+, \lambda_\psi^-)$ *be a BFHI of X. If* $(\ldots((x*a_1)*a_2)*\ldots)*a_n=0$ *for all* $x, a_1, a_2, \ldots, a_n \in X$, *then*

$$\lambda_\psi^+(x) \geq \min\{\lambda_\psi^+(a_1), \lambda_\psi^+(a_2), \ldots, \lambda_\psi^+(a_n)\} \text{ and } \lambda_\psi^-(x) \leq \max\{\lambda_\psi^-(a_1), \lambda_\psi^-(a_2), \ldots, \lambda_\psi^-(a_n)\}.$$

We provide a condition for a BFI to be a BFHI.

Theorem 3.12 *Let* $\Psi = (\lambda_\psi^+, \lambda_\psi^-)$ *be a BFI of X. If it satisfies the condition*

$$\lambda_\psi^+(x*y) \geq \lambda_\psi^+(x) \text{ and } \lambda_\psi^-(x*y) \leq \lambda_\psi^-(x)$$

for all $x, y \in X$, *then* Ψ *is a BFHI of X.*

Proof: Let $\Psi = (\lambda_\psi^+, \lambda_\psi^-)$ be a BFI of X. For any $x, y, z \in X$, we have

$$\lambda_\psi^+(x*z) \geq \min\{\lambda_\psi^+((x*z)*(y*z)), \lambda_\psi^+(y*z)\}$$
$$= \min\{\lambda_\psi^+((x*(y*z))*z), \lambda_\psi^+(y*z)\}$$
$$\geq \min\{\lambda_\psi^+(x*(y*z)), \lambda_\psi^+(y)\}$$

and

$$\begin{aligned}\lambda_\psi^-(x*z) &\le \max\{\lambda_\psi^-((x*z)*(y*z)), \lambda_\psi^-(y*z)\}\\ &= \max\{\lambda_\psi^-((x*(y*z))*z), \lambda_\psi^-(y*z)\}\\ &\le \max\{\lambda_\psi^-(x*(y*z)), \lambda_\psi^-(y)\}.\end{aligned}$$

Hence, $\Psi = (\lambda_\psi^+, \lambda_\psi^-)$ is a BFHI of X.

Theorem 3.13 *If* $\Psi = (\lambda_\psi^+, \lambda_\psi^-)$ *be a BFS of associative BCI-algebra X such that* Ψ *is a BFI of X, then* Ψ *is a BFHI of X.*

Proof: Let Ψ is a BFI of associative *BCI*-algebra *X*. For any $x,y,z \in X$, we have

$$\begin{aligned}\lambda_\psi^+(x*z) &\ge \min\{\lambda_\psi^+((x*z)*y), \lambda_\psi^+(y)\}\\ &= \min\{\lambda_\psi^+((x*y)*z), \lambda_\psi^+(y)\}\\ &= \min\{\lambda_\psi^+(x*(y*z)), \lambda_\psi^+(y)\}\end{aligned}$$

and

$$\begin{aligned}\lambda_\psi^-(x*z) &\le \max\{\lambda_{\text{Ë}}^-((x*z)*y), \lambda_\psi^-(y)\}\\ &= \max\{\lambda_\psi^-((x*y)*z), \lambda_\psi^-(y)\}\\ &= \max\{\lambda_\psi^-(x*(y*z)), \lambda_\psi^-(y)\}.\end{aligned}$$

Hence, Ψ is a BFHI of X.

Theorem 3.14 *If* $\Psi = (\lambda_\psi^+, \lambda_\psi^-)$ *is a BFHI of X, then the sets*

$T_{\lambda_A}^P = \{x \in X \mid \lambda_\psi^+(x) = \lambda_\psi^+(0)\}$ and $T_{\nu_A}^N = \{x \in X \mid \lambda_\psi^-(x) = \lambda_\psi^-(0)\}$

are H-ideals of X.

Proof: Let $\Psi = (\lambda_\psi^+, \lambda_\psi^-)$ is a BFHI of X. Obviously, $0 \in T_{\lambda_A}^P, T_{\nu_A}^N$. Let $x,y,z \in X$ be such that $x*(y*z) \in T_{\lambda_A}^P$ and $y \in T_{\lambda_A}^P$. Then

$$\lambda_\psi^+(x*(y*z)) = \lambda_\psi^+(0) = \lambda_\psi^+(y)$$

and so

$$\lambda_{\psi}^{+}(x*z) \geq \min\{\lambda_{\psi}^{+}(x*(y*z)), \lambda_{\psi}^{+}(y)\} = \lambda_{\psi}^{+}(0).$$

Again, let *a,b,c*∈*X* be such that $a*(b*c) \in T_{\lambda_A}^{N}$ and $b \in T_{\lambda_A}^{N}$. Then

$$\lambda_{\psi}^{-}(a*(b*c)) = \lambda_{\psi}^{-}(0) = \lambda_{\psi}^{-}(b)$$

and so

$$\lambda_{\psi}^{-}(a*c) \leq \max\{\lambda_{\psi}^{-}(a*(b*c)), \lambda_{\psi}^{-}(b)\} = \lambda_{\psi}^{-}(0).$$

Since $\Psi = (\lambda_{\psi}^{+}, \lambda_{\psi}^{-})$ is a BFHI of *X*, we conclude that

$$\lambda_{\text{Ë}}^{+}(x*z) = \lambda_{\psi}^{+}(0) \text{ and } \lambda_{\psi}^{-}(a*c) = \lambda_{\psi}^{-}(0)$$

i.e., $x*z \in T_{\lambda_A}^{P}$ and $a*c \in T_{\lambda_A}^{N}$. Hence, $T_{\lambda_A}^{P}$ and $T_{v_A}^{N}$ are *H*-ideals of *X*.

Proposition 3.15 *Let* $\Psi = (\lambda_{\psi}^{+}, \lambda_{\psi}^{-})$ *be a BFHI of X. If x≤y then*

$$\lambda_{\psi}^{+}(x) \geq \lambda_{\psi}^{+}(y) \text{ and } \lambda_{\psi}^{-}(x) \leq \lambda_{\psi}^{-}(y),$$

that is, λ_{ψ}^{+} *is order reversing and* λ_{ψ}^{-} *is order prereversing.*

Proof: Let *x,y*∈*X* be such that *x*≤*y*. Then *x***y*=0 and hence

$$\begin{aligned}\lambda_{\psi}^{+}(x) &= \lambda_{\psi}^{+}(x*0)\\ &\geq \min\{\lambda_{\psi}^{+}(x*(y*0)), \lambda_{\psi}^{+}(y)\}\\ &= \min\{\lambda_{\psi}^{+}(x*y), \lambda_{\psi}^{+}(y)\}\\ &= \min\{\lambda_{\psi}^{+}(0), \lambda_{\psi}^{+}(y)\} = \lambda_{\psi}^{+}(y)\end{aligned}$$

and

$$\begin{aligned}\lambda_{\psi}^{-}(x) &= \lambda_{\psi}^{-}(x*0)\\ &\leq \max\{\lambda_{\psi}^{-}(x*(y*0)), \lambda_{\psi}^{-}(y)\}\\ &= \max\{\lambda_{\psi}^{-}(x*y), \lambda_{\psi}^{-}(y)\}\\ &= \max\{\lambda_{\psi}^{-}(0), \lambda_{\psi}^{-}(y)\} = \lambda_{\psi}^{-}(y).\end{aligned}$$

The characterizations of BFI $\Psi = \left(\lambda_\psi^+, \lambda_\psi^-\right)$ is given by the following theorem.

Theorem 3.16 *Let* $\Psi = \left(\lambda_\psi^+, \lambda_\psi^-\right)$ *be a BFI of X. Then the following assertions are equivalent:*

(i) $\Psi = \left(\lambda_\psi^+, \lambda_\psi^-\right)$ is a BFHI of *X*,
(ii) $\lambda_\psi^+(x*y) \geq \lambda_\psi^+(x*(0*y))$ and $\lambda_\psi^-(x*y) \leq \lambda_\psi^-(x*(0*y))$ for all $x,y \in X$,
(iii) $\lambda_\psi^+((x*y)*z) \geq \lambda_\psi^+(x*(y*z))$ and $\lambda_\psi^-((x*y)*z) \leq \lambda_\psi^-(x*(y*z))$ for all $x,y,z \in X$.

Proof: Assume that $\Psi = \left(\lambda_\psi^+, \lambda_\psi^-\right)$ is a bipolar fuzzy ideal of *X*. (i)⇒(ii) Let $\Psi = \left(\lambda_\psi^+, \lambda_\psi^-\right)$ is a BFHI of *X*. Then for all $x,y \in X$ we have

$$\lambda_\psi^+(x*y) \geq \min\left\{\lambda_\psi^+(x*(0*y)), \lambda_\psi^+(0)\right\} = \lambda_\psi^+(x*(0*y))$$

and

$$\lambda_\psi^-(x*y) \leq \max\left\{\lambda_\psi^-(x*(0*\mathrm{y})), \lambda_\psi^-(0)\right\} = \lambda_\psi^-(x*(0*y)).$$

Therefore, the inequality (ii) is satisfied. (ii)⇒(iii) Assume that (ii) is satisfied. For all $x,y,z \in X$, we have

$$\begin{aligned}
&((x*y)*(0*z))*(x*(y*z)) \\
&=((x*y)*(x*(y*z)))*(0*z) \leq ((y*z)*y)*(0*z). \\
&=((y*y)*z)*(0*z) = (0*z)*(0*z) = 0
\end{aligned}$$

It follows from Proposition 3.15 that

$$\lambda_\psi^+((x*y)*(0*z))*(x*(y*z)) \geq \lambda_\psi^+(0) \text{ and } \lambda_\psi^-((x*y)*(0*z))*(x*(y*z)) \leq \lambda_\psi^-(0).$$

Since $\Psi = \left(\lambda_\psi^+, \lambda_\psi^-\right)$ is BFHI of *X*, therefore, we have

$$\lambda_\psi^+((x*y)*(0*z))*(x*(y*z)) = \lambda_\psi^+(0) \text{ and } \lambda_\psi^-((x*y)*(0*z))*(x*(y*z)) = \lambda_\psi^-(0).$$

Using (ii) we get

$$\begin{aligned}\lambda_\psi^+((x*y)*z) &\geq \lambda_\psi^+((x*y)*(0*z)) \\ &= \min\{\lambda_\psi^+(((x*y)*(0*z))*(x*(y*z))), \lambda_\psi^+(x*(y*z))\} \\ &= \min\{\lambda_\psi^+(0), \lambda_\psi^+(x*(y*z))\} = \lambda_\psi^+(x*(y*z))\end{aligned}$$

and

$$\begin{aligned}\lambda_\psi^-((x*y)*z) &\leq \lambda_\psi^-((x*y)*(0*z)) \\ &= \max\{\lambda_\psi^-(((x*y)*(0*z))*(x*(y*z))), \lambda_\psi^-(x*(y*z))\} \\ &= \max\{\lambda_\psi^-(0), \lambda_\psi^-(x*(y*z))\} = \lambda_\psi^-(x*(y*z)).\end{aligned}$$

Therefore, inequality (iii) is also satisfied. (iii)⇒(i) Assume that (iii) is valid. For all $x,y,z \in X$, we have

$$\begin{aligned}\lambda_\psi^+(x*z) &\geq \min\{\lambda_\psi^+((x*z)*y), \lambda_\psi^+(y)\} \\ &= \min\{\lambda_\psi^+((x*y)*z), \lambda_\psi^+(y)\} \\ &\geq \min\{\lambda_\psi^+(x*(y*z)), \lambda_\psi^+(y)\}\end{aligned}$$

and

$$\begin{aligned}\lambda_\psi^-(x*z) &\leq \max\{\lambda_\psi^-((x*z)*y), \lambda_\psi^-(y)\} \\ &= \max\{\lambda_\psi^-((x*y)*z), \lambda_\psi^-(y)\} \\ &\leq \max\{\lambda_\psi^-(x*(y*z)), \lambda_\psi^-(y)\}.\end{aligned}$$

Therefore, $\Psi = (\lambda_\psi^+, \lambda_\psi^-)$ is a BFHI of X. Hence, the assertion (i) holds.

Next we give another characterizations of BFI $\Psi = (\lambda_\psi^+, \lambda_\psi^-)$ in the following theorem.

Theorem 3.17 *Let $\Psi = (\lambda_\psi^+, \lambda_\psi^-)$ be a BFI of X. Then the following assertions are equivalent:*

(i) $\Psi = (\lambda_\psi^+, \lambda_\psi^-)$ is a BFHI of X,

(ii) $\lambda_\psi^+((x*z)*y) \geq \lambda_\psi^+((x*z)*(0*y))$ and $\lambda_\psi^-((x*z)*y) \leq \lambda_\psi^-((x*z)*(0*y))$ for all $x,y,z \in X$,

(iii) $\lambda_\psi^+(x*y) \geq \min\{\lambda_\psi^+((x*z)*(0*y)), \lambda_\psi^+(z)\}$ and $\lambda_\psi^-(x*y) \leq \max\{\lambda_\psi^-((x*z)*(0*y)), \lambda_\psi^-(z)\}$ for all $x,y,z \in X$.

Proof: (i)⇒(ii) Same as above theorem. (ii)⇒(iii) Assume that (ii) is valid. For all $x,y,z \in X$, we have

$$\begin{aligned} \lambda_\psi^+(x*y) &\geq \min\{\lambda_\psi^+((x*y)*z), \lambda_\psi^+(z)\} \\ &= \min\{\lambda_\psi^+((x*z)*y), \lambda_\psi^+(z)\} \\ &\geq \min\{\lambda_\psi^+((x*z)*(0*y)), \lambda_\psi^+(z)\} \end{aligned}$$

and

$$\begin{aligned} \lambda_\psi^-(x*y) &\leq \max\{\lambda_\psi^-((x*y)*z), \lambda_\psi^-(z)\} \\ &= \max\{\lambda_\psi^-((x*z)*y), \lambda_\psi^-(z)\} \\ &\leq \max\{\lambda_\psi^-((x*z)*(0*y)), \lambda_\psi^-(z)\}. \end{aligned}$$

Therefore, (iii) is satisfied. (iii)⇒(i) Assume that (iii) is valid. Therefore, for all $x,y,z \in X$, we have

$$\lambda_\psi^+(x*y) \geq \min\{\lambda_\psi^+((x*z)*(0*y)), \lambda_\psi^+(z)\} \text{ and } \lambda_\psi^-(x*y) \leq \max\{\lambda_\psi^-((x*z)*(0*y)), \lambda_\psi^-(z)\}.$$

Putting $z=0$ we get

$$\lambda_\psi^+(x*y) \geq \min\{\lambda_\psi^+((x*0)*(0*y)), \lambda_\psi^+(0)\} = \min\{\lambda_\psi^+(x*(0*y)), \lambda_\psi^+(0)\} = \lambda_\psi^+(x*(0*y))$$

and

$$\lambda_\psi^-(x*y) \leq \max\{\lambda_\psi^-((x*0)*(0*y)), \lambda_\psi^-(0)\} = \max\{\lambda_\psi^-(x*(0*y)), \lambda_\psi^-(0)\} = \lambda_\psi^-(x*(0*y)).$$

It follows from Theorem 3.16 that $\Psi = (\lambda_\psi^+, \lambda_\psi^-)$ is a BFHI of X.

For a BFS $\Psi = (\lambda_\psi^+, \lambda_\psi^-)$ and $(s,t) \in [-1,0] \times [0,1]$, we define

$$\psi_t^P = \{x \in X \mid \lambda_\psi^+(x) \geq t\} \text{ and } \psi_s^N = \{x \in X \mid \lambda_\psi^-(x) \leq s\}$$

which are called the positive t-cut of $\Psi = (\lambda_\psi^+, \lambda_\psi^-)$ and the negative s-cut of $\Psi = (\lambda_\psi^+, \lambda_\psi^-)$, respectively. The set $C(\psi;(s,t)) = \psi_t^P \cap \psi_s^N$ is called the (s,t)-cut of $\Psi = (\lambda_\psi^+, \lambda_\psi^-)$.

Theorem 3.18 *Let* $\psi = \left(\lambda_\psi^+, \lambda_\psi^-\right)$ *be a BFS in X and* $(s,t) \in [-1,0] \times [0,1]$. *Then* $\psi = \left(\lambda_\psi^+, \lambda_\psi^-\right)$ *is a BFHI of X if and only if the nonempty positive t-cut* ψ_t^P *and the nonempty negative s-cut* ψ_s^N *are H-ideals of X.*

Proof: Assume that $\psi = \left(\lambda_\psi^+, \lambda_\psi^-\right)$ is a BFHI of *X* and let $(s,t) \in [-1,0] \times [0,1]$ be such that

$\psi_t^P \neq \phi$ and $\psi_s^N \neq \phi$.

Obviously, $0 \in \psi_t^P \cap \psi_s^N$. Let $x,y,z \in X$ be such that

$x*(y*z) \in \psi_t^P$ and $z \in \psi_t^P$.

Then

$\lambda_\psi^+\left(x*(y*z)\right) \geq t$ and $\lambda_\psi^+(z) \geq t$.

It follows from (*F*6) that

$$\lambda_\psi^+(x*z) \geq \min\left\{\lambda_\psi^+\left(x*(y*z)\right), \lambda_\psi^+(y)\right\} \geq t$$

so that $x*z \in \psi_t^P$. Hence the nonempty positive *t*-cut ψ_t^P is a *H*-ideal of X. Now assume that

$x*(y*z) \in \psi_s^N$ and $z \in \psi_s^N$

for all $x,y,z \in X$. Then $\lambda_\psi^-\left(x*(y*z)\right) \leq s$ and $\lambda_\psi^-(z) \leq s$. Using (F7), we have

$$\lambda_\psi^-(x*z) \leq \max\left\{\lambda_\psi^-\left(x*(y*z)\right), \lambda_\psi^-(y)\right\} \leq s$$

and so $x*z \in \psi_s^N$. Therefore, the nonempty negative *s*-cut ψ_s^N is a *H*-ideals of *X*.

Conversely, assume that for every $(s,t) \in [-1,0] \times [0,1]$ the non-empty positive *t*-cut ψ_t^P and the non-empty negative *s*-cut ψ_s^N are *H*-ideals of *X*. If $a,b \in X$ be such that

$\lambda_\psi^+(0) < \lambda_\psi^+(a)$ and $\lambda_\psi^-(0) > \lambda_\psi^-(b)$,

then $0 \notin \psi_t^P$ and $0 \notin \psi_t^P$. This is a contradiction. Thus

$\lambda_\psi^+(0) \geq \lambda_\psi^+(x)$ and $\lambda_\psi^-(0) \leq \lambda_\psi^-(x)$

for all $x \in X$. Suppose that

$$\lambda_{\psi}^{+}(a*c) < min\{\lambda_{\psi}^{+}(a*(b*c)), \lambda_{\psi}^{+}(b)\} = t$$

for some $a,b,c \in X$, then $a*(b*c) \in \psi_t^P$ and $b \in \psi_t^P$ but $a*c \notin \psi_t^P$. This is not true since ψ_t^P is a *H*-ideal of *X*. Therefore,

$$\lambda_{\psi}^{+}(x*z) \geq \min\{\lambda_{\psi}^{+}(x*(y*z)), \lambda_{\psi}^{+}(y)\}$$

for all $x,y,z \in X$. Again, suppose that

$$\lambda_{\psi}^{-}(a*c) > max\{\lambda_{\psi}^{-}(a*(b*c)), \lambda_{\psi}^{-}(b)\} = s$$

for some $a,b,c \in X$, then $a*(b*c) \in \psi_s^N$ and $b \in \psi_s^N$ but $a*c \notin \psi_s^N$. This is impossible since ψ_s^N is a *H*-ideal of *X*. Therefore,

$$\lambda_{\psi}^{-}(x*z) \leq \max\{\lambda_{\psi}^{-}(x*(y*z)), \lambda_{\psi}^{-}(y)\}$$

for all $x,y,z \in X$. Consequently, $\psi = (\lambda_{\psi}^{+}, \lambda_{\psi}^{-})$ is a BFHI of *X*.

Corollary 3.19 *If* $\psi = (\lambda_{\psi}^{+}, \lambda_{\psi}^{-})$ *is a BFHI of X, then the nonempty (s,t) -cut C(Ψ;(s,t)) of* $\psi = (\lambda_{\psi}^{+}, \lambda_{\psi}^{-})$ *is a H-ideal of X.*

Theorem 3.20 *If every BFHI* $\psi = (\lambda_{\psi}^{+}, \lambda_{\psi}^{-})$ *of X has the finite image, then every descending chain of H-ideals of X terminates at finite step.*

Proof: Suppose there exists a strictly descending chain $S_0 \times S_1 \times S_2 \ldots$ of *H*-ideals of *X* which does not terminate at finite step. Define a BFS $\psi = (\lambda_{\psi}^{+}, \lambda_{\psi}^{-})$ in *X* by

$$\lambda_{\psi}^{+}(x) = \begin{cases} \dfrac{n}{n+1} & \text{if } x \in S_n \setminus S_{n+1} \\ 1 & \text{if } x \in \cap_{n=0}^{\infty} S_n \end{cases} \quad \text{and} \quad \lambda_{\psi}^{-}(x) = \begin{cases} -\dfrac{n}{n+1} & \text{if } x \in S_n \setminus S_{n+1} \\ -1 & \text{if } x \in \cap_{n=0}^{\infty} S_n \end{cases}$$

where $n=0,1,2,\ldots$ and S_0 stands for *X*. Since

$$0 \in \cap_{n=0}^{\infty} S_n,\ \lambda_{\psi}^{+}(0) = 1 \geq \lambda_{\psi}^{+}(x) \text{ and } \lambda_{\psi}^{-}(0) = -1 \leq \lambda_{\psi}^{-}(x)$$

for all $x \in X$. Let $x, y \in X$. Assume that $x*(y*z) \in S_n \backslash S_{n+1}$ and $y \in S_k \backslash S_{k+1}$ for $n=0,1,2,\ldots$; $k=0,1,2,\ldots$. Without loss of generality, we may assume that $n \leq k$. Then obviously $x*(y*z)$ and $y \in S_n$, so $x*z \in S_n$ because S_n is a H-ideal of X. Hence,

$$\lambda_\psi^+(x*z) \geq \frac{n}{n+1} = \min\left\{\lambda_\psi^+(x*(y*z)), \lambda_\psi^+(y)\right\}$$

$$\lambda_\psi^-(x*z) \leq -\frac{n}{n+1} = \max\left\{\lambda_\psi^-(x*(y*z)), \lambda_\psi^-(y)\right\}.$$

If $x*(y*z), y \in \cap_{n=0}^{\infty} S_n$, then $x*z \in \cap_{n=0}^{\infty} S_n$. Thus

$$\lambda_\psi^+(x*z) = 1 = \min\left\{\lambda_\psi^+(x*(y*z)), \lambda_\psi^+(y)\right\}$$

$$\lambda_\psi^-(x*z) = -1 = \max\left\{\lambda_\psi^-(x*(y*z)), \lambda_\psi^-(y)\right\}.$$

If $x*(y*z) \notin \cap_{n=0}^{\infty} S_n$ and $y \in \cap_{n=0}^{\infty} S_n$, then there exists a positive integer r such that $x*(y*z) \in S_r \backslash S_{r+1}$. It follows that $x*z \in S_r$ so that

$$\lambda_\psi^+(x*z) \geq \frac{r}{r+1} = \min\left\{\lambda_\psi^+(x*(y*z)), \lambda_\psi^+(y)\right\}$$

$$\lambda_\psi^-(x*z) \leq -\frac{r}{r+1} = \max\left\{\lambda_\psi^-(x*(y*z)), \lambda_\psi^-(y)\right\}.$$

Finally suppose that $x*(y*z) \in \cap_{n=0}^{\infty} S_n$ and $y \notin \cap_{n=0}^{\infty} S_n$. Then $y \in S_s \backslash S_{s+1}$ for some positive integer s. It follows that $x*z \in S_s$, and hence

$$\lambda_\psi^+(x*z) \geq \frac{s}{s+1} = \min\left\{\lambda_\psi^+(x*(y*z)), \lambda_\psi^+(y)\right\}$$

$$\lambda_\psi^-(x*z) \leq -\frac{s}{s+1} = \max\left\{\lambda_\psi^-(x*(y*z)), \lambda_\psi^-(y)\right\}.$$

This proves that $\psi = \left(\lambda_\psi^+, \lambda_\psi^-\right)$ is a BFHI with an infinite number of different values, which is a contradiction. This completes the proof.

4. CONCLUSION AND FUTURE WORK

In this paper, notion of BFHIs in *BCI*-algebra are introduced and investigated some of their useful properties. It is our hope that this work would other foundations for further study of the theory of *BCK/ BCI*-algebras. In our future study of fuzzy structure of *BCK/BCI* -algebra, may be the following topics should be considered: (i) to find bipolar fuzzy *p*-ideals in *BCI* -algebra, (i) to find the relationship between BFHIs, *a*-ideals and *p*-ideals in *BCI* -algebra, (iii) to find bipolar fuzzy implicative ideals, positive implicative ideals, commutative ideals in *BCI* -algebra and their relationships.

REFERENCES

Akram, M. (2011). Bipolar fuzzy graphs. *Information Sciences*, *181*(24), 5548–5564. doi:10.1016/j.ins.2011.07.037

Bhowmik, M., Senapati, T., & Pal, M. (2014). Intuitionistic L BG. *Afrika Matematika*, *25*(3), 577–590. doi:10.100713370-013-0139-5

Dubois, D., & Prade, H. (2008). An introduction to bipolar representations of information and preference. *International Journal of Intelligent Systems*, *23*(8), 866–877. doi:10.1002/int.20297

Hu, Q. P., & Iseki, K. (1980). On BCI-algebra satisfying (x*y)*z=x*(y*z). *Math Sem Notes*, *8*, 553–555.

Imai, Y., & Iseki, K. (1966). On axiom system of propositional calculi. *XIV Proc. Japan Academy*, *42*, 19-22.

Iseki, K. (1966). An algebra related with a propositional calculus. *Proceedings of the Japan Academy*, *42*(1), 26–29. doi:10.3792/pja/1195522171

Jun, Y. B. (1993). Closed fuzzy ideals in BCI-algebras. *Math Japonica*, *38*, 199–202.

Jun, Y. B., Kim, H. S., & Lee, K. J. (2009). Bipolar fuzzy translation in BCK/BCI-algebra. Journal of the Chungcheong Math. *Soc.*, *22*(3), 399–408.

Khalid, H. M., & Ahmad, B. (1999). Fuzzy H BCI. *Fuzzy Sets and Systems*, *101*(1), 153–158. doi:10.1016/S0165-0114(97)00042-0

Lee, K. J. (2009). Bipolar fuzzy subalgerbas and bipolar fuzzy ideals of BCK/BCI-algerbas. *Bulletin of the Malaysian Mathematical Sciences Society*, *32*(3), 361–373.

Lee, K. J., & Jun, Y. B. (2011). Bipolar fuzzy a BCI. *Communications of the Korean Mathematical Societ.*, *26*(4), 531–542. doi:10.4134/CKMS.2011.26.4.531

Mitra, S., & Pal, S. K. (2005). Fuzzy sets in pattern recognition and machine intelligence. *Fuzzy Sets and Systems*, *156*(3), 381–386. doi:10.1016/j.fss.2005.05.035

Senapati, T. (2015b). Bipolar fuzzy structure of BG-subalgebras. *J. Fuzzy Math.*, *23*(1), 209–220.

Senapati, T., Bhowmik, M., & Pal, M. (2012). Interval-valued intuitionistic fuzzy BG-subalgebras. *Journal of Fuzzy Mathematics*, *20*(3), 707–720.

Senapati, T., Bhowmik, M., & Pal, M. (2013). Atanassov's intuitionistic fuzzy translations of intuitionistic fuzzy H-ideals in BCK/BCI-algebras. *Notes Intuition. Fuzzy Sets*, *19*(1), 32–47.

Senapati, T., Bhowmik, M., & Pal, M. (2014). Fuzzy dot structure of BG. *Fuzzy Information & Engineering*, *6*(3), 315–329. doi:10.1016/j.fiae.2014.12.004

Senapati, T., Bhowmik, M., & Pal, M. (2016). Triangular norm based fuzzy BG-algebras. *Afrika Matematika*, *27*(1-2), 187–199. doi:10.100713370-015-0330-y

Senapati, T., Bhowmik, M., Pal, M., & Davvaz, B. (2015). Fuzzy translations of fuzzy H BCK/BCI. *Journal of the Indonesian Mathematical Society*, *21*(1), 45–58. doi:10.22342/jims.21.1.200.45-58

Senapati, T., Jana, C., Bhowmik, M., & Pal, M. (2015a). -fuzzy LGG. *Journal of the Egyptian Mathematical Society*, *23*(2), 219–223. doi:10.1016/j.joems.2014.05.010

Senapati, T., Kim, C. S., Bhowmik, M., & Pal, M. (2015c). Cubic subalgebras and cubic closed ideals of B. *Fuzzy Information & Engineering*, *7*(2), 129–149. doi:10.1016/j.fiae.2015.05.001

Xi, O. (1991). Fuzzy BCK-algebras. *Math Japonica*, *36*, 935–942.

Zadeh, L. A. (1965). Fuzzy sets. *Information and Control*, *8*(3), 338–353. doi:10.1016/S0019-9958(65)90241-X

Zadrozny, S., & Kacprzyk, J. (2012). Bipolar queries: An aggregation operator focused perspective. *Fuzzy Sets and Systems*, *196*, 69–81. doi:10.1016/j.fss.2011.10.013

Zhang, W. R. (1994). Bipolar fuzzy sets and relations: a computational framework for cognitive modeling and multiagent decision analysis. *Proceedings of IEEE Conf.*, 305-309.

Zhang, W. R., & Zhang, L. (2004). YinYang bipolar logic and bipolar fuzzy logic. *Information Sciences*, *165*(3-4), 265–287. doi:10.1016/j.ins.2003.05.010

Chapter 12
Lukaswize Triple-Valued Intuitionistic Fuzzy BCK/BCI-Subalgebras

Chiranjibe Jana
Vidyasagar University, India

Karping Shum
Yunnan University, China

ABSTRACT

The authors studied the notion of (α,β)-intuitionistic fuzzy BCK/BCI-subalgebras by applying the Lukasiewicz 3-valued implication operator, where $\alpha,\beta\in\{\in,q,\in\wedge q,\in\vee q\}$ for $\alpha\neq\in\wedge q$. In this chapter, an intuitionistic fuzzy set A is an $(\in,\in)$ (or $(\in\wedge q,\in)$ or $(\in,\in\vee q)$)-intuitionistic fuzzy subalgebras if and only if for any $p\in(0,1]$ (or $p\in(0,0.5]$ or $p\in(0.5,1]$), then A_p is a fuzzy subalgebras of X respectively. Next, the authors defined intuitionistic fuzzy subalgebras with thresholds (s,t) and then provided intuitionistic fuzzy subalgebras with thresholds $(0,1)$ (or $(0,0.5)$ or $(0.5,1)$) by the concept of quasi-coincidence of fuzzy point respectively. Also, A is an intuitionistic fuzzy subalgebras with thresholds (s,t) if and only if for any $p\in(s,t]$, then cut set A_p is a fuzzy subalgebras.

1 INTRODUCTION

The role of logic in mathematics and computer science is two-fold which acts as a tool for applications in theoretical areas as well as a technical areas for laying the foundations. Thus, non-classical logic including many-valued logic and fuzzy logic takes the advantage of the classical logic (Boms and Mack, 1975) to handle information with various facets of uncertainty (Zadeh, 2005, 2008) such as fuzziness, randomness and so on. Presently, non-classical logic has become a formal and useful tool for computer science which can be used to deal with uncertain information and fuzzy information. The semantical systems of non-classical logic systems have produced various logical algebras. In 1966. Imai and Iseki

DOI: 10.4018/978-1-7998-0190-0.ch012

(1966) introduced a notions of logical algebras called BCK/BCI.algebras based on the generalization of the concept of set-theoretic difference and propositional calculus.

The concept of fuzzy sets was proposed by Zadeh in his tremendous paper (Zadeh, 1965), provided a natural mathematical framework for generalizing some notions of classical algebraic structures. After the introduction of fuzzy sets many researchers conducted the researches on the generalizations of fuzzy sets that have successful applications in computer science, artificial intelligence, control engineering, expert, robotics, automat theory, finite state machine, graph theory, logics and many branches of pure and applied mathematics. In 1971. Rosenfeld (1971) introduced the notion of fuzzy group. Since then many enormous papers have been published in fuzzy algebra. Mordeson and Malik introduced fuzzy subring and fuzzy subalgebra in ring theory. Jana et al. (2015), Bej and Pal (2015) have been done lot of works on fuzzy BCK/BCI/G.algebras. Based on this concept Bhakat and Das (1996) introduced a new generalization of fuzzy subgroups. Then this concept extended to BCK/BCI.algebras by many researchers in (Jun, 2005, 2007; Ma et al., 2011, 2012; Ma et al. (2009)). In 1999. the intuitionistic fuzzy sets was proposed by Atanassov (1999), a nobel generalization of fuzzy sets. Then different types of fuzzy setting have been conducted on intuitionistic fuzzy sets by many researchers in (Jana et al., 2016; Jun & Kim, 2000; Larimi, 2013; Yuan, 2010; Zhan & Tan, 2004). Though fuzzy sets and intuitionistic fuzzy sets have a applications in different theoretical study as well as real life applications. Implication operator with intuitionistic fuzzy sets have a applications in logical point of view. These points are enough to motivated us and best of our knowledge there is no work available on 3.valued intuitionistic fuzzy BCK. algebra. For this reason, we developed theoretical study of 3.valued intuitionistic fuzzy subalgebras of BCK/BCI.algebras based on Lukasiewicz 3.valued implication operator.

In this paper, applying the Lukasiewicz implication operators, the notions of (α,β).intuitionistic fuzzy subalgebras of BCK/BCI.algebras has been defined by using the grades of a fuzzy point x_a on intuitionistic fuzzy sets A. Also, $(\in,\in)$, $(\in\wedge q)$, and $(\in\vee q)$ intuitionistic fuzzy subalgebra of X.are investigated and define intuitionistic fuzzy subalgebra with thresholds (s,t) of X.

2 PRELIMINARIES

In this section, some elementary aspects that are necessary for this paper are included.

By a BCI.algebra we mean an algebra $(X,*,0)$ of type (2,0) satisfying the following axioms for all $x,y,z\in X$.

(C1) $((x*y)*(x*z))*(z*y)=0$
(C2) $(x*(x*y))*y=0$
(C3) $x*x=0$
(C4) $x*y=0$ and $y*x=0$ imply $x=y$.

The partial ordering is defined as $x\leq y$ if and only if $x*y=0$.

If a BCI.algebra X satisfies $0*x=0$ for all $x\in X$ then X is a BCK.algebra. Any BCI.algebra X satisfies the following axioms for all $x,y,z\in X$.

(C5) $(x*y)*z=(x*z)*y$
(C6) $((x*z)*(y*z))*(x*y)=0$

(C7) $x*0=x$
(C8) $x*y=0 \Rightarrow (x*z)*(y*z)=0$, $(z*y)*(z*x)=0$
(C9) $x*(x*(x*y))=x*y$
(C10) $0*(x*y)=(0*x)*(0*y)$.

Throughout this paper, X always means a BCK/BCI algebra without any specification.
A non-empty subset S of X.is called a subalgebra of X if $x*y \in S$ for any $x,y \in S$.

Definition 2.1 (Zadeh, 1965) *Let X be a non-empty set. A fuzzy set A of X is defined as a mapping A:* $X \rightarrow [0,1]$, where $[0,1]$ is the usual interval of real numbers.

Definition 2.2 (Jun, 2005) *A fuzzy set* $A=\{\langle x, \mu_A(x)\rangle : x \in X\}$ *.in a BCK/BCI algebra X is called a fuzzy BCK/BCI subalgebras of X if it satisfies for all x,y∈X*

$\mu_A(x*y) \geq \mu_A(x) \wedge \mu_A(y)$.

Definition 2.3 (Pu and Liu, 1980) *A fuzzy set A in a set X is of the form*

$$\mu_A(y)=\begin{cases} a \in (0,1] & \text{if } y=x \\ 0 & \text{if } y \neq x. \end{cases}$$

is said to be a fuzzy point with support x and value a and is denoted by x_a.

Definition 2.4 (Bhakat and Das, 1992) *Let A be a fuzzy subsets of X and* x_a *be fuzzy point.*

1. If $A(x) \geq a$ then x_a belong to A denoted $x_a \in A$.
2. If $A(x)+a \geq 1$ then x_a is quasi-coincident with A and denoted by $x_a qA$.
3. $x_a \in \wedge qA \Leftrightarrow x_a \in A$ and $x_a qA$.
4. $x_a \in \vee qA \Leftrightarrow x_a \in A$ or $x_a qA$.

Definition 2.5 (Atanassov, 1999) *The intuitionistic fuzzy set (IFS) defined on a Universe of discourse S. is an object having the form*

$$A=(\mu_A, v_A)=\{\langle x, \mu_A(x), v_A(x)\rangle : x \in S\}.$$

where the functions $\mu_A: S \rightarrow [0,1]$ *and* $v_A: S \rightarrow [0,1]$ *denote the degree of membership (namely,* $\mu_A(x)$*) and the degree of nonmembership (namely,* $v_A(x)$*) of each element* $x \in S$ *to the set A respectively, and* $0 \leq \mu_A(x)+v_A(x) \leq 1$ *for all* $x \in S$*. By IFS(S) we denote the set of all IFSs of S. For every two intuitionistic fuzzy sets* $A=(\mu_A, v_A)$ *and* $B=(\mu_B, v_B)$ *in S.*

1. $A \subseteq B$ .iff $\mu_A(x) \leq \mu_B(x)$ and $v_A(x) \geq V_B(x)$ for all $x \in S$.
2. $A = B$ .iff $\mu_A(x)=\mu_B(x)$.and $v_A(x)=V_B(x)$.for all $x \in S$.

3. $(A \cup B)(x) = \{\max\{\mu_A(x), \mu_B(x)\}, \min\{v_A(x), v_B(x)\} \mid x \in S\}$.
4. $(A \cap B)(x) = \{\min\{\mu_A(x), \mu_B(x)\}, \max\{v_A(x), v_B(x)\} \mid x \in S\}$..

Definition 2.6 (Jun and Kim, 2000) *An intuitionistic fuzzy set* $A = \{\langle x, \mu_A(x), v_A(x)\rangle : x \in X\}$ *in X is called an intuitionistic fuzzy subalgebras of X if it satisfies the following two conditions*

1. $\mu_A(x*y) \geq \mu_A(x) \wedge \mu_A(y)$.
2. $v_A(x*y) \leq v_A(x) \vee v_A(y)$ for all $x,y \in X$.

Definition 2.7 (Yuan et al., 2009) *Let* $A = \langle \mu_A, v_A \rangle$ *.be an IFSs of X and* $a \in [0,1]$. *Then*

1. $A_a(x) = \begin{cases} 1, & \text{if } \mu_A(x) \geq a \\ \frac{1}{2}, & \text{if } \mu_A(x) < a \leq 1 - v_A(x) \\ 0, & \text{if } a > 1 - v_A(x) \end{cases}$ *and* $A_{\underline{a}}(x) = \begin{cases} 1, & \text{if } \mu_A(x) > a \\ \frac{1}{2}, & \text{if } \mu_A(x) \leq a < 1 - v_A(x) \\ 0, & \text{if } a \geq 1 - v_A(x) \end{cases}$

are called the *a* upper cut set and *a* stronger upper cut set of *A* respectively.

2. $A^a(x) = \begin{cases} 1, & \text{if } v_A(x) \geq a \\ \frac{1}{2}, & \text{if } v_A(x) < a \leq 1 - \mu_A(x) \\ 0, & \text{if } a > 1 - \mu_A(x) \end{cases}$ *and* $A^{\underline{a}}(x) = \begin{cases} 1, & \text{if } v_A(x) > a \\ \frac{1}{2}, & \text{if } v_A(x) \leq a < 1 - \mu_A(x) \\ 0, & \text{if } a \geq 1 - \mu_A(x) \end{cases}$

are called the *a* lower cut set and *a* strong lower cut set of *A* respectively.

3. $A_{[a]}(x) = \begin{cases} 1, & \text{if } \mu_A(x) + a \geq 1 \\ \frac{1}{2}, & \text{if } v_A(x) \leq a < 1 - \mu_A(x) \\ 0, & \text{if } a < v_A(x) \end{cases}$ *and* $A_{[\underline{a}]}(x) = \begin{cases} 1, & \text{if } \mu_A(x) + a > 1 \\ \frac{1}{2}, & \text{if } v_A(x) < a \leq 1 - \mu_A(x) \\ 0, & \text{if } a \leq v_A(x) \end{cases}$

are called the *a* upper *Q* cut set and *a* strong upper *Q* cut set of *A* respectively.

4. $A^{[a]}(x) = \begin{cases} 1, & \text{if } v_A(x) + a \geq 1 \\ \frac{1}{2}, & \text{if } \mu_A(x) \leq a < 1 - v_A(x) \\ 0, & \text{if } a < \mu_A(x) \end{cases}$ *and* $A^{[\underline{a}]}(x) = \begin{cases} 1, & \text{if } v_A(x) + a > 1 \\ \frac{1}{2}, & \text{if } \mu_A(x) < a \leq 1 - v_A(x) \\ 0, & \text{if } a \leq \mu_A(x) \end{cases}$

are called the *a* lower *Q* cut set and *a* strong lower *Q* cut set of *A* respectively.

Proposition 2.1 (Yuan et al., 2009)

1. $A_{[a]}(x) = A_{\underline{1-a}}(x)$.

2. $A_{\underline{a}} \subset A_a$.
3. $a < b \Rightarrow A_{\underline{a}} \supset A_{\underline{b}}$.

Definition 2.8 (Yuan et al., 2009)

1. The grade of membership of $x_a \in A$ is $[x_a \in A]$ and grade of membership of $x_a qA$ is $[x_a qA]$ and are given by the following relations:

$[x_a \in A] = A_a(x)$ and $\left[x_a qA\right] = A_{[\underline{a}](x)}$.

2. The grade of membership of $x_a \in A$ and $x_a qA$ is $[x_a \in \wedge qA]$ and grade of membership of $x_a \in A$ or $x_a qA$ is $[x_a \in \vee qA]$ and are given by the following relations:

$$\left[x_a \in \wedge qA\right] = \left[x_a \in A\right] \wedge \left[x_a qA\right] = A_a(x) \wedge A_{[\underline{a}](x)}$$

and

$$\left[x_a \in \vee qA\right] = \left[x_a \in A\right] \vee \left[x_a qA\right] = A_a(x) \vee A_{[\underline{a}](x)}.$$

3. The grade of nonmembership of $x_a \in A$ is $\left[x_a \overline{\in} A\right]$ and grade of nonmembership of $x_a qA$ is $\left[x_a \overline{q} A\right]$ and are given by the following relations:

$$\left[x_a \overline{\in} A\right] = A^a(x) \text{ and } \left[x_a \overline{q} A\right] = A^{[a](x)}.$$

4. The grade of nonmembership of $x_a \in A$ and $x_a qA$ is $\left[x_a \overline{\in \wedge qA}\right]$ and grade of nonmembership of $x_a \in A$ or $x_a qA$ is $\left[x_a \overline{\in \vee qA}\right]$ and are given by the following relations:

$$\left[x_a \overline{\in \wedge qA}\right] = \left[x_a \overline{\in} \vee \overline{q} A\right] = \left[x_a \overline{\in} A\right] \vee \left[x_a \overline{q} A\right] = A^a(x) \vee A^{[\underline{a}](x)}$$

and

$$\left[x_a \overline{\in \vee qA}\right] = \left[x_a \overline{\in} A \wedge x_a \overline{q} A\right] = \left[x_a \overline{\in} A\right] \wedge \left[x_a \overline{q} A\right] = A^a(x) \wedge A^{[\underline{a}](x)}.$$

Now, we have given the property

Property 2.2 (Yuan et al., 2009)

$\left[x_a \bar{\in} A\right]=\left[x_a \in A^c\right] \quad \left[x_a \bar{q} A\right]=\left[x_a q A^c\right]$

$\left[x_a \bar{\in} \wedge \bar{q} A\right]=\left[x_a \in \wedge q A^c\right] \quad \left[x_a \bar{\in} \vee \bar{q} A\right]=\left[x_a \in \vee q A^c\right]$

$\left[x_a \in \left(\bigcap_{t\in T} A_t\right)\right]=\wedge_{t\in T}\left[x_a \in A\right] \quad \left[x_a q\left(\bigcup_{t\in T} A_t\right)\right]=\vee_{t\in T}\left[x_a q A\right]$

$\left[x_a \bar{\in} \left(\bigcup_{t\in T} A_t\right)\right]=\wedge_{t\in T}\left[x_a \bar{\in} A\right] \quad \left[x_a \bar{q}\left(\bigcap_{t\in T} A_t\right)\right]=\vee_{t\in T}\left[x_a \bar{q} A\right]$

3 (α,β) INTUITIONISTIC FUZZY BCK/BCI SUBALGEBRAS

In this section, we study (α,β) intuitionistic fuzzy BCK/BCI subalgebras of NCK/BCI algebras. Let → denote the implication of Lukasiewicz in triple valued logic.

Let X be a BCK/BCI algebras and α,β∈{∈,q,∈∧q,∈∨q}. Then for a∈[0,1] x∈X x_a is a fuzzy point and $[x_a \alpha A]$, $[x_b \beta A]$∈{0,1/2,1}. Then we have

Definition 3.1 *Let X be a BCK/BCI algebras and* $A=\langle \mu_A,\nu_A\rangle$ *.be an IFS defined on the set X. If for any* α,β∈{∈,q,∈∧q,∈∨q} *and s,t*∈(0,1] *then following conditions are satisfied:* $[x_s \alpha A] \wedge [y_t \alpha A] \rightarrow [(x_s * y_t) \beta A]=1$ for all x,y∈X.

Then $A=\langle \mu_A,\nu_A\rangle$ is called an (α,β) intuitionistic fuzzy BCK/BCI subalgebras of X where $(x_s * y_t)=(x*y)_{s\wedge t}$.

Let X be a set. Then we define a mapping A: X→0,1/2,1 a 3 valued fuzzy set. It is noted that, for p,q∈{0,1/2,1} $(p\rightarrow q)=1 \Leftrightarrow q\geq p$. Therefore, Definition 3.1 is equivalent to the following definition.

Definition 3.2 *Let* $A=\langle \mu_A,\nu_A\rangle$ *.be an IFS defined on the set X. If for any* α,β∈{∈,q,∈∧q,∈∨q} *s,t*∈(0,1] *then the following conditions are satisfied:* $[(x_s * y_t)\beta A] \geq [x_s \alpha A] \wedge [y_t \alpha A]$ for all x,y∈X.

Then A is called an (α,β) intuitionistic fuzzy BCK/BCI subalgebras of X where $(x_s * y_t)=(x*y)_{s\wedge t}$.

Example 3.1 Consider a BCK/BCI algebras X={0,a,b,c,d} with the Caley table (*Table 2*).

Table 1. The truth table of Lukasiewicz implication

→	0	1/2	1
0	1	1	1
1/2	1/2	1	1
1	0	1/2	1

Table 2.

*	0	a	b	c	d
0	0	0	0	0	0
a	a	0	a	0	a
b	b	b	0	b	0
c	c	a	c	0	c
d	d	d	b	d	0

Let us consider the following IFS of *X*.

$$\mu_A(x)=\begin{cases}0.7 & \text{if } x=0\\ 0.4 & \text{if } x=a,c\\ 0.3 & \text{if } x=b,d.\end{cases}$$

$$\nu_A(x)=\begin{cases}0.4 & \text{if } x=0\\ 0.6 & \text{if } x=b,d\\ 0.7 & \text{if } x=a,c.\end{cases}$$

Then, it can be verified that $A=\langle\mu_A,\nu_A\rangle$ .is both an (∈,∈) and (∈,∈∨*q*). IF subalgebras of *X*. Also, if we consider the IFS of *X*.

$$\mu_A(x)=\begin{cases}0.6 & \text{if } x=0\\ 0.3 & \text{if } x=a,c\\ 0.2 & \text{if } x=b,d.\end{cases}$$

$$\nu_A(x)=\begin{cases}0.2 & \text{if } x=0\\ 0.3 & \text{if } x=a,c\\ 0.7 & \text{if } x=b,d.\end{cases}$$

Then it can be easily verified that $A=\langle\mu_A,\nu_A\rangle$ is an (∈∧*q*,∈) intuitionistic fuzzy subalgebra of *X*.

Here, the Definition 3.1 and Definition 4.7 are the generalization of the concepts of (α,β).fuzzy subalgebras of BCK/BCI algebras in (Jun, 2005). In Definition 4.7, α can be chosen one from four kinds of relations and β can be chosen from four kinds of relations. Therefore, there are 16 kinds of (α,β) intuitionistic fuzzy subalgebras of BCK/BCI algebras. Now, we will discuss 16 kinds of (α,β) intuitionistic fuzzy subalgebras of *X*.

Example 3.2 *Let us consider a BCI algebra X={0,a,b,c} with the Caley table (Table 3).*

Table 3.

*	0	a	b	c
0	0	a	b	c
a	a	0	c	b
b	b	c	0	a
c	c	b	a	0

Let $A=\langle\mu_A,\nu_A\rangle$ be a IFS defied in X such that $\mu_A(0)=0.6$, $\mu_A(a)=0.7$, and $\mu_A(b)=\mu_A(c)=0.3$ and $\nu_A(0)=0.4$, $\nu_A(a)=\nu_A(c)=0.5$ and $\nu_A(b)=0.4$. Then, it is examine that $A=\langle\mu_A,\nu_A\rangle$ is an $(\in,\in\vee q)$ intuitionistic fuzzy subalgebras of X. But,

1. $A=\langle\mu_A,\nu_A\rangle$ is not an $(\in,\in)$ intuitionistic fuzzy subalgebras of X because $a_{0.63}\in\mu_A$ and $a_{0.65}\in\mu_A$ but $(a*a)_{\min\{0.63,0.65\}}=0_{0.65}\,\overline{\in}\,\mu_A$. Also, $b_{0.32}\in\nu_A$ and $b_{0.36}\in\nu_A$ but $(b*b)_{\max\{0.32,0.36\}}=0_{0.36}\,\overline{\in}\,\nu_A$.
2. $A=\langle\mu_A,\nu_A\rangle$ is not an $(q,\in\vee q)$ intuitionistic fuzzy subalgebras of X because $a_{0.42}q\mu_A$ and $b_{0.75}q\mu_A$ but $(a*b)_{\min\{0.42,0.75\}}=c_{0.42}\overline{\in\vee q}\mu_A$ and $b_{0.6}q\nu_A$ and $c_{0.4}q\nu_A$ but $(b*c)_{\max\{0.6,0.4\}}=c_{0.6}\overline{q}\nu_A$.
3. $A=\langle\mu_A,\nu_A\rangle$ is not an $(\in\vee q,\in\vee q)$.intuitionistic fuzzy subalgebras of X because $a_{0.6}\in\vee q\mu_A$ and $c_{0.75}\in\vee q\mu_A$ but $(a*c)_{\min\{0.6,0.75\}}=b_{0.6}\overline{\in\vee q}\mu_A$.

Theorem 3.1 *Let $A=\langle\mu_A,\nu_A\rangle$ be a (α,β) intuitionistic fuzzy subalgebras of X. If $\alpha\notin\wedge q$ then A0.is a fuzzy subalgebras of X.*

Proof: To, show that for all $x,y\in X$ $A0(x*y)\geq A0(x)\wedge A0(y)$. First we prove that $A0(x)\wedge A0(y)=1\Rightarrow A0(x*y)=1$. Let $A0(x)\wedge A0(y)=1$. Then $A_t(x)=1$ and $A_t(y)=1$ and so $\mu_A(x)>0$ and $\mu_A(y)>0$. Let $t=\mu_A(x)\wedge\mu_A(y)$ imply $t>0$. Therefore, there exist $s\in(0,1)$ such that $0<1-s<t=\mu_A(x)\wedge\mu_A(y)$. Thus, $[x_t\in A]=A_t(x)=1$ and $[y_t\in A]=A_t(y)=1$.

$$[x_s qA]=A_{[\underline{s}]}(x)=1\quad [y_s qA]=A_{[\underline{s}]}(y)=1.$$

If $\alpha=\in$ or $\alpha=\in\vee q$ then $[x_t\alpha A]=[t_t\alpha A]=1$ and for $\beta\in\{\in,q,\wedge q,\vee q\}$. Now, by Definition 4.7,

$$1\geq[(x_t*y_t)\beta A]\geq[x_t\alpha A]\wedge[y_t\alpha A]=1$$

i.e. $[(x_t*y_t)\beta A]=1$ which imply that either $A_t(x*y)=1$ or $A_{[\underline{t}]}(x*y)=1$ and so $\mu_A(x*y)\geq t>0$ or $\mu_A(x*y)>1-t\geq 0$. Therefore, $A0(x*y)=1$.

If $\alpha=q$ then $[x_s\alpha A]=[y_s\alpha A]=1$. Thus, for $\beta\in\{\in,q,\wedge q,\vee q\}$. So, from Definition 4.7,

$$1\geq[(x_s*y_s)\beta A]\geq[x_s\alpha A]\wedge[y_s\alpha A]=1$$

i.e. $[(x_s*y_s)\beta A]=1$ which imply that $A_s(x*y)=1$ or $A_{[\underline{s}]}(x*y)=1$ and so $\mu_A(x*y)\geq s>0$ or $\mu_A(x*y)>1-s\geq 0$

$$\Rightarrow\mu_A(x*y)>0\Rightarrow A_{\underline{0}}(x*y)=1.$$

Next we show,

$$A_{\underline{0}}(x)\wedge A_{\underline{0}}(y)=1/2\Rightarrow A_{\underline{0}}(x*y)\geq 1/2.$$

Let $A0(x) \wedge A0(y)=1/2$. Then $A0(x) \geq 1/2$ and $A0(y) \geq 1/2$ which imply $v_A(x)<1$ and $v_A(y)<1$ so that $v_A(x) \vee v_A(y)<1$. Let $s,t \in (0,1)$ such that $v_A(x) \vee v_A(y)<1-t<s<1$. Then

$$0<t<1-v_A(x) \vee v_A(y)=(1-v_A(x)) \wedge (1-v_A(y))$$

implies that $1-v_A(x)>t$ and $1-v_A(y)>t$. Thus, $[x_t \in A]=A_t(x) \geq 1/2$ and $[y_t \in A]=A_t(y) \geq 1/2$. Again, $v_A(x) \vee v_A(y)<s<1$ implies that $v_A(x)<s$ and $v_A(y)<s$. Thus,

$$[x_s qA]=A_{[\underline{s}]}(x) \geq 1/2 \text{ and } [y_s qA]=A_{[\underline{s}]}(y) \geq 1/2.$$

If $\alpha = \in$ or $\alpha = \in \vee q$ then $[x_t \alpha A]=[y_t \alpha A]=1/2$ and for $\beta \in \{\in, q, \wedge q, \vee q\}$.

$$[(x_t * y_t)\beta A] \geq [x_t \alpha A] \wedge [y_t \alpha A] \geq 1/2$$

by Definition 4.7. Thus, $[(x_t * y_t)\beta A] \geq 1/2$ which implies $[(x_t * y_t) \in A]=A_t(x*y) \geq 1/2$ or

$$[(x_t * y_t) qA]=A_{[\underline{t}]}(x*y) \geq 1/2 \Rightarrow v_A(x*y) \leq 1-t<1-0$$

or

$$v_A(x*y)<1-t<1-0 \Rightarrow v_A(x*y)<1-0 \Rightarrow A_{\underline{0}}(x*y) \geq 1/2.$$

If $\alpha = q$ then $[x_s \alpha A] \wedge [y_s \alpha A] \geq 1/2$ and for $\beta \in \{\in, q, \wedge q, \vee q\}$. $[(x_s * y_s)\beta A] \geq 1/2$ which implies }. $[(x_s * y_s) \in A]=A_s(x*y) \geq 1/2$ or

$$[(x_s * y_s) qA]=A_{[\underline{s}]}(x*y) \geq 1/2 \Rightarrow v_A(x*y) \leq 1-s<1$$

or

$$v_A(x*y)<s<1 \Rightarrow v_A(x*y)<1 \Rightarrow A_{\underline{0}}(x*y) \geq 1/2.$$

Therefore, $A_{\underline{0}}(x) \wedge A_{\underline{0}}(y) \leq A_{\underline{0}}(x*y)$ for all $x,y \in X$. Hence, $A0$ is a fuzzy subalgebras of X.

Theorem 3.2 *An IFS* $A=\langle \mu_A, v_A \rangle$ *of X is an* $(\in,\in)$ *intuitionistic fuzzy subalgebras of X if and only if for any* $p \in [0,1]$ *then* A_p *is a 3 valued fuzzy subalgebras of X.*

Proof: Suppose, A is an $(\in,\in)$ intuitionistic fuzzy subalgebras of X. Let $x,y \in X$ and for $p \in [0,1]$ then $[(x_p * y_p) \in A] \geq [x_p \in A] \wedge [y_p \in A]$, i.e. $A_p(x*y) \geq A_p(x) \wedge A_p(y)$. Hence, A_p is a fuzzy subalgebras of X.

Conversely, assume for any $p \in [0,1]$. A_p is a fuzzy subalgebras of X. Let $x,y \in X$ and $s,t \in [0,1]$. We have to prove

$$[(x*y)_{s\wedge t} \in A] \geq [x_s \in A] \wedge [y_t \in A].$$

$$[(x*y)_{s\wedge t} \in A] \geq [x_s \in A] \wedge [y_t \in A].$$

Let $a=[x_s \in A] \wedge [y_t \in A]$.

Case 1. Let $a=1$. Then $[x_s \in A]=A_s(x)=1$ and $[y_t \in A]=A_t(y)=1$ then,

$$[(x*y)_{s\wedge t} \in A] = A_{s\wedge t}(x*y) \geq A_{s\wedge t}(x) \wedge A_{s\wedge t}(y) \geq A_s(x) \wedge A_t(y) = [x_s \in A] \wedge [y_t \in \mathrm{A}] = 1$$

which implies that

$$A_{s\wedge t}(x*y) = 1 \Rightarrow \mu_A(x*y) \geq s \wedge t \Rightarrow [(x*y)_{s\wedge t} \in A] = 1.$$

Case 2. At $a=1/2$. Then $[x_s \in A]=A_s(x)=1/2$ and $[y_t \in A]=A_t(y)=1/2$. Therefore,

$$[(x*y)_{s\wedge t} \in A] = A_{s\wedge t}(x*y) \geq A_{s\wedge t}(x) \wedge A_{s\wedge t}(y) \geq A_s(x) \wedge A_t(y) = [x_s \in A] \wedge [y_t \in A] = 1/2$$

so we get

$$A_{s\wedge t}(x*y) \geq 1/2 \Rightarrow s \wedge t \leq 1 - \nu_A(x*y).$$

Thus, $[(x*y)_{s\wedge t} \in A] \geq 1/2$. Hence, $[(x*y)_{s\wedge t} \in A] \geq [x_s \in A] \wedge [y_t \in A]$.■

Theorem 3.3 *An IFS* $A = \langle \mu_A, \nu_A \rangle$ *of X is an* $(\in, \in\vee q)$ *intuitionistic fuzzy subalgebras of X if and only if for any* $p \in [0,0.5]$. A_p *is a 3 valued fuzzy subalgebras of X.*

Proof: Suppose, A is an $(\in, \in\vee q)$ intuitionistic fuzzy subalgebras of X. Then for any $p \in (0,0.5]$ and $x,y \in X$ we have and

$$[(x_p * y_p) \in \vee qA] \geq [x_p \in A] \wedge [y_p \in A] \Rightarrow A_p(x*y) \vee A_{\lceil \underline{p} \rceil}(x*y) \geq A_p(x) \wedge A_p(y)..$$

As $0<p\leq 0.5$ $p \leq 0.5 \leq 1-p$. Then

$$A_{\lceil \underline{p} \rceil}(x*y) = A_{\underline{1-p}}(x*y) \leq A_{\underline{p}}(x*y) \leq A_p(x*y).$$

Therefore,

$$A_p(x)\wedge A_p(y)\leq A_p(x^*y)\vee A_{[\underline{p}]}(x^*y)\leq A_p(x^*y)\vee A_p(x^*y)=A_p(x^*y).$$

Hence, $A_p(x^*y)\geq A_p(x)\wedge A_p(x)$. Therefore, for any $p\in(0,0.5]$. A_p is a fuzzy subalgebras of X.
Conversely, we claim for any $p\in[0,0.5]$. A_p is a fuzzy subalgebras of X. Let $x,y\in X$ and $s,t\in[0,1]$. We prove, $[(x^*y)_{s\wedge t}\in\vee qA]\geq[x_s\in A]\wedge[y_t\in A]$. If $s\wedge t\leq 0.5$ and then let $a=[x_s\in A]\wedge[y_t\in A]$.

Case 1. For $a=1$. Then $[x_s\in A]=A_s(x)=1$ and $[y_t\in A]=A_t(y)=1$ from which

$$A_{s\wedge t}(x^*y)\geq A_{s\wedge t}(x)\wedge A_{s\wedge t}(y)\geq A_s(x)\wedge A_t(y)=1.$$

This implies that $A_{s\wedge t}(x^*y)=1\Rightarrow[(x^*y)_{s\wedge t}\in A]=1$. Thus,

$$[(x^*y)_{s\wedge t}\in\vee qA]=[(x^*y)_{s\wedge t}\in A]\vee[(x^*y)_{s\wedge t}qA]=1.$$

Case 2. For $a=1/2$. Then $A_s(x)\geq 1/2$ and $A_t(y)\geq 1/2$ so

$$A_{s\wedge t}(x^*y)\geq A_{s\wedge t}(x)\wedge A_{s\wedge t}(y)\geq A_s(x)\wedge A_t(y)\geq 1/2\Rightarrow A_{s\wedge t}(x^*y)\geq 1/2$$

which imply $[(x^*y)_{s\wedge t}\in A]\geq 1/2$. Therefore,

$$[(x^*y)_{s\wedge t}\in\vee qA]=[(x^*y)_{s\wedge t}\in A]\vee[(x^*y)_{s\wedge t}qA]\geq 1/2.$$

Hence,

$$[(x^*y)_{s\wedge t}\in\vee qA]\geq[x_s\in A]\wedge[y_t\in A].$$

If $s\wedge t>0.5$ then let $p\in(0,1)$ be such that $1-s\wedge t<p<0.5<s\wedge t$. Now,

$$A_{\underline{s\wedge t}}(x^*y)=A_{\underline{1-s\wedge t}}(x^*y)\geq A_{s\wedge t}(x^*y)\ \text{ and }\ A_{[\underline{s\wedge t}]}(x^*y)=A_{\underline{1-s\wedge t}}(x^*y)\geq A_p(x^*y).$$

Thus,

$$\begin{aligned}[(x^*y)_{s\wedge t}\in\vee qA]&=[(x^*y)_{s\wedge t}\in A]\vee[(x^*y)_{s\wedge t}qA]=A_{s\wedge t}(x^*y)\vee A_{[\underline{s\wedge t}]}(x^*y))\\&=A_{[\underline{s\wedge t}]}(x^*y)\geq A_p(x^*y)\geq A_p(x)\wedge A_p(y)\geq A_s(x)\wedge A_t(y)=[x_s\in A]\wedge[y_t\in A]\end{aligned}.$$

Hence, $[(x*y)_{s\wedge t} \in \vee qA] \geq [x_s \in A] \wedge [y_t \in A]$. ■

Theorem 3.4 *An IFS* $A = \langle \mu_A, \nu_A \rangle$ *of X is an* (∈∧*q*,∈) *intuitionistic fuzzy subalgebras of X if and only if for any p*∈(0.5,1]. A_p *is a 3 valued fuzzy subalgebras of X.*

Proof: Suppose, *A* is an (∈∧*q*,∈) intuitionistic fuzzy subalgebras of *X*. For *p*∈(0.5,1] and *x,y*∈*X* so $A_{\lceil p \rceil}(x) \geq A_p(x)$. Then for all *x,y*∈*X*

$$A_p(x*y) = [(x_p * y_p) \in A] \geq [x_p \in \wedge A] \wedge [y_p \in \wedge A] = A_p(x) \wedge A_{\lceil p \rceil}(x) \wedge A_p(y) \wedge A_{\lceil p \rceil}(y) = A_p(x) \wedge A_p(y)$$

Therefore, $A_p(x) \geq A_p(x*y) \wedge A_p(y)$.
Hence, A_p is a fuzzy subalgebras of *X*.
Conversely, we claim that for any *p*∈(0.5,1]. A_p is a fuzzy subalgebras of *X*. To prove,

$$[(x*y)_{s\wedge t} \in A] \geq [x_s \in \wedge A] \wedge [y_t \in \wedge A].$$

Let $a=[x_s \in \wedge A] \wedge [y_t \in \wedge A]$.

Case 1. Let *a*=1. Then $\mu_A(x) \geq s$, $\mu_A(x) > 1 - s$ and $\mu_A(y) \geq t$, $\mu_A(y) > 1 - t$. Therefore, $\mu_A(x) > 0.5$, $\mu_A(y) > 0.5$. Let $p = \mu_A(x) \wedge \mu_A(y)$ then $p > 0.5$ so we get $\mu_A(x) \geq p$ and $\mu_A(y) \geq p \Rightarrow A_p(x) = 1$ and $A_p(y) = 1$. Thus, $A_p(x*y) \geq A_p(x) \wedge A_p(y) = 1 \Rightarrow A_p(x*y) = 1$ and so $\mu_A(x*y) \geq p = \mu_A(x) \wedge \mu_A(y) = s \wedge t$. Hence,

$$[(x*y)_{s\wedge t} \in A] = 1.$$

Case 2. Let *a*=1/2. Then $1 - \nu_A(x) \geq s$, $s > \nu_A(x)$ and $1 - \nu_A(y) \geq t$, $t > \nu_A(y)$ which imply $\nu_A(x) < 0.5$ and $\nu_A(y) < 0.5$. Then $1 - \nu_A(x) > 0.5$ and $1 - \nu_A(y) > 0.5$. Let $p = (1 - \nu_A(x)) \wedge (1 - \nu_A(y))$ then $p > 0.5$. Therefore, $A_p(x*y) \geq A_p(x) \wedge A_p(y) = 1/2$ as $1 - \nu_A(x) \geq p$ and $1 - \nu_A(y) \geq p$. Thus, $1 - \nu_A(x*y) \geq p = (1 - \nu_A(x)) \wedge (1 - \nu_A(y)) \geq s \wedge t$. Hence,

$$[(x*y)_{s\wedge t} \in A] \geq 1/2 = [x_s \in \wedge A] \wedge [y_t \in \wedge A]. ■$$

Theorem 3.5 *Let* $A = \langle \mu_A, \nu_A \rangle$ *be a* (α,β) *intuitionistic fuzzy subalgebras of X. For x*∈*X and* $A(x) = (\mu_A(x), \nu_A(x))$. *If* $\mu_A(x) > 0$ *then* $A(x) = (1,0)$ *for*

$$\alpha, \beta \in \{(\in, q), (\in, \in \wedge q), (\in \vee q, \in), (\in \vee q, \in \wedge q), (q, \in), (q, \in \wedge q), (\in \vee q, \in).$$

Theorem 3.6 *If A is an* (∈∨*q*, ∈∨*q*) *intuitionistic fuzzy subalgebras of X then A is an* (∈,∈∨*q*) *intuitionistic fuzzy subalgebras of X.*

Theorem 3.7 *If A is a (q,q) intuitionistic fuzzy subalgebra of X then*

1. $\mu_A(x)>0 \Rightarrow A(x)=(\mu_A(0),v_A(0))$.
2. $\mu_A(x)=0,\ v_A(x)<1 \Rightarrow A(x)=(0,v_A(0))$.

Theorem 3.8 *Let A be a* $(q,\in\vee q)$ *intuitionistic fuzzy subalgebra of X and* $M=\{x|x\in X,\ \mu_A(x)>0\}$ and $N=\{x|x\in X,\ \mu_A(x)=0,\ v_A(x)<1\}$. then

1. If $\mu_A(x)$ is not a constant on M then for any $x\in M$, $A(x)\geq(0.5,0.5)$, i.e. $\mu_A(x)\geq 0.5$ and $v_A(x)\leq 0.5$.
2. If $v_A(x)$ is not constant on N then for any $x\in N$, $v_A(x)\leq 0.5$.

4 INTUITIONISTIC FUZZY *BCK/BCI* SUBALGEBRAS WITH THRESHOLDS

Definition 4.1 *Let* $A=\langle\mu_A,v_A\rangle$ *.be an intuitionistic fuzzy set of X. Then A is said to be intuitionistic fuzzy subalgebras of X with threshold (s,t) if and only if it satisfies the following conditions*

1. $\mu_A(x*y)\vee s\geq\mu_A(x)\wedge\mu_A(y)\wedge t$.
2. $v_A(x*y)\wedge(1-s)\leq v_A(x)\vee v_A(y)\vee(1-t)$ for all $x,y\in X$.

Example 4.1 *Consider a BCI algebra* $X=\{0,a,b,c\}$ *with the following Caley table given in Example 3.4. We consider* $\mu_A(0)=m$, $\mu_A(a)=m$, $\mu_A(b)=s$, $\mu_A(c)=s$, *and* $v_A(0)=1-v$, $v_A(a)=1-v$, $v_A(b)_z$, $v_A(c)=z$ *where* $0<s<v<1$, $m\in(0,s]$ *and* $z\in[0,1-v]$. *Then* $A=\langle\mu_A,v_A\rangle$ *.is an intuitionistic fuzzy subalgebra of X with thresholds (s,v). Whenever, if we consider* $x=a$, $y=a$, $\alpha=\in$, $\beta=\in$, *and let* $p,q\in[0,1]$ *be such that* $[x_p\alpha A]\wedge[y_q\alpha A]=1$ *then* $s\geq p$ *and* $s\geq q$. $s\geq p\wedge q$. *Again,* $\mu_A(x)=m<s$. *Now, if A is an* $(\in,\in)$ *intuitionistic fuzzy subalgebra of X then* $[(x*y)_{p\wedge q}\beta A]\geq[x_p\alpha A]\wedge[y_q\alpha A]$ *implies* $m\geq p\wedge q$ *which to be a contradiction if we choose* $m<p$, $q<s$. *Therefore, A is not an* $(\in,\in)$ *intuitionistic fuzzy subalgebras of X. It may be noted that, A is not an intuitionistic fuzzy subalgebra of X with thresholds* (0,1).

Theorem 4.1 *An IFS* $A=\langle\mu_A,v_A\rangle$ *.of a BCK/BCI algebra X is an* $(\in,\in)$ *intuitionistic fuzzy subalgebras of X if and only if A is an intuitionistic fuzzy subalgebras of X with thresholds* (0,1).

Proof: Suppose that $A=\langle\mu_A,v_A\rangle$ .is an $(\in,\in)$.intuitionistic fuzzy subalgebras of *X*. We have to show that *A* is an intuitionistic fuzzy subalgebras of *X* with thresholds (0,1), i.e. to show

1. $\mu_A(x*y)\geq\mu_A(x)\wedge\mu_A(y)$.
2. $v_A(x*y)\leq v_A(x)\vee v_A(y)$ for all $x,y\in X$.

(1) Let $t=\mu_A(x)\wedge\mu_A(y)$ then $\mu_A(x)\geq t$ and $\mu_A(y)\geq t$ which indicate $A_t(x)=1$ and $A_t(y)=1$ and so $[x_t\in A]=1$ and $[y_t\in A]=1$. Therefore, $1\geq[(x_t*y_t)\in A]\geq[x_t\in A]\wedge[y_t\in A]=1$. which implies that $\mu_A(x*y)\geq t=\mu_A(x*y)\wedge\mu_A(y)$ for all $x,y\in X$.

(2) If $v_A(x)=0$ then it is obvious. Let $s=v_A(x)>0$ and take $t\in[0,1]$ such that $t>1-s=1-v_A(x)$ then get $0=[(x_t*y_t)\in A]\geq[x_t\in A]\wedge[y_t\in A]$ then $[x_t\in A]=0$ or $[y_t\in A]=0$, i.e. either $v_A(x)>1-t$ or $v_A(y)>1-t$. Thus, $v_A(x)\vee v_A(y)>1-t$. Hence, $v_A(x)\vee v_A(y)\geq\vee\{1-t|s>1-t\}=s=v_A(x*y)$. Therefore, $v_A(x*y)\leq v_A(x)\vee v_A(y)$ for all $x,y\in X$.

Conversely, assume that $A=\langle\mu_A,\nu_A\rangle$ .is an intuitionistic fuzzy subalgebra of X with threshold (0,1). We have to show $A=\langle\mu_A,\nu_A\rangle$ .is an $(\in,\in)$ intuitionistic fuzzy subalgebra of X. For any $x,y\in X$ and $s,t\in(0,1]$. For any $x,y\in X$ we claim $[x_sy_t\in A]\geq[x_s\in A]\wedge[y_t\in A]$. Let $a=[x_s\in A]\wedge[y_t\in A]$.

Case 1. $a=1$. Then $[x_s\in A]=1$ and $[y_t\in A]=1$ which implies $\mu_A(x)\geq s$ and

$$\mu_A(y)\geq t\Rightarrow\mu_A(x*y)\geq\mu_A(x)\wedge\mu_A(y)\geq s\wedge t\Rightarrow[x_sy_t\in A]=1\geq1=[x_s\in A]\wedge[y_t\in A].$$

Case 2. $a=1/2$. Then $[x_s\in A]\geq1/2$ and $[y_t\in A]\geq1/2$ implies $1-v_A(x)\geq s$ and

$$\begin{aligned}&1-v_A(y)\geq t\\&\qquad\Rightarrow1-v_A(x*y)\geq1-v_A(x)\vee v_A(y)=(1-v_A(x))\wedge(1-v_A(y))\geq s\wedge t.\\&\qquad\Rightarrow[x_sy_t\in A]\geq1/2=[x_s\in A]\wedge[y_t\in A]\end{aligned}$$

Therefore, $[x_sy_t\in A]\geq[x_s\in A]\wedge[y_t\in A]$. Hence, $A=\langle\mu_A,\nu_A\rangle$ is an $(\in,\in)$ intuitionistic fuzzy subalgebra of X. ■

Theorem 4.2 *An IFS* $A=\langle\mu_A,\nu_A\rangle$ *of X is an* $(\in,\in\vee q)$ *intuitionistic fuzzy subalgebra of X if and only if A is an intuitionistic fuzzy subalgebra of X with thresholds* (0,0.5).

Proof: Suppose $A=\langle\mu_A,\nu_A\rangle$ .is an $(\in,\in\vee q)$.intuitionistic fuzzy subalgebra of X. We have to show that A is an intuitionistic fuzzy subalgebra of X with thresholds (0,0.5), i.e. to show

1. $\mu_A(x*y)\geq\mu_A(x)\wedge\mu_A(y)\wedge0.5$.
2. $v_A(x*y)\leq v_A(x)\vee v_A(y)\vee0.5$ for all $x,y\in X$.

To show (1) let $x,y\in X$ and $t=\mu_A(x)\wedge\mu_A(y)\wedge0.5$. Then $\mu_A(x)\geq t$ and $\mu_A(y)\geq t$ and so $[x_t\in A]=1$ and $[y_t\in A]=1$. Therefore, from Definition 4.7,

$$\begin{aligned}&1\geq[(x_t*y_t)\in\vee qA]\geq[x_t\in A]\wedge[y_t\in A]=1\\&\Rightarrow[(x_t*y_t)\in\vee qA]=1\\&\Rightarrow[(x_t*y_t)\in A]\vee[(x_t*y_t)qA]=1\\&\Rightarrow[(x_t*y_t)\in A]=1\end{aligned}$$

or $[(x_t * y_t)aA]=1 \Rightarrow \mu_A(x*y)\geq t$ or $\mu_A(x*y)+t>1 \Rightarrow \mu_A(x*y)\geq t$ or $\mu_A(x*y)>1 - t \geq 0.5\geq t$. Thus, $\mu_A(x)\geq t = \mu_A(x*y) \wedge \mu_A(y) \wedge 0.5$ for all $x,y\in X$.

Next (2) for $x,y\in X$ and let $v_A(x)\vee v_A(y)\vee 0.5=1 - s$ then $v_A(x)\leq 1 - s$ and $v_A(y)\leq 1 - s \Rightarrow s\leq 1 - v_A(x)$ and $s\leq 1 - v_A(y)$ which implies $[x_s\in A]\geq 1/2$ and $[y_s\in A]\geq 1/2$. Therefore, from Definition 4.7,

$$\begin{aligned}&\left[(x_s * y_s)\in \vee qA\right]\geq\left[x_s\in A\right]\vee\left[y_s\in A\right]\geq 1/2\\ &\Rightarrow\left[(x_s * y_s)\in\vee qA\right]\geq 1/2\\ &\Rightarrow\left[(x_s * y_s)\in A\right]\geq 1/2\end{aligned}$$

or $[x_s * y_s)qA]\geq 1/2 \Rightarrow s\leq 1 - v_A(x*y)$ or $v_A(x*y)<s\leq 1 - s$ [since $1 - s\geq 0.5$ so, $s\leq 0.5$].

Hence, $v_A(x*y)\leq 1-s = v_A(x)\vee v_A(y)\vee 0.5$.

Conversely, we let A be an intuitionistic fuzzy subalgebras of X with thresholds (0,0.5). We declare A is an $(\in,\in\vee q)$ -intuitionistic fuzzy subalgebras of X. Let $s,t\in[0,1]$ and for $x,y\in X$. Let $a=[x_s\in A]\wedge[y_t\in A]$.

Case 1. When $a=1$. Then $[x_s\in A]=1$ and $[y_t\in A]=1$, implies $\mu_A(x)\geq s$ and $\mu_A(y)\geq t$. If

$$[(x*y)_{s\wedge t}\in\vee qA]\leq 1/2,$$

then $\mu_A(x*y)<s\wedge t$ and $\mu_A(x*y)\leq 1 - s\wedge t$. Thus, $0.5> \mu_A(x*y)\geq \mu_A(x)\wedge \mu_A(y)\wedge 0.5$ and so, $\mu_A(x*y)\geq \mu_A(x)\wedge \mu_A(y)\geq s\wedge t$, is a contradiction to $\mu_A(x*y)<s\wedge t$. Thus,

$$[(x*y)_{s\wedge t}\in\vee qA]=1.$$

Case 2. When $a=1/2$. Then $[x_s\in A]=1/2$ and $[y_t\in A]\geq 1/2 \Rightarrow s\leq 1 - v_A(x)\geq s$ and $1 - v_A(y)\geq t$. Now, $1 - v_A(x)\vee v_A(y) = (1 - v_A(x))\wedge(1 - v_A(y))\geq s\wedge t$. If

$$[(x*y)_{s\wedge t}\in\vee qA]=0,$$

then $(1 - v_A(x*y))<s\wedge t$ and $v_A(x*y)\geq s\wedge t$. Now we have, $0.5<v_A(x*y)\leq v_A(x)\vee v_A(y)\vee 0.5$, and implies $v_A(x*y)\leq v_A(x)\vee v_A(y)$, so we have $1 - v_A(x*y)\geq 1 - v_A(x)\vee v_A(y) = (1 - v_A(x))\wedge 1 - v_A(y))\geq s\wedge t$, is a contradiction to take $(1 - v_A(x*y))<s\wedge t$. Thus,

$$[(x*y)_{s\wedge t}\in\vee qA]\geq 1/2=\left[x_s\in A\right]\wedge\left[y_t\in A\right].$$

Hence, $[(x_s * y_t)\in\vee qA]\geq[x_s\in A]\wedge[y_t\in A]$. Therefore, $A=\langle\mu_A,v_A\rangle$ is an $(\in,\in\vee q)$ -intuitionistic fuzzy subalgebras of X.■

Theorem 4.3 *An intuitionistic fuzzy set* $A=\langle\mu_A,v_A\rangle$ *of X is an* $(\in\wedge q,\in)$ *-intuitionistic fuzzy subalgebras of X if and only if A is an intuitionistic fuzzy subalgebras of X with thresholds* (0.5,1).

Proof: Suppose, $A=\langle \mu_A, \nu_A \rangle$ is an $(\in\wedge q,\in)$-intuitionistic fuzzy subalgebras of X. We have to show that A is an intuitionistic fuzzy subalgebas of X with thresholds (0.5,1), i.e. to show

1. $\mu_A(x*y) \vee 0.5 \geq \mu_A(x) \wedge \mu_A(y)$
2. $\nu_A(x*y) \wedge 0.5 \leq \nu_A(x) \vee \nu_A(y)$ for all $x,y \in X$.

To prove (1), let $t=\mu_A(x)\wedge\mu_A(y)$. Now, if $\mu_A(x*y) \vee 0.5 < t = \mu_A(x)\wedge\mu_A(y)$, and so $\mu_A(x)\geq t>0.5$ and $\mu_A(y)\geq t>0.5$, which implies $[x_t \in A]=1$, $[x_t qA]=1$ and $[y_t \in A]=1$, $[y_t qA]=1$ These gives

$[x_t \in\wedge qA]=1$ and $[y_t \in\wedge qA]=1$

$\Rightarrow [x_t \in\wedge qA]\wedge[y_t \in\wedge qA]=1$.

Therefore,

$$\left[(x_t * y_t)\in A\right] \geq \left[x_t \in \wedge qA\right] \wedge \left[y_t \in \wedge qA\right] \Rightarrow \left[(x_t * y_t)\in A\right]=1$$

imply $\mu_A(x*y)\geq t$, is a contradiction to our assumption $\mu_A(x*y) \vee 0.5 < t = \mu_A(x)\wedge\mu_A(y)$.
Hence, $\mu_A(x*y) \vee 0.5 \geq t = \mu_A(x)\wedge\mu_A(y)$.

Now, we claim (2), let $t=1-s=\nu_A(x)\vee\nu_A(y)$, then $1-s\geq\nu_A(x)$ and $1-s\geq\nu_A(y)$. But, if $\nu_A(x*y)\wedge 0.5\geq t=1-s=\nu_A(x)\vee\nu_A(y)$, then $s\leq 1-\nu_A(x)$, $s\leq 1-\nu_A(y)$ and $\nu_A(x*y)>t$ and then $s>0.5>t$. Thus, $\nu_A(x)\leq t<s$ and $\nu_A(y)\leq t<s \Rightarrow [x_s \in A]\geq 1/2$, $[x_s qA]\geq 1/2$, $[y_s \in A]\geq 1/2$ and $[y_s qA]\geq 1/2$. Therefore,

$$\left[(x_s * y_s)\in A\right] \geq \left[x_s \in \wedge qA\right] \wedge \left[y_s \in \wedge qA\right] \geq 1/2 \Rightarrow \left[(x_s * y_s)\in A\right] \geq 1/2,$$

which implies $s\leq 1-\nu_A(x*y)$, is a contradiction to our assumption $\nu_A(x*y)>t$. Hence, $\nu_A(x*y) \wedge 0.5 \leq t=\nu_A(x)\vee\nu_A(y)$.

Conversely, suppose A is an intuitionistic fuzzy subalgebras of X with thresholds (0.5,1). Let $x,y\in X$. Let $s,t\in[0,1]$. Next, we prove that $[(x*y)_{s\wedge t} \in A] \geq [x_s \in \wedge qA] \wedge [y_t \in \wedge qA]$. Let $a=\mu_A(x)\wedge\mu_A(y)$.

Case 1. At $a=1$. Then, $\mu_A(x)\geq s$, $\mu_A(x)+s>1$ and $\mu_A(y)\geq t$, $\mu_A(y)+t>1 \Rightarrow \mu_A(x)\geq 0.5$ and $\mu_A(y)\geq 0.5$. Thus, $\mu_A(x*y) \vee 0.5 \geq \mu_A(x)\wedge\mu_A(y)\geq s\wedge t$, we get from which $[(x*y)_{s\wedge t} \in A]=1$.

Case 2. At $a=1/2$. Then, $s\leq 1-\nu_A(x)$, $\nu_A(x)<s$ and $t\leq 1-\nu_A(y)$, $\nu_A(y)<t$, which implies $1-\nu_A(x)\geq s>\nu_A(x)$ and $1-\nu_A(y)\geq t>\nu_A(y) \Rightarrow \nu_A(x)<0.5$ and $\nu_A(y)<0.5$. Thus,

$$\nu_A(x*y)\wedge 0.5 \leq \nu_A(x)\vee\nu_A(y) \Rightarrow \nu_A(x*y)\leq\nu_A(x)\vee\nu_A(y).$$

Therefore, $1-\nu_A(x*y)\geq(1-\nu_A(x))\wedge(1-\nu_A(y))\geq s\wedge t$.

Thus, $[(x*y)_{s\wedge t} \in A]\geq 1/2$. Hence, $[(x*y)_{s\wedge t} \in A] \geq [x_s \in \wedge qA] \wedge [y_t \in \wedge qA]$. ■

Theorem 4.4 *An IFS* $A = \langle \mu_A, \nu_A \rangle$ *of X is an intuitionistic fuzzy subalgebras of X with thresholds (s,t) of X if and only if for any p∈(s,t], then* A_p *is a fuzzy subalgebras of X.*

Proof: Suppose that *A* is an intuitionistic fuzzy subalgebras of *X* with thresholds (*s*,*t*) of *X*. Let *x*,*y*∈*X* and *p*∈(*s*,*t*]. We have to prove $A_p(x^*y) \geq A_p(x) \wedge A_p(y)$. Let $a = A_p(x) \wedge A_p(y)$.

Case 1. For *a*=1. Then, $A_p(x)=1$ and $A_p(y)=1$. This implies $\mu_A(x) \geq p > s$ and $\mu_A(y) \geq p > s$. Now,

$$\mu_A(x^*y) \vee s \geq \mu_A(x) \vee \mu_A(y) \wedge t \geq (p \vee p) \wedge t = p,$$

which implies that $\mu_A(x^*y) \geq p$, i.e. $A_p(x^*y)=1$.

Case 2. For *a*=1/2. Then, $A_p(x)=1/2$ and $A_p(y)=1/2$, this implies that $1 - \nu_A(x) \geq p$ or $1 - \nu_A(y) \geq p$. Thus, $\nu_A(x) \wedge \nu_A(y) \leq 1 - p < 1 - s$. Now from which we get,

$$\nu_A(x^*y) \wedge (1-s) \leq (\nu_A(x) \wedge \nu_A(y)) \vee (1-t) \leq (1-p) \vee (1-t) = 1-p.$$

Since $t \geq p$ and $1 - s > 1 - p$. Therefore, $1 - \nu_A(x^*y) \geq p$ and so, $\nu_A(x^*y) \geq 1/2 = \nu_A(x) \vee \nu_A(y)$. Hence, $A_p(x^*y) \geq A_p(x) \vee A_p(y)$.

Conversely, we assume for any *p*∈(*s*,*t*], A_p is a fuzzy subalgebras of *X*. We have to show

$$\mu_A(x^*y) \vee s \geq \mu_A(x) \wedge \mu_A(y) \wedge t.$$

If

$$\mu_A(x^*y) \vee s < p = \mu_A(x) \wedge \mu_A(y) \wedge t,$$

then *p*∈(*s*,*t*] and $\mu_A(x) \geq p$ and $\mu_A(y) \geq p$. Thus from,

$$A_p(x^*y) \geq A_p(x) \wedge A_p(y) = 1,$$

which implies $A_p(x^*y)=1$, and so, $\mu_A(x^*y) \geq p$ which contradicts to $\mu_A(x^*y) < p$. Therefore,

$$\mu_A(x^*y) \vee s \geq \mu_A(x) \wedge \mu_A(y) \wedge t.$$

We have to show

$$\nu_A(x^*y) \wedge (1-s) \leq \nu_A(x) \vee \nu_A(y) \vee (1-t).$$

If

$$\nu_A(x*y)\wedge(1-s)>p=\nu_A(x)\vee\nu_A(y)\vee(1-t),$$

then

$$(1-\nu_A(x*y))\vee s<c=1-p=(1-\nu_A(x))\wedge(1-\nu_A(y))\wedge t,$$

and so, $c\in(s,t]$ and $1-\nu_A(x)\geq c$ and $1-\nu_A(y)\geq c$. Thus, from

$$A_p(x*y)\geq A_p(x)\wedge A_p(y)\geq 1/2,$$

we have $A_p(x*y)\geq 1/2$ and so, $1-\nu_A(x*y)\geq c=1-p$. Therefore, $\nu_A(x*y)\leq p$, which contradicts to $\nu_A(x*y)>p$. Hence, $\nu_A(x*y)\wedge(1-s)\leq\nu_A(x)\vee\nu_A(y)\vee(1-t)$.

Therefore, $A=\langle\mu_A,\nu_A\rangle$ is an intuitionistic fuzzy subalgebras of X with thresholds (s,t). ■

Definition 4.2 (Yuan et al., 2010) *Let* $A=\langle\mu_A,\nu_A\rangle$ *be a IFS of X, for a fuzzy point* x_a *and* $s\in(0,1)$. *Then,*

1. $$[x_a q_s A]=\begin{cases}1, & \text{if } a+\mu_A(x)>2s\\ \frac{1}{2}, & \text{if } \mu_A(x)\leq 2s-a<1-\nu_A(x)\\ 0, & \text{if } \nu_A(x)+2s\geq a+1\end{cases}$$
2. $[x_a\in\vee q_s A]=[x_a\in A]\vee[x_a q_s A]$
3. $[x_a\in\wedge q_s A]=[x_a\in A]\wedge[x_s q_s A]$

The above definition introduced the following theorems.

Theorem 4.5 *For* $s,t\in[0,1]$ *and* $0<s<t$. *Then,* $A=\langle\mu_A,\nu_A\rangle$ *is a* $(s,t]$ *-intuitionistic fuzzy subalgebra of X if and only if*

$$[x_m\in q_s A]\wedge[y_n\in q_s A]\Rightarrow[x_m y_n\in A)=1$$

i.e.,

$$[x_m y_n\in A]\geq[x_m\in\wedge q_s A]\wedge[y_n\in\wedge q_s A]$$

for all $x,y\in X$ and for all $m,n\in(0,t]$.

Proof: ⇒ Let us take $m,n\in(0,t]$ and $p=\left[x_m \in \wedge q_s A\right]\wedge\left[y_n \in \wedge q_s A\right]$.

Case 1. For p=1. Then $\mu_A(x)\geq m$, $m+\mu_A(x)\geq 2s$ and $\mu_A(y)\geq n$, $n+\mu_A(y)\geq 2s$. Thus, $\mu_A(x)\geq s$ and $\mu_A(y)\geq s$. Therefore,

$$\mu_A(x{*}y)\vee s\geq \mu_A(x)\wedge\mu_A(y)\wedge t.$$

Since $\mu_A(x{*}y)\geq s$. So,

$$\mu_A(x{*}y)\geq \mu_A(x)\wedge\mu_A(y)\wedge t\geq m\wedge n\wedge t=m\wedge n.$$

Hence, $[x_m y_n\in A]=1$.

Case 2. $p=1/2$. Then, $1-v_A(x)\geq m$, $1-v_A(y)\geq n$, $2s-m<1-v_A(x)$ and $2s-n<1-v_A(y)$. Thus, $1-\mu_A(x)>s$ and $1-v_A(y)>s$. By

$$v_A(x{*}y)\wedge(1-s)\leq v_A(x)\vee v_A(y)\vee(1-t).$$

But, it known that

$$\left(1-v_A(x{*}y)\right)\vee s\geq\left(1-v_A(x)\right)\vee\left(1-v_A(y)\right)\vee t.$$

Thus obtained, $1-v_A(x{*}y)\geq s$ and

$$1-v_A(x{*}y)\geq m\wedge n\wedge t=m\wedge n.$$

Hence, $[x_m y_n\in A]\geq 1/2$. Therefore, $\left[x_m y_n \in A\right]\geq\left[x_m \in \wedge q_s A\right]\wedge\left[y_n \in \wedge q_s A\right]$

⇐ (1) If

$$\mu_A(x{*}y)\vee s<m=\mu_A(x)\wedge\mu_A(y)\wedge t,$$

then $s<a\leq t$, $\mu_A(x)\geq m$, $\mu_A(y)\geq m$. So, $m+\mu_A(y)>2s$. Thus,

$$\left[x_m \in \wedge q_s A\right]=\left[y_n \in \wedge q_s A\right]=1.$$

Hence, $[x_m y_n\in A]=1$, i.e., $\mu_A(x{*}y)\geq m$ is a contradiction to assume $\mu_A(x{*}y)<m$. Therefore,

$$\mu_A(x*y)\vee s\geq\mu_A(x)\wedge\mu_A(y)\wedge t.$$

(2) If

$$v_A(x*y)\wedge(1-s)>1-m=v_A(x)\vee v_A(y)\vee(1-t),$$

then

$$1-v_A(x*y))\vee s<m=(1-v_A(x))\wedge(1-v_A(y))\wedge t.$$

Thus, $s<m\leq t$, $1-v_A(x)\geq m$, $1-v_A(y)\geq m$. Then, $1-v_A(x)>2s-m$ and $1-v_A(y)>2s-m$ implies that

$$[x_m y_n\in A]\geq[x_m\in\wedge q_s A]\wedge[y_n\in\wedge q_s A]\geq 1/2.$$

Thus, $1-v_A(x*y)\geq m$ is contradicts to $1-v_A(x*y)<m$. Hence,

$$v_A(x*y)\wedge(1-s)\geq v_A(x)\vee v_A(y)\vee(1-t).$$

Therefore, A is a $(s,t]$ -intuitionistic fuzzy subalgebras of X. ■

Theorem 4.6 *Let $s,t\in[0,1]$ and $s<t<1$. Then $A=\langle\mu_A,v_A\rangle$ is a $(s,t]$ -intuitionistic fuzzy subalgebras of X if and only if*

$$([x_m\in A]\wedge[y_n\in A]\Rightarrow[x_m y_n\in\vee q_t A])=1,$$

i.e.,

$$[x_m y_n\in\vee q_t A]\geq[x_m\in A]\wedge[y_n\in A]$$

for all $m,n\in\{s,1]$ and for all $x,y\in X$.

Proof: Theorem can be similarly proved by using the Definition 4.7 and Theorem 4.8. ■

5 CONCLUSION

The study of different algebraic structures plays an important role in pure mathematics. In this article, we defined new type of (α,β) - IF subalgebras of *BCK/BCI*-algebras X where, $\alpha,\beta\in\{\in,q,\in\wedge q,\in\vee q\}$ based on the cut set of a IF sets A. The following results are investigated here:

We study (∈,∈) (or (∈∧q,∈), (∈,∈∨q)) - IF subalgebras of *BCK/BCI* -algebras. It is investigated for IFS A is an (∈,∈) (or (∈∧q,∈) or (∈,∈∨q))- IF subalgebras of X if and only if A is an IF subalgebras with thresholds (0,1) (or (0,0.5] or (0.5,1] and A is an (∈,∈) (or (∈∧q,∈) or (∈,∈∨q))-IF subalgebras of X if and only if for any p∈(0,1) (or p∈{0,0.5] or p∈(0.5,1]), then A_p is a fuzzy subalgebras of X respectively. By the definition of IF subalgebras with {s,t] thresholds, then A_p is a fuzzy subalgebras of X if and only if for any p∈(s,t]. It is our view that this is an opening for study different types of (α,β)-intuitionistic fuzzy subalgebras.

REFERENCES

Atanassov, K. T. (1999). *Intuitionistic Fuzzy Sets: Theory and Applications, Studies in Fuzziness and Soft Computing* (Vol. 35). Heidelberg, Germany: Physica-Verlag. doi:10.1007/978-3-7908-1870-3

Bej, T., & Pal, M. (2015). Doubt Atanassovs intuitionistic fuzzy Sub-implicative ideals in *BCI*. *Int. J. Comput. Int. Sys.*, *8*(2), 240–249. doi:10.1080/18756891.2015.1001948

Bhakat, S. K., & Das, P. (1992). On the definition of a fuzzy subgroup. *Fuzzy Sets and Systems*, *51*(2), 235–241. doi:10.1016/0165-0114(92)90196-B

Bhakat, S. K., & Das, P. (1996). -fuzzy subgroup. *Fuzzy Sets and Systems*, *80*(3), 359–368. (∈,∈∨q) doi:10.1016/0165-0114(95)00157-3

Borns, D. W., & Mack, J. M. (1975). *An algebraic Introduction to Mathematical Logic*. Berlin: Springer. doi:10.1007/978-1-4757-4489-7

Huang, Y. S. (2006). BCI-algebra. Science Press.

Imai, Y., & Iseki, K. (1966). On axiom system of propositional calculi. XIV. *Proc. of the Japan Aceademy*, *42*(1), 19–22. doi:10.3792/pja/1195522169

Jana, C., Pal, M., & Saied, A. B. (2017). -bipolar fuzzy (∈,∈∨q)BCK/BCI-algebras. *Missouri J. of Math. Sci.*, *29*(2), 139–160.

Jana, C., Senapati, T., Bhowmik, M., & Pal, M. (2015). On intuitionistic fuzzy *G*. *Fuzzy Inf. Eng.*, *7*(2), 195–209. doi:10.1016/j.fiae.2015.05.005

Jana, C., Senapati, T., & Pal, M. (2016). -intuitionistic fuzzy(∈,∈∨q)BCI. *Journal of Intelligent & Fuzzy Systems*, *31*(1), 613–621. doi:10.3233/IFS-162175

Jun, Y. B. (2005). On (α,β)BCK/BCI. *Bulletin of the Korean Mathematical Society*, *42*(4), 703–711. doi:10.4134/BKMS.2005.42.4.703

Jun, Y. B. (2007). Fuzzy subalgebras with thresholds in *BCK/BCI*. *Commun. Korean Math. Soc.*, *22*(2), 173–181. doi:10.4134/CKMS.2007.22.2.173

Jun, Y. B., & Kim, K. H. (2000). Intuitionistic fuzzy ideals in *BCK*. *International Journal of Mathematics and Mathematical Sciences*, *24*(12), 839–849. doi:10.1155/S0161171200004610

Larimi, M.A. (2013). On $(\in,\in\vee q_k)$ -intuitionistic fuzzy ideals of hemirings. *World Applied Sciences, 21*, 54-67.

Ma, X., Zhan, J., & Jun, Y. B. (2009). Some types of $(\in,\in\vee q)$-interval-valued fuzzy ideals of BCI algebras. *Iranian Journal of Fuzzy Systems, 6*, 53–63.

Ma, X., Zhan, J., & Jun, Y. B. (2011). Some kinds of$(\in\gamma,\in\gamma\vee q\delta)$ BCI. *Computers & Mathematics with Applications (Oxford, England), 61*(4), 1005–1015. doi:10.1016/j.camwa.2010.12.049 PMID:21542996

Ma, X., Zhan, J., & Jun, Y. B. (2012). New types of fuzzy ideals of *BCI*. *Neural Computing & Applications, 21*(S1), S19–S27. doi:10.100700521-011-0558-x

Mordeson, J. N., & Malik, D. S. (1998). *Fuzzy commutative algebra*. World scientific. doi:10.1142/3929

Murali, V. (2004). Fuzzy points of equivalent fuzzy subsets. *Inform. Sci., 158*, 277–288. doi:10.1016/j.ins.2003.07.008

Pu, P., & Liu, Y. (1980). Fuzzy topology I. Neighbourhood structures of a fuzzy point and Moore-Smith convergence. *Journal of Mathematical Analysis and Applications, 76*(2), 571–579. doi:10.1016/0022-247X(80)90048-7

Rosenfeld, R. (1971). Fuzzy groups. *Journal of Mathematical Analysis and Applications, 35*(3), 512–517. doi:10.1016/0022-247X(71)90199-5

Yuan, X. H., Li, H. X., & Lee, E. S. (2010). On the definition of the intuitionistic fuzzy subgroups. *Computers & Mathematics with Applications (Oxford, England), 59*(9), 3117–3129. doi:10.1016/j.camwa.2010.02.033

Yuan, X. H., Li, H. X., & Sun, K. B. (2009). The cut sets, decomposition theorems and representation theorems on intuitionistic fuzzy sets and interval-valued fuzzy sets [in Chinese]. *Science in China Series F: Information Sciences, 39*(9), 933–945.

Zadeh, L. A. (1965). Fuzzy sets. *Information and Control, 8*(3), 338–353. doi:10.1016/S0019-9958(65)90241-X

Zadeh, L. A. (2005). Toward a generalized theory of uncertainty (GTU)-an outline. *Inform. Sci., 172*(1-2), 1–40. doi:10.1016/j.ins.2005.01.017

Zadeh, L. A. (2008). Is there a need for fuzzy logic? *Inform. Sci., 178*(13), 2751–2779. doi:10.1016/j.ins.2008.02.012

Zhan, J., & Tan, Z. (2004). Intuitionistic fuzzy *a*-ideals in BCI-algebras. *Soochow Math. J., 30*, 207–216.

Chapter 13
Interval-Valued Intuitionistic Fuzzy Subnear Rings

Amal Kumar Adak
https://orcid.org/0000-0002-3644-782X
Ganesh Dutt College, India

ABSTRACT

The theory of interval-valued intuitionistic fuzzy sets is a generalization of both interval-valued fuzzy sets and intuitionistic fuzzy sets. In this chapter, the notion of interval-valued intuitionistic fuzzy subnear-ring is introduced, and some interesting properties are discussed. Some relations on the family of all interval-valued intuitionistic fuzzy subnear-ring are presented, and some related properties are investigated. Also, the authors represent upper and lower level set of interval-valued intuitionistic fuzzy set.

1 INTRODUCTION

The theory of Instuitionistic fuzzy set has been found to be more useful to deal with vagueness and uncertainty in decision situations than that of the fuzzy set. Atanassov and Gargov (1989) further generalized the IFS on the spirit of ordinary interval-valued fuzzy sets (IVFS) and defined the notion of interval-valued intuitionistic fuzzy set (IVIFS). The relations, operations and operators related to IFS and IVIFS have been systematically studied in (1994, 2010). Adak *et al.* investigated some interesting properties of interval-valued intuitionistic fuzzy sets and matrices in (2011c, 2012a, 2012b, 2012c, 2013).

Algebraic structures play a prominent role in mathematics with wide ranging applications in many disciplines such as theoretical physics, computer sciences, control engineering, information sciences and so on. This provides sufficient motivations to researchers to review various concepts and results from the real abstract algebra in the broader framework of fuzzy setting. To study more about fuzzy algebraic structures like semigroups and subgroups see (2013). A near-ring satisfying all axioms of an associative ring, except for commutativity of addition and one of the distributive laws. Abou-Zaid (1989) introduced the concept of fuzzy subnear-ring and studied fuzzy ideals of near-rings. The concept was discussed further by many researchers in (1999, 2009 2019). Roh (2006) *et. al.* discussed interval-valued fuzzy subalgebras in BCK algebras. Jun (2002) discussed some properties of interval-valued fuzzy R-subgroup of near-rings.

DOI: 10.4018/978-1-7998-0190-0.ch013

The structure of this paper is organized as follows. In Section 2, the preliminaries and some definitions of interval-valued intuitionistic fuzzy set are given. In Section 3, interval-valued intuitionistic fuzzy R-subgroup of near-ring and interval-valued intuitionistic fuzzy subnear-ring of near-ring are defined. In Section 4, we represent upper and lower level set of interval-valued intuitionistic fuzzy set. Also, some relations on the family of all interval-valued intuitionistic fuzzy subnear-ring are presented. Finally, a conclusion is made in Section 5.

2 PRELIMINARIES AND DEFINITION

In this section, we recall the concept of interval arithmetics. Let $D(0,1)$ be the set of all closed subintervals of the interval [0,1]. Then according to Zadeh's extension principle, we can popularize these operations such as $^{\vee}$, $^{\wedge}$ and c (complement) to $D(0,1)$. Thus $([I], {}^{\vee}, {}^{\wedge}, c)$ is a complete lattice with a minimal element $\overline{0}=[0,0]$ and a maximal element $\overline{1}=[1,1]$. Furthermore let $\overline{a}=[a^-,a^+]$, $\overline{b}=[b^-,b^+]$, where a^- and a^+ are lower and upper limits of $\overline{a}$, be two intervals then we have $\overline{a}=\overline{b} \Leftrightarrow a^-=b^-$, $a^+=b^+$, $\overline{a}\leq\overline{b} \Leftrightarrow a^-\leq b^-$, $a^+\leq b^+$, and $\overline{a}<\overline{b} \Leftrightarrow a^-<b^-$, $a^+<b^+$ and $\overline{a}\neq\overline{b}$.

Definition 2.1 Let X be a fixed set. An intuitionistic fuzzy set (IFS) A in X is an expression having the form $A=\{\langle x,\mu_A(x),v_A(x)\rangle : x\in X\}$, where the functions $\mu_A(x)$ and $v_A(x)$ are the degree of membership and the degree of non-membership of the element $x\in X$ respectively. Also $\mu_A: X\to[0,1]$, $v_A: X\to[0,1]$ and $0\leq\mu_A(x)+v_A(x)\leq 1$, for all $x\in X$.

Definition 2.2 An interval-valued intuitionistic fuzzy set (IVIFS) A over X (universe of discourse) is an object having the form $A=\{\langle x,M_A(x),N_A(x)\rangle \mid x\in X\}$, where $M_A(x): X\to[I]$ and $N_A(x): X\to[I]$. The intervals $M_A(x)$ and $N_A(x)$ denote the intervals of the degree of membership and degree of non-membership of the element x to the set A, where $M_A(x)=[M_{AL}(x), M_{AU}(x)]$ and $N_A(x)=[N_{AL}(x), N_{AU}(x)]$, for all $x\in X$, with the condition $0\leq M_{AU}(x)+N_{AU}(x)\leq 1$. For simplicity, we denote

$$A=\{\langle x,[A^-(x),A^+(x)],[B^-(x),B^+(x)]\rangle \mid x\in X\}$$

and $F(X)$ and $P(X)$ as the set of all IVIFSs and crisp sets in X respectively.

2.1 Some Operations on IVIFSs

In (2010), Bhowmik and Pal define some relational operations on IVIFSs. Let A and B be two IVIFSs on X, where

$$A=\{\langle[M_{AL}(x),M_{AU}],[N_{AL}(x),N_{AU}(x)] \mid x\in X\rangle\}$$

and

$$B=\{\langle[M_{BL}(x),M_{BU}],[N_{BL}(x),N_{BU}(x)] \mid x\in X\rangle\}.$$

Then

$A=B \Leftrightarrow M_A(x)=M_B(x)$ and $N_A(x)=N_B(x)$, for all $x \in X$

$$A \subseteq B \text{ iff } \{(M_{AU}(x) \leq M_{BU}(x) \text{ and } M_{AL}(x) \leq M_{BL}(x))\}$$

and

$$\{(N_{AU}(x) \geq N_{BU}(x) \text{ and } N_{AL}(x) \geq N_{BL}(x))\},$$

for all $x \in X$

$$\overline{A} = \{\langle x, N_A(x), M_A(x)\rangle \mid x \in X\},$$

for all $x \in X$

$$A \cap B = \{\langle [min\{M_{AL}(x), M_{BL}(x)\}, min\{M_{AU}(x), M_{BU}(x)\}], \\ [max\{N_{AL}(x), N_{BL}(x)\}, max\{N_{AU}(x), N_{BU}(x)\}]\rangle \mid x \in X\}.$$

$$A \cup B = \{\langle [max\{M_{AL}(x), M_{BL}(x)\}, max\{M_{AU}(x), M_{BU}(x)\}], \\ [min\{N_{AL}(x), N_{BL}(x)\}, min\{N_{AU}(x), N_{BU}(x)\}]\rangle \mid x \in X\}.$$

$$\overline{A} = [\overline{M_A(x)}, \overline{N_A(x)}] \text{ where } \overline{M_A(x)} = [1-M_{AU}(x), 1-M_{AL}(x)]$$

and

$$\overline{N_A(x)} = [1-N_{AU}(x), 1-N_{AL}(x)]$$

represent the complemetn of x in A

$$A = \langle M_A, \overline{M_A}\rangle \text{ where } \overline{M_A(x)} = [1-M_{AU}(x), 1-M_{AL}(x)].$$

$$A = \langle \overline{N_A}, N_A\rangle \text{ where } \overline{N_A(x)} = [1-N_{AU}(x), 1-N_{AL}(x)].$$

3 INTERVAL-VALUED INTUITIONISTIC FUZZY SUBGROUP OF NEAR RING

The concept of interval-valued intuitionistic fuzzy R-subgroup of near-ring is an important and useful generalization of ordinary fuzzy R-subgroup of near-rings. In this section, some fundamental aspects of interval-valued intuitionistic fuzzy R-subgroup of near-ring will be discussed.

A near ring is a non-empty set R with two binary operation + and . satisfying the following axioms:

(i) $(R,+)$ is a group,
(ii) $(R,.)$ is a semigroup,
(iii) $x.(y+z)=x.y+x.z$ for all $x, y, z \in R$.

Precisely speaking, it is a left near-ring because it satisfied left distributive law. We will use the word "near ring " instead of "left near ring ". Note that $x.0=0$ and $x(-y)=-xy$ for all $x,y \in R$ but in general $0.x \neq 0$ for some $x \in R$

An R-subgroup of a near ring R is a subset H of R such that

(i) $(H,+)$ is a subgroup of $(R,+)$
(ii) $RH \subseteq H$
(iii) $HR \subseteq H$.

If H satisfies (i) and (ii), then it is called a left R-subgroup of R and if H if H satisfies (i) and (iii), then it is called a right R-subgroup of R.

Definition 3.1 *In(2002), Jun and Kim defined interval-valued fuzzy R-subgroup with respect to Min-norm. Let an interval valued fuzzy set A where A=[A^-,A^+] in a near ring R is called an interval-valued fuzzy right (resp. left) R-subgroup of R with respect to the interval Min-norm if*

1. $A(x-y) \geq \mathrm{Min}^i(A(x),A(y))$
2. $A(xr) \geq A(x)$ resp. $A(rx) \geq A(x)$ for all $x, y, r \in R$.

Definition 3.2 *Let* $A=\langle M_A, N_A \rangle$ be an IVIFS of R. Then A is called an interval-valued intuitionistic fuzzy subnear-ring of near-ring R if for all x,y∈*R,*

(1) $M_{AL}(x-y) \geq M_{AL}(x) \wedge M_{AL}(y)$, $M_{AU}(x-y) \geq M_{AU}(x) \wedge M_{AU}(y)$ and $N_{AL}(x-y) \leq N_{AL}(x) \vee N_{AL}(y)$, $N_{AU}(x-y) \leq N_{AU}(x) \vee N_{AU}(y)$
(2) $M_{AL}(xy) \geq M_{AL}(x) \wedge M_{AL}(y)$, $M_{AU}(xy) \geq M_{AU}(x) \wedge M_{AU}(y)$ and $N_{AL}(xy) \leq N_{AL}(x) \vee N_{AL}(y)$, $N_{AU}(xy) \leq N_{AU}(x) \vee N_{AU}(y)$.

Definition 3.3 *Let* $A=\langle M_A, N_A \rangle$ be an IVIFS of R. Then A is called an interval-valued intuitionistic fuzzy R-subgroup of near-ring R if for all x,y∈*R,*

(1) $M_{AL}(x-y) \geq M_{AL}(x) \wedge M_{AL}(y)$, $M_{AU}(x-y) \geq M_{AU}(x) \wedge M_{AU}(y)$ and $N_{AL}(x-y) \leq N_{AL}(x) \vee N_{AL}(y)$, $N_{AU}(x-y) \leq N_{AU}(x) \vee N_{AU}(y)$.

(2) $M_{AL}(rx) \geq M_{AL}(x)$, $M_{AU}(rx) \geq M_{AU}(x)$ and $N_{AL}(rx) \leq N_{AL}(x)$, $N_{AU}(rx) \leq N_{AU}(x)$.
(3) $M_{AL}(xr) \geq M_{AL}(x)$, $M_{AU}(xr) \geq M_{AU}(x)$ and $N_{AL}(xr) \leq N_{AL}(x)$, $N_{AU}(xr) \leq N_{AU}(x)$.

If $A = \langle M_A, N_A \rangle$ satisfies (1) and (2), then A is called an interval-valued intuitionistic fuzzy left R-subgroup of R and if $A = \langle M_A, N_A \rangle$ satisfies (1) and (3), then A is called an interval-valued intuitionistic fuzzy right R-subgroup of R.

Example 3.1 Let $R=\{a,b,c,d\}$ be a set with two binary operations + and . as follows:

+	*a*	*b*	*c*	*d*
a	*a*	*b*	*c*	*d*
b	*b*	*a*	*d*	*c*
c	*c*	*d*	*b*	*a*
d	*d*	*c*	*a*	*b*

+	*a*	*b*	*c*	*d*
a	*a*	*a*	*a*	*a*
b	*a*	*a*	*a*	*a*
c	*a*	*a*	*a*	*a*
d	*a*	*a*	*b*	*b*

Then $(R,+,.)$ is a near ring. We define an IVIFS $A = \langle M_A, N_A \rangle$ by

$M_A(a)=[0.4,0.6]$, $N_A(a)=0.1,0.4]$

$M_A(b)=[0.4,0.5]$, $N_A(b)=[0.3,0.5]$

$M_A(c)=[0.3,0.4]$, $N_A(c)[0.2,0.6]$

$M_A(d)=0.2,0.3]$, $N_A(d)=[0.1,0.6]$

Then A is both interval-valued intuitionistic fuzzy subnear-ring and interval-valued intuitionistic fuzzy R-subgroup of near-ring R.

Proposition 3.1 *If A is an IVIF R-subgroup of a near-ring R, then* $M_{AL}(0) \geq M_{AL}(x)$, $M_{AU}(0) \geq M_{AU}(x)$ *and* $N_{AL}(0) \leq N_{AL}(x)$, $N_{AU}(0) \leq N_{AU}(x)$.

Proof. For every $x \in A$, we have

$$M_{AL}(0) = M_{AL}(x-x) \geq M_{AL}(x) \wedge M_{AL}(x) = M_{AL}(x)$$

and

$$M_{AU}(0) = M_{AU}(x-x) \geq M_{AU}(x) \wedge M_{AU}(x) = M_{AU}(x).$$

Therefore, $M_{AL}(0) \geq M_{AL}(x)$, $M_{AU}(0) \geq M_{AU}(x)$. Again, For every $x \in A$, we get

$$N_{AL}(0) = N_{AL}(x - x) \le N_{AL}(x) \vee N_{AL}(x) = N_{AL}(x)$$

and

$$N_{AU}(0) = N_{AU}(x - x) \le N_{AU}(x) \vee N_{AU}(x) = N_{AU}(x).$$

Therefore, $N_{AL}(0) \le N_{AL}(x)$, $N_{AU}(0) \le N_{AU}(x)$.
This completes the proof.

Theorem 3.1 *If* $A = \langle M_A, N_A \rangle$ and $B = \langle M_B, N_B \rangle$ are two IVIF R-subgroup of R then so is $A \cap B$.

Proof. For any $x,y \in R$, we have

$$\begin{aligned}(M_{AL} \wedge M_{BL})(x - y) &= M_{AL}(x - y) \wedge M_{BL}(x - y)\\ &\ge (M_{AL}(x) \wedge M_{AL}(y)) \wedge (M_{AL}(x) \wedge M_{AL}(y))\\ &= (M_{AL}(x) \wedge M_{BL}(x)) \wedge (M_{AL}(y) \wedge M_{BL}(y))\\ &= (M_{AL} \wedge M_{BL})(x) \wedge (M_{AL} \wedge M_{BL})(y).\end{aligned}$$

Similarly, it can be shown that

$$(M_{AL} \wedge M_{BL})(x - y) \ge (M_{AL} \wedge M_{BL})(x) \wedge (M_{AL} \wedge M_{BL})(y).$$

Again,

$$\begin{aligned}(N_{AL} \vee N_{BL})(x - y) &= N_{AL}(x - y) \vee N_{BL}(x - y)\\ &\le (N_{AL}(x) \vee N_{AL}(y)) \vee (N_{AL}(x) \vee N_{AL}(y))\\ &= (N_{AL}(x) \vee N_{BL}(x)) \vee (N_{AL}(y) \vee N_{BL}(y))\\ &= (N_{AL} \vee N_{BL})(x) \vee (N_{AL} \vee N_{BL})(y),\end{aligned}$$

and

$$(N_{AL} \vee N_{BL})(x - y) \le (N_{AL} \vee N_{BL})(x) \vee (N_{AL} \vee N_{BL})(y).$$

Now, if $x,r \in R$, we have

$$(M_{AL} \wedge M_{BL})(xr) = M_{AL}(xr) \wedge M_{BL}(xr) \ge M_{AL}(x) \wedge M_{BL}(x) = (M_{AL} \wedge M_{BL})(x).$$

Similarly,

$(M_{AU} \wedge M_{BU})(xr) \geq (M_{AU} \wedge M_{BU})(x)$.

$(N_{AL} \vee N_{BL})(xr) = N_{AL}(xr) \vee N_{BL}(xr) \leq N_{AL}(x) \vee N_{BL}(x) = (N_{AL} \vee N_{BL})(x)$.

Similarly,

$(N_{AU} \vee N_{BU})(xr) \leq (N_{AU} \vee N_{BU})(x)$.

Hence $A \cap B$ is an IVIF R-subgroups of *R*.

Theorem 3.2 *Let* $A = \langle M_A, N_A \rangle$ be an IVIFS in R. Then A is an IVIF R-subgroup of R if and only if M_A and $\overline{N_A}$ are interval-valued fuzzy R-subgroup in R.

Proof. Let $A = \langle M_A, N_A \rangle$ be an IVIF R-subgroup of *R*. Then by the definition of IVIF R-subgroup of near-rings we have for all $x,y \in R$

(1) $M_A(x-y) \geq M_A(x) \wedge M_A(y)$ and $N_A(x-y) \leq N_A(x) \vee N_A(y)$
(2) $M_A(xy) \geq M_A(x) \wedge M_A(y)$ and $N_A(xy) \leq N_A(x) \vee N_A(y)$.

Hence M_A is an IVF R-subgroup of near-ring.
Since $N_A(x-y) \leq N_A(x) \vee N_A(y)$ and so $\overline{N_A}(x-y) \geq \overline{N_A(x) \vee N_A(y)}$.
Therefore, $\overline{N_A(x-y)} \geq \overline{N_A(x)} \wedge \overline{N_A}(y)$.
Again, $N_A(xy) \leq N_A(x) \vee N_A(y)$ so $\overline{N_A(xy)} \geq \overline{N_A(x) \vee N_A(y)}$.
Therefore, $\overline{N_A(xy)} \geq \overline{N_A(x)} \wedge \overline{N_A(y)}$.
Hence $\overline{N_A}$ is an interval-valued fuzzy R-subgroup of near-ring.
Converse part is follows from the definitions.

4 SOME PROPERTIES OF INTERVAL-VALUED INTUITIONISTIC FUZZY SUBNEAR-RING

In this section, we discuss some interesting properties of interval-valued intuitionistic fuzzy subnear-ring. First, we define upper and lower level interval cut-sets. Then using these definition level sets, we give some theorems and obtain some useful properties.

Definition 4.1 *Let* $A = \langle M_A, N_A \rangle$ be an IVIFS. For $[\alpha_1,\alpha_2]$, $[\beta_1,\beta_2] \in D[0,1]$, *where* $0 \leq \alpha_1 \leq M_{AL}(0)$, $0 \leq \alpha_2 \leq M_{AU}(0)$ *and* $N_{AL}(0) \leq \beta_1 \leq 1$, $N_{AU}(0) \leq \beta_2 \leq 1$ *the set* $U(M_A;[\alpha_1,\alpha_2]) = \{x \in X: M_A(x) \geq [\alpha_1,\alpha_2]\}$ *is called upper* α *interval-cut and* $L(N_A;[\beta_1,\beta_2])$ *is called lower* β *interval-cut set of A.*

Theorem 4.1 *Let* $A = \langle M_A, N_A \rangle$ be an IVIFS in R. Then A is an IVIF R-subgroup of R if and only if $U(M_A;[\alpha_1,\alpha_2])$ *and* $L(N_A;[\beta_1,\beta_2])$ *are IVIF R-subgroup in R for any* $0 \leq \alpha_1 \leq M_{AL}(0)$, $0 \leq \alpha_2 \leq M_{AU}(0)$ *and* $N_{AL}(0) \leq \beta_1 \leq 1$, $N_{AU}(0) \leq \beta_2 \leq 1$.

Proof. Let $A=\langle M_A, N_A\rangle$ be an IVIF R-subgroup in R and α in $D(0,1)$ such that $0\leq\alpha_1\leq M_{AL}(0)$ and $0\leq\alpha_2\leq M_{AU}(0)$. If $x,y\in U(M_A;\alpha)$, then $M_{AL}(x)\geq\alpha_1$, $M_{AU}(x)\geq\alpha_2$ and $M_{AL}(y)\geq\alpha_1$, $M_{AU}(y)\geq\alpha_2$.

Hence

$$M_{AL}(x-y)\geq M_{AL}(x)\wedge M_{AL}(y)\geq\alpha_1$$

and

$$M_{AU}(x-y)\geq M_{AU}(x)\wedge M_{AU}(y)\geq\alpha_2.$$

It follows that $(x-y)\in U(M_A;\alpha)$. If $r\in R$ and $x\in U(M_A;\alpha)$, then $M_{AL}(y)\geq\alpha_1$ and $M_{AU}(y)\geq\alpha_2$. Then we have $M_A(rx)\geq M_A(x)\geq\alpha$, i.e.,

$$RU(M_A;\alpha)\subseteq U(M_A;\alpha) \text{ and } U(M_A;\alpha)R\subseteq U(M_A;\alpha).$$

Similarly, it can be prove for $L(N_A;\beta)$.

Conversely, let $x,y\in R$ and $\alpha=[\alpha_1,\alpha_2]$, where

$$\alpha_1=M_{AL}(x)\wedge M_{AL}(y) \text{ and } \alpha_2=M_{AU}(x)\wedge M_{AU}(y).$$

Then $x,y\in U(M_A;\alpha)$ and $(x-y)\in U(M_A;\alpha)$ since $U(M_A;\alpha)$ is an IVF R-subgroup of R. Hence

$$M_A(x-y)\geq\alpha=M_A(x)\wedge M_A(y).$$

If $x,y\in R$ and $\alpha=M_A(x)$ then $x\in U(M_A;\alpha)$ and xr, $rx\in U(M_A;\alpha)$ since $U(M_A;\alpha)$ is an IVF R-subgroup in R. Hence $M_A(rx)\geq\alpha=M_A(x)$ and $M_A(rx)\geq\alpha=M_A(x)$.

Definition 4.2: Let IVIFSN(R) be the family of all interval-valued intuitionistic fuzzy subnear-rings of R and α be a fixed interval in $D(0,1)$. We define two binary relations U^α and L^α as follows

$$(A,B)\in U^\alpha\Leftrightarrow U(M_A;\alpha)=U(M_B;\alpha) \text{ and } (A,B)\in L^\alpha\Leftrightarrow L(M_A;\alpha)=L(M_B;\alpha)$$

respectively for any $A=\langle M_A, N_A\rangle$ and $B=\langle M_B, N_B\rangle\in IVIFSN(R)$. Then the two relations U^α and L^α are equivalence relations on IVIFSN(R). These equivalence relations give rise to partitions of IVIFSN(R) into the equivalence classes $[A]_{U^\alpha}$ and $[A]_{L^\alpha}$ for any $A=\langle M_A, N_A\rangle\in IVIFSN(R)$ respectively. And we will denote the quotient sets of IVIFSN(R) by U^α and L^α as IVIFSN(R)/ U^α and IVIFSN(R)/ L^α respectively.

If IVFSN(R) is the family of all subnear-rings of R and $\alpha\in[0,1]$ then we define two maps U_α and L_α from IVIFSN(R) to IVSN(R)$\cup\{\Phi\}$ as follows:

$$U_\alpha(A)=U_\alpha(M_A,N_A)=U(M_A;\alpha) \text{ and } L_\alpha(A)=L_\alpha(M_A,N_A)=L(M_A;\alpha)$$

respectively, for each $A=\langle M_A, N_A\rangle$. Then the maps U_α and L_α are well defined.

Theorem 4.3 *For any α∈D(0,1), the maps U^α and L^α are surjective from IVIFSN(R) onto* $IVSN(R)\cup\{\varphi\}$.

Proof. Let α∈*D*(0,1). If $\bar{0}=\langle[0,0],[1,1]\rangle$ is an IVIFSN(R) and $U_\alpha(\bar{0}=L_\alpha(\bar{0})=\varphi$. If *P* is a subnear-ring of *R*, then for the interval-valued intuitionistic fuzzy subnear-ring �$=\langle\chi_{\blacklozenge},\overline{\chi_{\blacklozenge}}\rangle$ then

$$U_\alpha(\mathrm{P})=U(\chi_{\mathrm{P}},\alpha)=P \text{ and } L_\alpha(\mathrm{P})=L(\overline{\chi_{\mathrm{P}}};\alpha)=P.$$

Hence U_α, L_α are surjective.

Theorem 4.4 *For any* α∈*D*(0,1), *if*

$$I_\alpha : IVIFSN(R)\rightarrow IVSN(R)\cup\{\varphi\}$$

is a map defined by

$$I_\alpha(A)=U_\alpha(A)\cap L_\alpha(A)$$

for each $A=\langle M_A, N_A\rangle\in IVIFSN(R)$ then the map I_α is surjective.

Proof. Let α∈*D*(0,1). For

$$\bar{0}=\langle[0,0],[1,1]\rangle\in IVIFSN(R),\ I_\alpha(\bar{0})=U_\alpha(\bar{0})\cap L_\alpha(\bar{0})=U([0,0];\alpha)\cap L([1,1];\alpha)=\varphi.$$

And for any *P*∈*IVFSN*(*R*), there exists

$$P=\langle\chi_P,\overline{\chi_P}\rangle\in IVIFSN(R)$$

such that

$$I_\alpha(P)=U(\chi_P,\alpha)\cap L(\overline{\chi_P};\alpha)=P.$$

Theorem 4.5 *For any α∈D(0,1), the maps U^α and L^α are surjective from IVIFSN(R) onto* $IVSN(R)\cup\{\varphi\}$.

Proof. Let α∈*D*(0,1) and

$$\overline{I_\alpha} : IVIFSN(R)/R^\alpha\rightarrow IVFSN(R)/R^\alpha\cup\{\varphi\}$$

be a map defined by

$$\overline{I_\alpha}([A]_{R^\alpha}) = I_\alpha(A)$$

for each $[A]_{R^\alpha} \in IVIFSN(R)$. If $\overline{I_\alpha}([A]_{R^\alpha}) = \overline{I_\alpha}([B]_{R^\alpha})$ for any

$[A]_{R^\alpha}$, $[B]_{R^\alpha} \in IVIFSN(R) / R^\alpha$,

then

$$U(M_A;\alpha) \cap L(N_A;\alpha) = U(M_B;\alpha) \cap L(N_B;\alpha),$$

hence $(A,B) \in R^\alpha$ and $[A]_{R^\alpha} = [B]_{R^\alpha}$. It follows that I_α is injective. For

$$\bar{0} = \langle [0,0],[1,1] \rangle \in IVIFSN(R)$$

$$\overline{I_\alpha}(\bar{0}) = I_\alpha(\bar{0}) = \{\varphi\}.$$

If $P \in SN(R)$, then for $P = (\chi_P, \overline{\chi_P}) \in IVIFS(R)$, $\overline{I_\alpha}(P) = I_\alpha(P) = P$.
Hence $\overline{I_\alpha}$ is a bijective map.

5 CONCLUSION

In this paper, we introduce some definitions and theorems on interval-valued intuitionistic fuzzy R-subgroup of near-ring *R*. It is shown that the upper and lower interval level set of IVIF R-subgroup of near-ring again form IVIF R-subgroup of near-ring. Also, we investigate some relations on interval-valued intuitionistic fuzzy subnear-ring of near-ring. These works can be used in setting up the basic theory of interval-valued fuzzy set. In the next paper we should try to make related properties of generalized intuitionistic fuzzy R-subgroup of near-ring.

ACKNOWLEDGMENT

The authors are very grateful and would like to express their sincere thanks to the anonymous referees and Editor for their valuable comments to improve the presentation of the paper.

REFERENCES

Abdullah, S., Naeem, M., & Imran, M. (2013). On intuitionistic fuzzy Γ-ideals of Γ-LA-semigroups. *Annals of Fuzzy Mathematics and Informatics*, *6*(1), 17–31.

Abou-Zaid, A. (1989). On fuzzy subnear-ring. *Fuzzy Sets and Systems*, *81*, 383–393.

Adak, A. K., & Bhowmik, M. (2011a). Interval cut-set of interval-valued intuitionistic fuzzy sets. *African Journal of Mathematics and Computer Sciences*, *4*(4), 192–200.

Adak, A. K., Bhowmik, M., & Pal, M. (2011b). Application of generalized intuitionistic fuzzy matrix in multi-criteria decision making problem. *Journal of Mathematical and Computational Science*, *1*(1), 19–31.

Adak, A. K., Bhowmik, M., & Pal, M. (2011c). Semiring of interval-valued intuitionistic fuzzy matrices. *Global Journal of Computer Application and Technology*, *1*(3), 340–347.

Adak, A. K., Bhowmik, M., & Pal, M. (2012a). Interval cut-set of generalized interval-valued intuitionistic fuzzy sets. *International Journal of Fuzzy System Applications*, *2*(3), 35–50.

Adak, A. K., Bhowmik, M., & Pal, M. (2012b). Intuitionistic fuzzy block matrix and its some properties. *Annals of Pure and Applied Mathematics*, *1*(1), 13–31.

Adak, A. K., Bhowmik, M., & Pal, M. (2012c). Some properties of generalized intuitionistic fuzzy nilpotent matrices over distributive lattice. *International Journal of Fuzzy Information and Engineering*, *4*(4), 371–387.

Adak, A. K., Bhowmik, M., & Pal, M. (2012d). Semiring of Generalized Interval-valued Intuitionistic Fuzzy Matrices. *World Applied Sciences Journal*, *16*, 7-16.

Adak, A. K., Bhowmik, M., & Pal, M. (2013). Distributive Lattice over Intuitionistic Fuzzy Matrices. *The Journal of Fuzzy Mathematics*, *21*(2), 401–416.

Atanassov, K. (1986). Intuitionistic fuzzy sets. *Fuzzy Sets and Systems*, *20*, 87–96.

Atanassov, K. (1994). Operations over interval valued fuzzy set. *Fuzzy Sets and Systems*, *64*, 159–174.

Atanassov, K., & Gargo, G. (1989). Interval-valued intuitionistic fuzzy sets. *Fuzzy Sets and Systems*, *31*(1), 343–349.

Bhowmik, M., & Pal, M. (2010). Generalized interval-valued intuitionistic fuzzy sets. *The Journal of Fuzzy Mathematics*, *18*(2), 357–371.

Bustince, H., & Burillo, P. (1995). Intuitionistic fuzzy relation (part-1). *Mathware and Soft Computing*, *2*, 5–38.

Ebrahimnejad, A., Adak, A. K., & Jamkhaneh, E. B. (2019). Eigenvalue of Intuitionistic Fuzzy Matrices Over Distributive Lattice. *International Journal of Fuzzy System Applications*, *8*(1), 1–18.

Jun, Y. B. (2002). Interval-valued fuzzy R-subgroups of near-rings. *Indian Journal of Pure and Applied Mathematics*, *33*(1), 71–80.

Jun, Y. B., Kim, K. H., & Yon, Y. H. (1999). Intuitionistic fuzzy ideals of near-rings. *J. Inst. Math. Comp. Sci.*, *12*(3), 221–228.

Kim, K. H., & Jun, Y. B. (2000). On fuzzy R-subgroups of near-rings. *J. Fuzzy Math.*, *8*, 549–558.

Ma, X., & Zhan, J. (2009). On generalized fuzzy R-subgroups of near-rings. *International Workshop on Intelligent Systems and Applications*, 1724-1727.

Prince Williams, D. R., Latha, K. B., & Chandrasekar, E. (2013). -interval-valued T_G $(\in, \in, \vee q)$ -fuzzy subgroups. *Annals of Fuzzy Mathematics and Informatics*, *5*(2), 283–300.

Roh, E. H., Kim, K. H., & Lee, J. G. (2006). Intuitionistic Q-Fuzzy Subalgebras of BCK/BCI-Algebras. *International Mathematical Forum*, *1*(24), 1167-1174.

Zadeh, L. A. (1965). Fuzzy sets. *Information and Control*, *8*, 338-353.

Chapter 14
Fuzzy Translation of Doubt Interval-Valued Fuzzy Ideals in BF-Algebras

Tripti Bej
Vidyasagar University, India

Young Bae Jun
Gyeongsang National University, South Korea

ABSTRACT

This chapter focuses on the basics of a theory for such an interval-valued fuzzy set becoming doubt interval-valued fuzzy subalgebra and doubt interval-valued fuzzy ideal in BF-algebras. After a detailed study of its properties, the authors come to the conclusion that in BF-algebras an interval-valued fuzzy set in BF-algebras is a doubt interval-valued fuzzy subalgebra and a doubt interval-valued fuzzy ideal if and only if the complement of this interval-valued fuzzy set is an interval-valued fuzzy subalgebra and an interval-valued fuzzy ideal.

1. INTRODUCTION

It is known that mathematical logic is a discipline used in sciences and humanities with different point of views. Non-classical logic takes the advantage of the classical logic (two-valued logic) to handle information with various facts of uncertainty. The non-classical logic has become a formal and useful tool for computer science to deal with fuzzy information and uncertain information. The notion of logical algebras: *BCK* algebras and *BCI* algebras were initiated by Imai and Iseki (1966) in 1966 as a generalization of both classical and non-classical positional calculus. It is known that the class of *BCK* algebras is a proper subclass of the class of *BCI* algebras. In 2007, Andrzej Walendziak defined the *BF* algebra (Walendziak, A., 2007).

DOI: 10.4018/978-1-7998-0190-0.ch014

The work of Zadeh (1975) made an extension of the concept of a fuzzy set by an interval-valued fuzzy set (i.e., a fuzzy set with an interval-valued membership function). An interval-valued fuzzy set is a fuzzy set whose membership function is many-valued and forms an interval with respect to the membership scale. Biswas (1994) defined interval-valued fuzzy sub groups.

To develope the theory of *BCK/BCI* algebras the ideal theory plays an important role. Jun, Meng, Liu and several researchers investigated properties of fuzzy subalgebras and ideals in *BCK/BCI* algebras (Jun, Y. B., 1993; Jun, Y. B., 1994; Meng, J., 1994; Liu, Y, L. and Meng, J. 2001; Meng, J. and Guo, X., 2005; Bej, T. and Pal, M., 2014a; Bej, T. and Pal, M. 2014b; Bej, T. and Pal, M. 2015). Zarandi and Saeid (Zarandi, A. and Saeid, A. B., 2009) applied the notion of interval-valued fuzzy to *BF* subalgebras.

Jun (1994) introduced the notion of doubt fuzzy subalgebras and doubt fuzzy ideals in *BCK/BCI* algebras. Biswas (1990) introduced the concept of anti fuzzy subgroup. Modifying their idea, Saeid (2011) introduced the notion of doubt fuzzy subalgebra of *BF* algebras and Barbhuiya (2014) introduced the notion of doubt fuzzy ideal of *BF* algebras.

The authors have studied doubt intuitionistic fuzzy ideals in *BCK/BCI* algebras, doubt intuitionistic fuzzy H-ideals in *BCK/BCI* algebras and doubt Atanassov's intuitionistic fuzzy subimplicative-ideals in *BCI* algebras in their earlier works (Bej, T. and Pal, M. 2014a; Bej, T. and Pal, M. 2014b; Bej, T. and Pal, M. 2015).

Following the discussion of Barbhuiya (2014); Senapati, T., Bhowmik, M. & Pal, M. (2012c); Senapati, T., Bhowmik, M. & Pal, M.. (2013) authors are going to introduce the concept of doubt interval-valued fuzzy ideals in *BF* algebras. After a detailed study of its properties, they come to this conclusion that in *BF* algebras, an interval-valued fuzzy set is a doubt interval-valued fuzzy ideal in *BF* algebras if and only if the complement of this interval-valued fuzzy set is an interval-valued fuzzy ideal. Finally, results based on doubt interval-valued fuzzy subalgebras of *BF* algebras and doubt interval-valued fuzzy ideals of *BF* algebras are established. At the same time the product of doubt interval-valued fuzzy ideals in *BF* algebras has been introduced. Besides the idea of fuzzy translation, fuzzy multiplication of a doubt interval-valued fuzzy subalgebra and doubt interval-valued fuzzy ideal of a BF-algebra are given along with several related characterisation theorems.

2. PRELIMINARIES

In this section, some elementary aspects that are necessary for this paper are included.

Definition 2.1 *An algebra* $(X;*,0)$ *of type (2,0) is called a BF algebra if it satisfies the following axioms for all* $x,y,z \in X$:

(I) $x*x=0$

(II) $x*0=x$

(III) $0*(x*y)=(y*x)$ for all $x,y \in X$.

A partial ordering "$\leq$ on a *BF* algebra X can be defined by $x \leq y$ if and only if $x*y=0$.

Example 2.2 *Let* $A=[0;\infty)$. *Define the binary operation* $*$ *on A as follows:* $x*y=|x-y|$ *for all* $x,y \in A$. *Then* $(A;*,0)$ *is a BF algebra.*

Throughout this paper, *X* always means a *BF* algebra without any specification. Walendziak (2007) give some properties of *BF*-algebra, which are given below:

(a) (0*(0*x))=x for all x∈X.
(b) if 0*x=0*y then x=y for any x,y∈X.
(c) if x*y=0 then y*x=0 for any x,y∈X.

Definition 2.3 *(BF subalbebra) A non-empty subset I of a BF algebra X is called a subalgebra of X if x*y∈I for any x,y∈I* (Walendziak, 2007).

Definition 2.4 *(Ideal) A nonempty subset I of a BF algebra X is to be an ideal* (Walendziak, 2007). *of X if*

(i) 0∈I
(ii) x*y∈I and Y∈I then x∈I for all x,y∈X.

Saeid and Rezvani (2009) extends the concepts of *BF* algebra to fuzzy set. In the fuzzy set, the membership values of the elements are written together along with the elements. The membership values lie between 0 and 1. The definition of this set is given below.

Definition 2.5 *(Fuzzy set) Let X be the collection of objects denoted generally by x then a fuzzy set A in X is defined as* $A=\{\langle x,\mu_A(x)\rangle : x\in X\}$*.where μA(x) is called the membership value of x in A and* 0≤μA(x)≤1 (Zadeh, 1965).

Definition 2.6 *(Fuzzy BF subalgebra) Let* μ *be a fuzzy set in a BF algebra. Then μ is called a fuzzy An IVFS A in X is called an interval-valued fuzzy BF subalgebra (algebra) of X if μ*(x*y) ≥ min{μ(x),μ(y)} for *all x,y∈X (*Saeid & Rezvani, 2000).

Definition 2.7 (*Fuzzy BF ideal) A fuzzy set μ* of *a BF algebra X is called a fuzzy ideal of X if it satisfied the following conditions. (i)μ*(0)≥μ(x) *and (ii) μ*(x) ≥ min{μ(x*y),*μ*(y)} *for all x,y∈X (*Barbhuiya, 2014). Zadeh (1975) defined another type of fuzzy set called interval-valued fuzzy set (IVSs*). T*he membership value of an element of this set is not a single number, it is an interval and this interval is a subinterval of the interval [0,1].

Definition 2.8 **(**IV*FS)* *An interval-valued fuzzy set A defined on X is given by* $A=\{(x,[\mu_A^L(x),\mu_A^U(x)]):x\in X\}$. *Briefly, denoted by* $A=[\mu_A^L,\mu_A^U]$ *where* μ_A^L and μ_A^U *are any two fuzzy sets in X such that* $\mu_A^L(x)\le\mu_A^U(x)$ *for all x∈X* (Zadeh, 1975).

Let $\bar{\mu}_A(x)=[\mu_A^L(x),\mu_A^U(x)]$ for all x∈X and let D[0,1] denotes the family of all closed sub-intervals of [0,1]. It is clear that if $\mu_A^L(x)=\mu_A^U(x)=p$, where 0≤p≤1 then $\bar{\mu}_A(x)=[p,p]$ is in D[0,1]. Thus $\bar{\mu}_A(x)\in D[0,1]$ for all x∈X. Therefore, the IVFS A is given by $A=\{(x,\bar{\mu}_A(x)):x\in X\}$ where $\bar{\mu}_A:X\to D[0,1]$.

For the sake of simplicity, we use the symbol form $A=\left(\bar{\mu}_A(x)\right)$ or $\left(\bar{\mu}_A\right)$ for the interval-valued fuzzy set $A=\left[\mu_A^L,\mu_A^U\right]$.

Biswas(1994) described a method (refinement of intervals) to find max/sup and min/inf between two intervals or a set of intervals.

Consider two elements D_1 $D_2 \in D[0,1]$. If $D_1=[a_1,b_1]$ and $D_2=[a_2,b_2]$ then

$r\min(D_1,D_2) = [\min(a_1,a_2), \min(b_1,b_2)]$.

and

$r\max(D_1,D_2) = [\max(a_1,a_2), \max(b_1,b_2)]$.

Now, $D_1 \leq D_2$ if and only if $a_1 \leq a_2$ and $b_1 \leq b_2$. Similarly, the relations $D_1 \geq D_2$ and $D_1 = D_2$ are defined.

Extending the idea of fuzzy BF-subalgebras, Zarandi and Saeid (2009) defined interval-valued fuzzy *BF* subalgebras

Definition 2.9 *(Interval-valued fuzzy BF subalgebra) An IVFS A in X is called an interval-valued fuzzy BF subalgebra of X if* $\bar{\mu}_A(x*y) \geq rmin\left\{\bar{\mu}_A(x),\bar{\mu}_A(y)\right\}$ *for all x,y∈X* (Azarandi & Saeid, 2009).

Jun (1994) introduced the notion of doubt fuzzy subalgebra and doubt fuzzy ideals in *BCK/BCI* algebras. Biswas (1990) introduced the concept of anti fuzzy subgroup. Modifying their idea, Saeid (2011) introduced the notion of doubt fuzzy subalgebra of *BF* algebras and Barbhuiya(2014) introduced the notion of doubt fuzzy ideal of *BF* algebras.which are as follows:

Definition 2.10 *(Doubt fuzzy subalgebra) A fuzzy set* $A=\left\{\langle x,\mu_A(x)\rangle : x\in X\right\}$ *in X is called a doubt fuzzy subalgebra of X if* $\mu A_(x*y) \leq \mu A_(x) \vee \mu A_(y)$ for all x,y∈X (Saeid, 2011).

Definition 2.11 *(Doubt fuzzy ideal) A fuzzy set* $A=\left\{\langle x,\mu_A(x)\rangle : x\in X\right\}$ *in X is called a doubt fuzzy ideal of X if*

(i) $\mu_A(0) \leq \mu_A(x)$
(ii) $\mu_A(x) \leq \mu_A(x*y) \vee \mu_A(y)$ for all *x,y*∈*X* (Barbhuiya, 2014).

3. DOUBT INTERVAL-VALUED FUZZY SUBALGEBRA IN *BF* ALGEBRAS

In this section, we define doubt interval-valued fuzzy subalgebra in *BF* algebras and proved some essential theorems of it.

Definition 3.1 *An interval-valued fuzzy set A in X is called a doubt interval-valued fuzzy BF subalgebras of X if* $\bar{\mu}_A(x*y) \leq rmax\left\{\bar{\mu}_A(x),\bar{\mu}_A(y)\right\}$ *for all x,y∈X.*

Example 3.2 *Let X*={1,2,3,4} *be a BF algebra with the Cayley table (table 1).*

Table 1.

*	0	1	2	3	4
0	0	4	3	2	1
1	1	0	4	3	2
2	2	1	0	4	3
3	3	2	1	0	4
4	4	3	2	1	0

Let $A=\{(x,\bar{\mu}_A(x)):x\in X\}$ be an IVFS defined as

$\bar{\mu}_A(0)=[0,0.2]$, $\bar{\mu}_A(1)=[0.2,0.4]$, $\bar{\mu}_A(2)=[0.3,0.4]$, $\bar{\mu}_A(3)=[0.5,0.6]$ and $\bar{\mu}_A(4)=[0.3,0.4]$.

Then *A* is a doubt interval-valued fuzzy *BF* subalgebra of *X*.

Theorem 3.3 *An interval-valued fuzzy set $A=(\mu_A^L,\mu_A^U)$ in X is a doubt interval-valued fuzzy subalgebra of a BF algebra X if and only if the fuzzy sets μ_A^L and μ_A^U are doubt fuzzy subalgebras of X.*

Proof: Let μ_A^L and μ_A^U are doubt fuzzy subalgebras of *X* and *x,y*∈*X* then

$$\begin{aligned}\bar{\mu}_A(x*y)&=[\mu_A^L(x*y),\mu_A^U(x*y)]\le[max(\mu_A^L(x),\mu_A^L(y)),max(\mu_A^U(x),\mu_A^U(y))]\\&=rmax([\mu_A^L(x),\mu_A^U(x)],[\mu_A^L(y),\mu_A^U(y)])\\&=rmax\{\bar{\mu}_A(x),\bar{\mu}_A(y)\}.\end{aligned}$$

Hence, $A=(\mu_A^L,\mu_A^U)$ in *X* is a doubt interval-valued fuzzy subalgebra of a *BF* algebra *X*.

Conversely, let $A=(\mu_A^L,\mu_A^U)$ in *X* be a doubt interval-valued fuzzy subalgebra of a *BF* algebra *X*. Then for any *x,y*∈*X*, we have

$$\begin{aligned}[\mu_A^L(x*y),\mu_A^U(x*y)]=\bar{\mu}_A(x*y)&\le rmax\{\bar{\mu}_A(x),\bar{\mu}_A(y)\}\\&=rmax([\mu_A^L(x),\mu_A^U(x)],[\mu_A^L(y),\mu_A^U(y)])\\&=[max(\mu_A^L(x),\mu_A^L(y)),max(\mu_A^U(x),\mu_A^U(y))].\end{aligned}$$

Therefore,

$\mu_A^L(x*y)\le max(\mu_A^L(x),\mu_A^L(y))$ and $\mu_A^U(x*y)\le max(\mu_A^U(x),\mu_A^U(y))$

hence μ_A^L and μ_A^U are doubt fuzzy subalgebras of *X*. This the proof ends.

Lemma 3.4 *If* $A=\{(x,\bar{\mu}_A(x)):x\in X\}$ *is a doubt interval-valued fuzzy subalgebra in X then for all x∈X,* $\bar{\mu}_A(0)\leq\bar{\mu}_A(x)$.

Proof:

$$\bar{\mu}_A(0)=\bar{\mu}_A(x*x)\leq rmax\{\bar{\mu}_A(x),\bar{\mu}_A(x)\}$$
$$=rmax([\mu_A^L(x),\mu_A^U(x)],[\mu_A^L(x),\mu_A^u(x)])=\bar{\mu}_A(x).$$

This completes the proof.

Corollary 3.5 *Let* $A=\{(x,\bar{\mu}_A(x)):x\in X\}$ *be a doubt interval-valued fuzzy subalgera of X. Then the set,* $D_{\bar{\mu}_A}=\{x\in X/\bar{\mu}_A(x)=\bar{\mu}_A(0)\}$ *is a BF subalgebra of X.*

Proof: Here $A=\{(x,\bar{\mu}_A(x)):x\in X\}$ is a doubt interval-valued fuzzy subalgera of *X*. Obviously, $0\in D_{\bar{\mu}_A}$. Now, let $x,y\in D_{\bar{\mu}_A}$ then $\bar{\mu}_A(x)=\bar{\mu}_A(0)=\bar{\mu}_A(y)$. Also,

$$\bar{\mu}_A(x*y)\leq rmax\{\bar{\mu}_A(x),\bar{\mu}_A(y)\}=rmax\{\bar{\mu}_A(0),\bar{\mu}_A(0)\}=\bar{\mu}_A(0)..$$

Again, by Lemma 3.4, $\bar{\mu}_A(0)\leq\bar{\mu}_A(x*y)$. So, $\bar{\mu}_A(x*y)=\bar{\mu}_A(0)$. This means that $x*y\in D_{\bar{\mu}_A}$. Hence, $D_{\bar{\mu}_A}$ is a *BF* subalgebra of *X*.

Definition 3.6 *Let A and B be two doubt interval-valued fuzzy BF subalgebras of X. Then union of A and B denoted by* $A\cup B$ *is defined as*

$$A\cup B=\bar{\mu}_{A\cup B}(x)=[\mu_{A\cup B}^L(x),\mu_{A\cup B}^U(x)]=[max\{\mu_A^L(x),\mu_B^L(x)\},max\{\mu_A^U(x),\mu_B^U(x)\}],$$

for all x∈X.

Theorem 3.7 *Union of any two doubt interval-valued fuzzy subalgebras of a BF algebra X is also a doubt interval-valued fuzzy subalgebra of X.*

Proof: Let *P* and *Q* be two doubt interval-valued fuzzy *BF* subalgebras of *X*. Again let $x,y\in P\cup Q$. Then *x,y*∈*P* and *x,y*∈*Q*. Now

$$\begin{aligned}\bar{\mu}_{P\cup Q}(x*y) &= \left[\mu^L_{P\cup Q}(x*y), \mu^U_{P\cup Q}(x*y)\right] \\ &= \left[max\left(\mu^L_P(x*y), \mu^L_Q(x*y)\right), max\left(\mu^U_P(x*y), \mu^U_Q(x*y)\right)\right] \\ &\leq \left[max\left(\mu^L_{P\cup Q}(x), \mu^L_{P\cup Q}(y)\right), max\left(\mu^U_{P\cup Q}(x), \mu^U_{P\cup Q}(y)\right)\right] \\ &= rmax\left\{\bar{\mu}_{P\cup Q}(x), \bar{\mu}_{P\cup Q}(y)\right\}\end{aligned}$$

This completes the proof.
Also, the above theorem is verified by the following example.

Example 3.8 *Let X={0,1,2,3} be a BF algebra with the Cayley table (table 2).*

Let $A=\left\{\left(x,\bar{\mu}_A(x)\right): x\in X\right\}$ be an IVFS defined as Table 3.
Then *A* is an doubt interval-valued fuzzy subalgebra of *X*
Again, let $B=\left\{\left(x,\bar{\mu}_B(x)\right): x\in X\right\}$ be an IVFS defined as Table 4.
Then *B* is a doubt interval-valued fuzzy subalgebra of *X*.
We also assume that $P=A\cup B$ and *P* is defined as table 5.
Then *P* is a doubt interval-valued fuzzy subalgebra of *X*.

Theorem 3.9 *Let* $A=\left\{\left(x,\bar{\mu}_A(x)\right): x\in X\right\}$ *be a doubt interval-valued fuzzy subalgera of X. Then for any x∈X, we have*

(i) $\bar{\mu}_A\left(x^n*x\right)\leq\bar{\mu}_A(x)$, if *n* is odd
(ii) $\bar{\mu}_A\left(x^n*x\right)=\bar{\mu}_A(x)$, if *n* is even.

Table 2.

*	0	1	2	3
0	0	3	0	1
1	1	0	1	3
2	2	3	0	1
3	3	1	3	0

Table 3.

X	0	1	2	3
$\bar{\mu}_A$	[0.1, 0.2]	[0.2, 0.3]	[0.3, 0.5]	[0.2, 0.3]

Table 4.

X	0	1	2	3
$\bar{\mu}_B$	[0, 0.1]	[0.1, 0.2]	[0.4, 0.5]	[0.1, 0.2]

Table 5.

X	0	1	2	3
$\bar{\mu}_P$	[0.1, 0.2]	[0.2, 0.3]	[0.4, 0.5]	[0.2, 0.3]

Proof: (i) Let $x \in X$, then $\bar{\mu}_A(x*x) = \bar{\mu}_A(0) \le \bar{\mu}_A(x)$ [by Lemma 3.4]. If n is odd then let $n=2p-1$ for some positive integer p. Now, assume that $\bar{\mu}_A(x^{2p-1}*x) \le \bar{\mu}_A(x)$, for some positive integer p. Then,

$$\begin{aligned}\bar{\mu}_A\left(x^{2(p+1)-1}*x\right) &= \bar{\mu}_A\left(x^{2p+1}*x\right)\\ &= \bar{\mu}_A\left(x^{2p-1}*\left(x*\left(x*x\right)\right)\right)\\ &= \bar{\mu}_A\left(x^{2p-1}*\left(x*0\right)\right)\\ &= \bar{\mu}_A\left(x^{2p-1}*x\right)\\ &\le \bar{\mu}_A(x).\end{aligned}$$

Hence, $\bar{\mu}_A(x^n*x) \le \bar{\mu}_A(x)$, if n is odd.

(ii) Again, let n be even, and $n=2q$.

Now for $q=1$, $\bar{\mu}_A(x^2*x) = \bar{\mu}_A(x*(x*x)) = \bar{\mu}_A(x*0) = \bar{\mu}_A(x)$.
Also assume that, $\bar{\mu}_A(x^{2q}*x) = \bar{\mu}_A(x)$, for some positive integer q. Then,

$$\begin{aligned}\bar{\mu}_A\left(x^{2(q+1)}*x\right) &= \bar{\mu}_A\left(x^{2q}*\left(x*\left(x*x\right)\right)\right)\\ &= \bar{\mu}_A\left(x^{2q}*x\right)\\ &= \bar{\mu}_A(x).\end{aligned}$$

Hence, $\bar{\mu}_A(x^n*x) = \bar{\mu}_A(x)$, if n is even.

4. INTERVAL-VALUED DOUBT FUZZY IDEAL IN *BF*-ALGEBRAS

In this section, we define doubt interval-valued fuzzy ideal in *BF*-algebras and investigated its properties.

Definition 4.1 *An interval-valued fuzzy set A in X is called a doubt interval-valued fuzzy ideal (briefly, DIVF-ideal) of BF-algebra X if*

(i) $\bar{\mu}_A(0) \le \bar{\mu}_A(x)$
(ii) $\bar{\mu}_A(x) \le rmax\{\bar{\mu}_A(x*y), \bar{\mu}_A(y)\}$ for all $x,y \in X$.

Example 4.2 *Consider a BF-algebra* $X=\{0,1,2,3\}$ *with the table 6.*

Let A be an interval-valued fuzzy set of X as defined by table 7.

Table 6.

*	0	1	2
0	0	1	2
1	1	0	1
2	2	2	0

Table 7.

X	0	1	2
$\bar{\mu}_A$	[0.2, 0.3]	[0.4, 0.6]	[0.6, 0.8]

Then *A* is a doubt interval-valued fuzzy ideal of *X*.

Proposition 4.3 *Let $\bar{\mu}_A(x)$ be a DIVF-ideal of X. Then the following hold.*

(a) If $x \leq y$ then $\bar{\mu}_A(x) \leq \bar{\mu}_A(y)$.
(b) If $x*y \leq z$ then $\bar{\mu}_A(x) \leq rmax\{\bar{\mu}_A(y), \bar{\mu}_A(z)\}$ for all $x,y,z \in X$.

Proof: (a) Let $x \leq y$ then $x*y=0$. Now

$$\begin{aligned}\bar{\mu}_A(x) &\leq r\max\{\bar{\mu}_A(x*y), \bar{\mu}_A(y)\} && [\text{Since } \bar{\mu}_A(x) \text{ is an } DIVF-ideal] \\ &= r\max\{\bar{\mu}_A(0), \bar{\mu}_A(y)\} = \bar{\mu}_A(y) && [\text{Since } \bar{\mu}_A(0) \leq \bar{\mu}_A(y) \text{ for } DIVF-ideal].\end{aligned}$$

So, $\bar{\mu}_A(x) \leq \bar{\mu}_A(y)$.

(b) Here $x*y \leq z$ therefore $(x*y)*z=0$. Now,

$$\begin{aligned}\bar{\mu}_A(x*y) &\leq r\max\{\bar{\mu}_A((x*y)*z), \bar{\mu}_A(z)\} && [\text{Since } \bar{\mu}_A(x) \text{ is an } DIVF-ideal] \\ &= r\max\{\bar{\mu}_A(0), \bar{\mu}_A(z)\} = \bar{\mu}_A(z) && [\text{Since } \bar{\mu}_A(0) \leq \bar{\mu}_A(z) \text{ for } DIVF-ideal].\end{aligned}$$

Therefore, $\bar{\mu}_A(x*y) \leq \bar{\mu}_A(z)$.
Again, as $\bar{\mu}_A(x)$ is an DIVF-ideal, so

$$\bar{\mu}_A(x) \leq r\max\{\bar{\mu}_A(x*y), \bar{\mu}_A(y)\} \leq r\max\{\bar{\mu}_A(z), \bar{\mu}_A(y)\} \quad [\text{Since } \bar{\mu}_A(x*y) \leq \bar{\mu}_A(z)].$$

Hence, $\bar{\mu}_A(x) \leq rmax\{\bar{\mu}_A(y), \bar{\mu}_A(z)\}$ for all $x,y,z \in X$.

Theorem 4.4 *Let $A = \{(x, \bar{\mu}_A(x)) : x \in X\}$ be a doubt interval-valued fuzzy ideal of a BF algebra X. Then the set, $D_{\bar{\mu}_A} = \{x \in X / \bar{\mu}_A(x) = \bar{\mu}_A(0)\}$ is an ideal of X.*

Proof: $A=\{(x,\bar{\mu}_A(x)):x\in X\}$ is a doubt interval-valued fuzzy ideal of *X*. Obviously, $0\in D_{\bar{\mu}_A}$. Now, let $x*y, y\in D_{\bar{\mu}_A}$ then $\bar{\mu}_A(x*y)=\bar{\mu}_A(0)=\bar{\mu}_A(y)$.

Now,

$$\bar{\mu}_A(x)\le rmax\{\bar{\mu}_A(x*y),\bar{\mu}_A(y)\}=rmax\{\bar{\mu}_A(0),\bar{\mu}_A(0)\}=\bar{\mu}_A(0).$$

Therefore, $\bar{\mu}_A(x)\le\bar{\mu}_A(0)$. Also, $\bar{\mu}_A(0)\le\bar{\mu}_A(x)$ [Since *A* is an DIVF-ideal]. Hence, $\bar{\mu}_A(x)=\bar{\mu}_A(0)$. So, $x\in D_{\bar{\mu}_A}$. That is, $x*y, y\in D_{\bar{\mu}_A}$ implies that $x\in D_{\bar{\mu}_A}$. So, $D_{\bar{\mu}_A}$ is an ideal. This completes the proof.

Theorem 4.5 *Let* $A=\{(x,\bar{\mu}_A(x)):x\in X\}$ *be a doubt interval-valued fuzzy set of a BF algebra X then A is a DIVF-ideal of X if and only if its complement is DIVF-ideal of X.*

Proof: Let *A* is an interval-valued fuzzy ideal of *X* to prove A^c (complement of *A* is DIVF-ideal, let *x,y,z*∈*X*. Now,

$$\bar{\mu}_A^c(0)=1-\bar{\mu}_A(0)\le 1-\bar{\mu}_A(x)\quad\left[\text{As }\bar{\mu}_A(0)\ge\bar{\mu}_A(x)\right]=\bar{\mu}_A^c(x).$$

So, $\bar{\mu}_A^c(0)\le\bar{\mu}_A^c(x)$ for all *x*∈*X*. Again,

$$\begin{aligned}\bar{\mu}_A^c(x)&=1-\bar{\mu}_A(x)\le 1-r\min\{\bar{\mu}_A(x*y),\bar{\mu}_A(y)\}\quad\left[\text{As }\bar{\mu}_A(x)\ge r\min\{\bar{\mu}_A(x*y),\bar{\mu}_A(y)\right]\\&=1-r\min\{1-\bar{\mu}_A^c(x*y),1-\bar{\mu}_A^c(y)\}=r\max\{\bar{\mu}_A^c(x*y),\bar{\mu}_A^c(y)\}\end{aligned}.$$

This implies that, A^c is an DIVF-ideal.

Conversely, let A^c is DIVF-ideal of *X*. Then,

$$\bar{\mu}_A^c(0)\le\bar{\mu}_A^c(x)\text{ and }\bar{\mu}_A^c(x)\le rmax\{\bar{\mu}_A^c(x*y),\bar{\mu}_A^c(y)\}.$$

Now, $\bar{\mu}_A^c(0)\le\bar{\mu}_A^c(x)$ which implies that, $1-\bar{\mu}_A(0)\le 1-\bar{\mu}_A(x)$ that is, $\bar{\mu}_A(0)\ge\bar{\mu}_A(x)$.

Again, $\bar{\mu}_A^c(x)\le rmax\{\bar{\mu}_A^c(x*y),\bar{\mu}_A^c(y)\}$ then $1-\bar{\mu}_A(x)\le rmax\{1-\bar{\mu}_A(x*y),1-\bar{\mu}_A(y)\}$ next $\bar{\mu}_A(x)\ge rmin\{\bar{\mu}_A(x*y),\bar{\mu}_A(y)\}$ which implies that, *A* is an interval-valued fuzzy ideal of *X*.

Theorem 4.6 *Let* $A=\{(x,\bar{\mu}_A(x)):x\in X\}$ *be an DIVF-subalgebra of X if*

$$\bar{\mu}_A(x)\le rmax\{\bar{\mu}_A(y),\bar{\mu}_A(z)\}$$

with $x*y\le z$, then *A* is an DIVF-ideal of *X*.

Proof: Here $A=\{(x,\bar{\mu}_A(x)):x\in X\}$ is a doubt interval-valued fuzzy subalgebra of *X*. We have, $\bar{\mu}_A(0)\le\bar{\mu}_A(x)$. Also since, $x*(x*y)\le y$ and from the hypothesis,

$$\bar{\mu}_A(x)\le rmax\{\bar{\mu}_A(x*y),\bar{\mu}_A(y)\}.$$

So, *A* is an DIVF-ideal of *X*.

Theorem 4.7 *An interval-valued fuzzy set* $A=(\mu_A^L,\mu_A^U)$ *in X is a doubt interval-valued fuzzy ideal of a BF algebra X if and only if the fuzzy sets* μ_A^L and μ_A^U *are doubt fuzzy ideals of X.*

Proof: Let μ_A^L and μ_A^U are doubt fuzzy ideals of *X* and *x,y*∈*X* then

$$\mu_A^L(0)\le\mu_A^L(x) \text{ and } \mu_A^U(0)\le\mu_A^U(x).$$

Also,

$$\mu_A^L(x)\le max\left[\mu_A^L(x*y),\mu_A^L(y)\right] \text{ and } \mu_A^U(x)\le max\left[\mu_A^U(x*y),\mu_A^U(y)\right].$$

Now,

$$\begin{aligned}\bar{\mu}_A(x)&=\left[\mu_A^L(x),\mu_A^U(x)\right]\\&\le\left[max\left(\mu_A^L(x*y),\mu_A^L(y)\right),max\left(\mu_A^U(x*y),\mu_A^U(y)\right)\right].\\&=rmax\{\bar{\mu}_A(x*y),\bar{\mu}_A(y)\}\end{aligned}$$

Therefore, $\bar{\mu}_A(x)\le rmax\{\bar{\mu}_A(x*y),\bar{\mu}_A(y)\}$. Hence, $A=(\mu_A^L,\mu_A^U)$ in *X* is a doubt interval-valued fuzzy ideal of a *BF* algebra *X*.

Conversely, let $A=(\mu_A^L,\mu_A^U)$ in *X* be a doubt interval-valued fuzzy ideal of a *BF* algebra *X*. Then for any *x,y*∈*X*, we have

$$\begin{aligned}\left[\mu_A^L(x),\mu_A^U(x)\right]&=\bar{\mu}_A(x)\le rmax\{\bar{\mu}_A(x*y),\bar{\mu}_A(y)\}\\&=\left[max\left(\mu_A^L(x*y),\mu_A^L(y)\right),max\left(\mu_A^U(x*y),\mu_A^U(y)\right)\right].\end{aligned}$$

Therefore, $\mu_A^L(x)\le max\left[\mu_A^L(x*y),\mu_A^L(y)\right]$ and $\mu_A^U(x)\le max\left[\mu_A^U(x*y),\mu_A^U(y)\right]$. Hence we get μ_A^L and μ_A^U are doubt fuzzy ideal of *X*. This completes the proof.

Definition 4.8 *Let A and B be two doubt interval-valued fuzzy BF ideals of X. Then union of A and B denoted by* $A \cup B$. *is defined as*

$$A \cup B = \bar{\mu}_{A\cup B}(x) = \left[\mu^L_{A\cup B}(x), \mu^U_{A\cup B}(x)\right] = \left[max\left\{\mu^L_A(x), \mu^L_B(x)\right\}, max\left\{\mu^U_A(x), \mu^U_B(x)\right\}\right],$$

for all x∈X.

Theorem 4.9 *Union of any two doubt interval-valued fuzzy ideals of a BF algebra X is also an doubt interval-valued fuzzy ideal of X.*

Proof: Let *A* and *B* be two doubt interval-valued fuzzy ideals of *X*. Now let $x, y \in A \cup B$. Then *x,y*∈*A* and *x,y*∈*B*. Again let, $C = A \cup B = (\bar{\mu}_C)$ where $\bar{\mu}_C = \bar{\mu}_A \vee \bar{\mu}_B$. Let *x,y*∈*X,* then,

$$\bar{\mu}_C(0) = \bar{\mu}_A(0) \vee \bar{\mu}_B(0) \le \bar{\mu}_A(x) \vee \bar{\mu}_B(x) = \bar{\mu}_C(x).$$

Also,

$$\begin{aligned}\bar{\mu}_{A\cup B}(x) &= \left[\mu^L_{A\cup B}(x), \mu^U_{A\cup B}(x)\right] \\ &= \left[max\left(\mu^L_A(x), \mu^L_B(x)\right), max\left(\mu^U_A(x), \mu^U_B(x)\right)\right] \\ &\le \left[max\left(\mu^L_{A\cup B}(x*y), \mu^L_{A\cup B}(y)\right), max\left(\mu^U_{A\cup B}(x*y), \mu^U_{A\cup B}(y)\right)\right]. \\ &= rmax\left\{\bar{\mu}_{A\cup B}(x*y), \bar{\mu}_{A\cup B}(y)\right\}\end{aligned}$$

This completes the proof.

Example 4.10 *Let X*={0,1,2,3} *be a BF algebra with the Cayley table (table 8).*

Let *A* and *B* be two doubt interval-valued fuzzy ideals of *X* as defined by Tables 9 and 10.
We also assume that $C = A \cup B$. and *C* is defined as table 11.
Then clearly, *C* is a doubt interval-valued fuzzy ideal of *X*.
The above theorem can be generalised as follows:

Table 8.

*	0	1	2	3
0	0	1	2	3
1	1	0	3	2
2	2	3	0	1
3	3	2	1	0

Table 9.

X	0	1	2	3
$\bar{\mu}_A$	[0.1, 0.2]	[0.2, 0.3]	[0.5, 0.6]	[0.5, 0.6]

Table 10.

X	0	1	2	3
$\bar{\mu}_B$	[0.2, 0.2]	[0.3, 0.4]	[0.6, 0.7]	[0.6, 0.7]

Table 11.

X	0	1	2	3
$\bar{\mu}_C$	[02, 02]	[0.3, 0.4]	[0.6, 0.7]	[]0.6, 0.7

Corollary 4.11 *Let* $\{A_i|i=1,2,3,\ldots\}$ *be a family of DIVF-ideal of BF-algebra X, then* $\bigcup_{i=1}^{n} A_i$ *is also a DIVF-ideal of BF-algebra X, where* $\cup(A_i) = max\{A_i(x) : i = 1,2,3,....\}$.

5. PRODUCT OF DOUBT INTERVAL-VALUED FUZZY IDEAL IN *BF*-ALGEBRAS

Definition 5.1 *Let* $\bar{\mu}_A$ and $\bar{\mu}_B$ *be two DIVF-ideals of a BF-algebra X. Then the Cartesian product* $\bar{\mu}_A \times \bar{\mu}_B$ *is* $(\bar{\mu}_A \times \bar{\mu}_B)$: *X×X→[0,1], and is defined by*

$$(\bar{\mu}_A \times \bar{\mu}_B)(x, y) = r\max\{\bar{\mu}_A(x), \bar{\mu}_B(y)\},$$

for all $x,y \in X$.

Theorem 5.2 *Let* $\bar{\mu}_A$ and $\bar{\mu}_B$ *be two DIVF-ideals of a BF-algebra X. Then* $\bar{\mu}_A \times \bar{\mu}_B$ *is also a DIVF-ideal of X×X.*

Proof: Let $(x,y) \in X \times X$, then by definition,

$$\begin{aligned}
(\bar{\mu}_A \times \bar{\mu}_B)(0,0) &= rmax\{\bar{\mu}_A(0), \bar{\mu}_B(0)\} \\
&= rmax\{(\mu_A^L(0), \mu_A^U(0)), (\mu_B^L(0), \mu_B^U(0))\} \\
&= max[(\mu_A^L(0), \mu_B^L(0)), (\mu_A^U(0), \mu_B^U(0))] \\
&\le max[(\mu_A^L(x), \mu_B^L(y)), (\mu_A^U(x), \mu_B^U(y))] \\
&= rmax\{(\mu_A^L(x), \mu_A^U(x)), (\mu_B^L(y), \mu_B^U(y))\} \\
&= rmax\{\bar{\mu}_A(x), \bar{\mu}_B(y)\} \\
&= (\bar{\mu}_A \times \bar{\mu}_B)(x, y).
\end{aligned}$$

Therefore, $(\bar{\mu}_A \times \bar{\mu}_B)(0,0) \le (\bar{\mu}_A \times \bar{\mu}_B)(x, y)$. (i)

Again let, (x_1,x_2), $(y_1,y_2) \in X \times X$. Then,

$$\begin{aligned}(\bar{\mu}_A \times \bar{\mu}_B)(x_1,x_2) &= rmax\{\bar{\mu}_A(x_1),\bar{\mu}_B(x_2)\}\\ &\le rmax\{rmax\{\bar{\mu}_A(x_1*y_1),\bar{\mu}_A(y_1)\},rmax\{\bar{\mu}_B(x_2*y_2),\bar{\mu}_B(y_2)\}\}\\ &= rmax\{[max[(\mu_A^L(x_1*y_1),\mu_A^L(y_1))],max[(\mu_A^U(x_1*y_1),\mu_A^U(y_1))]],\\ &[max[(\mu_B^L(x_2*y_2),\mu_B^L(y_2))],max[(\mu_B^U(x_2*y_2),\mu_B^U(y_2))]]\}\\ &= [max\{max[(\mu_A^L(x_1*y_1),(\mu_B^L(x_2*y_2)],max[\mu_A^L(y_1),\mu_B^L(y_2)]\},\\ &max\{max[(\mu_A^U(x_1*y_1),(\mu_B^U(x_2*y_2)],max[\mu_A^U(y_1),\mu_B^U(y_2)]\}\\ &= rmax\{(\bar{\mu}_A \times \bar{\mu}_B)[(x_1*y_1,x_2*y_2),(\bar{\mu}_A \times \bar{\mu}_B)(y_1,y_2)\}.\end{aligned}$$

Hence, $(\bar{\mu}_A \times \bar{\mu}_B)(x_1,x_2) \le rmax\{(\bar{\mu}_A \times \bar{\mu}_B)[(x_1*y_1,x_2*y_2),(\bar{\mu}_A \times \bar{\mu}_B)(y_1,y_2)\}$. (ii)

From (i) and (ii) see that $\bar{\mu}_A \times \bar{\mu}_B$ is also a DIVF-ideal of $X \times X$.

6. FUZZY-TRANSLATION AND FUZZY-MULTIPLICATION OF DIVF-SUBALGEBRA AND DIVF-IDEAL IN *BF*-ALGEBRAS

This section deals with the notion of fuzzy-translation and fuzzy-multiplication and fuzzy magnified translation of a DIVF-subalgebra/ DIVF-ideal in *BF*-algebras. For any IVFS

$$M = \{(v_1,\bar{\zeta}_M(v_1)) : v_1 \in X\},$$

let $\bar{\delta} = [\delta^L,\delta^U]$ where $\delta^U = 1 - sup\{\bar{\zeta}_M^U(v_1) : v_1 \in X\}$ and $\delta^L \le \delta^U$.

Definition 6.1 *Let* $M = (\zeta_M^L,\zeta_M^U)$ *be an IVFS in X and* $0 \le \beta^U \le \delta^U$, *where* $\bar{\beta} = [\beta^L,\beta^U]$. *An object of the form* $M_{\bar{\beta}}^T = [\zeta_{M\bar{\beta}}^{LT},\zeta_{M\bar{\beta}}^{UT}]$ *is identified as fuzzy β-translation of M if* $M_{\bar{\beta}}^T(v_1) = \bar{\zeta}_M(v_1) + \bar{\beta}$ *for all* $v_1 \in X$.

Definition 6.2 *Let* $M = (\zeta_M^L,\zeta_M^U)$ *be an IVFS in X and* $\bar{\rho} \in [0,1]$. *An object having the form* $M_{\bar{\rho}}^m = [\zeta_{M\bar{\rho}}^{Lm},\zeta_{M\bar{\rho}}^{Um}]$ *is called fuzzy ρ-multiplication of M if* $M_{\bar{\rho}}^m(v_1) = \bar{\zeta}_M(v_1).\bar{\rho}$ *for all* $v_1 \in X$.

Theorem 6.3 *Let* $M = (\zeta_M^L,\zeta_M^U)$ *be a DIVF-subalgebra in X. Then the fuzzy β-translation* $M_{\bar{\beta}}^T(v_1)$ *of M is also a DIVF-subalgebra in X for all* $\bar{\beta} \in [\bar{0},\bar{\delta}]$ *and vice-versa.*

Proof: Let $M=\left(\zeta_M^L,\zeta_M^U\right)$ be a DIVF-subalgebra in *X* and let $\beta\in\left[0,\delta\right]$. Then

$$\bar{\zeta}_M\left(v_1*v_2\right)\leq rmax\left\{\bar{\zeta}_M\left(v_1\right),\bar{\zeta}_M\left(v_2\right)\right\}$$

for all $v_1,v_2\in X$. Now,

$$\begin{aligned}M_{\bar{\beta}}^T\left(v_1*v_2\right)&=\bar{\zeta}_M\left(v_1*v_2\right)+\bar{\beta}\\&\leq rmax\left\{\bar{\zeta}_M\left(v_1\right),\bar{\zeta}_M\left(v_2\right)\right\}+\bar{\beta}\\&=rmax\left\{\bar{\zeta}_M\left(v_1\right)+\bar{\beta},\bar{\zeta}_M\left(v_2\right)+\bar{\beta}\right\}\\&=rmax\left\{M_{\bar{\beta}}^T\left(v_1\right),M_{\bar{\beta}}^T\left(v_2\right)\right\}\end{aligned}$$

Hence, $M_{\bar{\beta}}^T$ is a DIVF-subalgebra.

Conversely, let $M_{\bar{\beta}}^T$ is a DIVF-subalgebra. Then

$$\begin{aligned}M_{\bar{\beta}}^T\left(v_1*v_2\right)&\leq rmax\left\{M_{\bar{\beta}}^T\left(v_1\right),M_{\bar{\beta}}^T\left(v_2\right)\right\}\\&=rmax\left\{\bar{\zeta}_M\left(v_1\right)+\bar{\beta},\bar{\zeta}_M\left(v_2\right)+\bar{\beta}\right\}\\&=\bar{\zeta}_M\left(v_1*v_2\right)+\bar{\beta}\\&=rmax\left\{\bar{\zeta}_M\left(v_1\right),\bar{\zeta}_M\left(v_2\right)\right\}+\bar{\beta}\end{aligned}$$

So,

$$\bar{\zeta}_M\left(v_1*v_2\right)+\bar{\beta}\leq rmax\left\{\bar{\zeta}_M\left(v_1\right),\bar{\zeta}_M\left(v_2\right)\right\}+\bar{\beta},$$

implies that

$$\bar{\zeta}_M\left(v_1*v_2\right)\leq rmax\left\{\bar{\zeta}_M\left(v_1\right),\bar{\zeta}_M\left(v_2\right)\right\},$$

for all $v_1,v_2\in X$. Hence, *M* is a DIVF-subalgebra in *X*.

Theorem 6.4 *Let* $M=\left(\zeta_M^L,\zeta_M^U\right)$ *be a DIVF-ideal in X. Then the fuzzy* β*-translation* $M_{\bar{\beta}}^T\left(v_1\right)$ *of M is also a DIVF-ideal in X for all* $\bar{\beta}\in\left[\bar{0},\bar{\delta}\right]$ *and vice-versa.*

Proof: Let *M* is a DIVF-ideal of a *BCK*-algebra *X* and let $\bar{\beta}\in\left[\bar{0},\bar{\delta}\right]$. Then for all $v_1,v_2\in X$,

$$M_{\bar{\beta}}^T\left(0\right)=\bar{\zeta}_M\left(0\right)+\bar{\beta}\leq\bar{\zeta}_M\left(v_1\right)+\bar{\beta}=M_{\bar{\beta}}^T\left(v_1\right).$$

That is $M_{\bar{\beta}}^{T}(0) \leq M_{\bar{\beta}}^{T}(v_1)$. And,

$$\begin{aligned} M_{\bar{\beta}}^{T}(v_1) &= \bar{\zeta}_M(v_1) + \bar{\beta} \\ &\leq rmax\{\bar{\zeta}_M(v_1 * v_2), \bar{\zeta}_M(v_2)\} + \bar{\beta} \\ &= rmax\{\bar{\zeta}_M(v_1 * v_2) + \bar{\beta}, \bar{\zeta}_M(v_2) + \bar{\beta}\} \\ &= rmax\{M_{\bar{\beta}}^{T}(v_1 * v_2), M_{\bar{\beta}}^{T}(v_2)\} \end{aligned}$$

Hence, $M_{\bar{\beta}}^{T}$ is a DIVF-ideal .

Conversely, let $M_{\bar{\beta}}^{T}$ is a DIVF-ideal. Then $M_{\bar{\beta}}^{T}(0) \leq M_{\bar{\beta}}^{T}(v_1)$, implies that $\bar{\zeta}_M(0) \leq \bar{\zeta}_M(v_1)$. Also,

$$\begin{aligned} M_{\bar{\beta}}^{T}(v_1) &= \bar{\zeta}_M(v_1) + \bar{\beta} \\ &\leq rmax\{M_{\bar{\beta}}^{T}(v_1 * v_2), M_{\bar{\beta}}^{T}(v_2)\} \\ &= rmax\{\bar{\zeta}_M(v_1 * v_2) + \bar{\beta}, \bar{\zeta}_M(v_2) + \bar{\beta}\} \\ &= rmax\{\bar{\zeta}_M(v_1 * v_2), \bar{\zeta}_M(v_2)\} + \bar{\beta} \end{aligned}$$

So,

$$\bar{\zeta}_M(v_1) + \bar{\beta} \leq rmax\{\bar{\zeta}_M(v_1 * v_2), \bar{\zeta}_M(v_2)\} + \bar{\beta},$$

implies that $\bar{\zeta}_M(v_1) \leq rmax\{\bar{\zeta}_M(v_1 * v_2), \bar{\zeta}_M(v_2)\}$, for all $v_1, v_2 \in X$ Hence, *M* is a DIVF-ideal in *X*.

Theorem 6.5 *Let $M = (\zeta_M^L, \zeta_M^U)$ be a DIVF-subalgebra in X. Then the fuzzy* ρ-multiplication $M_{\bar{\rho}}^{m}$ *of M is also a DIVF-subalgebra in X for all $\bar{\rho} \in [0,1]$ and vice-versa.*

Proof: Let $M = (\zeta_M^L, \zeta_M^U)$ be a DIVF-subalgebra in *X* and let $\bar{\rho} \in [0,1]$. Then

$$\bar{\zeta}_M(v_1 * v_2) \leq rmax\{\bar{\zeta}_M(v_1), \bar{\zeta}_M(v_2)\}$$

for all $v_1, v_2 \in X$. Now,

$$\begin{aligned} M_{\bar{\rho}}^{m}(v_1 * v_2) &= \bar{\zeta}_M(v_1 * v_2).\bar{\rho} \\ &\leq r\max\{\bar{\zeta}_M(v_1), \bar{\zeta}_M(v_2)\}.\bar{\rho} \\ &= r\max\{\bar{\zeta}_M(v_1).\bar{\rho}, \bar{\zeta}_M(v_2).\bar{\rho}\} \\ &= r\max\{M_{\bar{\rho}}^{m}(v_1), M_{\bar{\rho}}^{m}(v_2)\} \end{aligned}$$

Hence, $M_{\bar{\rho}}^{m}$ is a DIVF-subalgebra.

Conversely, let $M_{\bar{\rho}}^{m}$ is a DIVF-subalgebra. Then

$$\begin{aligned} M_{\bar{\rho}}^{m}(v_1 * v_2) &\leq r\max\{M_{\bar{\rho}}^{m}(v_1), M_{\bar{\rho}}^{m}(v_2)\} \\ &= r\max\{\bar{\zeta}_M(v_1).\bar{\rho}, \bar{\zeta}_M(v_2).\bar{\rho}\} \\ &= r\max\{\bar{\zeta}_M(v_1), \bar{\zeta}_M(v_2)\}.\bar{\rho} \\ &= \bar{\zeta}_M(v_1 * v_2).\bar{\rho} \end{aligned}$$

So,

$$\bar{\zeta}_M(v_1 * v_2).\bar{\rho} \leq rmax\{\bar{\zeta}_M(v_1), \bar{\zeta}_M(v_2)\}.\bar{\rho},$$

implies that

$$\bar{\zeta}_M(v_1 * v_2) \leq r\max\{\bar{\zeta}_M(v_1), \bar{\zeta}_M(v_2)\},$$

for all $v_1, v_2 \in X$ Hence, M is a DIVF-subalgebra in X.

Theorem 6.6 *Let* $M = (\zeta_M^L, \zeta_M^U)$ *be a DIVF-ideal in X. Then the fuzzy* ρ-multiplication $M_{\bar{\rho}}^{m}$ *of M is also a DIVF-ideal in X for all* $\bar{\rho} \in [0,1]$ *and vice-versa.*

Proof: The proof is straightforward.

7. CONCLUSION

In this present chapter, the authors have introduced the concept of doubt interval-valued fuzzy subalgebras and doubt interval-valued fuzzy ideals in *BF*-algebras and investigated some of its essential properties. Besides fuzzy translation and fuzzy multiplication of an interval-valued doubt fuzzy subalgebra/ideal in BCK-algebras is introduced and the product of interval-valued doubt fuzzy ideals in BF-algebras is defined. The results obtained here probably be applied in different fields such as automata theory, pattern recognition, automata theory, robotics, genetic algorithm and medical diagnosis. In terms of the above notion it is obvious and apprehensible that the research in this direction can be fruitful for further developments of fuzzy BF-algebra.

REFERENCES

Barbhuiya, S. R., (2014). Doubt fuzzy ideals of *BF*-algebra. *IOSR journal of Mathematics, 10*(2), 65-70.

Bej, T., & Pal, M. (2014a). Doubt intuitionistic fuzzy H-ideals in BCK/BCI-algebras. *Annals of Fuzzy Mathematics and Informatics*, *8*(4), 593–605.

Bej, T., & Pal, M. (2014b). Doubt Atanassov's intuitionistic fuzzy sub-implicative ideals in *BCI. Computational Intelligence Systems*, *8*(2), 240–249. doi:10.1080/18756891.2015.1001948

Bej, T., & Pal, M. (2015). Doubt intuitionistic fuzzy ideals in *BCK/BCI. Int. J. Fuzzy Logic Systems*, *5*(1), 1–13. doi:10.5121/ijfls.2015.5101

Biswas, R. (1990). Fuzzy subgroups and antifuzzy subgroups. *Fuzzy Sets and Systems*, *35*(1), 121–124. doi:10.1016/0165-0114(90)90025-2

Biswas, R. (1994). Rosenfeld fuzzy subgroups with interval-valued membership function. *Fuzzy Sets and Systems*, *63*(1), 87–90. doi:10.1016/0165-0114(94)90148-1

Imai, Y., & Iseki, K. (1966). On axiom systems of propositional calculi. *Proceedings of the Japan Academy*, *42*(1), 19–22. doi:10.3792/pja/1195522169

Jun, Y. B. (1993). Closed fuzzy ideals in *BCI*-algebras. *Math. Japonica*, *38*, 199–202.

Jun, Y. B. (1994). Doubt fuzzy BCK/BCI-algebras. *Soochow Journal of Mathematics*, *20*(3), 351–358.

Liu, Y. L., & Meng, J. (2001). Fuzzy ideals in BCI. *Fuzzy Sets and Systems*, *123*(2), 227–237. doi:10.1016/S0165-0114(00)00047-6

Meng, J. (1994). On ideals in BCK-algebras. *Math. Japonica*, *40*, 143–154.

Meng, J., & Guo, X. (2005). On fuzzy ideals in BCK. *Fuzzy Sets and Systems*, *149*(3), 509–525. doi:10.1016/j.fss.2003.11.014

Meng, J., & Jun, Y. B. (1994). BCK-algebras. Kyung Moon Sa Co.

Saeid, A. B. (2011). Some results in Doubt fuzzy *BF*-algebras. *Analele Universitatii de Vest, Timi soara Seria Mathematica Informatica*, *1*, 125-134.

Saeid, A. B., & Rezvani, M. A. (2009). On fuzzy *BF*-algebras. *Int. Mathematical Forum*, *4*(1), 13-25.

Senapati, T., Bhowmik, M., & Pal, M. (2012c). Interval-valued intuitionistic fuzzy BG-subalgebras. *The Journal of Fuzzy Mathematics*, *20*(3), 707–720.

Senapati, T., Bhowmik, M., & Pal, M. (2013). Atanassov's intuitionistic fuzzy translations of intuitionistic fuzzy H-ideals in BCK/BCI-algebras. *Notes on Intuitionistic Fuzzy Sets*, *19*(1), 32–47.

Walendziak, A. (2007). On BF-algebras. *Mathematica Slovaca*, *57*(2), 119–128.

Zadeh, L. A. (1965). Fuzzy sets. *Information and Control*, *8*(3), 338–353. doi:10.1016/S0019-9958(65)90241-X

Zadeh, L. A. (1975). The concept of a linguistic variable and its application to approximate reasoning-I. *Information Sciences*, *8*(3), 199–249. doi:10.1016/0020-0255(75)90036-5

Zarandi, A., & Saeid, A. B. (2009). Interval-valued fuzzy BF-algebras. *Int. J. Contemp. Math. Sciences*, *4*(17), 803–814.

Chapter 15
Interval-Valued Fuzzy H-Ideals on ß-Algebra

Prakasam Muralikrishna
Muthurangam Government Arts College, India

Tapan Senapati
Southwest University, China

Perumal Hemavathi
Sri Venkateswara College of Engineering and Technology, India

ABSTRACT

The notion of interval valued fuzzy set was first introduced by Zadeh as a generalization of fuzzy sets. Using interval valued fuzzy set, various algebraic structures and related topics were discussed. This chapter extends fuzzy H-ideal into interval valued fuzzy H-ideals of β-algebra and deals some related results. It also provides the study on homomorphic images of an interval valued fuzzy H-ideals of β-algebra and the idea of Cartesian product of interval valued fuzzy H-ideals of β-algebra.

1. INTRODUCTION

Neggers (Neggers, 2002a) and Kim (Neggers, & Kim., 2002) initiated the class of B-algebras which is extended from different classes of algebras such as BCH/BCI/BCK-algebras. They also proposed the notion of β – algebras in which two operations were used. Young Bae Jun et al. (Jun, & Kim., 2012) are also explored the notion of β – algebras and related topics.

Abu Ayub Ansari et al. (Ansari, & Chandramouleeswaran., 2013),((Ansari, & Chandramouleeswaran., 2014),(Ansari, & Chandramouleeswaran., 2014a) defined the notion of fuzzy β – subalgebras of β – algebras and inverstigated some of their properties. Further, the authors extended their new ideas towards fuzzy β – ideals of β – algebras. By introducing the concepts of Cartesian product of fuzzy β – ideals of β – algebras. In 2014 the authors of (Ansari, & Chandramouleeswaran., 2014b) introduced the notion of T-fuzzy of β – subalgebras of β – algebras.

DOI: 10.4018/978-1-7998-0190-0.ch015

After Zadeh's (Zadeh, 1965) presentation of fuzzy sets, there have been various speculations of this crucial idea. In (Zadeh, 1975) likewise the authors presented the idea of interval valued fuzzy subsets where the estimations of the membership functions are intervals of numbers rather than the numbers.

Biswas and Rosenfeld (Biswas, & Rosenfeld., 1994) proposed the notion of fuzzy subgroups with interval valued membership functions. Fuzzy BCI-subalgebras with interval valued membership functions was studied by Sung Min Houng et al. (Hong, Jun, Kim, & Kim., 1993). Further the new idea of interval valued fuzzy subalgebras/ideals in BCK-algebras was propounded by Jun (Jun, 2000) and also the author (Young, & Min., 2001) redefined the concept of fuzzy subalgebras of BCK/BCI-algebras.

Khalid et al. (Khalid, & Ahmad., 1999) introduced the notion of Fuzzy H-ideals in BCI-algebras. The concept of intuitionistic L-fuzzy H-ideals using BF-algebras has been presented by Muralikrishna et al. (Muralikrishna, & Chandramouleeswaran., 2011). In 2017, Sabarinathan et al. (Sabarinathan, Muralikrishna, & Kumar., 2017) studied the concepts of Bipolar interval valued fuzzy H-ideals of BF-algebras. The new thought of fuzzy H-ideals in BCK-algebras was proposed by Zhan et al. (Zhan, & Tan., 2005).

In 2015, Rajam et al. (Rajam, & Chandramouleeswaran., 2015) proposed the concept of L-fuzzy T-ideals of β – algebras. Sujatha et al. (Sujatha, Chandramouleeswaran, & Muralikrishna., 2014) opened the notion of product on intuitionistic fuzzy β – subalgebras of β – algebras. Kordi et al. (Kordi, Moussavi, & Ahmad., 2008) defined the notion of fuzzy H-ideals of BCI-algebras with interval valued membership functions. Recently, the concept of interval valued fuzzy β – subalgebras were initiated by Hemavathi et al.(Hemavathi, Muralikrishna, Palanivel., 2015). In 2017 (Hemavathi, Muralikrishna, Palanivel., 2017) the authors extended the notion of interval valued fuzzy β – ideals of β – algebras. With all these ideas, in this chapter, we discuss the ideas of interval valued fuzzy H-ideals on β – algebras.

This work entitled as "Interval valued fuzzy H-ideals on β-algebras which is divided into four subsections as follows:

Section1 Introduction
Section 2 Preliminaries
Section 3 discusses about the interval valued fuzzy H-ideals on β-algebra.
Section 4 deals about the product on interval valued fuzzy H-ideals on β-algebra .
Section 5 describes $\bar{T}$ - interval valued fuzzy H-ideals on β-algebra which provides the interpretation of product, homomorphism.
Section 6 Conclusion

2. Preliminaries

This section provides the basic definition of fuzzy sets and some operations.

Definition 1: A fuzzy set in a universal set X is defined as σ: X→[0,1]. For each element x∈X, σ(x) is called the membership value of x.

Definition 2: If σ_1 and σ_2 are two fuzzy sets in X, then the intersection of σ_1 and σ_2, denoted by $\sigma_1 \cap \sigma_2$ is defined by

$$\sigma_1 \cap \sigma_2 = \min\left\{\sigma_1\left(x\right), \sigma_2\left(x\right)\right\} \forall x \in X.$$

Definition 3: If σ_1 and σ_2 are two fuzzy sets in X, then the union of σ_1 and σ_2, denoted by $\sigma_1 \cap \sigma_2$ is defined by

$$\sigma_1 \cup \sigma_2 = \max\left\{\sigma_1\left(x\right), \sigma_2\left(x\right)\right\} \forall x \in X.$$

Definition 4: Let σ_1 and σ_2 be two fuzzy sets of X_1 and X_2 respectively. Then the direct product $\sigma_1 \times \sigma_2$ of σ_1 and σ_2 is defined by

$$\left(\sigma_1 \times \sigma_2\right)\left(x_1, x_2\right) = \min\left\{\sigma_1\left(x_1\right), \sigma_2\left(x_2\right)\right\} \forall \left(x_1, x_2\right) \in X_1 \times X_2$$

and $\sigma_1 \times \sigma_2$ is a fuzzy set of $X_1 \times X_2$.

Definition 5: A triangular norm (*T*-norm) is a function T: $[0,1]\times[0,1]\rightarrow[0,1]$ $\forall x,y,z \in [0,1]$, if it satisfies the following conditions,

1. T(x,1)=x (boundary condition)
2. T(x,y)=T(y,x) (commutativity)
3. T(T(x,y)z)=T(x,T(y,z)) (associativity)
4: $T(x,y)\leq T(x,z)$ if $y\leq z$ (monotonicity)

The following are some *T*-norms used in many times.

1. Standard *T*-norm TM: T(x,y)=min(x,y)
2. Bounded difference (T_L): T(x,y)=max(0,x+y−1)
3. Algebraic product (T_P): T(x,y)=xy
4. Drastic intersection:

$$T_D : T\left(x, y\right) = \begin{cases} x & when\ y = 1 \\ y & when\ x = 1 \\ 0 & otherwise \end{cases}$$

Definition 6: Let D[0,1] denote the family of all closed sub intervals of [0,1]. Consider two elements $D_1D_2 \in D[0,1]$ If $D_1=[a_1,b_1]$ and $D_2=[a_2,b_2]$, then $\mathrm{rmax}(D_1,D_2)=[\max\{a_1,a_2\}, \max\{b_1,b_2\}]$ which is denoted by $D_1 \vee^r D_2$ and $\mathrm{rmin}(D_1,D_2) = [\min\{a_1,a_2\}, \min\{b_1,b_2\}]$ which is denoted by $D_1 \wedge^r D_2$. Thus if $D_i=[a_1,b_1]$ $\in D[0,1]$ for i=1,2,3,…

$\mathrm{rsup}_i(D_i)=[\sup_i(a_i), \sup(b_i)]$,

i.e. $\vee_i^r D_i = [\vee_i a_i, \vee_i b_i]$

Similarly

$rinf_i(D_i)=[inf_i(a_i, inf_i(b_i)]$,

i.e. $\wedge_i^r D_i = [\wedge_i a_i, \wedge_i b_i]$.

Now $D_1 \geq D_2$ iff $a_1 \geq a_2$ and $b_1 \geq b_2$.
Similarly the relations $D_1 \leq D_2$ and $D_1 = D_2$ are defined.

Remark: Let $D_1=[a_1,b_1]$ and $D_2=[a_2,b_2] \in D[0,1]$ Then

1. $D_1 \leq D_2 \Rightarrow a_1 \leq a_2$ and $b_1 \leq b_2$
2. $D_1 = D_2 \Rightarrow a_1 = a_2$ and $b_1 = b_2$
3. $D_1 + D_2 = [a_1+a_2, b_1+b_2]$ whenever $a_1+a_2 \leq 1$ and $b_1+b_2 \leq 1$
4. $D_1 - D_2 = [a_1-a_2, b_1-b_2]$ whenever $a_1-a_2 \leq 1$ and $b_1-b_2 \leq 1$

Definition 7: An interval valued fuzzy set (briefly i-v fuzzy set) A defined on X is given by

$$A = \{(x, [\sigma_A^L(x), \sigma_A^U(x)])\} \forall x \in X$$

(briefly denoted by $A = [\sigma_A^L, \sigma_A^U]$), where σ_A^L and σ_A^U are two fuzzy sets in X such that $\sigma_A^L(x) \leq \sigma_A^U(x)$ $\forall x \in X$.

For $\bar{\mu}_A(x) = [\mu_A^L(x), \mu_A^U(x)]$ $\forall x \in X$ and let D[0,1] denotes the family of all closed sub intervals of [0,1]. If $\sigma_A^L(x) = \sigma_A^U(x) = c$, say, where $0 \leq c \leq 1$, then $\bar{\sigma}_A(x) = [c, c]$ which we also assume, for the sake of convenience, to belong to D[0,1].

Thus $\bar{\sigma}_A(x) \in D[0,1]$ $\forall x \in X$, and therefore the i-v fuzzy set A is given by

$$A = \{(x, \bar{\sigma}_A(x))\} \forall x \in X,$$

where $\bar{\sigma}_A : X \to D[0,1]$. Now let us define what is known as refined minimum (briefly rmin) of two elements in D[0,1]. We also define the symbols "≥", "≤"and "=" in case of two elements in D[0,1].

Consider two elements

$D_1=[a_1,b_1]$ and $D_2=[a_2,b_2] \in D[0,1]$.

Then

$rmin(D_1,D_2) = [min\{a_1,a_2\}, min\{b_1,b_2\}]$;

$D_1 \geq D_2$ if and only if $a_1 \geq a_2$, $b_1 \geq b_2$;

Similarly, we may have $D_1 \leq D_2$ and $D_1 = D_2$.

Definition 8: Let (X,+,-,0) and (Y,+,-,0) be two sets. Let

$$A = \left\{x, \bar{\sigma}_A\left(x\right) : x \in X\right\} \text{ and } B = \left\{y, \bar{\sigma}_B\left(y\right) : y \in Y\right\}$$

be i-v fuzzy subsets in X and Y respectively. The Cartesian product of A and B denoted by A×B is defined to be the set

$$A \times B = \left\{\left(x, y\right), \bar{\sigma}_{A\times B}\left(x, y\right) : \left(x, y\right) \in X \times Y\right\},$$

where

$$\bar{\sigma}_{A\times B} : X \times Y \rightarrow D\left[0,1\right]$$

is given by

$$\bar{\sigma}_{A\times B}\left(x, y\right) = r\min\left\{\bar{\sigma}_A\left(x\right), \bar{\sigma}_B\left(y\right)\right\}.$$

Definition 9: A β-algebra is a non-empty set X with a constant 0 and two binary operations + and - satisfying the following axioms:

(i) $x - 0 = x$
(ii) $(0 - x) + x = 0$
(iii) $(x - y) - z = x - (z + y)$ $\forall x,y,z \in X$

Example: Let X={0,1,2,3} be a set with constant 0 and binary operations + and – are defined on X by Cayley's table (Table 1).

Table 1. Cayley's table

+	0	1	2	3		-	0	1	2	3
0	0	1	2	3		0	0	3	2	1
1	1	2	3	0		1	1	0	3	2
2	2	3	0	1		2	2	1	0	3
3	3	0	1	2		3	3	2	1	0

Then (X,+,-,0) is a β-algebra.

Definition 10: A non empty subset A of a β-algebra (X,+,-,0) is called a β -subalgebra of X, if

(i) $x + y \in A$
(ii) $x - y \in A$ $\forall x,y \in A$

Example: In the above example, the subset A={0,2} is a β -subalgebra of X.

Definition 11: Let(X,+,-,0) and (Y,+,-,0) be two β -algebras. A mapping f: $X \rightarrow Y$ is said to be a β-homomorphismif $\forall x,y \in X$

(i) $f(x+y) = f(x)+f(y)$
(ii) $f(x+y) = f(x)+f(y)$

Definition 12: Let σ be a fuzzy set in a β -algebra of X. Then σ is called a fuzzy β -subalgebra of X, if $\forall x,y \in X$

(i) $\sigma(x+y) \geq \min\{\sigma(x), \sigma(y)\}$ and
(ii) $\sigma(x-y) \geq \min\{\sigma(x), \sigma(y)\}$

Definition 13: Let σ be a fuzzy set in a β -algebra of X. For a given t-norm T, σ is called a T–fuzzy β -subalgebra of X, if $\forall x,y \in X$

(i) $\sigma(x+y) \geq T\{\sigma(x),\sigma(y)\}$ and
(ii) $\sigma(x-y) \geq T\{\sigma(x),\sigma(y)\}$

Definition 14: A non empty subset I of a β-algebra $(X,+,-,0)$ is called a β -ideals of X, if

(i) $0 \in I$
(ii) $x+y \in I$ $\forall x,y \in I$
(iii) if $x - y$ and $y \in I$ then $x \in I$ $\forall x,y \in X$

Example 2.6: Consider the β -algebra X in example. Then the subset I_1={0,1} is a β -ideals of X. But I_2={0,1,3} is not a β -ideal of X, (since $1+3=2 \notin I_2$)

Definition 15: Let σ be a fuzzy set in a β -algebra of X. Then σ is called a fuzzy β -ideal of X, if $\forall x,y \in X$

(i) $\sigma(0) \geq \sigma(x)$
(ii) $\sigma(x+y) \geq \min\{\sigma(x),\sigma(y)\}$
(iii) $\sigma(x) \geq \min\{\sigma(x-y),\sigma(y)\}$

Definition 16: Let σ be a fuzzy set in a β -algebra of X. Then σ is called a T- fuzzy β -ideal of X, if $\forall x,y \in X$

(i) $\sigma(0) \geq \sigma(x)$
(ii) $\sigma(x+y) \geq T\{\sigma(x),\sigma(y)\}$
(iii) $\sigma(x) \geq T\{\sigma(x-y),\sigma(y)\}$

Definition 17: A non empty subset I of a β -algebra $(X,+,-,0)$ is called a H-ideals of X, if $\forall x,y,z \in X$

(i) $0 \in I$
(ii) $(x+y)+z \in I$ and $y \in I \Rightarrow x+z \in I$
(iii) $(x-y)-z \in I$ and $y \in I \Rightarrow x - z \in I$

Definition 18: Let σ be a fuzzy set in a β -algebra of X. Then σ is called a fuzzy H-ideal of X, if $\forall x,y \in X$

(i) $\sigma(0) \geq \sigma(x)$
(ii) $\sigma(x+z) \geq \min\{\sigma((x+y)+z),\sigma(y)\}$
(iii) $\sigma(x - z) \geq \min\{\sigma((x - y) - z),\sigma(y)\}$

Definition 19: Let σ be a fuzzy set in a β -algebra of X. Then σ is called a T- fuzzy H-ideal of X, if $\forall x,y \in X$

(i) $\sigma(0) \geq \sigma(x)$
(ii) $\sigma(x+z) \geq T\{\sigma((x+y)+z),\sigma(y)\}$
(iii) $\sigma(x - z) \geq T\{\sigma((x - y) - z),\sigma(y)\}$

Definition 20: Let$(X,+,-,0)$ be a β -algebra. Then the interval valued fuzzy subset $A = \left\{x, \bar{\sigma}_A\left(x\right) : x \in X\right\}$ is called an interval valued fuzzy (i-v fuzzy) β -subalgebra of X if

(i) $\bar{\sigma}_A\left(x+y\right) \geq r\min\left\{\bar{\sigma}_A\left(x\right), \bar{\sigma}_A\left(y\right)\right\}$
(ii) $\bar{\sigma}_A\left(x-y\right) \geq r\min\left\{\bar{\sigma}_A\left(x\right), \bar{\sigma}_A\left(y\right)\right\} \ \forall x, y \in X$

Definition 21: Let $A = \left\{x, \bar{\sigma}_A\left(x\right) : x \in X\right\}$ be an interval valued fuzzy set in a β -algebra X. Then A is called an interval valued fuzzy(i-v-fuzzy) β -ideal of X, if $\forall x,y \in X$

(i) $\bar{\sigma}_A\left(0\right) \geq \bar{\sigma}_A\left(x\right)$
(ii) $\bar{\sigma}_A\left(x+y\right) \geq r\min\left\{\bar{\sigma}_A\left(x\right), \bar{\sigma}_A\left(y\right)\right\}$
(iii) $\bar{\sigma}_A\left(x\right) \geq r\min\left\{\bar{\sigma}_A\left(x-y\right), \bar{\sigma}_A\left(y\right)\right\}$

Example: The i-v fuzzy set defined in the β -algebra X in the above example as, $\bar{\sigma}_A : X \rightarrow D\left[0,1\right]$ such that

$$\bar{\sigma}_A(x) = \begin{cases} [0.3, 0.7] \; x = 0 \\ [0.1, 0.5] \; x = 1, 3 \\ [0.2, 0.6] \; x = 2 \end{cases}$$

is a i-v fuzzy β -ideal of *X*.

Definition 22: Let A be an i-v fuzzy subset of X, $\bar{\alpha} \in D[0,1]$. Then $A_{\bar{\alpha}} = \{x \in X : \bar{A}_{\alpha} \geq \bar{\alpha}\ \}$ is called a i-v-level set of A.

3. INTERVAL VALUED FUZZY H-IDEALS ON β –ALGEBRA

This section introduces the notion of interval valued fuzzy H-ideals on β -algebras and deals some simple results.

Definition 23: Let $\bar{\sigma}$ be an interval valued fuzzy (i-v-fuzzy) set in a β –algebra of X. Then $\bar{\sigma}$ is called an i-v-fuzzy H-ideals of X if

(i) $\bar{\sigma}(0) \geq \bar{\sigma}(x) \quad \forall x \in X$

$$\text{(ii)}\,\bar{\sigma}(x+z) \geq r\min\{\bar{\sigma}((x+y)+z), \bar{\sigma}(y)\}$$
$$\text{(ii)}\,\bar{\sigma}(x-z) \geq r\min\{\bar{\sigma}((x-y)-z), \bar{\sigma}(y)\} \quad \forall x, y \in X.$$

Example: Consider a β –algebra *X*={0,1,2,3} with constant 0 and binary operations + and – are defined on X by Cayley's table (Table 2).

$$\bar{\sigma}(x) = \begin{cases} [0.4, 0.6] & x = 0 \\ [0.3, 0.5] & x = 1 \\ [0.2, 0.4] & x = 3, 2 \end{cases}$$

Table 2. Cayley's table

-	0	1	2	3		+	0	1	2	3
0	0	1	3	2		0	0	1	2	3
1	1	0	2	3		1	1	0	3	2
2	2	3	0	1		2	2	3	1	0
3	3	2	1	0		3	3	2	0	1

where

$$\bar{0} \leq [0.2, 0.4] \leq [0.3, 0.5] \leq [0.4, 0.6] \leq \bar{1}.$$

It is easy to verify that $\bar{\sigma}(x)$ is an i-v fuzzy H-ideal of X.

Lemma: Let $A = \{x, \bar{\sigma}_A(x) : x \in X\}$ be an i-v-fuzzy H-ideal of a β –algebra of X. If x≤y then $\bar{\sigma}(x) \geq \bar{\sigma}(y)$. ie., $\bar{\sigma}$ is order reversing.

Proof: For x,y∈X, x≤y ⇒ x – y=0. Then

$$\begin{aligned} \Rightarrow \bar{\sigma}(x) = \bar{\sigma}(x-0) &\geq r\min\{\bar{\sigma}((x-y)-0), \bar{\sigma}(y)\} \\ &= r\min\{\bar{\sigma}(0-0), \bar{\sigma}(y)\} \\ &= r\min\{\bar{\sigma}(0), \bar{\sigma}(y)\} \\ &= \bar{\sigma}(y) \end{aligned}$$

Hence $\bar{\sigma}(x) \geq \bar{\sigma}(y)$.

Theorem 1: Let A be a subset of X. Define an i-v fuzzy set $\bar{\sigma} : X \to D[0,1]$ such that

$$\bar{\sigma}_A(x) = \begin{cases} [t_0, t_1] & \text{if } x \in A \\ [s_0, s_1] & \text{if } x \notin A \end{cases}$$

where $[t_0,t_1],[s_0,s_1] \in D[0,1]$ with $[t_0,t_1] \geq [s_0,s_1]$. Then $\bar{\sigma}$ is an i-v fuzzy H-ideal of a β–algebra X if and only if A is a H-ideal of X.

Proof: Assume that $\bar{\sigma}$ is an i-v fuzzy H-ideal of X. If $x \in A, \bar{\sigma} \geq [t_0, t_1]$ and if $x \notin A, \bar{\sigma} \geq [s_0, s_1]$ since $[t_0,t_1] > [s_0,s_1]$

$$\begin{aligned} &\Rightarrow \bar{\sigma}_A(0) \geq [t_0, t_1] > [s_0, s_1] \\ &\Rightarrow \bar{\sigma}_A(0) = [t_0, t_1] \\ &\Rightarrow 0 \in A \end{aligned}$$

For x,y,z∈A

$$\Rightarrow \bar{\sigma}_A(x) = [t_0, t_1], \bar{\sigma}_A(y) = [t_0, t_1] and \bar{\sigma}_A(z) = [t_0, t_1]$$

$$\begin{aligned}\bar{\sigma}_A(x+z) &\geq r\min\{\bar{\sigma}_A((x+y)+z),\bar{\sigma}_A(y)\}\\ &\geq r\min\left\{r\min\{\bar{\sigma}_A(x+y),\bar{\sigma}_A(z)\},\bar{\sigma}_A(y)\right\}\\ &= r\min\left\{r\min\left\{r\min\{\bar{\sigma}_A(x),\bar{\sigma}_A(y)\},\bar{\sigma}_A(z)\right\},\bar{\sigma}_A(y)\right\}\\ &= r\min\left\{r\min\left\{r\min\left\{[t_0,t_1],[t_0,t_1]\right\},[t_0,t_1]\right\},[t_0,t_1]\right\}\\ &= r\min\left\{r\min\left\{[t_0,t_1],[t_0,t_1]\right\},[t_0,t_1]\right\}\end{aligned}$$

$=[t_0,t_1]$

$$\therefore \bar{\sigma}_A(x+z)\geq[t_0,t_1]\Rightarrow x+z\in A$$

For x,y,z∈A

$$\Rightarrow \bar{\sigma}_A((x-y)-z)=[t_0,t_1] \text{ and } \bar{\sigma}_A(y)=[t_0,t_1].$$

Now

$$\begin{aligned}\bar{\sigma}_A(x-z) &\geq r\min\{\bar{\sigma}_A((x-y)-z),\bar{\sigma}_A(y)\}\\ &= r\min\left\{[t_0,t_1],[t_0,t_1]\right\}\\ &= [t_0,t_1]\\ &\Rightarrow \bar{\sigma}_A(x-z)\geq[t_0,t_1]\Rightarrow x-z\in A\end{aligned}$$

Hence A is a H-ideal of X.
Suppose A is a H-ideal of X. Now

$$0\in A\Rightarrow \bar{\sigma}_A(0)=[t_0,t_1]$$

Also

$$\forall x\in X,\ \operatorname{Im}\bar{\sigma}=\left\{[t_0,t_1],[s_0,s_1]\right\}$$

and

$$[t_0,t_1]>[s_0,s_1]\Rightarrow \bar{\sigma}_A(0)\geq\bar{\sigma}_A(x)$$

Since A is a H-ideal of X.

∀x,y,z∈S, (x+y)+z∈A and y∈A, and y∈A ⇒ x+z, x – z∈A

Then

$$\bar{\sigma}_A\left(x+z\right)=\left[t_0,t_1\right]\geq r\min\left\{\bar{\sigma}_A\left(\left(x+y\right)+z\right),\bar{\sigma}_A\left(y\right)\right\}$$

Similarly,

$$\bar{\sigma}_A\left(x-z\right)\geq r\min\{\bar{\sigma}_A\left((x-y\right)-z),\bar{\sigma}_A\left(y\right)\}$$

Hence $\bar{\sigma}$ is an i-v fuzzy H-ideal of X.

Theorem 2: Let A be an i-v fuzzy set in a β-algebra X. Then A is an i-v fuzzy H-ideal of X if and only if the non empty set $A_{\bar{\alpha}}=\{x\in X/\bar{\sigma}_A\left(x\right)\geq\bar{\alpha}\}$ is a H-ideal of X for every $\bar{\alpha}\in D\left[0,1\right]$. We call $A_{\bar{\alpha}}$ the i-v-level H-ideal of X.

Proof: Assume that A is an i-v-fuzzy H-ideal of X. Since for all $x\in A_{\bar{\alpha}}$ we have

$$\bar{\sigma}_A\left(0\right)\geq\bar{\sigma}_A\left(x\right)\geq\bar{\alpha}$$
$$\therefore 0\in A_{\bar{\alpha}}$$

Now, let x,y,z∈X

$$\begin{aligned}\bar{\sigma}_A\left(x+z\right)&\geq r\min\{\bar{\sigma}_A\left(x+y\right)+z),\bar{\sigma}_A\left(y\right)\}\\&\geq r\min\left\{\bar{\alpha},\bar{\alpha}\right\}\\&=\bar{\alpha}\end{aligned}$$

And so

$$x+z\in A_{\bar{\alpha}}$$

Similarly

$$\bar{\sigma}_A\left(x-z\right)\geq\bar{\alpha}$$

Thus $A_{\bar{\alpha}}$ is a H-ideal of β-algebra of X.

Conversely, Assume that $A_{\bar{\alpha}}$ is a H-ideal of X for every $\bar{\alpha}\in D\left[0,1\right]$.

Suppose that there exist $x_0,y_0,z_0\in X$ such that

$$\bar{\sigma}_A(x_0 + z_0) < r \min\left\{\bar{\sigma}_A((x_0 + y_0) + z_0), \bar{\sigma}_A(y_0)\right\}$$

Let

$$\bar{\sigma}_A((x_0 + y_0) + z_0) = \bar{t}_1, \bar{\sigma}_A(y_0) = \bar{t}_2 \quad and \quad \bar{\sigma}_A(x_0 + z_0) = \bar{\alpha},$$

then

$$\begin{aligned} \bar{\alpha} &< r \min\left\{\bar{t}_{1,}\bar{t}_2\right\} \\ &< r \min\left[t_1^L, t_1^U\right], \left[t_2^L, t_2^U\right] \end{aligned}$$

$$\left[\alpha_1^L, \alpha_1^U\right] < \left[\min\left\{\left[t_1^L, t_2^L\right], \left[t_1^U, t_2^U\right]\right\}\right]$$

Taking

$$\bar{S} = \frac{1}{2}(\bar{\sigma}_A(x_0 + z_0) + r \min\{\bar{\sigma}_A((x_0 + y_0) + z_0), \bar{\sigma}_A(y_0)\}$$

We obtain

$$\begin{aligned} \bar{S} &= \frac{1}{2}([\alpha^L, \alpha^U] + [\min\left\{t_1^L, t_2^L\right\}, \min\left\{t_1^U, t_2^U\right\}]) \\ &= \left[\frac{1}{2}(\alpha^L + \min\left\{t_1^L, t_2^L\right\}\right), \frac{1}{2}(\alpha^U + \min\left\{t_1^U, t_2^U\right\})] \end{aligned}$$

It follows that,

$$\min\left\{t_1^L, t_2^L\right\} > t_1^L = \frac{1}{2}(\alpha^L + \min\left\{t_1^L, t_2^L\right\}) > \alpha^L$$

$$\min\left\{t_1^U, t_2^U\right\} > t_1^U = \frac{1}{2}(\alpha^U + \min\left\{t_1^U, t_2^U\right\}) > \alpha^U$$

So that

$$\left[\min\left\{t_1^L, t_2^L\right\}, \min\left\{t_1^U, t_2^U\right\}\right] > \left[t_1^L, t_1^U\right] > \bar{\alpha} = \bar{\sigma}_A(x_0 + z_0)$$

$$\therefore (x_0 + z_0) \notin A_{\bar{\alpha}}$$

On the other hand,

$$\bar{\sigma}_A\left(\left(x_0+y_0\right)+z_0\right)=\left[t_1^L,t_1^U\right]\geq[\min\left\{t_1^L,t_2^L\right\},\min\left\{t_1^U,t_2^U\right\}]>\bar{t}$$

and

$$\bar{\sigma}_A\left(y_0\right)=\left[t_2^L,t_2^U\right]\geq[\min\left\{t_1^L,t_2^L\right\},\min\left\{t_1^U,t_2^U\right\}]>\bar{t}$$

and so

$$\left(x_0+y_0\right)+z_0,y_0\in A_{\bar{\alpha}}$$

But this contradiction the fact that $A_{\bar{\alpha}}$ is a *H*-ideal of *X*. Hence

$$\bar{\sigma}_A\left(x+z\right)\geq r\min\{\bar{\sigma}_A\left(x+y\right)+z),\bar{\sigma}_A\left(y\right)\}\forall x,y,z\in X.$$

Similarly,

$$\bar{\sigma}_A\left(x-z\right)\geq r\min\{\bar{\sigma}_A\left(x-y\right)-z),\bar{\sigma}_A\left(y\right)\}$$

Theorem 3: Every *H*-ideal of a β-algebras *X* can be realized as an i-v level *H*-ideal of an i-v fuzzy *H*-ideal of *X*.

Proof: Let *J* be a *H*-ideal and *A* be an i-v fuzzy set on *X* defined by,

$$\bar{\sigma}_A=\begin{cases}[t_1^L,t_1^U]:x\in J\\ \left[0,0\right]:otherwise\end{cases}$$

where

$$\bar{t}_1\in\left[0,1\right]\text{ with }t_1^L<t_1^U.$$

We show that *A* is an i-v fuzzy *H*-ideal of *X*. Let (x+y)+z, y∈*J* then *x*+*z*∈*J* and so

$$\begin{aligned}\bar{\sigma}_A\left(x-z\right)&=\bar{t}_1=r\min\left\{\bar{t}_1,\bar{t}_1\right\}\\&=r\min\ \{[t_1^L,t_1^U],[t_1^L,t_1^U]\ \}\\&=r\min\{\bar{\sigma}_A\left(\left(x+y\right)+z\right),\bar{\sigma}_A\left(y\right)\}\end{aligned}$$

If at least one of (x+y)+z and y is not in *A*, then at least one of $\bar{\sigma}_A\left(\left(x+y\right)+z\right)$ & $\bar{\sigma}_A\left(y\right)$ is 0.

$$\therefore \bar{\sigma}_A\left(x+z\right) \geq \left[0,0\right] = r\min\left\{\left[0,0\right],\left[0,0\right]\right\}$$
$$= r\min\{\bar{\sigma}_A\left(x+y\right)+z),\bar{\sigma}_A\left(y\right)\}$$

This means that $\bar{\sigma}_A$ satisfies the condition.

On the other hands, since 0∈*J*, $\bar{\sigma}_A\left(0\right) = \bar{t}_1 \geq \bar{\sigma}_A\left(x\right)$, ∀x∈X and so $\bar{\sigma}_A$ satisfies condition. Thus $\bar{\sigma}_A$ is an i-v fuzzy *H*-ideal of *X*. It is clear that $A_{\bar{\alpha}} = J$.

Theorem 4: Let f: X→X be an endomorphism on a β-algebra. Let $\bar{\sigma}$ be an i-v-fuzzy H-ideal of X. Define a i-v-fuzzy set $\bar{\sigma}_f : X \to D\left[0,1\right]$ defined by $\bar{\sigma}_f\left(x\right) = \bar{\sigma}\left(f\left(x\right)\right)$ ∀*x*∈*X*. Then $\bar{\sigma}_f$ is an i-v fuzzy H-ideal of X.

Proof: Let x∈X. Then

$$\bar{\sigma}_f = \bar{\sigma}\left(f\left(x\right)\right) \leq \bar{\sigma}\left(0\right) = \bar{\sigma}\left(f\left(0\right)\right) = \bar{\sigma}_f\left(0\right)$$

Let x,y∈X

$$\begin{aligned}\bar{\sigma}_f\left(x+z\right) &= \bar{\sigma}\left(f\left(x+z\right)\right)\\ &= \bar{\sigma}\left(f\left(x\right)+f(y)\right)\\ &\geq r\min\{\bar{\sigma}\left(\left(f\left(x\right)+f(y)\right)+f\left(z\right)\right),\bar{\sigma}\left(f\left(y\right)\right)\}\\ &= r\min\{\bar{\sigma}\left(\left(f\left(x+y\right)\right)+f\left(z\right)\right),\bar{\sigma}\left(f\left(y\right)\right)\}\\ &= r\min\{\bar{\sigma}\left(\left(f\left(x+y\right)+z\right)\right),\bar{\sigma}\left(f\left(y\right)\right)\}\\ &= r\min\{\bar{\sigma}_f\left(\left(x+y\right)+z\right)\right),\bar{\sigma}_f\left(y\right)\}\end{aligned}$$

Similarly,

$$\bar{\sigma}_f\left(x-z\right) \geq r\min\{\bar{\sigma}_f\left(\left(x-y\right)-z\right)\right),\bar{\sigma}_f\left(y\right)\}$$

Hence $\bar{\sigma}_f$ is an i-v fuzzy H-ideal of X.

Theorem 5: Let:X→Y be an homomorphism of β-algebra. If A is an i-v-fuzzy H-ideal of Y, then the pre image f^{-1}(A) is an i-v-fuzzy H-ideal of X.

Proof: Let A be an i- v-fuzzy H-ideal of Y. For any x∈X.

$$f^{-1}\left(\bar{\sigma}_A\left(0\right)\right) = \bar{\sigma}_A\left(f\left(0\right)\right) = \bar{\sigma}_A\left(0\right) \geq \bar{\sigma}_A\left(x\right)$$

For x,y∈X,

$$\begin{aligned} f^{-1}\left(\bar{\sigma}_A\left(x+z\right)\right) &= \bar{\sigma}_A\left(f\left(x+z\right)\right) \\ &= \bar{\sigma}_A\left(f\left(x\right)+f(z)\right) \\ &\geq r\min\left\{\bar{\sigma}_A\left(\left(f\left(x\right)+f\left(y\right)\right)+f\left(z\right)\right),\bar{\sigma}_A\left(f\left(y\right)\right)\right\} \\ &\geq r\min\left\{\bar{\sigma}_A\left(\left(f(x+y\right)+z\right)\right),\bar{\sigma}_A\left(f\left(y\right)\right)\right\} \\ &\geq r\min\left\{f^{-1}(\bar{\sigma}_A\left(\left(\left(x+y\right)+z\right)\right)),f^{-1}(\bar{\sigma}_A\left(\left(y\right)\right)\right\} \end{aligned}$$

$\therefore f^{-1}(\bar{\sigma}_A)$ is a fuzzy H-ideal of X.

Similarly,

$$f^{-1}\left(\bar{\sigma}_A\left(x-z\right)\right) \geq r\min\left\{f^{-1}(\bar{\sigma}_A\left(\left(\left(x-y\right)-z\right)\right)),f^{-1}(\bar{\sigma}_A\left(\left(y\right)\right)\right\}$$

Theorem 6: Intersection of any two i-v-fuzzy H-ideal of a β-algebra is also a i-v-fuzzy H-ideal.

Proof: Let

$$\begin{aligned} \left(\bar{\sigma}_1 \cap \bar{\sigma}_2\right)(0 &\geq r\min\left\{\bar{\sigma}_1\left(0\right),\bar{\sigma}_2\left(0\right)\right\} \\ &\geq r\min\left\{\bar{\sigma}_1\left(x\right),\bar{\sigma}_2\left(x\right)\right\} \\ &= \left(\bar{\sigma}_1 \cap \bar{\sigma}_2\right)\left(x\right) \end{aligned}$$

$$\begin{aligned} \left(\bar{\sigma}_1 \cap \bar{\sigma}_2\right)\left(x+z\right) &\geq r\min\left\{\bar{\sigma}_1\left(x+z\right),\bar{\sigma}_2\left(x+z\right)\right\} \\ &\geq r\min\left\{r\min\left\{\bar{\sigma}_1\left(\left(x+y\right)+z\right),\bar{\sigma}_1\left(y\right)\right\},r\min\left\{\bar{\sigma}_2\left(\left(x+y\right)+z\right),\bar{\sigma}_2\left(y\right)\right\}\right\} \\ &\geq r\min\left\{r\min\left\{\bar{\sigma}_1\left(\left(x+y\right)+z\right),\bar{\sigma}_1\left(y\right)\right\},r\min\left\{\bar{\sigma}_2\left(\left(x+y\right)+z\right),\bar{\sigma}_2\left(y\right)\right\}\right\} \\ &\geq r\min\left\{r\min\left\{\bar{\sigma}_1\left(\left(x+y\right)+z\right),\bar{\sigma}_2\left(\left(x+y\right)+z\right)\right\},r\min\left\{\bar{\sigma}_1\left(y\right),\bar{\sigma}_2\left(y\right)\right\}\right\} \\ &= r\min\left\{\left(\bar{\sigma}_1 \cap \bar{\sigma}_2\right)\left(x+y\right)+z\right),\left(\bar{\sigma}_1 \cap \bar{\sigma}_2\right)\left(y\right)\right\} \end{aligned}$$

Similarly,

$$\left(\bar{\sigma}_1 \cap \bar{\sigma}_2\right)\left(x-z\right) \geq r\min\left\{\left(\bar{\sigma}_1 \cap \bar{\sigma}_2\right)\left(x-y\right)-z\right),\left(\bar{\sigma}_1 \cap \bar{\sigma}_2\right)\left(y\right)\right\}$$

Theorem 7: An i-v fuzzy set $A=\left[\sigma_A^L,\sigma_A^U\right]$ in X is an i-v fuzzy H-ideal of X if and only if σ_A^L and σ_A^U are fuzzy H-ideals of X.

Proof: Suppose that σ_A^L and σ_A^U are fuzzy H-ideal of X.

$$\therefore \sigma_A^L(0) \geq \sigma_A^L(x) \text{ and } \sigma_A^U(0) \geq \sigma_A^U(x)$$
$$\Rightarrow \bar{\sigma}_A(0) \geq \bar{\sigma}_A(x)$$

Let x,y,z∈X. Then

$$\begin{aligned}\bar{\sigma}_A(x+z) &= \left[\sigma_A^L(x+z), \sigma_A^U(x+z)\right] \\ &\geq \left[\min\left\{\sigma_A^L((x+y)+z), \sigma_A^L(y)\right\}, \min\left\{\sigma_A^U((x+y)+z), \sigma_A^U(y)\right\}\right] \\ &= r\min\left\{\left[\sigma_A^L((x+y)+z), \sigma_A^U((x+y)+z)\right], \left[\sigma_A^L(y), \sigma_A^U(y)\right]\right\} \\ &= r\min\left\{\bar{\sigma}_A((x+y)+z), \bar{\sigma}_A(y)\right\}\end{aligned}$$

Similarly,

$$\bar{\sigma}_A(x-z) \geq r\min\left\{\bar{\sigma}_A((x-y)-z), \bar{\sigma}_A(y)\right\}$$

Hence A is an i-v fuzzy H-ideal of X.

Conversely, Suppose A is an i-v fuzzy H-ideal of X. Then for any x,y∈X, we have

$$\begin{aligned}\left[\sigma_A^L((x+y)+z), \sigma_A^U((x+y)+z)\right] &= \bar{\sigma}_A((x+y)+z) \\ &\geq r\min\left\{\bar{\sigma}_A((x+y)+z), \bar{\sigma}_A(y)\right\} \\ &\geq r\min\left\{\left[\sigma_A^L((x+y)+z), \sigma_A^U((x+y)+z)\right], \left[\sigma_A^L(y), \sigma_A^U(y)\right]\right\} \\ &= \left[\min\left\{\sigma_A^L((x+y)+z), \sigma_A^L(y)\right\}, \min\left\{\sigma_A^U((x+y)+z), \sigma_A^U(y)\right\}\right]\end{aligned}$$

It follows that

$$\sigma_A^L(x+z) \geq \min\left\{\sigma_A^L((x+y)+z), \sigma_A^L(y)\right\}$$

and

$$\sigma_A^U(x+z) \geq \min\left\{\sigma_A^U((x+y)+z), \sigma_A^U(y)\right\}$$

Similarly,

$$\sigma_A^L(x-z) \geq \min\left\{\sigma_A^L((x-y)-z), \sigma_A^L(y)\right\}$$

and

$$\sigma_A^U\left(x-z\right)\geq \min\left\{\sigma_A^U\left(x-y\right)-z\right),\sigma_A^U\left(y\right)\}$$

Hence σ_A^L and σ_A^U are fuzzy H-ideals of X.

Theorem 8: Let A be an i-v fuzzy H-ideal of X. If there is a sequence x_n in X such that $\lim_{n\to\infty}\bar{\sigma}_A(x_n)$ =[1,1] then $\bar{\sigma}_A\left(0\right)=\left[1,1\right]$.

Proof: Since $\bar{\sigma}_A\left(0\right)\geq\bar{\sigma}_A\left(x\right)$ for all x in X. We have $\bar{\sigma}_A\left(0\right)\geq\bar{\sigma}_A\left(x_n\right)$ for every positive integer n. Note that

$$\left[1,1\right]\geq\bar{\sigma}_A\left(0\right)\geq\lim_{n\to\infty}\bar{\sigma}_A\left(x_n\right)=\left[1,1\right]$$

Hence $\bar{\sigma}_A\left(0\right)=\left[1,1\right]$.

Theorem 9: Let *I* be a subset of a β-algebras *X* such that 0∈I and let *A* be an i-v fuzzy set on *X* which is given in the proof of theorem 2 if *A* is an i-v fuzzy *H*-ideal of *X*, then I is a *H*-ideal of *X*.

Proof: Assume that *A* is an i-v fuzzy *H*-ideal of *X*. Let $(x+y)+z$, $y\in I$, then

$$\bar{\sigma}_A\left(\left(x+y\right)+z\right)=\bar{\alpha}=\bar{\sigma}_A\left(y\right)$$

and so,

$$\begin{aligned}\bar{\sigma}_A\left(x+z\right)&\geq r\min\{\bar{\sigma}_A\left(x+y\right)+z),\bar{\sigma}_A\left(y\right)\}\\&=r\min\left\{\bar{\alpha},\bar{\alpha}\right\}\\&=\bar{\alpha}\end{aligned}$$

Similarly,

$$\bar{\sigma}_A\left(x-z\right)\geq\bar{\alpha}\Rightarrow x+z\ \text{ and }\ x-z\in I.$$

Theorem 10: If *A* is an i-v fuzzy *H*-ideal of a β-algebras *X*, then the set

$$X_{\bar{\sigma}_A}=\left\{x\in X:\bar{\sigma}_A\left(x\right)=\bar{\sigma}_A\left(0\right)\right\}$$

is a *H*-ideal of *X*.

Proof: Let $\left(x+y\right)+z\in X_{\bar{\sigma}_A}$. Then

$$\overline{\sigma}_A(y) = \overline{\sigma}_A(0) = \overline{\sigma}_A((x+y)+z)$$

and so

$$\begin{aligned}\overline{\sigma}_A(x+z) &\geq r\min\{\overline{\sigma}_A(x+y)+z), \overline{\sigma}_A(y)\} \\ &= r\min\{\overline{\sigma}_A(0), \overline{\sigma}_A(0)\} \\ &= \overline{\sigma}_A(0)\end{aligned}$$

ie, $x+z \in X_{\overline{\sigma}_A}$.

Similarly,

$$x - z \in X_{\overline{\sigma}_A}$$

Hence $X_{\overline{\sigma}_A}$ is a *H*-ideal of *X*.

4. PRODUCT ON INTERVAL VALUED FUZZY H-IDEALS ON β-ALGEBRA

This section presents the notion of the product on interval valued fuzzy H-ideals on β-algebras.

Theorem 11: Let $\overline{\sigma}_1$ and $\overline{\sigma}_2$ be two i-v-fuzzy H-ideals in a β-algebras of X. Then the direct product $\overline{\sigma}_1 \times \overline{\sigma}_2$ is an i-v-fuzzy H-ideal of $X_1 \times X_2$.

Proof: For $(x_1,x_2) \in X_1 \times X_2$. Then

$$\begin{aligned}(\overline{\sigma}_1 \times \overline{\sigma}_2)(0,0) &= r\min\{\overline{\sigma}_1(0), \overline{\sigma}_2(0)\} \\ &\geq r\min\{\overline{\sigma}_1(x_1), \overline{\sigma}_2(x_2)\} \\ &= (\overline{\sigma}_1 \times \overline{\sigma}_2)(x_1, x_2)\end{aligned}$$

Take $a,b,c \in X_1 \times X_2$, where $a=(x_1,x_2)$, $b=(y_1,y_2)$ and $c=(z_1,z_2)$. Then

$$
\begin{aligned}
(\bar{\sigma}_1 \times \bar{\sigma}_2)((x_1, x_2) + (z_1, z_2)) &= \{(\bar{\sigma}_1 \times \bar{\sigma}_2)((x_1, z_1), (x_2, z_2))\} \\
&= r\min\{\bar{\sigma}_1(x_1 + z_1), \bar{\sigma}_2(x_2 + z_2)\} \\
&\geq r\min\{r\min\{\bar{\sigma}_1((x_1 + y_1) + z_1), \bar{\sigma}_1(y_1)\}, r\min\{\bar{\sigma}_2((x_2 + y_2) + z_2), \bar{\sigma}_2(y_2)\} \\
&= r\min\{r\min\{\bar{\sigma}_1((x_1 + y_1) + z_1), \bar{\sigma}_2((x_2 + y_2) + z_2)\}, r\min\{\bar{\sigma}_1(y_1), \bar{\sigma}_2(y_2)\} \\
&\geq r\min\{r\min\{(\bar{\sigma}_1 \times \bar{\sigma}_2)((x_1 + y_1) + z_1), ((x_2 + y_2) + z_2))(\bar{\sigma}_1 \times \bar{\sigma}_2)(y_1, y_2)\} \\
&= r\min\{(\bar{\sigma}_1 \times \bar{\sigma}_2)((x_1 + y_1), (x_2 + y_2)) + (z_1, z_2))(\bar{\sigma}_1 \times \bar{\sigma}_2)(y_1, y_2)\} \\
&= r\min\{(\bar{\sigma}_1 \times \bar{\sigma}_2)((x_1, x_2), (y_1, y_2)) + (z_1, z_2))((\bar{\sigma}_1 \times \bar{\sigma}_2)(y_1, y_2)\}
\end{aligned}
$$

Similarly,

$$(\bar{\sigma}_1 \times \bar{\sigma}_2)((x_1, x_2) - (z_1, z_2) \geq r\min(\bar{\sigma}_1 \times \bar{\sigma}_2)((x_1, x_2), (y_1, y_2)) - (z_1, z_2))((\bar{\sigma}_1 \times \bar{\sigma}_2)(y_1, y_2)\}$$

Hence $\bar{\sigma}_1 \times \bar{\sigma}_2$ is an i-v fuzzy H-ideal of a β-algebra in $X_1 \times X_2$.

Theorem 12: Let $\bar{\sigma}_1$ and $\bar{\sigma}_2$ be two i-v fuzzy sets in a β-algebra X such that $\bar{\sigma}_1 \times \bar{\sigma}_2$ is an i-v fuzzy H-ideal of $X_1 \times X_2$. Then

(1) Either $\bar{\sigma}_1(0) \geq \bar{\sigma}_1(x)$ and $\bar{\sigma}_2(0) \geq \bar{\sigma}_2(x)$ $\forall x \in X$
(2) If $\bar{\sigma}_1(0) \geq \bar{\sigma}_1(x)$ $\forall x \in X$ then either $\bar{\sigma}_2(0) \geq \bar{\sigma}_1(x)$ or $\bar{\sigma}_2(0) \geq \bar{\sigma}_2(x)$
(3) If $\bar{\sigma}_2(0) \geq \bar{\sigma}_2(x)$ $\forall x \in X$ then either $\bar{\sigma}_1(0) \geq \bar{\sigma}_1(x)$ or $\bar{\sigma}_1(0) \geq \bar{\sigma}_2(x)$
(4) Either $\bar{\sigma}_1$ or $\bar{\sigma}_2$ is a i-v-fuzzy H-ideal of X.

Proof: Let $(\bar{\sigma}_1 \times \bar{\sigma}_2)$ is an i-v-fuzzy H-ideal of $X_1 \times X_2$. Suppose that

$$\bar{\sigma}_1(0) < \bar{\sigma}_1(x) \text{ and } \bar{\sigma}_2(0) < \bar{\sigma}_1(y) \quad \forall x, y \in X.$$

Then

$$
\begin{aligned}
(\bar{\sigma}_1 \times \bar{\sigma}_2)(x, y) &= r\min\{\bar{\sigma}_1(x), \bar{\sigma}_2(y)\} \\
&\geq r\min\{\bar{\sigma}_1(0), \bar{\sigma}_2(0)\} \\
&= (\bar{\sigma}_1 \times \bar{\sigma}_2)(0, 0)
\end{aligned}
$$

This contradiction yields that either

$$\bar{\sigma}_1(0) \geq \bar{\sigma}_1(x) \text{ or } \bar{\sigma}_2(0) \geq \bar{\sigma}_2(x) \quad \forall x \in X.$$

Given

$$\bar{\sigma}_1(0) \geq \bar{\sigma}_1(x) \ \forall x \in X$$

and assume that there exists x,y∈X such that

$$\bar{\sigma}_2(0) < \bar{\sigma}_1(x) \text{ and } \bar{\sigma}_2(0) < \bar{\sigma}_2(y) \ \forall x, y \in X.$$

Now

$$\bar{\sigma}_2(0) < \bar{\sigma}_1(x) \leq \bar{\sigma}_1(0) \Rightarrow \bar{\sigma}_2(0) < \bar{\sigma}_1(0)$$

Then

$$\begin{aligned}(\bar{\sigma}_1 \times \bar{\sigma}_2)(0,0) &= r\min\{\bar{\sigma}_1(0), \bar{\sigma}_2(0)\} \\ &= \bar{\sigma}_2(0)\end{aligned}$$

$$\begin{aligned}(\bar{\sigma}_1 \times \bar{\sigma}_2)(x,y) &= r\min\{\bar{\sigma}_1(x), \bar{\sigma}_2(y)\} \\ &\geq r\min\{\bar{\sigma}_2(0), \bar{\sigma}_2(0)\} \\ &= \bar{\sigma}_2(0) \\ &= (\bar{\sigma}_1 \times \bar{\sigma}_2)(0,0)\end{aligned}$$

Thus

$$(\bar{\sigma}_1 \times \bar{\sigma}_2)(x,y) \geq (\bar{\sigma}_1 \times \bar{\sigma}_2)(0,0)$$

which is a contradiction. Hence if

$$\bar{\sigma}_1(0) \geq \bar{\sigma}_1(x) \ \forall x \in X$$

then either

$$\bar{\sigma}_2(0) \geq \bar{\sigma}_1(x) \text{ or } \bar{\sigma}_2(0) \geq \bar{\sigma}_2(x).$$

Similarly, if

$$\bar{\sigma}_2(0) \geq \bar{\sigma}_2(x) \quad \forall x \in X$$

then either

$$\bar{\sigma}_1(0) \geq \bar{\sigma}_1(x) \text{ or } \bar{\sigma}_1(0) \geq \bar{\sigma}_2(x)$$

First to prove that $\bar{\sigma}_2$ is a i-v-fuzzy H-ideal of X. Assume that

$$\bar{\sigma}_2(0) \geq \bar{\sigma}_2(x) \quad \forall x \in X$$

Then it follows that either

$$\bar{\sigma}_2(0) \geq \bar{\sigma}_1(x) \text{ or } \bar{\sigma}_1(0) \geq \bar{\sigma}_2(x)$$

If

$$\bar{\sigma}_1(0) \geq \bar{\sigma}_2(x) \quad \forall x \in X.$$

Then

$$(\bar{\sigma}_2)(x) = r\min\{\bar{\sigma}_1(0), \bar{\sigma}_2(x)\} = (\bar{\sigma}_1 \times \bar{\sigma}_2)(0, x)$$

$$\begin{aligned}(\bar{\sigma}_2)(x+z) &\geq r\min\{\bar{\sigma}_1(0), \bar{\sigma}_2(x+z)\} \\ &= (\bar{\sigma}_1 \times \bar{\sigma}_2)(0, x+z) \\ &= (\bar{\sigma}_1 \times \bar{\sigma}_2)(0+0, x+z) \\ &= (\bar{\sigma}_1 \times \bar{\sigma}_2)((0,x)+(0+z)\end{aligned}$$

$$\begin{aligned}&\geq r\min\left\{(\bar{\sigma}_1 \times \bar{\sigma}_2)\left(\left((0,x)+(0,y)\right)+(0,z)\right), (\bar{\sigma}_1 \times \bar{\sigma}_2)(0,y)\right\} \\ &= r\min\left\{(\bar{\sigma}_1 \times \bar{\sigma}_2)\left(\left((0+0),(x+y)\right)+(0,z)\right), (\bar{\sigma}_1 \times \bar{\sigma}_2)(0,y)\right\} \\ &= r\min\left\{(\bar{\sigma}_1 \times \bar{\sigma}_2)\left(\left((0+0)+0,(x+y)+z\right)\right), (\bar{\sigma}_1 \times \bar{\sigma}_2)(0,y)\right\} \\ &= r\min\left\{(\bar{\sigma}_1 \times \bar{\sigma}_2)\left((0,(x+y)+z)\right), (\bar{\sigma}_1 \times \bar{\sigma}_2)(0,y)\right\} \\ &= r\min\left\{\bar{\sigma}_2\left((x+y)+z\right), \bar{\sigma}_2(y)\right\}\end{aligned}$$

Similarly,

$$\bar{\sigma}_2(x-z) \geq r\min\{\bar{\sigma}_2((x-y)-z), \bar{\sigma}_2(y)\}$$

Hence $\bar{\sigma}_2$ is a i-v-fuzzy H-ideal of X.

Theorem 13: Let $A = [\sigma_A^L, \sigma_A^U]$ & $B = [\sigma_B^L, \sigma_B^U]$ be two i-v fuzzy subsets in a set X, then A×B is an i-v fuzzy H-ideals of X×X.

Proof: Let $(x,y) \in$ X×X, then by definition

$$\begin{aligned}
(\bar{\sigma}_A \times \bar{\sigma}_B)(0,0) &= r\min\{\bar{\sigma}_A(0), \bar{\sigma}_B(0)\\
&= r\min\{[\sigma_A^L(0), \sigma_A^U(0)], [\sigma_B^L(0), \sigma_B^U(0)]\}\\
&= [\min\{\sigma_A^L(0), \sigma_B^L(0)\}, \min\{\sigma_A^U(0), \sigma_B^U(0)\}]\\
&\geq [\min\{\sigma_A^L(x), \sigma_B^L(y)\}, \min\{\sigma_A^U(x), \sigma_B^U(y)\}]\\
&= r\min\{[\sigma_A^L(x), \sigma_A^U(x)], [\sigma_B^L(y), \sigma_B^U(y)]\}\\
&= r\min\{\bar{\sigma}_A(x), \bar{\sigma}_B(y)\}\\
&= (\bar{\sigma}_A \times \bar{\sigma}_B)(x,y)
\end{aligned}$$

Now, ∀x,y,z∈X
Take a,b,c ∈ X×X where $a=(x_1,x_2)$, $b=(y_1,y_2)$ and $c=(z_1,z_2)$. Then

$$\begin{aligned}
(\bar{\sigma}_A \times \bar{\sigma}_B)(a+c) &= (\bar{\sigma}_A \times \bar{\sigma}_B)((x_1,x_2)\times(z_1,z_2))\\
&= (\bar{\sigma}_A \times \bar{\sigma}_B)((x_1,z_1)\times(x_2,z_2))\\
&\geq r\min\{\bar{\sigma}_A(x_1,z_1), \bar{\sigma}_B(x_2,z_2)\}\\
&\geq r\min\{r\min\{\bar{\sigma}_A((x_1+y_1)+z_1), \bar{\sigma}_A(y_1)\}, r\min\{\bar{\sigma}_B((x_2+y_2)+z_2), \bar{\sigma}_B(y_2)\}\\
&\geq r\min\{[\min\{\sigma_A^L((x_1+y_1)+z_1), \sigma_A^L(y_1)\}, \min\{\sigma_A^U((x_1+y_1)+z_1), \sigma_A^U(y_1)\}]\\
&\quad [\min\{\sigma_B^L((x_2+y_2)+z_2), \sigma_B^L(y_2)\}, \min\{\sigma_B^U((x_2+y_2)+z_2), \sigma_B^U(y_2)\}]\}\\
&= [\min\{\min\{\sigma_A^L((x_1+y_1)+z_1), \sigma_B^L((x_2+y_2)+z_2)\}, \min\{\sigma_A^L(y_1), \sigma_B^L(y_2)\}\},\\
&\quad \min\{\min\{\sigma_A^U((x_1+y_1)+z_1), \sigma_B^U((x_2+y_2)+z_2)\}, \min\{\sigma_A^U(y_1), \sigma_B^U(y_2)\}\}]\\
&= r\min\{(\bar{\sigma}_A \times \bar{\sigma}_B)((x_1+y_1)+z_1), ((x_2+y_2)+z_2)), (\bar{\sigma}_A \times \bar{\sigma}_B)(y_1,y_2)\}
\end{aligned}$$

Similarly

$$(\bar{\sigma}_A \times \bar{\sigma}_B)(x-z) \geq r\min\{(\bar{\sigma}_A \times \bar{\sigma}_B)(x-y)-z), (\bar{\sigma}_A \times \bar{\sigma}_B)(y)\}$$

Hence *A*×*B* is an i-v fuzzy H-ideal of *X*×*X*.

Definition 24: Let $B=\left[\sigma_B^L,\sigma_B^U\right]$ be an i-v subset in a set *X*. Then the strongest i-v fuzzy relation on *X* that is a i-v *A* or *B* is A_B and defined by $A_B=\left[\sigma_{A_B}^L,\sigma_{A_B}^U\right]$.

Theorem 14: Let $B=\left[\sigma_B^L,\sigma_B^U\right]$ be an i-v subset in a set *X* and $A_B=\left[\sigma_{A_B}^L,\sigma_{A_B}^U\right]$ be the strongest i-v fuzzy relation on X . Then *B* is an i-v fuzzy H-ideals of *X* if and only if A_B is i-v fuzzy H-ideals of *X*×*X*.

Proof: Let *B* is an i-v fuzzy H-ideals of *X*.

$$\begin{aligned}\bar{\sigma}_{A_B}(0,0)&=r\min\{\bar{\sigma}_B(0),\bar{\sigma}_B(0)\}\\&\geq r\min\{\bar{\sigma}_B(x),\bar{\sigma}_B(y)\}\\&=\bar{\sigma}_{A_B}(x,y)\forall x,y\in X\times X.\end{aligned}$$

On the other hand,

$$\begin{aligned}\bar{\sigma}_{A_B}((x_1,x_2)+(z_1,z_2))&=\bar{\sigma}_{A_B}((x_1+z_1),(x_2+z_2))\\&=r\min\{\bar{\sigma}_B(x_1+z_1),\bar{\sigma}_B(x_2+z_2)\}\\&\geq r\min\{r\min\{\bar{\sigma}_B(x_1+y_1)+z_1),\bar{\sigma}_B(y_1)\},r\min\{\bar{\sigma}_B(x_2+y_2)+z_{2,}\bar{\sigma}_B(y_2)\}\}\\&\geq r\min\{r\min\{\bar{\sigma}_B(x_1+y_1)+z_1),\bar{\sigma}_B(x_2+y_2)+z_2)\},r\min\{\bar{\sigma}_B(y_1),\bar{\sigma}_B(y_2)\}\}\\&=r\min\{\bar{\sigma}_B(x_1+y_1)+z_1),\bar{\sigma}_B(x_2+y_2)+z_2)\},r\min\{\bar{\sigma}_B(y_1),\bar{\sigma}_B(y_2)\}\}\\&=r\min\{\bar{\sigma}_{A_B}((x_1+y_1)+z_1),(x_2+y_2)+z_2)),\bar{\sigma}_{A_B}(y_1,y_2)\}\end{aligned}$$

Hence A_B is an i-v fuzzy H-ideal of *X*×*X*.
Conversely, Let A_B is an i-v fuzzy H-ideal of *X*×*X*.
Then ∀(x,x) ∈ *X*×*X*, we have

$$r\min\{\bar{\sigma}_B(0),\bar{\sigma}_B(0)\}=\bar{\sigma}_{A_B}(0,0)\geq\bar{\sigma}_{A_B}(x,x)=r\min\{\bar{\sigma}_B(x),\bar{\sigma}_B(x)\}$$

(or)

$$\bar{\sigma}_B(0)\geq\bar{\sigma}_B(x)\quad\forall x\in X\times X.$$

Now, let (x_1,x_2), (y_1,y_2), (z_1,z_2) ∈ X×X, then

$$\begin{aligned} r\min\{\bar{\sigma}_B(x_1+z_1),\bar{\sigma}_B(x_2+z_2)\} &= \bar{\sigma}_{A_B}((x_1+z_1,x_2+z_2)) \\ &= \bar{\sigma}_{A_B}\left((x_1,x_2)+(z_1,z_2)\right) \\ &\geq r\min\{\bar{\sigma}_{A_B}((x_1,x_2)+(y_1,y_2)+(z_1,z_2)),\bar{\sigma}_{A_B}(y_1,y_2)\} \\ &= r\min\{r\min\{\bar{\sigma}_B(x_1+y_1)+z_1),\bar{\sigma}_B(y_1)\},r\min\{\bar{\sigma}_B(x_2+y_2)+z_2,\bar{\sigma}_B(y_2)\}\} \end{aligned}$$

If $x_2=y_2=z_2=0$,

$$r\min\{\bar{\sigma}_B(x_1+z_1),\bar{\sigma}_B(0)\} \geq r\min\{r\min\{\bar{\sigma}_B((x_1+y_1)+z_1),\bar{\sigma}_B(y_1)\},\bar{\sigma}_B(0)\}$$

(or)

$$\bar{\sigma}_B(x_1+z_1) \geq r\min\{\bar{\sigma}_B(x_1+y_1)+z_1),\bar{\sigma}_B(y_1)\}$$

Therefore *B* is an i-v fuzzy H-ideal of *X*.

5. $\bar{\mathrm{T}}$-INTERVAL VALUED FUZZY H-IDEALS ON β-ALGEBRA

This section introduces the notion of $\bar{\mathrm{T}}$ – interval valued fuzzy H-ideals on β-algebras and deals some simple results.

Definition 25: Let $\bar{\sigma}$ be an interval valued fuzzy (i-v-fuzzy) set in a β–algebra of X. Then $\bar{\sigma}$ is called an i-v-fuzzy H-ideals of X if

(i) $\bar{\sigma}(0) \geq \bar{\sigma}(x) \quad \forall x \in X$

(ii) $\bar{\sigma}(x+z) \geq \bar{\mathrm{T}}\{\bar{\sigma}((x+y)+z),\bar{\sigma}(y)\}$

(ii) $\bar{\sigma}(x-z) \geq \bar{\mathrm{T}}\{\bar{\sigma}((x-y)-z),\bar{\sigma}(y)\} \quad \forall x,y \in X.$

Theorem 15: Intersection of any two $\bar{\mathrm{T}}$ - i-v-fuzzy H-ideal of a β -algebra is also a $\bar{\mathrm{T}}$ - i-v-fuzzy H-ideal.

Proof: Let $\bar{\sigma}_1,\bar{\sigma}_2$ are two $\bar{\mathrm{T}}$ - i-v-fuzzy H-ideal of a -algebra X.

$$\begin{aligned} (\bar{\sigma}_1 \cap \bar{\sigma}_2)(0) &\geq \bar{\mathrm{T}}\{\bar{\sigma}_1(0),\bar{\sigma}_2(0)\} \\ &\geq \bar{\mathrm{T}}\{\bar{\sigma}_1(x),\bar{\sigma}_2(x)\} \\ &= (\bar{\sigma}_1 \cap \bar{\sigma}_2)(x) \end{aligned}$$

$$(\bar{\sigma}_1 \cap \bar{\sigma}_2)(x+z) \geq \bar{\mathrm{T}}\{\bar{\sigma}_1(x+z), \bar{\sigma}_2(x+z)\}$$
$$\geq \bar{\mathrm{T}}\{\bar{\mathrm{T}}\{\bar{\sigma}_1((x+y)+z), \bar{\sigma}_1(y)\}, \bar{\mathrm{T}}\{\bar{\sigma}_2((x+y)+z), \bar{\sigma}_2(y)\}\}$$
$$\geq \bar{\mathrm{T}}\{\bar{\mathrm{T}}\{\bar{\sigma}_1((x+y)+z), \bar{\sigma}_1(y)\}, \bar{\mathrm{T}}\{\bar{\sigma}_2((x+y)+z), \bar{\sigma}_2(y)\}\}$$
$$\geq \bar{\mathrm{T}}\{\bar{\mathrm{T}}\{\bar{\sigma}_1((x+y)+z), \bar{\sigma}_2((x+y)+z)\}, \bar{\mathrm{T}}\{\bar{\sigma}_1(y), \bar{\sigma}_2(y)\}\}$$
$$= \bar{\mathrm{T}}\{(\bar{\sigma}_1 \cap \bar{\sigma}_2)(x+y)+z), (\bar{\sigma}_1 \cap \bar{\sigma}_2)(y)\}$$

Similarly,

$$(\bar{\sigma}_1 \cap \bar{\sigma}_2)(x-z) \geq \bar{\mathrm{T}}\{(\bar{\sigma}_1 \cap \bar{\sigma}_2)(x-y)-z), (\bar{\sigma}_1 \cap \bar{\sigma}_2)(y)\}$$

Hence $(\bar{\sigma}_1 \cap \bar{\sigma}_2)$ is an $\bar{\mathrm{T}}$ - i-v-fuzzy H-ideal of a -algebra X.
It can be generalized as,

Theorem16: The intersection of any collection of $\bar{\mathrm{T}}$ - i-v-fuzzy H-ideal of a -algebra is also $\bar{\mathrm{T}}$ - i-v-fuzzy H-ideal.

Lemma: Let $A = \{x, \bar{\sigma}_A(x) : x \in X\}$ be an $\bar{\mathrm{T}}$ - i-v-fuzzy H-ideal of a β –algebra of X. If x≤y then $\bar{\sigma}(x) \geq \bar{\sigma}(y)$.

Proof: For x,y∈X, x≤y ⇒ x – y=0. Then

$$\Rightarrow \bar{\sigma}(x) = \bar{\sigma}(x-0) \geq \bar{\mathrm{T}}\{\bar{\sigma}((x-y)-0), \bar{\sigma}(y)\}$$
$$= \bar{\mathrm{T}}\{\bar{\sigma}(0-0), \bar{\sigma}(y)\}$$
$$= \bar{\mathrm{T}}\{\bar{\sigma}(0), \bar{\sigma}(y)\}$$
$$= \bar{\sigma}(y)$$

Hence $\bar{\sigma}(x) \geq \bar{\sigma}(y)$.

Theorem 17: An i-v fuzzy set $A = [\sigma_A^L, \sigma_A^U]$ in X is an $\bar{\mathrm{T}}$-i-v fuzzy H-ideal of X if and only if σ_A^L and σ_A^U are fuzzy H-ideals of X.

Proof: Suppose that σ_A^L and σ_A^U are fuzzy H-ideal of X.

$$\therefore \sigma_A^L(0) \geq \sigma_A^L(x) \text{ and } \sigma_A^U(0) \geq \sigma_A^U(x)$$
$$\Rightarrow \bar{\sigma}_A(0) \geq \bar{\sigma}_A(x)$$

Let x,y,z∈X. Then

$$=\overline{\mathrm{T}}\left\{\left[\sigma_A^L\left(x+y\right)+z\right),\sigma_A^U\left(x+y\right)+z\right)\right],\left[\sigma_A^L\left(y\right),\sigma_A^U\left(y\right)\right]\right\}$$
$$=\overline{\mathrm{T}}\left\{\bar{\sigma}_A\left(x+y\right)+z\right),\bar{\sigma}_A\left(y\right)\right\}$$

Similarly,

$$\bar{\sigma}_A\left(x-z\right)\geq\overline{\mathrm{T}}\left\{\bar{\sigma}_A\left(x-y\right)-z\right),\bar{\sigma}_A\left(y\right)\right\}$$

Hence A is an $\overline{\mathrm{T}}$ - i-v fuzzy H-ideal of X.
Conversely, Suppose A is an $\overline{\mathrm{T}}$ -i-v fuzzy H-ideal of X
Then for any x,y∈X, we have

$$\begin{aligned}\left[\sigma_A^L\left(x+z\right),\sigma_A^U\left(x+z\right)\right]&=\bar{\sigma}_A\left(x+z\right)\\&\geq\overline{\mathrm{T}}\left\{\bar{\sigma}_A\left(x+y\right)+z\right),\bar{\sigma}_A\left(y\right)\right\}\\&\geq\overline{\mathrm{T}}\left\{\left[\sigma_A^L\left(x+y\right)+z\right),\sigma_A^U\left(x+y\right)+z\right)\right],\left[\sigma_A^L\left(y\right),\sigma_A^U\left(y\right)\right]\right\}\\&=\left[\min\left\{\sigma_A^L\left(x+y\right)+z\right),\sigma_A^L\left(y\right)\right\},\min\left\{\sigma_A^U\left(x+y\right)+z\right),\sigma_A^U\left(y\right)\right\}\right]\end{aligned}$$

It follows that

$$\sigma_A^L\left(x+z\right)\geq\min\left\{\sigma_A^L\left(x+y\right)+z\right),\sigma_A^L\left(y\right)\right\}$$

and

$$\sigma_A^U\left(x+z\right)\geq\min\left\{\sigma_A^U\left(x+y\right)+z\right),\sigma_A^U\left(y\right)\right\}$$

Similarly,

$$\sigma_A^L\left(x-z\right)\geq\min\left\{\sigma_A^L\left(x-y\right)-z\right),\sigma_A^L\left(y\right)\right\}$$

and

$$\sigma_A^U\left(x-z\right)\geq\min\left\{\sigma_A^U\left(x-y\right)-z\right),\sigma_A^U\left(y\right)\right\}$$

Hence σ_A^L and σ_A^U are fuzzy H-ideals of X.

Theorem 18: Let A be a subset of X. Define an i-v fuzzy set $\bar{\sigma}: X \to D[0,1]$ such that

$$\bar{\sigma}_A(x) = \begin{cases} [t_0, t_1] & \text{if } x \in A \\ [s_0, s_1] & \text{if } x \notin A \end{cases}$$

where $[t_0,t_1]$, $[s_0,s_1] \in D[0,1]$ with $[t_0,t_1] \geq [s_0,s_1]$. Then $\bar{\sigma}$ is an $\overline{\mathrm{T}}$-i-v fuzzy H-ideal of a β–algebra X if and only if Ais a H-ideal of X.

Proof: Assume that $\bar{\sigma}$ is an $\overline{\mathrm{T}}$-i-v fuzzy H-ideal of X. If $x \in A, \bar{\sigma} \geq [t_0, t_1]$ and if $x \notin A, \bar{\sigma} \geq [s_0, s_1]$ since $[t_0,t_1] > [s_0,s_1]$

$$\Rightarrow \bar{\sigma}_A(0) \geq [t_0, t_1] > [s_0, s_1]$$
$$\Rightarrow \bar{\sigma}_A(0) = [t_0, t_1]$$
$$\Rightarrow 0 \in A$$

For

$$x, y, z \in A \Rightarrow \bar{\sigma}_A(x) = [t_0, t_1], \bar{\sigma}_A(y) = [t_0, t_1] \text{ and } \bar{\sigma}_A(z) = [t_0, t_1]$$

$$\begin{aligned} \bar{\sigma}_A(x+z) &\geq \overline{\mathrm{T}}\{\bar{\sigma}_A((x+y)+z), \bar{\sigma}_A(y)\} \\ &\geq \overline{\mathrm{T}}\{\overline{\mathrm{T}}\{\bar{\sigma}_A(x+y), \bar{\sigma}_A(z)\}, \bar{\sigma}_A(y)\} \\ &= \overline{\mathrm{T}}\{\overline{\mathrm{T}}\{\overline{\mathrm{T}}\{\bar{\sigma}_A(x), \bar{\sigma}_A(y)\}, \bar{\sigma}_A(z)\}, \bar{\sigma}_A(y)\} \\ &= \overline{\mathrm{T}}\{\overline{\mathrm{T}}\{\overline{\mathrm{T}}\{[t_0,t_1],[t_0,t_1]\},[t_0,t_1]\},[t_0,t_1]\} \\ &= \overline{\mathrm{T}}\{\overline{\mathrm{T}}\{[t_0,t_1],[t_0,t_1]\},[t_0,t_1]\} \end{aligned}$$

$=[t_0,t_1]$

$$\therefore \bar{\sigma}_A(x+z) \geq [t_0, t_1] \Rightarrow x+z \in A$$

For

$$x, y, z \in A \Rightarrow \bar{\sigma}_A((x-y)-z) = [t_0, t_1] \text{ and } \bar{\sigma}_A(y) = [t_0, t_1]$$

Now

$$\bar{\sigma}_A(x-z) \geq \bar{\mathrm{T}}\{\bar{\sigma}_A((x-y)-z), \bar{\sigma}_A(y)\}$$
$$= \bar{\mathrm{T}}\{[t_0,t_1],[t_0,t_1]\}$$

$=[t_0,t_1]$

$$\Rightarrow \bar{\sigma}_A(x-z) \geq [t_0,t_1] \Rightarrow x-z \in A$$

Hence A is a H-ideal of X.

Conversely, Suppose A is a H-ideal of X. Now

$$0 \in A \Rightarrow \bar{\sigma}_A(0) = [t_0,t_1]$$

Also

$$\forall x \in X,\ \ \mathrm{Im}\,\bar{\sigma} = \{[t_0,t_1],[s_0,s_1]\}$$

and $[t_0,t_1]>[s_0,s_1]$

$$\Rightarrow \bar{\sigma}_A(0) \geq \bar{\sigma}_A(x)$$

Since A is a H-ideal of X.

$\forall x,y,z \in S$, $(x+y)+z \in A$ and $y \in A$, and $y \in A \Rightarrow x+z$, $x-z \in A$

Then

$$\bar{\sigma}_A(x+z) = [t_0,t_1] \geq \bar{\mathrm{T}}\{\bar{\sigma}_A((x+y)+z), \bar{\sigma}_A(y)\}$$

Similarly,

$$\bar{\sigma}_A(x-z) \geq \bar{\mathrm{T}}\{\bar{\sigma}_A((x-y)-z), \bar{\sigma}_A(y)\}$$

Hence $\bar{\sigma}$ is an i-v fuzzy H-ideal of X.

Theorem 19: Let f: X→X be an endomorphism on a β-algebra. Let $\bar{\sigma}$ be an $\bar{\mathrm{T}}$ - i-v-fuzzy H-ideal of X. Define a $\bar{\mathrm{T}}$ -i-v-fuzzy set $\bar{\sigma}_f : X \to D[0,1]$ defined by $\bar{\sigma}_f(x) = \bar{\sigma}(f(x))$ $\forall x \in X$. Then $\bar{\sigma}_f$ is an $\bar{\mathrm{T}}$ - i-v fuzzy H-ideal of X.

Proof: Let $x \in X$. Then

$$\bar{\sigma}_f = \bar{\sigma}\left(f\left(x\right)\right) \leq \bar{\sigma}\left(0\right) = \bar{\sigma}\left(f\left(0\right)\right) = \bar{\sigma}_f\left(0\right)$$

Let x,y∈X

$$\begin{aligned}\bar{\sigma}_f\left(x+z\right) &= \bar{\sigma}\left(f\left(x+z\right)\right)\\ &= \bar{\sigma}\left(f\left(x\right)+f(y)\right)\\ &\geq \bar{\mathrm{T}}\{\bar{\sigma}\left((f\left(x\right)+f(y)\right)+f\left(z\right)),\bar{\sigma}\left(f\left(y\right)\right)\}\\ &= \bar{\mathrm{T}}\{\bar{\sigma}\left((f\left(x+y\right)\right)+f\left(z\right)),\bar{\sigma}\left(f\left(y\right)\right)\}\\ &= \bar{\mathrm{T}}\{\bar{\sigma}\left((f\left(x+y\right)+z\right)),\bar{\sigma}\left(f\left(y\right)\right)\}\\ &= \bar{\mathrm{T}}\{\bar{\sigma}_f\left(\left(x+y\right)+z\right)),\bar{\sigma}_f\left(y\right)\}\end{aligned}$$

Similarly,

$$\bar{\sigma}_f\left(x-z\right) \geq \bar{\mathrm{T}}\{\bar{\sigma}_f\left(\left(x-y\right)-z\right)),\bar{\sigma}_f\left(y\right)\}.$$

Hence $\bar{\sigma}_f$ is an $\bar{\mathrm{T}}$-i-v fuzzy H-ideal of X.

Theorem 20: Let:X→Y be an homomorphism of β-algebra. If A is an $\bar{\mathrm{T}}$-i-v-fuzzy H-ideal of Y, then the pre image $f^{-1}(A)$ is an $\bar{\mathrm{T}}$- i-v-fuzzy H-ideal of X.

Proof: Let A be an $\bar{\mathrm{T}}$- i- v-fuzzy H-ideal of Y. For any x∈X,

$$f^{-1}\left(\bar{\sigma}_A\left(0\right)\right) = \bar{\sigma}_A\left(f\left(0\right)\right) = \bar{\sigma}_A\left(0\right) \geq \bar{\sigma}_A\left(x\right)$$

For x,y∈X,

$$\begin{aligned}f^{-1}\left(\bar{\sigma}_A\left(x+z\right)\right) &= \bar{\sigma}_A\left(f\left(x+z\right)\right)\\ &= \bar{\sigma}_A\left(f\left(x\right)+f(z)\right)\\ &\geq \bar{\mathrm{T}}\left\{\bar{\sigma}_A\left(\left(f\left(x\right)+f\left(y\right)\right)+f\left(z\right)\right),\bar{\sigma}_A\left(f\left(y\right)\right)\right\}\\ &\geq \bar{\mathrm{T}}\left\{\bar{\sigma}_A\left(\left(f(x+y\right)+z)\right),\bar{\sigma}_A\left(f\left(y\right)\right)\right\}\\ &\geq \bar{\mathrm{T}}\left\{f^{-1}(\bar{\sigma}_A\left(\left((x+y\right)+z)\right)),f^{-1}(\bar{\sigma}_A\left(\left(y\right)\right)\right\}\end{aligned}$$

$\therefore f^{-1}(\bar{\sigma}_A)$ is an $\bar{\mathrm{T}}$- fuzzy H-ideal of X.

Similarly,

$$f^{-1}\left(\bar{\sigma}_A\left(x-z\right)\right)\geq \bar{\mathrm{T}}\left\{f^{-1}(\bar{\sigma}_A\left(((x-y)-z)\right)), f^{-1}(\bar{\sigma}_A\left((y)\right)\right\}$$

Proposition: Let $A=\left\{x,\bar{\sigma}_A\left(x\right):x\in X\right\}$ be an $\bar{\mathrm{T}}$-i-v fuzzy β-ideal of a β-algebra X. If x≤z+y then

$$\bar{\sigma}_A\left(x\right)\geq \bar{\mathrm{T}}\left\{\bar{\sigma}_A\left(z\right),\bar{\sigma}_A\left(y\right)\right\}$$

Proof: For x,y,z∈X

$$\begin{aligned}\bar{\sigma}_A\left(x\right) &\geq \bar{\mathrm{T}}\left\{\bar{\sigma}_A\left(x-y\right),\bar{\sigma}_A\left(y\right)\right\}\\ &=\bar{\mathrm{T}}\left\{\bar{\mathrm{T}}\left\{\bar{\sigma}_A\left((x-y)-z\right),\bar{\sigma}_A\left(z\right)\right\},\bar{\sigma}_A\left(y\right)\right\}\\ &=\bar{\mathrm{T}}\left\{\bar{\mathrm{T}}\left\{\bar{\sigma}_A\left(x-(z+y\right),\bar{\sigma}_A\left(z\right)\right\},\bar{\sigma}_A\left(y\right)\right\}\\ &=\bar{\mathrm{T}}\left\{\bar{\mathrm{T}}\left\{\bar{\sigma}_A\left(0\right),\bar{\sigma}_A\left(z\right)\right\},\bar{\sigma}_A\left(y\right)\right\}\\ &=\bar{\mathrm{T}}\left\{\bar{\sigma}_A\left(z\right),\bar{\sigma}_A\left(y\right)\right\}\end{aligned}$$

6. CONCLUSION

This work shows the theoretical development and proof of H-ideals in β-algebras using IVF set and $\bar{\mathrm{T}}$–norms. This research work has not only show that the new direction in the development of IVF β-sub algebra also created the large scope for further research works primarily in terms of extending these concepts into fuzzy topological settings.

REFERENCES

Ansari, M. A. A., & Chandramouleeswaran, M. (2013). Fuzzy β– subalgebras of β– algebras. *International Journal of Mathematical Sciences and Engineering applications*, *5*(7), 239-249.

Ansari, M. A. A., & Chandramouleeswaran, M. (2014). Fuzzy β–ideals of β– algebras. *International Journal of Mathematical Sciences and Engineering Applications, 8*(1), 1-10.

Ansari, M. A. A., & Chandramouleeswaran, M. (2014a). Cartesian product of fuzzy β–ideals of β–algebras. *Global Journal of Pure and Applied Mathematics*, *10*(1), 35–42.

Ansari, M. A. A., & Chandramouleeswaran, M. (2014b). T- Fuzzy of β–subalgebras of β–algebras, *International Journal of Mathematical Sciences and Engineering Applications, 8*(1), 177-187.

Biswas, R. (1994). Rosenfeld's fuzzy subgroups with interval-valued membership functions. *Fuzzy Sets and Systems*, *63*(1), 87–90. doi:10.1016/0165-0114(94)90148-1

Hemavathi, P., Muralikrishna, P., & Palanivel, K. (2015). A note on interval valued fuzzy β-subalgebras. *Global Journal of Pure and Applied Mathematics*, *11*(4), 2553–2560.

Hemavathi, P., Muralikrishna, P., & Palanivel, K. (2017). ivf β--Ideals of β-Algebras. Materials Science and Engineering Conference Series, 263(4), 1-10.

Hong, S. M., Jun, Y. B., Kim, S. J., & Kim, G. I. (1993).. . *Fuzzy BCI-Subalgebras with Interval Valued Membership Functions Math Japonica*, *40*(2), 199–202.

Jun, Y. B. (2000). Interval-valued fuzzy subalgebras/ideals in BCK-algebras. *Scientiae Mathematicae*, *3*(3), 435–444.

Jun, Y. B., & Kim. (2012). β–subalgebras and related topics. *Commun. Korean. Math. Soc*, *27*(2), 243-255.

Khalid, H. M., & Ahmad, B. (1999). Fuzzy H-ideals in BCI-algebras. *Fuzzy Sets and Systems*, *101*(1), 153–158. doi:10.1016/S0165-0114(97)00042-0

Kordi, A., Moussavi, A., & Ahmadi, A. (2008). Fuzzy H-Ideals of BCI-Algebras with Interval Valued Membership Functions. *International Mathematical Forum, 3*(27), 1327-1338.

Muralikrishna, P., & Chandramouleeswaran, M. (2011). A Note on Intuitionistic L-Fuzzy H-ideals of BF-Algebras. *International Review of Fuzzy Mathematics*, *6*(2), 63–68.

Neggers, J., & Kim, H. S. (2002). On β-algebras. *Mathematica Slovaca*, *52*(5), 517–530.

Neggers, J., & Kim, H. S. (2002a). On B-algebras. *Mat. Vesnik*, *54*, 21–29.

Rajam, K., & Chadramouleeswaran, M. (2015). L-Fuzzy T-ideals in β– Algebras. *Applied Mathematical Sciences*, *9*(145), 7221–7228. doi:10.12988/ams.2015.59581

Sabarinathan, S., Muralikrishna, P., & Kumar, D. C. (2017). Bipolar valued fuzzy H-ideals of BF-algebras. *International Journal of Pure and Applied Mathematics*, *112*(5), 87–92.

Sujatha, K., Chandramouleeswaran, M., & Muralikrishna, P. (2014). Product on Intuitionistic Fuzzy β-sub Algebras of β-algebras. *Indian Journal of Science and Technology*, *7*(3), 318–322.

Young, B. J., & Min, H. S. (2001). Fuzzy subalgebras of BCK/BCI algebras redefined. *Scientiae. Mathematicae Japanicae,* 4, 769-775.

Zadeh, L. A. (1965). Fuzzy sets. *Information and Control*, *8*, 338-353.

Zadeh, L. A. (1975). The concept of a linguistic variable and its application to approximate reasoning—I. *Information Sciences*, *8*(3), 199–249. doi:10.1016/0020-0255(75)90036-5

Zhan, J., & Tan, Z. (2005). Fuzzy H-ideals in BCK-algebras. *Southeast Asian Bulletin of Mathematics*, *29*(6), 1165–1173.

Chapter 16
Some Types of Ideals in QI-Algebras

Ravikumar Bandaru
GITAM University (Deemed), Hyderabad, India

ABSTRACT

In this chapter, ideals of QI-algebra are considered. Given a subset of a right distributive QI-algebra, the smallest ideal containing it is constructed. Also, the notions of implicative ideal, fantastic ideal, and normal ideal in a right distributive QI-algebra are introduced, and the authors proved that these notions are equivalent.

1. INTRODUCTION

BCK-algebras and BCI-algebras were introduced by Imai and Iseki(1966), Iseki and Tanaka(1978) and Iseki(1980). Since their introduction, several generalizations of BCK-algebras were introduced and extensively studied by many researchers. Abbott (1967) introduced a concept of an implication algebra in the sake to formalize the logical connective implication in the classical propositional logic. Recently, Saeid et al.(2017) introduced the concept of a BI-algebra as a generalization of (dual) implication algebra and studied its properties. Bandaru(2017) introduced the concept of a QI-algebra as a generalization of BI-algebra and studied the concept of ideals, congruences in a QI-algebra. Also given connection between ideals and congruence kernels whenever a QI-algebra is right distributive. In this paper, we introduce the notions of implicative ideal, fantastic ideal and normal ideal in a right distributive QI-algebra and we show that these notions are equivalent.

DOI: 10.4018/978-1-7998-0190-0.ch016

2. PRELIMINARIES

First, we recall certain definitions and properties from Saeid et al.(2017), Abbott (1967) & Chen et.al. (1995) that are required in the paper.

Definition 2.1 (Iseki, 1980) *A BCI-algebra is an algebra* $(X,*,0)$ *of type (2,0) satisfying the following conditions:*

(1) $(x*y)*(x*z) \leq (z*y)$
(2) $x*(x*y) \leq y$
(3) $x \leq x$
(4) $x \leq y$ and $y \leq y$ imply $x=y$
(5) $x \leq 0$ implies $x=0$

where $x \leq y$ is defined by $x*y=0$.

If (5) is replaced by (6) $0 \leq x$, then the algebra is called a BCK-algebra. It is known that every BCK-algebra is a BCI-algebra but not conversely. A BCK-algebra satisfying the property $x*(y*x)= x$ for all $x,y \in X$ is called an implicative BCK-algebra.

Several generalizations of a BCK-algebra, in the form of definitions, one can see in the paper (Saeid et al., 2017).

Definition 2.2 (Abbott, 1967) *A groupoid* $(X,*)$ *is called an implication algebra if it satisfies the following identities:*

(a) $(x*y)*x=x$
(b) $(x*y)*y=(y*x)*x$
(c) $X*(y*z)=y*(x*z)$

for all $x,y,z \in X$.

Definition 2.3 (Abbott, 1967) *Let* $(X,*)$ *be an implication algebra and binary operation "* $\circ$ *" on X be defined by* $x*y = y \circ x$. Then $(X,\circ)$ is said to be a dual implication algebra. In fact, the axioms of that are as follows:

(a) $x \circ (y \circ x) = x$
(b) $x \circ (x \circ y) = y \circ (y \circ x)$
(c) $(x \circ y) \circ z = (x \circ z) \circ y$

for all $x,y,z \in X$.

Chen & Oliveira (1995) proved that in any implication algebra $(X,*)$ the identity $x*y=y*y$ holds for all $x,y \in X$. We denote the identity $x*x=y*y$ by the constant 0. The notion of BI-algebras comes from the (dual) implication algebra.

Definition 2.4 (Saeid et al., 2017) *An algebra* $(X,*,0)$ *of type* $(2,0)$ *is called a BI-algebra if*

(*BI* 1) $x*x=0$
(*BI* 2) $x*(y*x)=x$ for all $x,y\in X$.

It can be observed that every dual implication algebra is a BI-algebra but converse need not be true.

3. QI-ALGEBRA

In this section, we give the defintion of a QI-algebra and its properties.

Definition 3.1 (Bandaru, 2017). *A QI-algebra is a non-empty set X with a constant 0 and a binary operation * satisfying axioms:*

(QI1) $x*x=0$
(QI2) $x*0=x$
(QI3) $x*(y*(x*y))=x*y$

for all $x,y\in X$.

Let $(X,*,0)$ be a QI-algebra. We introduce a relation "$\leq$"on X by $x\leq y$ if and only if $X*y=0$. A relation $\leq$ is not a partially order, but it is only reflexive.

Note that every BI-algebra is a QI-algebra but converse need not be true.

Example 3.2 (Bandaru, 2017). *Let* $X=\{0,1,2,3\}$ *be a set with table 1.*

Then $(X,*,0)$ is a QI-algebra but not a BI-algebra because $3*(2*3) = 3*2 = 1 \neq 3$. Also, every implicative BCK-algebra is a QI-algebra but converse need not be true.

Example 3.3 (Bandaru, 2017) *Let* $X=\{0,a,b,c\}$ *be a set with table 2.*

Then $(X,*,0)$ is a QI-algebra but not an implicative BCK-algebra because $a*c=0$ and $c*a=0$ but $a\neq c$ and $c*(b*c)=c*b=a\neq c$.

Table 1.

*	0	1	2	3
0	0	2	1	0
1	1	0	1	0
2	2	2	0	2
3	3	2	1	0

Table 2.

*	0	a	b	c
0	0	0	0	0
a	a	0	a	0
b	b	b	0	b
c	c	0	a	0

Proposition 3.4 (Bandaru, 2017) *Let* $(X,*,0)$ *be a QI-algebra. Then*

(i) $x*(0*x)=x$
(ii) if $x\leq 0$, then $x=0$
(iii) if $x*y=y$, then $x=y$
(iv) if $x*y=x$, then $x*(y*x)=x$
(v) if $(x*y)*(z*u) = (x*z)*(u*y)$, then $X=\{0\}$

for all $x,y,z,u\in X$.

Definition 3.5 (Bandaru, 2017) *A QI-algebra X is said to be right distributive (or left distributive, resp.) if*

(*QI*4) $(x*y)*z = (x*z)*(y*z)$, ($z*(x*y)=(z*x)*(z*y)$, resp.) for all $x,y,z\in X$.

Example 3.6 (i) Example 3.3 is a right distributive QI-algebra. (ii) Example 3.2 is not a right distributive QI-algebra, since

$(3*1)*3 = 2*3 = 2 \neq 0 = 0*0 = (3*3)*(1*3)$

Proposition 3.7 *If X is a left distributive QI-algebra, then* $X=\{0\}$.

Proposition 3.8 (Bandaru, 2017) *If X is a right distributive QI-algebra, then*

(*QI*5) $0*x=0$
(*QI*6) $(x*y)*y=x*y$,

for any $x,y\in X$.

Proposition 3.9 (Bandaru, 2017) *In a right distributive QI-algebra X, for all* $x,y,z\in X$, *the following conditions hold:*

(1) $y*x\leq y$
(2) $(y*x)*x\leq y$
(3) $(x*z)*(y*z)\leq x*y$
(4) $x\leq y$ implies $x*z\leq y*z$
(5) $(x*y)*z\leq x*(y*z)$
(6) If $x\leq y$ and $y\leq z$, then $x\leq z$
(7) $x\leq y$ implies $z*y\leq z*x$
(8) $(x*y)*z\leq (x*z)*y$
(9) $(z*x)*(z*y)\leq (y*x)$
(10) If $x*y=z*y$, then $(x*z)*y=0$.

4. ON IDEALS OF QI-ALGEBRA

In this section, we study some properties of ideals in a right distributive QI-algebra and give the smallest ideal containing a subset of a right distributive QI-algebra.

Definition 4.1 (Bandaru, 2017) *Let $(X,*,0)$ be a QI-algebra and $I \subseteq X$. Then I is called an ideal of X if it satisfies the following:*

(*I*1) $0 \in I$
(*I*2) if $x*y \in I$ and $y \in I$, then $x \in I$.

Clearly, $\{0\}$ and X are ideals of X and we call them as zero ideal and trivial ideal respectively. An ideal I is said to be proper if $I \neq X$

Example 4.2 (Bandaru, 2017) *Let $X=\{0,a,b,c\}$ be a set with table 3.*

Then $(X,*,0)$ is a QI-algebra. Clearly, $I_1=\{0,a\}$ and $I_2=\{0,a,c\}$ are ideals of X. But $I_3=\{0,a,b\}$ is not an ideal of X.

Lemma 4.3 *Let X be a QI-algebra and I a non-empty subset of X satisfying the following conditions:*

(*I*3) $x \in X$ and $y \in I$ imply $y*x \in I$
(*I*4) $x \in X$, $a,b \in I$ imply $x*((x*a)*b) \in I$.

Then I is an ideal of X.
The converse of the above lemma does not hold in general.

Example 4.4 *Let $X=\{0,a,b,c\}$ be a set with table 4.*

Then $(X,*,0)$ is a QI-algebra. Clearly, $I=\{0,a\}$ is an ideal of X but it doesn't satisfy (*I*3) and (*I*4). However, for right distributive QI-algebras we have

Theorem 4.5 (Bandaru, 2017) *If X is a right distributive QI-algebra and I is an ideal of X, Then I satisfies (I3) and (I4).*

Table 3.

*	0	a	b	c
0	0	b	a	0
a	a	0	a	0
b	b	b	0	b
c	c	b	a	0

Table 4.

*	0	a	b	c
0	0	0	0	0
a	a	0	a	b
b	b	b	0	b
c	c	b	c	0

Proof. Let I be an ideal of X and $a \in I$, $x \in X$. Then $(a*x)*a=0 \in I$ and, applying (I2), we conclude $a*x \in I$, i.e., I satisfies (*I*3). Now, suppose $a,b \in I$ and $x \in X$. Then

$$\begin{aligned}\left(x*\left(\left(x*a\right)*b\right)\right)*b &= \left(x*b\right)*\left[\left(\left(x*a\right)*b\right)*b\right] = \left(x*b\right)*\left[\left(\left(x*a\right)*b\right)*\left(b*b\right)\right] \\ &= \left(x*b\right)*\left(\left(x*a\right)*b\right) = \left(x*\left(x*a\right)\right)*b \le \left(x*\left(x*a\right)\right) \le a\end{aligned}$$

and hence $\left[\left(x*\left(\left(x*a\right)*b\right)\right)*b\right]*a = 0 \in I$ and applying (*I*2) twice, we get $x*((x*a)*b) \in I$ proving (*I*4).

Theorem 4.6 (Bandaru, 2017) *If X is a right distributive QI-algebra and I a non-empty subset of X. Then I is an ideal of X if and only if I satisfies* (*I*3) *and* (*I*4).

Lemma 4.7 *Let X be a right distributive QI-algebra and $x,y \in X$ such that $x*y=0$. Then, for all $a_1,a_2,\ldots,a_n \in X$, $(\ldots(y*a_1)*\ldots)*a_n=0$ implies $(\ldots(x*a_1)*\ldots)*a_n=0$.*

Proof. We have $x*y=0$; that is, $x \le y$, and from Proposition 3.9(4), we can see that

$$\left(\ldots\left(x*a_1\right)*\ldots\right)*a_n \le \left(\ldots\left(y*a_1\right)*\ldots\right)*a_n = 0.$$

Therefore, from Proposition 3.4(ii), $(\ldots(x*a_1)*\ldots)*a_n=0$
We denote the set of all ideals of X by $\mathcal{I}\left(X\right)$. Clearly, $\left\{0\right\}, X \in \mathcal{I}\left(X\right)$.

Theorem 4.8 *If $I_i (i \in \Delta)$ are ideals of a QI-algebra X, then $\bigcap_{i\in\Delta} I_i$ is an ideal of X.*

The following is a characterization of ideals.

Theorem 4.9 *Let I be a subset of a right distributive QI-algebra X containing 0. Then $I \in \mathcal{I}\left(X\right)$ if and only if for any $a,b \in I$ and $x \in X$, $(x*a)*b=0$ implies $x \in I$.*

Proof. Let $I \in \mathcal{I}\left(X\right)$. Assume $a,b \in I$ and $x \in X$ such that $(x*a)*b=0$ Since I is an ideal of X, we have $x \in I$. Conversely, assume that the condition holds. Suppose $x \in I$ and $a*x \in I$. Then

$$\left(a*\left(a*x\right)\right)*x = \left(a*x\right)*\left(\left(a*x\right)*x\right) = \left(a*x\right)*\left(a*x\right) = 0.$$

Therefore $(a*(a*x))*x \in I$ and hence $a \in I$. Thus $I \in \mathcal{I}\left(X\right)$.

Corollary 4.10 *Let X be a subset of a right distributive QI-algebra X containing 0. Then $I \in \mathcal{I}\left(X\right)$ if and only if for any $a_1,a_2,\ldots,a_n \in I$ and $x \in X$, $(\ldots(x*a_1)*\ldots)*a_n=0$ implies $x \in I$.*

Definition 4.11 *For every subset $A \subseteq X$, the smallest ideal of X which contains A, that is, the intersection of all ideals $I \supseteq A$, is said to be the ideal generated by A, and will be denoted by* (*A*]. *Obviously,* $\left(\varnothing\right] = \left\{0\right\}$.

Lemma 4.12 *Let X be a right distributive QI-algebra and $x,y,z \in X$. Then $(x*y)*z=0$ if and only if $(x*z)*y=0$.*

Proof. Let $(x*y)*z=0$. Then $((x*y)*z)*y=0$ and hence $((x*y)*y)*(z*y)=0$. Therefore, $(x*y)*(z*y)=0$. Thus $(x*z)*y=0$ Similarly, we can prove the converse.

Theorem 4.13 *Let X be a right distributive implication groupiod and* $A(\neq \varnothing) \subseteq X$. *Then*

$$(A] = \{x \in X : x = 0 \text{ or } (\ldots(x*a_1)*\ldots)*a_n = 0 \text{ for some } a_1, a_2, \ldots, a_n \in A\}.$$

Proof. Let

$$I = \{x \in X : x = 0 \text{or} (\ldots(x*a_1)*\ldots)*a_n = 0 \text{forsome} a_1, a_2, \ldots, a_n \in A\}.$$

Since $a*a=0$ for all $a\in A$, we obtain $A \subseteq I$. Obviously $0\in I$. Let $x*y\in I$ and $y\in I$. To prove $x\in I$, we will consider three cases.

Case 1: $y=0$. Then $x=x*0\in I$.
Case 2: $x*y=0$ and $y\neq 0$.

Since $y\in I$ and $y\neq 0$, we conclude that $(\ldots(y*a_1)*\ldots)*a_n=0$ for some $a_1,a_2,\ldots,a_n\in A$. From Lemma 4.7, $(\ldots(x*a_1)*\ldots)*a_n=0$. Therefore, $x\in I$.

Cases 3: $x*y\neq 0$ and $y\neq 0$. Then there are $a_1,a_2,\ldots,a_n, b_1,b_2,\ldots,b_m\in A$ such that $(\ldots((x*y)*a_1)*\ldots)*a_n=0$ and $(\ldots(y*b_1)*\ldots)*b_m=0$. Applying Lemma 4.12, we deduce that $(\ldots(x*a_1)*\ldots)*a_n\leq y$ and by Lemma 3.9(4), we see that

$$(\ldots(((\ldots(x*a_1)*\ldots)*a_n)*b_1)*\ldots)*b_m \leq (\ldots(y*b_1)*\ldots)*b_m = 0$$

Consequently,

$$(\ldots(((\ldots(x*a_1)*\ldots)*a_n)*b_1)*\ldots)*b_m = 0$$

This implies that $x\in I$. Hence I is an ideal of X. Obviously, $A \subseteq I$. Let J be an ideal of X containing A. Let $x\in I$. Then, there are $a_1,a_2,\ldots,a_n\in A$ such that $(\ldots(x*a_1)*\ldots)*a_n=0$

Since $0\in J$, we have $(\ldots(x*a_1)*\ldots)*a_n\in J$

Since J is an ideal and each $a_i\in J$, it follows that $x\in J$. Hence $I \subseteq J$ which shows that $I=(A]$.

Let $I_1, I_2 \in \mathcal{I}(X)$; we define the meet of I_1 and I_2, denoted by $I_1 \wedge I_2$, by $I_1 \wedge I_2 = I_1 \cap I_2$ and the join of I_1 and I_2, denoted by $I_1 \vee I_2$, by $I_1 \vee I_2 = (I_1 \cup I_2]$. We can observe that $(\mathcal{I}(X), \wedge, \vee)$ is a lattice.

Theorem 4.14 $(\mathcal{I}(X), \wedge, \vee)$ *is a complete lattice.*

5. IMPLICATIVE IDEAL

In this section, we introduce the concept of implicative ideals and study their properties. A necessary and sufficient condition for an ideal to be an implicative ideal is given. Throughout this section, X denote a right distributive QI-algebra unless otherwise specified. We give below the definition of an implicative ideal.

Definition 5.1 *A subset I of X is called an implicative ideal of X if it satisfies the following conditions.*

(II_1) $0 \in I$.
(II_2) $(x*(y*x))*z \in I$ and $z \in I$ implies that $x \in I$ for all $x,y,z \in X$.

Example 5.2 *Let* $X=\{0,a,b,c,d\}$ *in which * is defined by table 5.*

Then $(X,*,0)$ is a right distributive QI-algebra. The subset $I=\{0,a,c,d\}$ of X is an implicative ideal of X. In a right distributive QI-algebra, the following theorem is obvious.

Theorem 5.3 *In a right distributive QI-algebra, every implicative ideal is an ideal.*

Proof. Put $y=0$ in (II_2). Then $x*z \in I$ and $z \in I$ implies that $x \in I$.

It is noted that the converse of the above theorem is not true. We can observe that, in Example 5.2, $I=\{0,a,c\}$ is an ideal of X but not an implicative ideal of X.

In the following, we give a necessary and sufficient condition for an ideal to become an implicative ideal.

Theorem 5.4 *Let I be an ideal of X. Then the following are equivalent.*

(i) I is an implicative ideal.
(ii) $x*(y*x) \in I \Rightarrow x \in I$.

Proof. (i) ⇒(ii): Assume (i). Let $x*(y*x) \in I$. Since $0 \in I$ and $(x*(y*x))*0=x*$, $(y*x) \in I$, we have $x \in I$.
(ii) ⇒(i): Assume (ii). Let $x*(y*x)*z \in I$ and $z \in I$. Then $x*(y*x) \in I$ and hence $x \in I$.

Table 5.

*	0	a	b	c	d
0	0	0	0	0	0
a	a	0	a	0	a
b	b	b	0	b	b
c	c	0	a	0	c
d	d	d	0	d	0

Theorem 5.5 *Let I be a subset of X containing 0. Then I is an implicative ideal of X if and only if for any a,b∈I and* $x, y \in X, \left(\left(x * \left(y * x\right)\right) * a\right) * b = 0$ *implies x∈I.*

Proof. Let *I* be an implicative ideal of *X*. Then *I* is an ideal of *X*. Assume *a,b*∈*I* and *x,y*∈*X* such that $\left(\left(x^*\left(y^*x\right)\right)^*a\right)^*b = 0$. Then, by Theorem 4.9, *x**(*y**x)∈*I*. Therefore *x*∈*I*.

Conversely, assume that the condition holds. Let *a,b*∈*I* and *x*∈*X* such that (*x**a)*b)=0.

Then $\left(\left(x^*\left(0^*x\right)\right)^*a\right)^*b = \left(x^*a\right)^*b = 0$ and hence *x*∈*I*. So that *I* is an ideal of *X*. Let *x**(*y**x)∈*I*
Then $\left(\left(x^*\left(y^*x\right)\right)^*0\right)^*\left(x^*\left(y^*x\right)\right) = 0$ and hence *x*∈*I*. Therefore *I* is an implicative ideal of *X*.

We now give an extension property for the implicative ideals.

Theorem 5.6 *Let I be an implicative ideal of X. If J is an ideal of X containing I, then J is an implicative ideal of X.*

Proof. Let *J* be an ideal of *X* containing *I* and (*x**(*y**x))*z∈*J*, *z*∈*J*. Then *x**(*y**x)∈*J*. Now

$$\left[x^*\left(y^*x\right)\right]^*\left[x^*\left(y^*x\right)\right] = 0 \in I$$

implies that

$$\left[x^*\left(x^*\left(y^*x\right)\right)\right]^*\left[\left(y^*x\right)^*\left(x^*\left(y^*x\right)\right)\right] \in I$$

and hence

$$\left[x^*\left(x^*\left(y^*x\right)\right)\right]^*\left\{\left[y^*\left(x^*\left(y^*x\right)\right)\right]^*\left[x^*\left(x^*\left(y^*x\right)\right)\right]\right\} \in I.$$

Then by Theorem 5.4, we get

$$x^*\left(x^*\left(y^*x\right)\right) \in I \subseteq J.$$

Since *J* is an ideal of *X*, we have *x*∈*J*. Therefore *J* is an implicative ideal of *X*.

6. FANTASTIC AND NORMAL IDEALS

In this section, we introduce the concept of fantastic ideals and normal ideals in a right distributive QI-algebra and we show that implicative ideals, fantastic ideals and normal ideals are equivalent. In what follows, X denotes a right distributive QI-algebra unless otherwise specified.

Definition 6.1 *A non-empty subset I of X is said to be fantastic ideal of X if it is satisfies the following conditions:*

(FI_1) $0 \in I$
(FI_2) $(x*y)*z \in I$ and $z \in I$ implies $x*(y*(y*x)) \in I$, for all $x,y,z \in X$.

Example 6.2 *Let* $X=\{0,a,b,c,d\}$ *in which * is defined by* table 6.

Then $(X,*,0)$ is a right distributive QI-algebra. The subset $I=\{0,a,c\}$ of X is a fantastic ideal of X. We can observe that every fantastic ideal of X is an ideal of X but converse need not be true. From Example 5.2, we can see that, $I=\{0,a,c\}$ is an ideal of X but not fantastic ideal of X.

Theorem 6.3 *Let I be an ideal of X. Then I is a fantastic ideal of X if and only if* $x*y \in I$ *implies* $x*(y*(y*x)) \in I$*. for all* $x,y \in X$.

Proof. Let I be a fantastic ideal of X and $x*y \in I$. Then $(x*y)*0=x*y$ and hence $x*(y*(y*x)) \in I$. Conversely assume that I satisfies the condition. Let $x,y,z \in X$ such that $(x*y)*z \in I$ and $z \in I$. Then $x*y \in I$ and hence $x*(y*(y*x)) \in I$. Therefore I is a fantastic ideal of X.

Theorem 6.4 *Let I be an ideal of X. Then the following are equivalent:*

(1) I is an implicative ideal.
(2) I is a fantastic ideal.

Proof. (1)⇒(2) Assume (1) and $x*y \in I$. We know that

$$\left(x*\left(y*\left(y*x\right)\right)\right)*\left(y*x\right)=\left(x*\left(y*x\right)\right)*\left(y*\left(y*x\right)\right)=\left(x*y\right)*\left(y*x\right)\leq x*y.$$

Therefore $(x*(y*(y*x)))*(y*x) \leq x*y$. Now $x*(y*(y*x)) \leq x$ implies that $y*x \leq y*(x*(y*(y*x)))$ and hence

$$\left[x*\left(y*\left(y*x\right)\right)\right]*\left[y*\left(x*\left(y*\left(y*x\right)\right)\right)\right]\leq\left[x*\left(y*\left(y*x\right)\right)\right]*\left(y*x\right)\leq x*y.$$

Table 6.

*	0	a	b	c	d
0	0	0	0	0	0
a	a	0	a	0	c
b	b	b	0	b	0
c	c	0	a	0	c
d	d	d	0	d	0

That implies

$$\left[x*\left(y*\left(y*x\right)\right)\right]*\left[y*\left(x*\left(y*\left(y*x\right)\right)\right)\right]\in I.$$

Hence $x*(y*(y*x))\in I$ since I is an implicative ideal of X.

(2) ⇒ (1) Assume (2) and $x*(y*x)\in I$. Hence $x*((y*x)*((y*x)*x))\in I$, since I is a fantastic ideal. That implies $x*((y*x)*(y*x)) = x*0 = x\in I$. Therefore $x\in I$. Hence, by Theorem 5.4, I is an implicative ideal of X.

Definition 6.5 *An ideal I of X is called normal ideal of X if, for any $x,y\in X$, $y*(y*x)\in I$ implies $x*(x*y)\in I$.*

Example 6.6 *Let X={0,a,b,c,d} be a set with table 7.*

Then $(X,*,0)$ is a right distributive QI-algebra and $I=\{0,a,c\}$ is a normal ideal of X.

Theorem 6.7 *Let I be an ideal of X. Then the following are equivalent: [(1)] I is an implicative ideal [(2)] $a\in I$, $(x*a)*(y*x)\in I$ implies $x\in I$.*

Proof. Suppose that I is an implicative ideal of X and $a\in I$, $(x*a)*(y*x)\in I$. We know that $a*(y*x) \leq a$ which implies that

$$\left[x*\left(y*x\right)\right]*a\leq\left[x*\left(y*x\right)\right]*\left[a*\left(y*x\right)\right].$$

So that

$$\left[x*\left(y*x\right)\right]*a\leq\left(x*a\right)*\left(y*x\right)$$

and hence $[x*(y*x)]*a\in I$, since I is an ideal of X.

Therefore $x*(y*x)\in I$ and hence $x\in I$ since I is an implicative ideal. Conversely, assume that the condition holds. Let $x*(y*x)\in I$. Then

$$\left(x*\left(x*\left(y*x\right)\right)\right)*\left(y*x\right)=0\in I.$$

Table 7.

*	0	a	b	c	d
0	0	0	0	0	0
a	a	0	a	0	a
b	b	b	0	b	b
c	c	0	a	0	c
d	d	d	d	d	0

Hence, by assumption, $x \in I$. Therefore I is an implicative ideal.

Theorem 6.8 *Let I be an ideal of X. Then the following are equivalent.* [(1)] I is an implicative ideal [(2)] I is a normal ideal.

Proof. (1)⇒(2) Assume (1) and $y*(y*x) \in I$. By Proposition 3.9(7), $x*(x*y) \leq x$ implies that $y*x \leq y*(x*(x*y))$ and hence

$$\begin{aligned} \left[x*\left(x*y\right)\right]*\left[y*\left(x*\left(x*y\right)\right)\right] &\leq \left[x*\left(x*y\right)\right]*\left(y*x\right) && \left(\text{by Proposition3.9}\left(7\right)\right) \\ &\leq \left(x*\left(y*x\right)\right)*\left(x*y\right) && \left(\text{by Proposition3.9}\left(8\right)\right) \\ &\leq y*\left(y*x\right) && \left(\text{by Proposition3.9}\left(9\right)\right) \end{aligned}$$

Since I is an ideal of X, we have $\left[x*\left(x*y\right)\right]*\left[y*\left(x*\left(x*y\right)\right)\right] \in I$. Hence, by Theorem 5.4, $x*(x*y) \in I$. Thus (2) follows.

(2)⇒(1) Assume (2) and $x*(y*x) \in I$. We know that $y*(y*x) \leq x$ Then, by Lemma 3.9(4), $(y*(y*x)) * (y*x) \leq x*(y*x)$ and hence $(y*(y*x))*(y*x) \in I$, since I is an ideal. That implies $y*(y*x) \in I$. Therefore $x*(x*y) \in I$. Also, $y*x \leq y$ which implies that $x*y \leq x*(y*x)$ Hence $x*y \in I$. Now $x*(x*y) \in I$ and $x*y \in I$ implies that $x \in I$. Thus (1) follows.

We conclude this section with the following theorem.

Theorem 6.9 *Let I be an ideal of X. Then the following are equivalent.*

(1) I is an implicative ideal
(2) I is a normal ideal
(3) I is a fantastic ideal.

REFERENCES

Abbott, J. C. (1967). Semi-bolean algebra. *Matem. Vestnik*, *4*, 177-198. http://eudml.org/doc/258960

Bandaru, R. K. (2017). On QI-algebras. *Discussiones Mathematicae. General Algebra and Applications*, *37*(2), 137–145. doi:10.7151/dmgaa.1269

Bandaru, R. K., Shum, K. P., & Rafi, N. (2017). Some types of ideals in distributive implication groupoids. *International Journal of Pure and Applied Mathematics*, *116*(2), 377–384. doi:10.12732/ijpam.v116i2.9

Chen, W. Y., & Oliveira, J. S. (1995). Implication algebras and the metropolis rota axioms for cubic lattices. *Journal of Algebra*, *171*(2), 383–396. doi:10.1006/jabr.1995.1017

Imai, Y., & Iseki, K. (1966). On axiom systems of propositional calculi, XIV. *Proceedings of the Japan Academy*, *42*(1), 19–22. doi:10.3792/pja/1195522169

Iseki, K. (1980). On BCI-algebras. *Mathematics Seminar Notes*, *8*, 125-130. http://www.math.kobe-u.ac.jp/jmsj/kjm/

Iseki, K., & Tanaka, S. (1978). An introduction to the theory of BCK-algebras. *Math. Japon*, *23*(1), 1–26.

Saeid, A. B., Kim, H. S., & Rezaei, A. (2017). On BI-algebras, *An.St. Univ.Ovidius Constana*, *25*(1), 177–194. doi:t doi:10.1515/auom-2017-0014

Saeid, A. B., Rezaei, A., & Borzooei, R. A. (2013). Some types of filters in BE-algebras. *Mathematics in Computer Science*, *7*(3), 341–352. doi:10.100711786-013-0157-6

Chapter 17
Regular and Intra-Regular Neutrosophic Left Almost Semihypergroups

Muhammad Gulistan
https://orcid.org/0000-0002-6438-1047
Hazara University, Pakistan

Rashid Ullah
Hazara University, Pakistan

ABSTRACT

In this chapter, the authors characterized regular and intra-regular neutrosophic LA-semihypergroups by giving some examples with different neutrosophic identities and prove a various important result about it with the help of different neutrosophic identities. The aim in this chapter is to characterize the idea of regular and intra-regular neutrosophic LA-semihypergroups, (weakly, strongly, completely) regular neutrosophic LA-semihypergroups using neutrosophuic sets. They also define different substructures of regular neutrosophic LA-semihypergroup namely (weakly, left, right, strongly, completely) regular neutrosophic LA-semihypergroup with some basic results. Further, invertible neutrosophic LA-semihypergroup are also define in the last.

1. INTRODUCTION

Zadeh in 1965 has introduced the fundamental concept of a fuzzy set which provide a natural framework for generalizing several basic notions of algebra (Zadeh, 1965). As a generalization of fuzzy set the intuitionistic fuzzy set was introduced by Atanassov (Atanassov, 1986). Inspired from the reallities of natural phenomena like from sport games (winning/tie/defeating), from votes decision (yes/NA/no), from decision making (conform decision/ hesitating/not making), from statement (accepted /pending /rejected) etc. conducted by the fact that the law of excluded middle did not work any longer in the modern logics. Smarandache in 1995 for the first time introduced the concept of philosophy (Sma-

DOI: 10.4018/978-1-7998-0190-0.ch017

randache, 2005) and combined the non-standard analysis (Broumi, Smarandache, & Maji. 2014) with a tri-component logic/set/probability theory which represents the main distinction between fuzzy and intuitionistic fuzzy set. (Broumi, & Smarandachi. 2014). Here he included the middle component i.e. the indeterminate besides the membership and non-membership components that both seem in fuzzy logic to separate between 'absolute membership and relative membership' or 'absolute non-membership and relative non-membership' (Maji., 2013). There are also some authors who have enhanced the theory of neutrosophic sets to another extended form namely soft set (Cutland, 1988).

At instant Kazim et al. introduced the concept of left almost semigroups and right almost semigroups (Golan, 1999). Mushtaq et al. explored this structure and added various useful results to the theory of LA-semigroups (Clifford, & Preston. 1961). Later, Madad et al. work on intra-regular LA-semigroups characterized by their anti-fuzzy ideals (Khan, & Ahmad., 2010) and an ideal in intra-regular LA-semigroup (Protic, & Stevanovic., 1995) etc.

Hyper structure theory was introduced in 1934 when Marty defined hyper groups, began to analyze their properties and applied them to groups (Mushtaq, & Yusuf., 1979). A lot of papers and several books have been written on hyper structure theory (Kazim, Naseeruddin, Mushtaq, Khan., et al. 1972). Many authors studied on different aspects of hyper structure theory, for instance, (Davvaz, et al., 2007), (Aslam, & Howie., 1976), (Hasankhani, 1999) and (Yaqoob, et al., 2013), (Hilla, et al., 2011) etc introduced a semihypergroups and studied the quasi hyper ideals in semihypergroups.

Recently Hilla, et al. in 2011 introduced the notion of LA-semihypergroups as a generalization of semi groups, semihypergroups, and LA-semigroups. Yousafzai et al. works on some characterization problems in LA-semihypergroup (Yousafzai, & Corsini., 2013). Naveed et al. display a solid result on intra-regular LA-semihypergroup with pure left identity (Yaqoob, Corsini, & Yousafzai., 2013) also they present a partially ordered LA-semihypergroup.

Using neutrosophic theory, Kandasamy, et al. introduced the concept of neutrosophic algebraic structures. Soon M. Khan et al. write a book on neutrosophic set approaches to algebraic structure and a theory of Abel Grossmann's groupoids. Agboola, et al. studied neutrosophic groups and subgroups and also the structure of neutrosophic polynomial.

Let $(H,*)$ be an LA-semihypergroup and $H \cup I = \{(k, lI) : k, l \in S\}$. The neutrosophic left almost semihypergroup is generated by H and I under the binary hyper operation 'o' represented by $N(H) = (H \cup I, o)$ the part k is non-neutrosophic and the part l is called neutrosophic part of generating set. Then the couple $(H \cup I, o)$ becomes a neutrosophic hypergroupoid if $\ell o \mathcal{g} = \mathcal{g} o \ell$ hold for all $\forall \mathcal{g}, \ell \in H \cup I$. A neutrosophic hyper groupoid satisfying the neutrosophic left invertive law i.e

$$(\hbar o \ell) o \mathcal{g} = (\mathcal{g} o \ell) o \hbar \quad \forall \mathcal{g}, \hbar, \ell \in H \cup I$$

is called a neutrosophic left almost semihypergroup shortly abbreviated as neutrosophic LA-semihypergroup (Gulistan, Yaqoob, Smarandachi, & Rashid., 2017).

Our aim in this paper is to characterize the idea of regular and intra-regular neutrosophic LA-semihypergroup and is the extended form of neutrosophic LA-semihypergroups.

2. BASIC DEFINITIONS

In this section, we give some basic definitions from the hyper structure's theory.

Definition 2.1 (Hila, & Dine. 2011). A map $o:\mathcal{S}\times\mathcal{S}\to P^*(\mathcal{S})$ is called hyperoperation or join operation on the non-empty set $\mathcal{S}$, and $P^*(\mathcal{S})=P(\mathcal{S})\setminus\phi$ denotes the set of all non-empty subsets of $\mathcal{S}$.

Definition 2.2 (Hila, & Dine. 2011). A hyper groupoid is called the pair $(\mathcal{S},o)$, where 'o' is a hyperoperation on the set $\mathcal{S}$.

If $f\in\mathcal{S}$ and $\mathcal{A},\mathcal{B}$ are non-empty subsets of $\mathcal{S}$ then

$$\mathcal{A}o\mathcal{B}=\bigcup_{f\in\mathcal{A},g\in\mathcal{B}} fog,\ fo\mathcal{A}=\{f\}o\mathcal{A},\ \mathcal{B}of=\mathcal{B}o\{f\}$$

Definition 2.3 (Hila, & Dine. 2011). A hypergroupoid (H,o) is called an LA-semihypergroup if for all $l,m,n\in H$

$$(lom)on=(nom)ol$$

The law $(lom)on=(nom)ol$ is called a left invertive law.

Definition 2.4 (Gulistan, et al. 2019). Let $(H,*)$ be an LA-semihypergroup and $H\cup I=\{(k,lI):k,l\in\mathcal{S}\}$. The neutrosophic LA-semihypergroup is generated by H and I under binary hyperoperation "o" is represented by $N(H)=(H\cup I,o)$, the part k is called non-neutrosophic part and the part l is called neutrosophic part of generating set.

If $\hbar=(n,mI)$ and $\ell=(e,zI)$ are any two arbitrary elements of $H\cup I$ then we defined a hyper operation as:

$$\begin{aligned}\hbar o\ell&=(n,mI)o(e,zI)\\&=\{(r,sI)\mid r\in n*e,s\in(n*z\cup m*e\cup m*z)\}\\&=\{(n*e,(n*z\cup m*e\cup m*z)I)\}\in H\cup I\end{aligned}$$

Then $(H\cup I,o)$ becomes a neutrosophic LA-semihypergroup if $(\hbar o\ell)og=(go\ell)o\hbar$ hold $\forall g,\hbar,\ell\in N(H)$. It is since called the neutrosophic left invertive law.

A neutrosophic hypergroupoid satisfy the neutrosophic left invertive law is called a neutrosophic LA-semihypergroup.

Definition 2.5 (Gulistan, et al. 2019). In a neutrosophic LA-semihypergroup the following laws hold:

(a) with neutrosophic right identity a medial law:

$$[\mathcal{g} o\hbar]o[\kappa o\ell]=[\mathcal{g} o\kappa]o[\hbar o\ell] \quad (1)$$

(b) and with neutrosophic left identity a paramedial law hold i.e.

$$[\mathcal{g} o\hbar]o[\kappa o\ell]=[\ell o\hbar]o[\kappa o\mathcal{g}] \quad (2)$$

$$[\mathcal{g} o\hbar]o[\kappa o\ell]=[\ell o\kappa]o[\hbar o\mathcal{g}] \quad (3)$$

$\forall \mathcal{g},\hbar,^{o},\ell \in N(\mathcal{S})$.

Definition 2.6 (Onipchuk. 1992). An element $\hbar$ in an LA-semihypergroup$(\mathcal{S},*)$ is called a regular element if $\exists \mathcal{g} \in \mathcal{S}$ such that $\hbar \in (\hbar o\mathcal{g})o\hbar$. However if all the element of $(\mathcal{S},*)$ is regular then it is called a regular LA-semihypergroup.

Definition 2.7 (Yaqoob. et al. 2013). An element $\hbar \in \mathcal{S}$ is called an intra-regular element if $\exists \mathcal{g},\ell \in \mathcal{S}$ such that $\hbar \in (\mathcal{g}*\hbar^2)*\ell$. If all the element of $(\mathcal{S},*)$ is intra-regular then it is called intra-regular LA-semihypergroup.

Definition 2.8 (Onipchuk. 1992). An element $\hbar$ of an LA-semihypergroup $(\mathcal{S},*)$ is called a weakly regular element if $\exists \mathcal{g},\ell \in \mathcal{S}$ such that $\hbar \in (\hbar*\mathcal{g})*(\hbar*\ell)$. If all the element of $(\mathcal{S},*)$ is weakly regular then it is called weakly regular LA-semihypergroup.

Definition 2.9 (Onipchuk. 1992). An element $\hbar$ of an LA-semihypergroup $(\mathcal{S},*)$ is called a strongly regular element if $\exists \mathcal{g},\ell \in \mathcal{S}$ such that $\hbar = (\hbar*\mathcal{g})*(\hbar*\ell)$. If all the element of $(\mathcal{S},*)$ is strongly regular then it is called strongly regular LA-semihypergroup.

3. REGULAR AND INTRA-REGULAR NEUTROSOPHIC LA-SEMIHYPERGROUP

In this section, we presented regular and intra-regular neutrosophic LA-semihypergroups with some important results.

Definition 3.1: A neutrosophic element $\{(\mathcal{g},\ell I)\}$ of a neutrosphic left almost semihypergroup *N*(*H*) is called a regular neutrosophic element if there exist $(\mathsf{r},\mathsf{s}I) \in H \cup I$ such that

$$\{(g,\ell I)\} \in [(g,\ell I)o(r,sI)]o(g,\ell I), \text{ where } g,\ell,r,s \in H.$$

The neutrosophic left almost semihypergroup is called a regular neutrosophic left almost semihypergroup if every neutrosophic element is regular neutrosophic element.

And $\{(g,\ell I)\}$ is called intra-regular neutrosophic element if there exist $(r,sI),(\hbar,\mathcal{I}I) \in H \cup I$, such that:

$$\{(g,\ell I)\} \in \left[(r,sI)o(g,\ell I)^2\right]o(\hbar,\mathcal{I}I), \text{ where } g,\ell,r,s \in H.$$

N(H) is an intra-regular neutrosophic LA-semihypergroup if every neutrosophic element isintra-regular neutrosophic element.

Example of Regular and Intra-regular neutrosophic LA-semihypergroup with neutrosophic right identity:

Example 3.1: Let $H = \{g,\hbar,\ell\}$ is an LA-semihypergroup with right identity g under "*o*" with Table 1.

Then

$$H \cup I = \{(g,gI),(g,\hbar I),(g,\ell I),(\hbar,\hbar I),(\hbar,gI),(\hbar,\ell I),(\ell,\ell I),(\ell,gI),(\ell,\hbar I)\}$$

is regular and intra-regular neutrosophic LA-semihypergroup with neutrosophic three right identities $(g,gI),(g,\hbar I)$, and $(g,\ell I)$ and has binary hyperoperation "*o*" is defined as shown in Table 2:

Clearly *N(H)* is neutrosophic LA-semihypergroup also

$$[(\hbar,\hbar I)o(\hbar,\hbar I)]o(\hbar,\hbar I) \neq (\hbar,\hbar I)o[(\hbar,\hbar I)o(\hbar,\hbar I)],$$

Table 1.

*	g	$\hbar$	ℓ
g	$\{g,\ell\}$	ℓ	$\{\hbar,\ell\}$
$\hbar$	$\{\hbar,\ell\}$	ℓ	ℓ
ℓ	$\{\hbar,\ell\}$	$\{\hbar,\ell\}$	$\{\hbar,\ell\}$

Table 2.

o	(g,gI)	$(g,\hbar I)$	$(g,\ell I)$	$(\hbar,\hbar I)$	$(\hbar,gI)$	$(\hbar,\ell I)$	$(\ell,\ell I)$	(ℓ,gI)	$(\ell,\hbar I)$
(g,gI)	(g,gI) $(g,\ell I)$ $(\ell,\ell I)$ (ℓ,gI)	(g,gI) $(g,\ell I)$ $(\ell,\ell I)$ (ℓ,gI)	(g,gI) $(g,\hbar I)$ $(g,\ell I)$ $(\ell,\ell I)$ (ℓ,gI) $(\ell,\hbar I)$	$(\ell,\ell I)$	(ℓ,gI) $(\ell,\ell I)$	$(\ell,\ell I)$ $(\ell,\hbar I)$	$(\hbar,\hbar I)$ $(\hbar,gI)$ $(\hbar,\ell I)$ $(\ell,\ell I)$ (ℓ,gI) $(\ell,\hbar I)$	$(\hbar,\hbar I)$ $(\hbar,\ell I)$ $(\ell,\ell I)$ $(\ell,\hbar I)$	$(\hbar,\hbar I)$ $(\hbar,\ell I)$ $(\ell,\ell I)$ $(\ell,\hbar I)$
$(g,\hbar I)$	(g,gI) $(g,\hbar I)$ $(g,\ell I)$ $(\ell,\ell I)$ (ℓ,gI) $(\ell,\hbar I)$	$(g,\hbar I)$ $(g,\ell I)$ $(\ell,\ell I)$ $(\ell,\hbar I)$	$(g,\hbar I)$ $(g,\ell I)$ $(\ell,\ell I)$ $(\ell,\hbar I)$	$(\ell,\ell I)$	$(\ell,\ell I)$ (ℓ,gI) $(\ell,\hbar I)$	$(\ell,\ell I)$ $(\ell,\hbar I)$	$(\hbar,\hbar I)$ $(\hbar,gI)$ $(\hbar,\ell I)$ $(\ell,\ell I)$ (ℓ,gI) $(\ell,\hbar I)$	$(\hbar,\ell I)$ $(\ell,\ell I)$	$(\hbar,\hbar I)$ $(\hbar,\ell I)$ $(\ell,\ell I)$ $(\ell,\hbar I)$
$(g,\ell I)$	(g,gI) $(g,\hbar I)$ $(g,\ell I)$ $(\ell,\ell I)$ (ℓ,gI) $(\ell,\hbar I)$	$(g,\hbar I)$ $(g,\ell I)$ $(\ell,\ell I)$ $(\ell,\hbar I)$	$(g,\hbar I)$ $(g,\ell I)$ $(\ell,\ell I)$ $(\ell,\hbar I)$	$(\ell,\ell I)$ $(\ell,\hbar I)$		$(\ell,\ell I)$ $(\ell,\hbar I)$	$(\hbar,\hbar I)$ $(\hbar,gI)$ $(\hbar,\ell I)$ $(\ell,\ell I)$ (ℓ,gI) $(\ell,\hbar I)$	$(\hbar,\hbar I)$ $(\hbar,\ell I)$ $(\ell,\ell I)$ $(\ell,\hbar I)$	$(\hbar,\hbar I)$ $(\hbar,\ell I)$ $(\ell,\ell I)$ $(\ell,\hbar I)$
$(\hbar,gI)$	$(\ell,\ell I)$ (ℓ,gI) $(\ell,\hbar I)$ (g,gI) $(g,\hbar I)$ $(g,\ell I)$	$(\ell,\ell I)$ (ℓ,gI) $(\ell,\hbar I)$ (g,gI) $(g,\hbar I)$ $(g,\ell I)$	$(\ell,\ell I)$ (ℓ,gI) $(\ell,\hbar I)$ (g,gI) $(g,\hbar I)$ $(g,\ell I)$	$(\ell,\ell I)$	$(\ell,\ell I)$ (ℓ,gI) $(\ell,\hbar I)$	$(\ell,\ell I)$ $(\ell,\hbar I)$	$(\ell,\ell I)$ $(\ell,\hbar I)$	$(\ell,\ell I)$ (ℓ,gI) $(\ell,\hbar I)$	$(\ell,\ell I)$ $(\ell,\hbar I)$
$(\hbar,\hbar I)$	$(\hbar,\hbar I)$ $(\hbar,gI)$ $(\hbar,\ell I)$ $(\ell,\ell I)$ (ℓ,gI) $(\ell,\hbar I)$	$(\hbar,\hbar I)$ $(\hbar,gI)$ $(\hbar,\ell I)$ $(\ell,\ell I)$ (ℓ,gI) $(\ell,\hbar I)$	$(\hbar,\hbar I)$ $(\hbar,gI)$ $(\hbar,\ell I)$ $(\ell,\ell I)$ (ℓ,gI) $(\ell,\hbar I)$	$(\ell,\ell I)$	$(\ell,\ell I)$ (ℓ,gI) $(\ell,\hbar I)$	$(\ell,\ell I)$	$(\ell,\ell I)$	$(\ell,\ell I)$ (ℓ,gI) $(\ell,\hbar I)$	$(\ell,\ell I)$

continues on following page

Table 2. Continued

o	(g,gI)	$(g,\hbar I)$	$(g,\ell I)$	$(\hbar,\hbar I)$	$(\hbar,gI)$	$(\hbar,\ell I)$	$(\ell,\ell I)$	(ℓ,gI)	$(\ell,\hbar I)$
$(\hbar,\ell I)$	$(\hbar,\hbar I)$ $(\hbar,gI)$ $(\hbar,\ell I)$ $(\ell,\ell I)$ (ℓ,gI) $(\ell,\hbar I)$	$(\hbar,\hbar I)$ $(\hbar,gI)$ $(\hbar,\ell I)$ $(\ell,\ell I)$ (ℓ,gI) $(\ell,\hbar I)$	$(\hbar,\hbar I)$ $(\hbar,gI)$ $(\hbar,\ell I)$ $(\ell,\ell I)$ (ℓ,gI) $(\ell,\hbar I)$	$(\ell,\ell I)$ $(\ell,\hbar I)$	$(\ell,\ell I)$ $(\ell,\hbar I)$	$(\ell,\ell I)$ $(\ell,\hbar I)$	$(\ell,\ell I)$ $(\ell,\hbar I)$	$(\ell,\ell I)$ $(\ell,\hbar I)$	$(\ell,\ell I)$ $(\ell,\hbar I)$
$(\ell,\ell I)$	$(\hbar,\hbar I)$ $(\hbar,\ell I)$ $(\ell,\ell I)$ $(\ell,\hbar I)$	$(\hbar,\hbar I)$ $(\hbar,\ell I)$ $(\ell,\ell I)$ $(\ell,\hbar I)$	$(\hbar,\hbar I)$ $(\hbar,\ell I)$ $(\ell,\ell I)$ $(\ell,\hbar I)$	$(\hbar,\hbar I)$ $(\hbar,\ell I)$ $(\ell,\ell I)$ $(\ell,\hbar I)$	$(\hbar,\hbar I)$ $(\hbar,\ell I)$ $(\ell,\ell I)$ $(\ell,\hbar I)$	$(\hbar,\hbar I)$ $(\hbar,\ell I)$ $(\ell,\ell I)$ $(\ell,\hbar I)$	$(\hbar,\hbar I)$ $(\hbar,\ell I)$ $(\ell,\ell I)$ $(\ell,\hbar I)$	$(\hbar,\hbar I)$ $(\hbar,\ell I)$ $(\ell,\ell I)$ $(\ell,\hbar I)$	$(\hbar,\hbar I)$ $(\hbar,\ell I)$ $(\ell,\ell I)$ $(\ell,\hbar I)$
(ℓ,gI)	$(\hbar,\hbar I)$ $(\hbar,gI)$ $(\hbar,\ell I)$ $(\ell,\ell I)$ (ℓ,gI) $(\ell,\hbar I)$	$(\hbar,\hbar I)$ $(\hbar,gI)$ $(\hbar,\ell I)$ $(\ell,\ell I)$ (ℓ,gI) $(\ell,\hbar I)$	$(\hbar,\hbar I)$ $(\hbar,gI)$ $(\hbar,\ell I)$ $(\ell,\ell I)$ (ℓ,gI) $(\ell,\hbar I)$	$(\hbar,\hbar I)$ $(\hbar,\ell I)$ $(\ell,\ell I)$ $(\ell,\hbar I)$	$(\hbar,\hbar I)$ $(\hbar,gI)$ $(\hbar,\ell I)$ $(\ell,\ell I)$ (ℓ,gI) $(\ell,\hbar I)$	$(\hbar,\hbar I)$ $(\hbar,\ell I)$ $(\ell,\ell I)$ $(\ell,\hbar I)$	$(\hbar,\hbar I)$ $(\hbar,\ell I)$ $(\ell,\ell I)$ $(\ell,\hbar I)$	$(\hbar,\hbar I)$ $(\hbar,gI)$ $(\hbar,\ell I)$ $(\ell,\ell I)$ (ℓ,gI) $(\ell,\hbar I)$	$(\hbar,\hbar I)$ $(\hbar,\ell I)$ $(\ell,\ell I)$ $(\ell,\hbar I)$
$(\ell,\hbar I)$	$(\hbar,\hbar I)$ $(\hbar,\ell I)$ $(\ell,\ell I)$ $(\ell,\hbar I)$	$(\hbar,\hbar I)$ $(\hbar,\ell I)$ $(\ell,\ell I)$ $(\ell,\hbar I)$	$(\hbar,\hbar I)$ $(\hbar,\ell I)$ $(\ell,\ell I)$ $(\ell,\hbar I)$	$(\hbar,\hbar I)$ $(\hbar,\ell I)$ $(\ell,\ell I)$ $(\ell,\hbar I)$	$(\hbar,\hbar I)$ $(\hbar,\ell I)$ $(\ell,\ell I)$ $(\ell,\hbar I)$	$(\hbar,\hbar I)$ $(\hbar,\ell I)$ $(\ell,\ell I)$ $(\ell,\hbar I)$	$(\hbar,\hbar I)$ $(\hbar,\ell I)$ $(\ell,\ell I)$ $(\ell,\hbar I)$	$(\hbar,\hbar I)$ $(\hbar,\ell I)$ $(\ell,\ell I)$ $(\ell,\hbar I)$	$(\hbar,\hbar I)$ $(\hbar,\ell I)$ $(\ell,\ell I)$ $(\ell,\hbar I)$

So *N*(*H*) is non-associative and is regular because

$$\{(g,gI)\}\in[(g,gI)o(g,\hbar I)]o(g,gI),$$
$$\{(\hbar,\hbar I)\}\in[(\hbar,\hbar I)o(\ell,gI)]o(\hbar,\hbar I),\{(\ell,\ell I)\}\in[(\ell,\ell I)o(g,\hbar I)]o(\ell,\ell I) \text{ etc.}$$

Therefore *N*(*H*) is regular neutrosopohic LA-semihypergroup. Also *N*(*H*) is intra-regular neutrosopohic LA-semihypergroup because

$$\{(g,gI)\} \in \left[(g,\hbar I)o(g,gI)^2\right]o(g,\ell I),$$
$$\{(\hbar,\hbar I)\} \in \left[(\ell,\hbar I)o(\hbar,\hbar I)^2\right]o(g,\ell I), \{(\ell,\ell I)\} \in \left[(g,\hbar I)o(\ell,\ell I)^2\right]o(\hbar,\ell I) \text{ etc.}$$

identity:

Example 3.2: Let $H = \{g, \hbar, \ell\}$ is an LA-semihypergroup with the Table 3*:*

Then

$$H \cup I = \{(g,gI),(g,\hbar I),(g,lI),(\hbar,\hbar I),(\hbar,gI),(\hbar,lI),(l,lI),(l,gI),(l,\hbar I)\}$$

is non- associative because of

$$[(g,gI)o(g,gI)]o(\hbar,\hbar I) \neq (g,gI)o[(g,gI)o(\hbar,\hbar I)].$$

Also, it is regular and intra-regular neutrosophic LA-semihypergroup under hyperoperation "*o*" with neutrosophic left identities $(g,gI),(g,\hbar I)$ and $(g,\ell I)$. Thus

$$\{(g,gI)\} \in [(g,gI)o(g,\hbar I)]o(g,gI),$$
$$\{(\hbar,\hbar I)\} \in [(\hbar,\hbar I)o(\ell,gI)]o(\hbar,\hbar I), \{(\ell,\ell I)\} \in [(\ell,\ell I)o(\hbar,lI)]o(\ell,\ell I) \text{ etc.}$$

And

$$\{(g,gI)\} \in \left[(\ell,\ell I)o(g,gI)^2\right]o(g,\hbar I),$$
$$\{(\hbar,\hbar I)\} \in \left[(\hbar,\ell I)o(\hbar,\hbar I)^2\right]o(\hbar,\hbar I), \{(\ell,\ell I)\} \in \left[(g,gI)o(\ell,\ell I)^2\right]o(\hbar,gI) \text{ etc.}$$

Table 3.

*	g	$\hbar$	ℓ
g	g	$\{\hbar,\ell\}$	$\{g,\ell\}$
$\hbar$	$\{g,\ell\}$	$\{g,\ell\}$	$\{g,\ell\}$
ℓ	$\{g,\ell\}$	$\{g,\ell\}$	$\{g,\ell\}$

Example of Regular and Intra-regular neutrosophic LA-semihypergroup with Neutrosophic identity:

Example 3.3: Let $H=\{g,\hbar,\ell\}$ is an LA-semihypergroup with the following Table 4:

Obviously (*H*,*) is neither associative and nor commutative. Thus the generating set

$$H\cup I=\{(g,gI),(g,\hbar I),(g,lI),(\hbar,\hbar I),(\hbar,gI),(\hbar,lI),(l,lI),(l,gI),(l,\hbar I)\}$$

is regular and intra-regular neutrosophic LA-semihypergroup under hyperoperation "*o*" with neutrosophic identity (*g,gI*). Thus

$$\{(g,gI)\}\in[(g,gI)o(g,\hbar I)]o(g,gI),$$
$$\{(\hbar,lI)\}\in[(\hbar,lI)o(l,gI)]o(\hbar,lI),\{(l,lI)\}\in[(l,lI)o(\hbar,\hbar I)]o(l,lI) \text{ etc.}$$

one can easily check similarly all other regular neutrosophic element of $H\cup I$. Now for intra-regular we have

$$\{(g,gI)\}\in[(g,\hbar I)o(g,gI)^2]o(g,gI),$$
$$\{(\hbar,lI)\}\in[(l,gI)o(\hbar,lI)^2]o(\hbar,\hbar I),\{(l,lI)\}\in[(l,lI)o(l,lI)^2]o(\hbar,gI) \text{ etc.}$$

Consequently $(H\cup I,o)$ is regular and intra-regular neutrosophic LA-semihypergroup of $H\cup I$.

Example 3.4: Consider a set of integers $\mathsf{Z}=H$, hence a neutrosophic generating set of integers is $\mathsf{Z}\cup I=H\cup I$, where *I* is a neutrosophic element so the binary hyperoperation "*o*" on $\mathsf{Z}\cup I$ is define as:

$$[(g,\hbar I)o(\ell,\ell I)]o(\mathrm{r},\mathrm{s}I)=\{(g,\hbar I),(\ell,\ell I),(\mathrm{r},\mathrm{s}I)\}\cup\mathsf{Z}\cup I.$$

Table 4.

*	g	$\hbar$	ℓ
g	$\{g,\ell\}$	$\hbar$	$\{\hbar,\ell\}$
$\hbar$	$\{\hbar,\ell\}$	$\{\hbar,\ell\}$	$\{\hbar,\ell\}$
ℓ	$\{\hbar,\ell\}$	$\{\hbar,\ell\}$	$\{\hbar,\ell\}$

for all

$$\left(g,\hbar I\right),\left(\ell,\ell I\right),\left(r,sI\right)\in Z\cup I,$$

where $g,\hbar,\ell,r,s\in\left(Z,o\right)$. Then $\left(H\cup I,o\right)$ is regular and intra-regular neutrosophic LA-semihypergroup.

Example 3.5: Let $Z\cup I=H\cup I$ denote the neutrosophic generating set of integers. Assume the binary hyper operation "o" on $Z\cup I$ is defined in the following manner:

$$\left(g,\hbar I\right)o\left(\ell,\ell I\right)=\left(\ell,\ell I\right)-\left(g,\hbar I\right)$$

for all

$$\left(g,\hbar I\right),\left(\ell,\ell I\right)\in Z\cup I,$$

where $g,\hbar,\ell\in\left(Z,*\right)$. and "-" denotes the ordinary subtraction. Then $\left(Z\cup I,o\right)$ is regular and intra-regular neutrosophic LA-semihypergroup.

Remark 3.1: It is important to note that for a regular and intra-regular neutrosophic LA-semihypergroup it is enough but not necessary that a neutrosophic LA-semihypergroup must have a neutrosophic identities it may or may not exist.

Here we begin with Example of regular and intra-regular neutrosophic LA-semihypergroup with no neutrosophic identity.

Example 3.6: Let $H=\left\{g,\hbar,\ell\right\}$ is an LA-semihypergroup under the binary hyper operation "*"is defined as Table 5.

Table 5.

*	g	$\hbar$	ℓ
g	g	g	g
$\hbar$	g	g	ℓ
ℓ	g	$\hbar$	g

Clearly

$$H \cup I = \{(g, gI), (g, \hbar I), (g, lI), (\hbar, \hbar I), (\hbar, gI), (\hbar, lI), (l, lI), (l, gI), (l, \hbar I)\}$$

is regular and intra-regular neutrosophic LA-semihypergroup without any neutrosophic identities under binary hyperoperation "o".

Remark 3.2: Every neutrosophic LA-semihypergroup is regular and intra-regular iff every neutrosophic element of neutrosophic LA-semihypergroup is regular and intra-regular.

Let us see the following Example given below.

Example 3.7: Let $H = \{g, \hbar, \ell, \mathcal{I}\}$ with the binary hyperoperation "*" is defined as Table 6:

Clearly $(H,*)$ is an LA-semihypergroup and

$$H \cup I = \begin{Bmatrix} (g, gI), (g, \hbar I), (g, lI), (g, \mathcal{I}I), (\hbar, \hbar I), (\hbar, gI), (\hbar, lI), (\hbar, \mathcal{I}I) \\ (l, lI), (l, gI), (l, \hbar I), (g, \mathcal{I}I), (\mathcal{I}, lI), (\mathcal{I}, gI), (\mathcal{I}, \hbar I), (\mathcal{I}, \mathcal{I}I) \end{Bmatrix}$$

is neutrosophic LA-semihypergroup with hyper operation "o" but not a regular and intra-regular neutrosophic LA-semihypergroup because of $(l, lI), (l, gI), (l, \hbar I), (g, \mathcal{I}I)$ is not a regular and intra-regular neutrosophic element with respect to binary hyperoperation "o".

Lemma 3.1: Every neutrosophic hyper ideal of a regular neutrosophic LA-semihypergroup is neutrosophic idempotent.

Proof: Let $N(\ell)$ be any neutrosophic hyper ideal of a regular neutrosophic LA-semihypergroup $N(H)$. Then by definition:

$$N(\ell)oN(H) \subseteq N(\ell) \text{ and } N(H)oN(\ell) \subseteq N(\ell)$$

Table 6.

*	g	$\hbar$	ℓ	$\mathcal{I}$
g	$\hbar$	$\hbar$	$\{g, \hbar\}$	$\mathcal{I}$
$\hbar$	$\hbar$	$\hbar$	$\{g, \hbar\}$	$\mathcal{I}$
ℓ	$\{g, \hbar\}$	$\{g, \hbar\}$	g	$\mathcal{I}$
$\mathcal{I}$	$\mathcal{I}$	$\mathcal{I}$	$\mathcal{I}$	$\mathcal{I}$

So,

$$N(\ell)^2 = N(\ell)oN(\ell) \subseteq N(\ell)oN(H) \subseteq N(\ell) \tag{4}$$

Similarly,

$$N(\ell)^2 = N(\ell)oN(\ell) \subseteq N(H)oN(\ell) \subseteq N(\ell) \tag{5}$$

Assume $\{(g, \hbar I)\} \in N(\ell)$. Since *N(H)* is regular so there exists an element $(\mathrm{r}, \mathrm{s}I)$ such that

$$\{(g, \hbar I)\} \in [(g, \hbar I)o(\mathrm{r}, \mathrm{s}I)]o(g, \hbar I) \in [N(\ell)oN(H)]oN(\ell) \subseteq N(\ell)oN(\ell) = N(\ell)^2$$

Implies that

$$N(\ell) \subseteq N(\ell)^2 \tag{6}$$

Hence, from equation (5) and (6) we have, $N(\ell) = N(\ell)^2$.

As $N(\ell)$ is chosen arbitrary neutrosophic hyper ideal of *N(H)* So, every hyper ideal s of a regular neutrosophic LA-semihypergroup is neutrosophic idempotent.

Lemma 3.2: Every neutrosophic right hyper ideal of a regular neutrosophic LA-semihypergroup is neutrosophic hyper ideal .

Proof: The proof is easy to follow.

Lemma 3.3: A neutrosophic LA-semihypergroup with neutrosophic left identity is intra-regular if

$N(H)o(g, \ell I) \subseteq N(H)$ or $(g, \ell I)oN(H) \subseteq N(H)$ for all $\{(g, \ell I)\} \in H \cup I$.

Proof: Assume that *N(H)* is neutrosophic LA-semihypergroup and

$$N(H)o(g, \ell I) \subseteq N(H)$$

hold for all $(g, \ell I) \in H \cup I$. Now we need to show that *N(H)* is intra-regular. For this let $\{(g, \ell I)\} \in N(H)$ such that

$$\{(g,\ell I)\} \in [(\mathcal{I},\hbar I)o(\mathrm{r},\mathrm{s}I)] \forall (\mathcal{I},\hbar I),(\mathrm{r},\mathrm{s}I) \in H \cup I$$
$$\in [[N(H)o(g,\ell I)]o[N(H)o(g,\ell I)]]oN(H)$$

for all

$$\{(g,\ell I)\} \in H \cup I \in [[N(H)oN(H)]o[(g,\ell I)o(g,\ell I)]]oN(H)$$
$$=[[N(H)oN(H)]o(g,\ell I)^2]oN(H)=[N(H)o(g,\ell I)^2]oN(H).$$

This prove that *N(H)* is intra-regular neutrosophic LA-semihypergroup.
Second part is like prove if $(g,\ell I)oN(H) \subseteq N(H)$ for all $(g,\ell I) \in H \cup I$.

Lemma 3.4: Every neutrosophic left hyper ideal of an intra-regular neutrosophic LA-semihypergroup with neutrosophic pure left identity is neutrosophic idempotent.

Proof: *Easy to prove.*

Lemma 3.5: If $N(\mathcal{B})$ is a neutrosophic bi-hyper ideal (generalized bi-hyper ideal) of a regular and intra-regular neutrosophic LA-semihypergroup *N(H)* with neutrosophic (left, or pure left) identity, then

$$[N(\mathcal{B})oN(H)]oN(\mathcal{B})=N(\mathcal{B})$$

Proof: Let $N(\mathcal{B})$ be a neutrosophic bi-hyper ideal (neutrosophic generalized bi-hyper ideal) of *N(H)*, then by definition

$$[N(\mathcal{B})oN(H)]oN(\mathcal{B})=N(\mathcal{B}) \tag{7}$$

Now let $\{(g,\ell I)\} \in N(\mathcal{B})$, since *N(H)* is regular and intra-regular so there exists an element $(\mathcal{I},\hbar I) \in H \cup I$ such that

$$\{(g,\ell I)\} \in [(g,\ell I)o(\mathcal{I},\hbar I)]o(g,\ell I)$$

Therefore,

$$\{(g,\ell I)\} \in [N(\mathcal{B})oN(H)]oN(\mathcal{B})$$

This implies that

$$N(\mathcal{B}) \subseteq [N(\mathcal{B}) o N(H)] o N(\mathcal{B}) \tag{8}$$

Thus, from (7) and (8) we get $[N(\mathcal{B}) o N(H)] o N(\mathcal{B}) = N(\mathcal{B})$.

Corollary 3.1: If *N*(*H*) is intra-regular neutrosophic LA-semihypergroup with neutrosophic left identity, then

$$[N(\mathcal{B}) o N(H)] o N(\mathcal{B}) = N(\mathcal{B}) \cap N(H)$$

where, $N(\mathcal{B})$ is neutrosophic bi- (generalized bi-) hyper ideal of *N*(*H*). But the converse is not true in general.

Theorem 3.1: If *N*(*H*) is intra-regular neutrosophic LA-semihypergroup with neutrosophic left identity, then

$$[N(H) o N(\mathcal{I})] o N(H) = N(H) \cap N(\mathcal{I})$$

where $N(\mathcal{I})$ is neutrosophic interior hyper ideal of *N*(*H*).

Proof: Let *N*(*H*) be an intra-regular neutrosophic LA-semihypergroup with neutrosophic left identity, then clearly

$$[N(H) o N(\mathcal{I})] o N(H) \subseteq N(H) \cap N(\mathcal{I}) \tag{9}$$

Now let $\{(g, \hbar I)\} \in N(H) \cap N(\mathcal{I})$ then $\{(g, \hbar I)\} \in N(H)$ and $\{(g, \hbar I)\} \in N(\mathcal{I})$. Since *N*(*H*) is an intra-regular by definition there exist $(\mathrm{r}, \mathrm{s}I), (\mathcal{I}, \hbar I) \in H \cup I$ such that

$$\{(g, \hbar I)\} \in \left[(\mathrm{r}, \mathrm{s}I) o (g, \hbar I)^2\right] o (\mathcal{I}, \hbar I)$$

Now by using neutrosophic paramedial and neutrosophic left invertive law, we have

$$\{(g,\hbar I)\}\in\left[(\mathrm{r},\mathrm{s}I)o(g,\hbar I)^2\right]o(\mathcal{I},\hbar I)\in\left[(5,5I)o(\mathrm{r},\mathrm{s}I)\right]o\left[(g,\hbar I)o(g,\hbar I)\right]$$
$$o(\mathcal{I},\hbar I)\in\left[(g,\hbar I)o(g,\hbar I)\right]o\left[(\mathrm{r},\mathrm{s}I)o(5,5I)\right]o(\mathcal{I},\hbar I)\in\left[(g,\hbar I)o(\mathrm{r},\mathrm{s}I)\right]$$
$$o\left[(g,\hbar I)o(5,5I)\right]o(\mathcal{I},\hbar I)\in\left[(5,5I)o(g,\hbar I)\right]o\left[(\mathrm{r},\mathrm{s}I)o(g,\hbar I)\right]o(\mathcal{I},\hbar I)\in(g,\hbar I)$$
$$o\left[(\mathrm{r},\mathrm{s}I)o(g,\hbar I)\right]o(\mathcal{I},\hbar I)\in\left[N(H)oN(\mathcal{I})\right]oN(H)$$
$$\Rightarrow N(H)\cap N(\mathcal{I})\subseteq\left[N(H)oN(\mathcal{I})\right]oN(H) \tag{10}$$

Hence from (9) and (10) we get

$$\left[N(H)oN(\mathcal{I})\right]oN(H)=N(H)\cap N(\mathcal{I}).$$

The converse is not true in general. Which is follow from Example 6.3.

Corollary 3.2: If *N(H)* is intra-regular neutrosophic LA-semihypergroup with neutrosophic left identity, then

$$\left[N(H)oN(\mathcal{I})\right]oN(H)=N(\mathcal{I})$$

where, $N(\mathcal{I})$ is neutrosophic interior hyper ideal of *N(H)*.

Lemma 3.6: Every neutrosophic bi-hyper ideal of a regular neutrosophic LA-semihypergroup *N(H)* with neutrosophic left identity is neutrosophic quasi hyper ideal .

Proof: Let $N(\mathcal{B})$ be a neutrosophic bi-hyper ideal of *N(H)* and

$$\{(g,\hbar I)\}o\{(\ell,\mathcal{I}I)\}\in N(H)oN(\mathcal{B})$$

for some $\{(g,\hbar I)\}\in N(H)$ and $\{(\ell,\mathcal{I}I)\}\in N(\mathcal{B})$. Thus by supposition $\exists(\mathrm{r},\mathrm{s}I)\in H\cup I$ such that

$$\{(g,\hbar I)\}\in\left[(g,\hbar I)o(\mathrm{r},\mathrm{s}I)\right](g,\hbar I)$$

Then by using

$\left[g o\hbar\right]o\left[\mathcal{I}o\ell\right]=\left[\ell o\mathcal{I}\right]o\left[\hbar o g\right]$ for all $g,\hbar,\mathcal{I},\ell\in H\cup I$

We have

$$\begin{aligned}\{(g,\hbar I)\}o\{(\ell,\mathcal{I}I)\} &\in \{(g,\hbar I)\}o\left[\left[(\ell,\mathcal{I}I)o(\mathrm{r},\mathrm{s}I)\right]o(\ell,\mathcal{I}I)\right]\\ &\in \left[(\ell,\mathcal{I}I)o(\mathrm{r},\mathrm{s}I)\right]o\left[(g,\hbar I)o(\ell,\mathcal{I}I)\right]\\ &\in \left[(\ell,\mathcal{I}I)o(g,\hbar I)\right]o\left[(\mathrm{r},\mathrm{s}I)o(\ell,\mathcal{I}I)\right]\\ &\in \left[N(\mathcal{B})oN(H)\right]o\left[N(H)oN(\mathcal{B})\right]\\ &\subseteq \left[N(\mathcal{B})oN(H)\right]oN(\mathcal{B})\subseteq N(\mathcal{B})\end{aligned}$$

Hence

$$N(H)oN(\mathcal{B})\subseteq N(\mathcal{B}) \tag{11}$$

Similarly, we can also prove that

$$N(\mathcal{B})oN(H)\subseteq N(\mathcal{B}) \tag{12}$$

Hence from (11) and (12) we get $N(H)oN(\mathcal{B})\cap N(\mathcal{B})oN(H)\subseteq N(\mathcal{B})$

Theorem 3.2: In a neutrosophic LA-semihypergroup *N*(*H*) with neutrosophic pure left identity, the following parts are equivalent:

(i) *N*(*H*) is intra-regular.
(ii) $N(\ell)=\left[N(H)oN(\ell)\right]^2$, where *N*(ℓ) is neutrosophic right hyper ideal of *N*(*H*).

Proof: (i)⇒(ii) assume that *N*(ℓ) be a neutrosophic right hyper ideal of an intra-regular neutrosophic LA-semihypergroup *N*(*H*) with neutrosophic pure right identity. Then by definition

$$N(H)oN(\ell)\subseteq N(\ell) \tag{13}$$

And by Lemma 3.4 we have

$$\left[N(H)oN(\ell)\right]^2=N(H)oN(\ell)\subseteq N(\ell)\ \text{ by (13)} \tag{14}$$

Now

$$N(\ell)=N(\ell)oN(\ell)\subseteq N(H)oN(\ell)=\left[N(H)oN(\ell)\right]^2 \tag{15}$$

Hence equation (14) and (15) implies that

$$N(\ell)=\left[N(H)oN(\ell)\right]^{2} \tag{16}$$

(ii)⇒(i) Let $N(\ell)$ be a neutrosophic right hyper ideal of *N*(*H*). Then by definition

$$N(H)oN(\ell)\subseteq N(\ell)$$

Clearly

$$\left[N(\ell)\right]^{2}=N(\ell)oN(\ell)\subseteq N(H)oN(\ell)\subseteq N(\ell) \tag{17}$$

Now from (ii)

$$N(\ell)=\left[N(H)oN(\ell)\right]^{2}\subseteq\left[N(\ell)\right]^{2}\text{ by (16)} \tag{18}$$

Hence (17) and (18) complete the proof.

Theorem 3.3: For a neutrosophic LA-semihypergroup *N*(*H*) with neutrosophic pure left identity, the following are equivalent.

(i) *N*(*H*) is intra-regular neutrosophic LA-semihypergroup.
(ii) For neutrosophic quasi hyper ideal s $N(\mathcal{Q})$ and $N(\mathcal{K})$,

$$N(\mathcal{Q})\cap N(\mathcal{K})=\left[N(\mathcal{Q})oN(\mathcal{K})\right]oN(\mathcal{Q})$$

Proof: (i)⇒(ii) Let $N(\mathcal{Q})$ and $N(\mathcal{K})$ are two neutrosophic quasi hyper ideal s of *N*(*H*). Then obviously by definition it is neutrosophic hyper ideal s so

$$\left[N(\mathcal{Q})oN(\mathcal{K})\right]oN(\mathcal{Q})\subseteq\left[N(\mathcal{Q})oN(H)\right]oN(H)\subseteq N(\mathcal{Q})oN(H)\subseteq N(\mathcal{Q}) \tag{19}$$

And

$$\left[N(\mathcal{Q})oN(\mathcal{K})\right]oN(\mathcal{Q})\subseteq\left[N(H)oN(\mathcal{K})\right]oN(H)\subseteq N(\mathcal{K})oN(H)\subseteq N(\mathcal{K}) \tag{20}$$

Thus, from equation (19) and (20) we get

$$\left[N(\mathcal{Q})oN(\mathcal{K})\right]oN(\mathcal{Q})\subseteq N(\mathcal{Q})\cap N(\mathcal{K}) \tag{21}$$

Also, we know that the intersection of neutrosophic hyper ideal s is also a neutrosophic hyper ideal. Therefore $N(\mathcal{Q})\cap N(\mathcal{K})$ becomes a neurosophic hyper ideal. Now by using Lemma 3.4, we have

$$\begin{aligned} N(\mathcal{Q})\cap N(\mathcal{K}) &= \left[N(\mathcal{Q})\cap N(\mathcal{K})\right]^2 \\ &= \left[N(\mathcal{Q})\cap N(\mathcal{K})\right]o\left[N(\mathcal{Q})\cap N(\mathcal{K})\right] \\ &= \left[\left[N(\mathcal{Q})\cap N(\mathcal{K})\right]o\left[N(\mathcal{Q})\cap N(\mathcal{K})\right]\right]o\left[N(\mathcal{Q})\cap N(\mathcal{K})\right] \\ &= \left[N(\mathcal{Q})oN(\mathcal{K})\right]oN(\mathcal{Q}) \end{aligned} \tag{22}$$

Hence from (21) and (22) we get the required result.

(ii)⇒(i) Let $N(\mathcal{K})$ be a neutrosophic quasi hyper ideal of *N*(*H*) then by (ii), we get

$$\begin{aligned} N(\mathcal{K}) &= N(\mathcal{K})\cap N(\mathcal{K}) \\ &= \left[N(\mathcal{K})oN(\mathcal{K})\right]oN(\mathcal{K}) \\ &= N(\mathcal{K})^2 oN(\mathcal{K}) \\ &= N(\mathcal{K})oN(\mathcal{K}) \\ &= N(\mathcal{K})^2 \end{aligned}$$

Hence $N(\mathcal{K})$ is idempotent.

Let $\{(\mathcal{g},\hbar I)\}\in N(H)$. Since $N(H)o\{(\mathcal{g},\hbar I)\}$ is a neutrosophic quasi hyper ideal, therefore

$$\begin{aligned} \{(\mathcal{g},\hbar I)\} &\in N(H)o\{(\mathcal{g},\hbar I)\} \\ &= \left[N(H)o\{(\mathcal{g},\hbar I)\}\right]^2 \end{aligned}$$

Hence *N*(*H*) is intra-regular neutrosophic LA-semihypergroup.

Theorem 3.4: For a neutrosophic LA-semihypergroup with neutrosophic (left, right) identity the following parts are equivalent:

(i) *N*(*H*) is regular.

(ii) $N(\mathcal{R})oN(\mathcal{L})=N(\mathcal{R})\cap N(\mathcal{L})$, where $N(\mathcal{R})$ and $N(\mathcal{L})$ are any neutrosophic right and left hyper ideal s of *N*(*H*).

(iii) If $N(\mathcal{R})$ and $N(\mathcal{L})$ are any neutrosophic right and left hyper ideal s of *N*(*H*). Then

  (a) $[N(\mathcal{R})]^2=N(\mathcal{R})$.

  (b) $[N(\mathcal{L})]^2=N(\mathcal{L})$.

  (c) $N(\mathcal{R})oN(\mathcal{L})$ is a neutrosophic quasi hyper ideal of *N*(*H*) .

(iv) The set of all quasi hyper ideal s of *N*(*H*) is a regular neutrosophic LA-semihypergroup.

(v) $[N(\mathcal{Q})oN(H)]oN(\mathcal{Q})=N(\mathcal{Q})$, where $N(\mathcal{Q})$ is any neutrosophic quasi hyper ideal of *N*(*H*).

Proof: (i)⇒(ii) Let $N(\mathcal{R})$ and $N(\mathcal{L})$ are neutrosophic right and neutrosophic left hyper ideal s of *N*(*H*), then

$$N(\mathcal{R})oN(\mathcal{L})\subseteq N(\mathcal{R})oN(H)\subseteq N(\mathcal{R}) \tag{23}$$

And

$$N(\mathcal{R})oN(\mathcal{L})\subseteq N(H)oN(\mathcal{L})\subseteq N(\mathcal{L}) \tag{24}$$

This implies that

$$N(\mathcal{R})oN(\mathcal{L})\subseteq N(\mathcal{R})\cap N(\mathcal{L}) \tag{25}$$

Again let

$$\{(\mathcal{g},\hbar I)\}\in N(\mathcal{R})\cap N(\mathcal{L})$$

Then $\{(\mathcal{g},\hbar I)\}\in N(\mathcal{R})$ and $\{(\mathcal{g},\hbar I)\}\in N(\mathcal{L})$. Since *N*(*H*) is regular so there exist $(\mathcal{I},\hbar I)\in H\cup I$. Such that

$$\{(\mathcal{g},\hbar I)\}\in[(\mathcal{g},\hbar I)o(\mathcal{I},\hbar I)]o(\mathcal{g},\hbar I)$$

Since

$$\{(\mathcal{g},\hbar I)\}\in N(\mathcal{L})$$

And

$$(\mathcal{g},\hbar I)o(\mathcal{I},\hbar I)\in N(\mathcal{R})$$

Thus

$$\{(\mathcal{g},\hbar I)\}\in[(\mathcal{g},\hbar I)o(\mathcal{I},\hbar I)]o(\mathcal{g},\hbar I)\in N(\mathcal{R})oN(\mathcal{L})$$

This implies that

$$N(\mathcal{R})\cap N(\mathcal{L})\subseteq N(\mathcal{R})oN(\mathcal{L}) \tag{26}$$

Hence from (25) and (26) we get

$$N(\mathcal{R})oN(\mathcal{L})=N(\mathcal{R})\cap N(\mathcal{L})$$

So $N(\mathcal{R})oN(\mathcal{L})$ is a neutrosophic quasi hyper ideal . The two-sided neutrosophic hyper ideal of *N*(*H*) generated by $N(\mathcal{R})$ is $N(\mathcal{R})\cup N(\mathcal{R})oN(H)$ so from (ii) it follows that

$$\begin{aligned} N(\mathcal{R}) &= N(\mathcal{R})o[N(\mathcal{R})\cup N(\mathcal{R})oN(H)] \\ &= N(\mathcal{R})oN(\mathcal{R})\cup N(\mathcal{R})o[N(\mathcal{R})oN(H)] \\ &= N(\mathcal{R})^2\cup N(\mathcal{R})o[N(\mathcal{R})oN(H)] \\ &\subseteq N(\mathcal{R})^2\cup[N(\mathcal{R})oN(\mathcal{R})] \\ &= N(\mathcal{R})^2\cup N(\mathcal{R})^2=N(\mathcal{R})^2 \end{aligned}$$

This implies that

$$N(\mathcal{R})\subseteq N(\mathcal{R})^2 \tag{27}$$

Obviously

$$\begin{aligned} N(\mathcal{R})^2 &= N(\mathcal{R})oN(\mathcal{R})\subseteq N(\mathcal{R})oN(H)\subseteq N(\mathcal{R}) \\ &\Rightarrow N(\mathcal{R})^2\subseteq N(\mathcal{R}) \end{aligned} \tag{28}$$

Hence from (27) and (28) we get $N(\mathcal{R})=N(\mathcal{R})^2$

Similarly, we can also prove that $N(\mathcal{L}) = N(\mathcal{L})^2$.

(iii)⇒(v) If $N(\mathcal{Q})$ is neutrosophic quasi hyper ideal of *N*(*H*) then $N(\mathcal{Q}) \cup N(H) o N(\mathcal{Q})$ generated by $N(\mathcal{Q})$ is a neutrosophic left hyper ideal of *N*(*H*). By (iii) part (b) we have

$$\begin{aligned} N(\mathcal{Q}) &\subseteq N(\mathcal{Q}) \cup N(H) o N(\mathcal{Q}) = \left[N(\mathcal{Q}) \cup N(H) o N(\mathcal{Q})\right]^2 \\ &= \left[N(\mathcal{Q}) \cup N(H) o N(\mathcal{Q})\right] o \left[N(\mathcal{Q}) \cup N(H) o N(\mathcal{Q})\right] \\ &= N(\mathcal{Q}) o \left[N(\mathcal{Q}) \cup N(H) o N(\mathcal{Q})\right] \cup N(H) o N(\mathcal{Q}) o \left[N(\mathcal{Q}) \cup N(H) o N(\mathcal{Q})\right] \\ &= N(\mathcal{Q})^2 \cup \left[N(H) o N(\mathcal{Q})\right]^2 \\ &= N(\mathcal{Q}) \cup \left[N(H) o N(\mathcal{Q})\right] \subseteq N(H) o N(\mathcal{Q}) \end{aligned}$$

In the same way we can also show that $N(\mathcal{Q}) \subseteq N(\mathcal{Q}) o N(H)$

So, from these two relations we get $N(\mathcal{Q}) \subseteq N(\mathcal{Q}) o N(H) \cup N(H) o N(\mathcal{Q})$

(v)⇒(iii) part (c) directly from (i) and (iii)

$$N(\mathcal{R}) o N(\mathcal{L}) = N(H) o \left[N(\mathcal{R}) o N(\mathcal{L})\right] \cap \left[N(\mathcal{R}) o N(\mathcal{L})\right] o N(H)$$

for every neutrosophic right hyper ideal $N(\mathcal{R})$ and neutrosophic left hyper ideal $N(\mathcal{L})$ of *N*(*H*) respectively. Furthermore from (iii) part (a) and part (b) we get the following relations

$$N(H) o \left[N(\mathcal{Q}) o N(\mathcal{K})\right] = \left[N(H) o \left[N(\mathcal{Q}) o N(\mathcal{K})\right]\right] o \left[N(H) o \left[N(\mathcal{Q}) o N(\mathcal{K})\right]\right]$$

And

$$\left[N(\mathcal{Q}) o N(\mathcal{K})\right] o N(H) = \left[\left[N(\mathcal{Q}) o N(\mathcal{K})\right] o N(H)\right] o \left[\left[N(\mathcal{Q}) o N(\mathcal{K})\right] o N(H)\right]$$

Thus $N(\mathcal{Q}) o N(\mathcal{K})$ is neutrosophic quasi hyper ideal of *N*(*H*).

(iv)⇒(v) Let $N(\mathcal{Q})$ be a neutrosophic quasi hyper ideal of *N*(*H*). Then by (iv) there exist a neutrosophic quasi hyper ideal $N(\mathcal{K})$ of *N*(*H*) such that

$$\begin{aligned} N(\mathcal{Q}) &= \left[N(\mathcal{Q}) o N(\mathcal{K})\right] o N(\mathcal{Q}) \\ &\subseteq \left[N(\mathcal{Q}) o N(H)\right] o N(\mathcal{Q}) \\ &\subseteq \left[N(\mathcal{Q}) o N(H)\right] o \left[N(\mathcal{Q}) o N(H)\right] \end{aligned}$$

From supposition and these relations, we have

$$N(\mathcal{Q})=\left[N(\mathcal{Q})oN(H)\right]oN(\mathcal{Q})$$

(v)⇒(i) The intersection $(g,\hbar I)\cap(\mathrm{r},\mathrm{s}I)$ of neutrosophic principal right hyper ideal $(g,\hbar I)$ and left hyper ideal $(\mathrm{r},\mathrm{s}I)$ by a single neutrosophic element $(g,\hbar I)$ is a neutrosophic quasi hyper ideal of *N*(*H*). So by (v) we have

$$\begin{aligned}(g,\hbar I)\cap(\mathrm{r},\mathrm{s}I)&=\left[\left[(g,\hbar I)\cap(\mathrm{r},\mathrm{s}I)\right]oN(H)\right]o\left[N(H)o\left[(g,\hbar I)\cap(\mathrm{r},\mathrm{s}I)\right]\right]\\&\in\left[(g,\hbar I)oN(H)\right]o(\mathrm{r},\mathrm{s}I)\end{aligned}$$

Also

$$\begin{aligned}(\mathrm{r},\mathrm{s}I)\cap(g,\hbar I)&=\left[\left[(\mathrm{r},\mathrm{s}I)\cap(g,\hbar I)\right]oN(H)\right]o\left[N(H)o\left[(\mathrm{r},\mathrm{s}I)\cap(g,\hbar I)\right]\right]\\&\in\left[(\mathrm{r},\mathrm{s}I)oN(H)\right]o(g,\hbar I)\end{aligned}$$

Since $(g,\hbar I),(\mathrm{r},\mathrm{s}I)\in(g,\hbar I)\cap(\mathrm{r},\mathrm{s}I)$ so

$$(g,\hbar I)\in\left[(g,\hbar I)oN(H)\right]o(\mathrm{r},\mathrm{s}I)=\left[(g,\hbar I)oN(H)\right]o(g,\hbar I)$$

So, any neutrosophic element of *N*(*H*) is regular. Hence *N*(*H*) is reglular neutrosophic LA-Semihypergroup.

Theorem 3.5: For a neutrosophic LA-smeihypergroup *N*(*H*) with neutrosophic left identity the following assertion are equivalent:

(i) *N*(*H*) is regular neutrosophic LA-semihypergroup.
(ii) For all neutrosophic hyper ideals $N(\mathcal{I})$ and neutrosophic bi-hyper ideals $N(\mathcal{B})$ of *N*(*H*).

$$N(\mathcal{I})\cap N(\mathcal{B})=\left[N(\mathcal{B})oN(\mathcal{I})\right]oN(\mathcal{B})$$

(iii) For all neutrosophic hyper ideals $N(\mathcal{I})$ and neutrosophic quasi hyper ideal s $N(\mathcal{Q})$ of *N*(*H*).

$$N(\mathcal{I})\cap N(\mathcal{Q})=\left[N(\mathcal{Q})oN(\mathcal{I})\right]oN(\mathcal{Q})$$

Proof: (i)⇒(ii) Suppose $N(\mathcal{I})$ and $N(\mathcal{B})$ are neutrosophic hyper ideal and bi-hyper ideal of a $N(H)$ respectively, then

$N(\mathcal{I})oN(H)\subseteq N(\mathcal{I})$ and $N(H)oN(\mathcal{I})\subseteq N(\mathcal{I})$

Also

$$\begin{aligned}\left[N(\mathcal{B})oN(\mathcal{I})\right]oN(\mathcal{B}) &\subseteq \left[N(H)oN(\mathcal{I})\right]oN(\mathcal{B})\\ &\subseteq N(\mathcal{I})oN(\mathcal{B})\\ &\subseteq N(H)oN(\mathcal{I})\\ &\subseteq N(\mathcal{I})\end{aligned} \tag{29}$$

And

$$\left[N(\mathcal{B})oN(\mathcal{I})\right]oN(\mathcal{B})\subseteq\left[N(\mathcal{B})oN(H)\right]oN(\mathcal{B})\subseteq N(\mathcal{B}) \tag{30}$$

From equation (29) and (30) we get

$$\left[N(\mathcal{B})oN(\mathcal{I})\right]oN(\mathcal{B})\subseteq N(\mathcal{I})\cap N(\mathcal{B}) \tag{31}$$

Now Let $\{(g,\hbar I)\}\in N(\mathcal{I})\cap N(\mathcal{B})$ then $\{(g,\hbar I)\}\in N(\mathcal{I})$ and $\{(g,\hbar I)\}\in N(\mathcal{B})$. So $\{(g,\hbar I)\}\in\left[(g,\hbar I)o(\mathrm{r},\mathrm{s}I)\right]o(g,\hbar I)$ for some $(\mathrm{r},\mathrm{s}I)\in H\cup I$ where $g,\hbar,\mathrm{r},\mathrm{s}\in(H,*)$

$$\{(g,\hbar I)\}\in\left[(g,\hbar I)o(\mathrm{r},\mathrm{s}I)\right]o(g,\hbar I)\in\left[N(\mathcal{B})oN(H)\right]oN(\mathcal{B}) \tag{32}$$

Now

$$\begin{aligned}\{(g,\hbar I)\}&\in\left[(g,\hbar I)o(\mathrm{r},\mathrm{s}I)\right]o(g,\hbar I)\in\left[(g,\hbar I)o(\mathrm{r},\mathrm{s}I)\right]o\left[(g,\hbar I)o(\mathrm{r},\mathrm{s}I)\right]o(g,\hbar I)\\ &=\left[\left[(g,\hbar I)o(\mathrm{r},\mathrm{s}I)\right]o(\mathrm{r},\mathrm{s}I)\right]o(g,\hbar I)]o(g,\hbar I)\\ &\in\left[\left[N(\mathcal{I})oN(H)\right]oN(H)\right]oN(\mathcal{B})]oN(\mathcal{B})\\ &=\left[N(\mathcal{I})oN(H)\right]oN(\mathcal{B})]oN(\mathcal{B})=\left[N(\mathcal{B})oN(\mathcal{B})\right]o\left[N(\mathcal{I})oN(H)\right]\\ &=\left[N(\mathcal{B})oN(\mathcal{I})\right]o\left[N(\mathcal{B})oN(H)\right]=\left[N(\mathcal{B})oN(\mathcal{I})\right]oN(\mathcal{B})\\ &\Rightarrow N(\mathcal{I})\cap N(\mathcal{B})\subseteq\left[N(\mathcal{B})oN(\mathcal{I})\right]oN(\mathcal{B})\end{aligned} \tag{33}$$

From (32) and (33) we get $N(\mathcal{I})\cap N(\mathcal{B})=[N(\mathcal{B})oN(\mathcal{I})]oN(\mathcal{B})$.

(ii)⇒(iii) Let $N(\mathcal{I})$ and $N(\mathcal{Q})$ be a neutrosophic hyper ideal and quasi hyper ideal of $N(H)$. Since every neutrosophic quasi hyper ideal is a neutrosophic bi-hyper ideal so therefore

$$[N(\mathcal{Q})oN(H)]oN(\mathcal{Q})\subseteq N(\mathcal{Q}) \tag{34}$$

And

$$\begin{aligned}&N(\mathcal{I})oN(H)\subseteq N(\mathcal{I}) \text{ also } N(H)oN(\mathcal{I})\subseteq N(\mathcal{I})\\ &\Rightarrow[N(\mathcal{Q})oN(\mathcal{I})]oN(\mathcal{Q})\subseteq[N(\mathcal{Q})oN(H)]oN(\mathcal{Q})\subseteq N(\mathcal{Q}) \text{ by}(34)\end{aligned} \tag{35}$$

Also

$$\begin{aligned}&[N(\mathcal{Q})oN(\mathcal{I})]oN(\mathcal{Q})\subseteq[N(H)oN(\mathcal{I})]oN(\mathcal{Q})\\ &\qquad\subseteq N(\mathcal{I})oN(\mathcal{Q})\subseteq N(\mathcal{I})oN(H)\subseteq N(\mathcal{I}) \text{ by}(33)\end{aligned} \tag{36}$$

So

$$[N(\mathcal{Q})oN(\mathcal{I})]oN(\mathcal{Q})\subseteq N(\mathcal{I}) \tag{37}$$

Hence from (34) and (36) we have

$$[N(\mathcal{Q})oN(\mathcal{I})]oN(\mathcal{Q})\subseteq N(\mathcal{I})\cap N(\mathcal{Q}) \tag{38}$$

From (ii)

$$N(\mathcal{I})\cap N(\mathcal{Q})\subseteq[N(\mathcal{Q})oN(\mathcal{I})]oN(\mathcal{Q}) \tag{39}$$

(37) and (38) implies

$$N(\mathcal{I})\cap N(\mathcal{Q})=[N(\mathcal{Q})oN(\mathcal{I})]oN(\mathcal{Q})$$

(iii)⇒(i) Let $N(\mathcal{Q})$ be a neutrosophic quasi hyper ideal of *N*(*H*). Therefore by (iii),

$$N(H)\cap N(\mathcal{Q})=\left[N(\mathcal{Q})oN(H)\right]oN(\mathcal{Q})\Rightarrow N(\mathcal{Q})=\left[N(\mathcal{Q})oN(H)\right]oN(\mathcal{Q})$$

because $N(\mathcal{Q})\subseteq N(H)$

So, by Theorem 3.4-part v *N*(*H*) is regular neutroophic LA-semihypergroup.

Theorem 3.6: For a neutrosophic LA-semihypergroup *N*(*H*) with neutrosophic pure identity the following parts are equivalent:

(i) *N*(*H*) is both neutrosophic regular and intra-regular.
(ii) For any neutrosophic bi-hyper ideal $N(\mathcal{B})$ of *N*(*H*), we have $\left[N(\mathcal{B})\right]^2=N(\mathcal{B})$.
(iii) For any neutrosophic quasi hyper ideal $N(\mathcal{Q})$ of *N*(*H*), we have $\left[N(\mathcal{Q})\right]^2=N(\mathcal{Q})$

Proof: (i)⇒(ii) Assume that $N(\mathcal{B})$ is a neutrosophic bi-hyper ideal of *N*(*H*).

So, by using theorem 3.5 part (iii), we have

$$N(\mathcal{B})\subseteq\left[N(\mathcal{B})oN(H)\right]oN(\mathcal{B}) \tag{40}$$

Now

$$\begin{aligned}\left[N(\mathcal{B})\right]^2&=N(\mathcal{B})oN(\mathcal{B})\\&=N(\mathcal{B})o\left[N(\mathcal{B})oN(H)\right]oN(\mathcal{B})\ \text{by}(37)\\&\subseteq N(\mathcal{B})\end{aligned} \tag{41}$$

Consider $\{(g,\hbar I)\}\in N(\mathcal{B})$. Since *N*(*H*) is regular therefore

$$\{(g,\hbar I)\}\in\left[(g,\hbar I)o(\mathsf{r},\mathsf{s}I)\right]o(g,\hbar I)$$

for some $(\ell,5I),(\mathsf{r},\mathsf{s}I)\in H\cup I$ where $\ell,5,\mathsf{r},\mathsf{s}\in(H,*)$. Also *N*(*H*) is intra regular so $\{(g,\hbar I)\}\in\left[(\mathsf{r},\mathsf{s}I)o(g,\hbar I)^2\right]o(\ell,5I)$

Now

$$\begin{aligned}
\{(g,\hbar I)\} &\in [(g,\hbar I)o(\mathrm{r},\mathrm{s}I)]o(g,\hbar I)\\
&\in [[(g,\hbar I)o(\mathrm{r},\mathrm{s}I)]o(g,\hbar I)]o(\mathrm{r},\mathrm{s}I)]o(g,\hbar I)\\
&= [(\mathrm{r},\mathrm{s}I)o(g,\hbar I)]o[(g,\hbar I)o(\mathrm{r},\mathrm{s}I)]]o(g,\hbar I)\\
&= [(g,\hbar I)o[(g,\hbar I)o(\mathrm{r},\mathrm{s}I)]]o[(\mathrm{r},\mathrm{s}I)o(g,\hbar I)]\\
&\in N(\mathcal{B})o[N(\mathcal{B})oN(H)]o[N(H)oN(\mathcal{B})] \qquad (42)\\
&\subseteq N(\mathcal{B})o[N(\mathcal{B})oN(H)]oN(\mathcal{B})\\
&\subseteq N(\mathcal{B})oN(\mathcal{B})\\
&= N(\mathcal{B})^2\\
&\Rightarrow N(\mathcal{B}) \subseteq N(\mathcal{B})^2
\end{aligned}$$

Consequently, from equation (41) and (42) we get $N(\mathcal{B}) = N(\mathcal{B})^2$
Similarly, (i)⇒(iii) and (iii)⇒(i) are easily follow from theorem 3.5 part (iii).

Theorem 3.7: For neutrosophic intra-regular neutrosophic LA-semihypergroup *N(H)* with neutrosophic left identity, the following statements are coinciding.

(i) $N(\mathcal{L})$ is a neutrosophic left hyper ideal of *N(H)*.
(ii) $N(\mathcal{L})$ is a neutrosophic right hyper ideal of *N(H)*.
(iii) $N(\mathcal{L})$ is a neutrosophic hyper ideal of *N(H)*.
(iv) $N(\mathcal{L})$ is a neutrosophic bi hyper ideal of *N(H)*.
(v) $N(\mathcal{L})$ is a neutrosophic generalized bi hyper ideal of *N(H)*.
(vi) $N(\mathcal{L})$ is a neutrosophic interior hyper ideal of *N(H)*.
(vii) $N(\mathcal{L})$ is a neutrosophic quasi hyper ideal of *N(H)*.

Proof: Easy to prove.

4. SUBSTRUCTURES OF REGULAR NEUTROSOPHIC LA-SEMIHYPERGROUPS

In this section, we define different Substructures of regular neutrosophic LA-semihypergroup namely (weakly, left, right, strongly, completely) regular neutrosophic LA-semihypergroup with some basic results. Further, invertible neutrosophic LA-semihypergroup are also define in the last.

Definition 4.1: A neutrosophic element $\{(g,\hbar I)\}$ of a neutrosophic LA-semihypergroup $N(H) = (\langle H \cup I\rangle, o)$ is called a weakly regular neutrosophic element if

$$\{(\mathscr{g},\hbar I)\} \in [(\mathscr{g},\hbar I) o (\mathsf{r},\mathsf{s}I)] o (\mathscr{g},\hbar I)$$

for some $(\mathsf{r},\mathsf{s}I) \in H \cup I$, where $\mathscr{g},\hbar,\mathsf{r},\mathsf{s} \in (H,*)$ and *N*(*H*) is called a weakly regular neutrosophic LA-semihypergroup if all of its neutrosophic element is weakly regular neutrosophic element.

Definition 4.2: A neutrosophic element $\{(\mathscr{g},\hbar I)\}$ of a neutrosophic LA-semihypergroup *N*(*H*) is called a right regular neutrosophic element if

$$\{(\mathscr{g},\hbar I)\} \in [(\mathsf{r},\mathsf{s}I) o (\mathscr{g},\hbar I)] o (\mathscr{g},\hbar I)$$

For some $(\mathsf{r},\mathsf{s}I) \in \langle H \cup I \rangle$ and *N*(*H*) is called a right regular neutrosophic LA-semihypergroup if all of its neutrosophic elements is right regular neutrosophic element.

Definition 4.3: A neutrosophic element $\{(\mathscr{g},\hbar I)\}$ of a neutrosophic LA-semihypergroup *N*(*H*) is called a left regular neutrosophic element if

$$\{(\mathscr{g},\hbar I)\} \in (\mathscr{g},\hbar I) o [(\mathscr{g},\hbar I) o (\mathsf{r},\mathsf{s}I)]$$

For some $(\mathsf{r},\mathsf{s}I) \in \langle H \cup I \rangle$ and *N*(*H*) is called a left regular neutrosophic LA-semihypergroup if all of its neutrosophic elements is left regular neutrosophic element.

Definition 4.4: A neutrosophic element of a neutrosophic LA-semihypergroup *N*(*H*) is called a completely regular neutrosophic element if it is regular, left regular and right regular neutrosophic element of *N*(*H*) and neutrosophic LA-semihypergroup is called a completely regular neutrosophic LA-semihypergroup if all of its elements are completely regular neutrosophic element.

Definition 4.5: A neutrosophic element $\{(\mathscr{g},\hbar I)\}$ of a neutrosophic LA-semihypergroup $N(H) = (\langle H \cup I \rangle, o)$ is called a strongly regular neutrosophic element if

$$\{(\mathscr{g},\hbar I)\} = [(\mathscr{g},\hbar I) o (\mathsf{r},\mathsf{s}I)] o (\mathscr{g},\hbar I)$$

For some $(\mathsf{r},\mathsf{s}I) \in \langle H \cup I \rangle$ where $\mathscr{g},\hbar,\mathsf{r},\mathsf{s} \in (H,*)$, and *N*(*H*) is called a strongly regular neutrosophic LA-semihypergroup if all of its neutrosophic element is strongly regular neutrosophic element.

Definition 4.6: A neutrosophic element $\{(\mathscr{g},\hbar I)\}$ of a neutrosophic LA-semihypergroup *N*(*H*) is called a strongly right regular neutrosophic element if

$$\{(\mathscr{g},\hbar I)\} = [(\mathsf{r},\mathsf{s}I) o (\mathscr{g},\hbar I)] o (\mathscr{g},\hbar I)$$

For some $(r, sI) \in \langle H \cup I \rangle$ and *N(H)* is called a strongly right regular neutrosophic LA-semihypergroup if all of its neutrosophic elements is strongly right regular neutrosophic element.

Definition 4.7: A neutrosophic element $\{(g, \hbar I)\}$ of a neutrosophic LA-semihypergroup *N(H)* is called a strongly left regular neutrosophic element if

$$\{(g, \hbar I)\} = (g, \hbar I) o \left[(g, \hbar I) o (r, sI) \right]$$

For some $(r, sI) \in \langle H \cup I \rangle$ and *N(H)* is called a strongly left regular neutrosophic LA-semihypergroup if all of its neutrosophic elements is strongly left regular neutrosophic element.

Definition 4.8: A neutrosophic element $\{(g, \hbar I)\}$ of a neutrosophic LA-semihypergroup *N(H)* with neutrosophic left identity *(e,eI)* is called a left (right) invertible neutrosophic element if

$$(5, 5I) \in (r, sI) o (g, \hbar I) \left((5, 5I) \in (g, \hbar I) o (r, sI) \right)$$

For some $(r, sI) \in \langle H \cup I \rangle$ and *N(H)* is called a left (right) invertible neutrosophic LA-semihypergroup if all of its neutrosophic elements is left (right) invertible neutrosophic element.

N(H) is said to be an invertible neutrosophic LA-semihypergroup if it is both a left invertible and a right invertible neutrosophic LA-semihypergroup.

Theorem 4.1: Every strongly regular neutrosophic LA-semihypergroup is regular, but converse is not true.

Proof: Assume that *N(H)* is strongly regular neutrosophic LA-semihypergroup then clearly by definition it is regular neutrosophic LA-semihypergroup. Now for converse consider the following example.

Example 4.1: Let $H = \{g, \hbar, \ell\}$ is an LA-semihypergroup with the binary hyperoperation "*"is given as Table 7:

Table 7.

*	g	$\hbar$	ℓ
g	g	g	g
$\hbar$	g	g	ℓ
ℓ	g	$\hbar$	g

Then

$$\langle H \cup I \rangle = \{(g, gI),(g, \hbar I),(g, lI),(\hbar, \hbar I),(\hbar, gI),(\hbar, lI),(l, lI),(l, gI),(l, \hbar I)\}$$

is regular neutrosophic LA-semihypergroup under binary hyperoperation “o” follow by the following table given as Table 8:

Clearly $(\langle H \cup I \rangle, o)$ is regular Neutrosophic LA-semihypergroup but not strongly regular neutrosophic LA-semihypergroup because of

$$\{(g, \ell I)\} \neq \{(g, gI),(g, \ell I)\} = [(g, \ell I) o (\hbar, \ell I)] o (g, \ell I) \text{etc.}$$

Table 8.

o	(g,gI)	$(g,\hbar I)$	$(g,\ell I)$	$(\hbar,gI)$	$(\hbar,\hbar I)$	$(\hbar,\ell I)$	(ℓ,gI)	$(\ell,\hbar I)$	$(\ell,\ell I)$
(g,gI)	(g,gI)	(g,gI)	(g,gI)	(g,gI)	(g,gI)	(g,gI)	(g,gI)	(g,gI)	(g,gI)
$(g,\hbar I)$	(g,gI)	(g,gI)	(g,gI) $(g,\ell I)$	(g,gI)	(g,gI)	(g,gI) $(g,\ell I)$	(g,gI) $(g,\ell I)$	(g,gI) $(g,\ell I)$	(g,gI) $(g,\ell I)$
$(g,\ell I)$	(g,gI) $(g,\hbar I)$	(g,gI) $(g,\hbar I)$	(g,gI) $(g,\hbar I)$	(g,gI) $(g,\hbar I)$	(g,gI) $(g,\hbar I)$	(g,gI) $(g,\hbar I)$	(g,gI) $(g,\hbar I)$	(g,gI) $(g,\hbar I)$	(g,gI) $(g,\hbar I)$
$(\hbar,gI)$	(g,gI)	(g,gI)	(g,gI) $(g,\ell I)$	(g,gI)	(g,gI)	(g,gI) $(g,\ell I)$	(ℓ,gI)	(ℓ,gI)	(ℓ,gI) $(\ell,\ell I)$
$(\hbar,\hbar I)$	(g,gI)	(g,gI)	(g,gI) $(g,\ell I)$	(g,gI)	(g,gI)	(g,gI) $(g,\ell I)$	(ℓ,gI) $(\ell,\ell I)$	(ℓ,gI) $(\ell,\ell I)$	$(\ell,\ell I)$
$(\hbar,\ell I)$	(g,gI)	(g,gI) $(g,\hbar I)$	(g,gI) $(g,\ell I)$	(g,gI) $(g,\hbar I)$	(g,gI) $(g,\hbar I)$	(g,gI) $(g,\ell I)$	(ℓ,gI)	(ℓ,gI) $(\ell,\hbar I)$	(ℓ,gI) $(\ell,\ell I)$
(ℓ,gI)	(g,gI)	(g,gI) $(g,\hbar I)$	(g,gI)	$(\hbar,gI)$	$(\hbar,gI)$ $(\hbar,\hbar I)$	(g,gI)	(g,gI)	(g,gI) $(g,\hbar I)$	(g,gI)
$(\ell,\hbar I)$	(g,gI)	(g,gI) $(g,\hbar I)$	(g,gI) $(g,\ell I)$	$(\hbar,gI)$	$(\hbar,gI)$ $(\hbar,\hbar I)$	$(\hbar,gI)$ $(\hbar,\ell I)$	(g,gI) $(g,\ell I)$	(g,gI) $(g,\hbar I)$ $(g,\ell I)$	(g,gI) $(g,\ell I)$
$(\ell,\ell I)$	(g,gI)	(g,gI) $(g,\hbar I)$	(g,gI)	$(\hbar,gI)$ $(\hbar,\hbar I)$	$(\hbar,gI)$ $(\hbar,\hbar I)$	$(\hbar,\hbar I)$	(g,gI)	(g,gI) $(g,\hbar I)$	(g,gI)

Theorem 4.2: Every strongly regular neutrosophic LA-semihypergroup is intra-regular, but converse is not true.

Proof: Let *N(H)* be a strongly regular neutrosophic LA-semihypergroup then by definition for every $\{(g, \hbar I)\} \in N(H)$ there exists some $(r, sI) \in \langle H \cup I \rangle$ such that

$$\{(g, \hbar I)\} = [(g, \hbar I) o (r, sI)] o (g, \hbar I)$$

And

$$[(g, \hbar I) o (r, sI)] = [(r, sI) o (g, \hbar I)]$$

Then by using neutrosophic laws we have

$$\begin{aligned}\{(g, \hbar I)\} &= [(g, \hbar I) o (r, sI)] o (g, \hbar I) \\ &= [(g, \hbar I) o (r, sI)] o [[(g, \hbar I) o (r, sI)] o (g, \hbar I)] \\ &= [(g, \hbar I) o (r, sI)] o [[(r, sI) o (g, \hbar I)] o (g, \hbar I)] \\ &= [(g, \hbar I) o (r, sI)] o [[(g, \hbar I) o (g, \hbar I)] o (r, sI)] \\ &= [(g, \hbar I) o (r, sI)] o [[(g, \hbar I)]^2 o (r, sI)] \\ &= [(r, sI) o (g, \hbar I)] o [[(g, \hbar I)]^2 o (r, sI)] \\ &= [(r, sI) o [(g, \hbar I)]^2] o [(g, \hbar I) o (r, sI)] \\ &= [(r, sI) o [(g, \hbar I)]^2] o (\ell, \mathcal{I} I)\end{aligned}$$

where $(\ell, \mathcal{I} I) = (g, \hbar I) o (r, sI)$

Hence $(\langle H \cup I \rangle, o)$ is intra regular neutrosophic LA-semihypergroup. For converse let us see the Example 3.1 which is an intra-regular neutrosophic LA-semihypergroup but not strongly regular neutrosophic LA-semihypergroup.

Theorem 4.3: For a neutrosophic LA-semihypergroup *N(H)* with neutrosophic left identity *(e,eI)* the following parts are equivalent:

(i) *N(H)* is strongly regular neutrosophic LA-semihypergroup.
(ii) $N(H)o(g,\hbar I)$ is strongly regular for all $(g,\hbar I)\in\langle H\cup I\rangle$ where $g,\hbar\in H$.

Proof: (i)⇒(ii) Let $\{(g,\hbar I)\}\in N(H)o(g,\hbar I)$. Since *N(H)* is strongly regular so by definition there exists some $(\mathrm{r},\mathrm{s}I)\in\langle H\cup I\rangle$ such that

$$\{(g,\hbar I)\}=[(g,\hbar I)o(\mathrm{r},\mathrm{s}I)]o(g,\hbar I)\text{and }[(g,\hbar I)o(\mathrm{r},\mathrm{s}I)]=[(\mathrm{r},\mathrm{s}I)o(g,\hbar I)]$$

Suppose $\{(\hbar,lI)\}=[(\mathrm{r},\mathrm{s}I)o(g,\hbar I)]o(\mathrm{r},\mathrm{s}I)$ for some $\{(\hbar,lI)\}\in\langle H\cup I\rangle$. Now using neutrosophic left invertive law we get

$$\begin{aligned}\{(\hbar,lI)\}&=[(\mathrm{r},\mathrm{s}I)o(g,\hbar I)]o\\&=[(g,\hbar I)o(\mathrm{r},\mathrm{s}I)]o(\mathrm{r},\mathrm{s}I)\\&=[(\mathrm{r},\mathrm{s}I)o(\mathrm{r},\mathrm{s}I)]o(g,\hbar I)\\&=[(\mathrm{r},\mathrm{s}I)]^2o(g,\hbar I)\\&\in N(H)o(g,\hbar I)\end{aligned}$$

Again, using neutrosophic property (1), (3), and (i) we get

$$\{(g,\hbar I)\}=[(g,\hbar I)o(\text{r},\text{s}I)]o(g,\hbar I)$$
$$\begin{aligned}\{(g,\hbar I)\}&=[(\text{r},\text{s}I)o(g,\hbar I)]o(g,\hbar I)\\
&=[(\text{r},\text{s}I)o[[(g,\hbar I)o(\text{r},\text{s}I)]o(g,\hbar I)]]o(g,\hbar I)\\
&=[(\text{r},\text{s}I)o[[(g,\hbar I)o(\text{r},\text{s}I)]o[(5,5I)o(g,\hbar I)]]]o(g,\hbar I)\\
&=[(\text{r},\text{s}I)o[[(g,\hbar I)o(5,5I)]o[(\text{r},\text{s}I)o(g,\hbar I)]]]o(g,\hbar I)\\
&=[(g,\hbar I)o[[(g,\hbar I)o(5,5I)]o[(\text{r},\text{s}I)o(g,\hbar I)]]]o(\text{r},\text{s}I)\\
&=[(g,\hbar I)o[[(g,\hbar I)o(5,5I)]o[(g,\hbar I)o(\text{r},\text{s}I)]]]o(\text{r},\text{s}I)\\
&=[(g,\hbar I)o[[(g,\hbar I)o(g,\hbar I)]o[(5,5I)o(\text{r},\text{s}I)]]]o(\text{r},\text{s}I)\\
&=[(g,\hbar I)o[[(g,\hbar I)o(g,\hbar I)]o(\text{r},\text{s}I)]]o(\text{r},\text{s}I)\\
&=[(g,\hbar I)o[[(\text{r},\text{s}I)o(g,\hbar I)]o(g,\hbar I)]]o(\text{r},\text{s}I)\\
&=[(\text{r},\text{s}I)o[[(\text{r},\text{s}I)o(g,\hbar I)]o(g,\hbar I)]]o(g,\hbar I)\\
&=[(g,\hbar I)o[[(\text{r},\text{s}I)o(g,\hbar I)]o(\text{r},\text{s}I)]]o(g,\hbar I)\\
&=[(g,\hbar I)o(\hbar,lI)]o(g,\hbar I)\end{aligned}$$

And

$$\begin{aligned}\{(g,\hbar I)\}o\{(\hbar,lI)\}&=\{(g,\hbar I)\}o[[(\text{r},\text{s}I)o(g,\hbar I)]o(\text{r},\text{s}I)]\\
&=[(\text{r},\text{s}I)o(g,\hbar I)]o[(g,\hbar I)o(\text{r},\text{s}I)]\\
&=[(\text{r},\text{s}I)o[(g,\hbar I)o(\text{r},\text{s}I)]]o(g,\hbar I)\\
&=\{(\hbar,lI)\}o\{(g,\hbar I)\}\end{aligned}$$

which implies that $N(H)o(g,\hbar I)$ is strongly regular neutrosophic LA-semihypergroup. Now assume that (ii)⇒(i), **consider** $\{(g,\hbar I)\}\in N(H)$ so by (ii) $\{(g,\hbar I)\}\in N(H)o(g,\hbar I)$ and $N(H)o(g,\hbar I)$ is strongly regular neutrosophic LA-semihypergroup which implies *N*(*H*) is strongly regular.

Theorem 4.4: Let *N*(*H*) be a strongly regular neutrosophic LA-semihypergroup with neutrosophic left identity. Then for every $\{(g,\hbar I)\}\in N(H)$ there exists $\{(\hbar,lI)\}\in N(H)$ such that

$$\{(g,\hbar I)\}=[(g,\hbar I)o(\hbar,lI)]o(g,\hbar I),$$

$$\{(\hbar,lI)\}=[(\hbar,lI)o(g,\hbar I)]o(\hbar,lI)$$

and

$$(g,\hbar I)o(\hbar,lI)=(g,\hbar I)o(\hbar,lI).$$

Proof: The proof is similar to Theorem 4.3.

Theorem 4.5: If *N*(*H*) is neutrosophic LA-semihypergroup with neutrosophic right identity, then the following are equivalent.

(i) *N*(*H*) is weakly regular.
(ii) *N*(*H*) is right regular.

Proof: (i)⇒(ii) Assume that *N*(*H*) be a weakly regular neutrosophic LA-semihypergroup then for any $\{(g,\hbar I)\}\in N(H)$ there exist $(\hbar,lI)\in\langle II\cup I\rangle$ such that

$$\{(g,\hbar I)\}\in[(g,\hbar I)o(\hbar,lI)]o(g,\hbar I)$$

And suppose $(\hbar,lI)o(e,eI)=(\hbar,lI)$ by hypothesis. So, by using neutrosophic properties. We have

$$\begin{aligned}\{(g,\hbar I)\}&\in[(g,\hbar I)o(\hbar,lI)]o(g,\hbar I)\\&=[(g,\hbar I)o(\hbar,lI)]o[(g,\hbar I)o(e,eI)]\\&=[(g,\hbar I)o(g,\hbar I)]o[(\hbar,lI)o(e,eI)]\\&=[(g,\hbar I)]^2o(\hbar,lI)\end{aligned}$$

Thus *N*(*H*) is right regular neutrosophic LA-semihypergroup.
Let (ii) hold and implies (i). Now follows from (ii) and neutrosophic property (1) we have

$$\begin{aligned}\{(\mathcal{g},\hbar I)\} &\in [(\mathcal{g},\hbar I)]^2 o(\hbar,lI)\\ &=[(\mathcal{g},\hbar I)o(\mathcal{g},\hbar I)]o[(\hbar,lI)o(e,eI)]\\ &=[(\mathcal{g},\hbar I)o(\hbar,lI)]o[(\mathcal{g},\hbar I)o(e,eI)]\\ &=[(\mathcal{g},\hbar I)o(\hbar,lI)]o(\mathcal{g},\hbar I)\end{aligned}$$

Hence *N*(*H*) is weakly regular neutrosophic LA-semihypergroup.

Theorem 4.6: If *N*(*H*) is neutrosophic LA-semihypergroup with neutrosophic left identity, then the following are equivalent.

(i) *N*(*H*) is weakly regular.
(ii) *N*(*H*) is left regular.
(iii) *N*(*H*) is completely regular.

Proof: (i)⇒(ii) Let (i) and implies (ii). As *N*(*H*) is weakly regular neutrosophic LA-semihypergroup with neutrosophic left identity then we need to prove that *N*(*H*) is left regular for this assume for any arbitrary $\{(\mathcal{g},\hbar I)\}\in N(H)$ there exist $(5,5I),(\hbar,lI)\in\langle H\cup I\rangle$, such that

$$\{(\mathcal{g},\hbar I)\}\in[(\mathcal{g},\hbar I)o(\hbar,lI)]o(\mathcal{g},\hbar I)$$

Now by using properties (1), (2) and (3) we have

$$\begin{aligned}\{(\mathcal{g},\hbar I)\} &\in [(\mathcal{g},\hbar I)o(\hbar,lI)]o(\mathcal{g},\hbar I)\\ &=[(\mathcal{g},\hbar I)o(\hbar,lI)]o[(5,5I)o(\mathcal{g},\hbar I)]\\ &=[(\mathcal{g},\hbar I)o(5,5I)]o[(\hbar,lI)o(\mathcal{g},\hbar I)]\\ &=(\hbar,lI)o[[(\mathcal{g},\hbar I)o(5,5I)]o(\mathcal{g},\hbar I)]\\ &=(\hbar,lI)o[(\mathcal{g},\hbar I)o[(5,5I)o(\mathcal{g},\hbar I)]]\\ &=(\hbar,lI)o[(\mathcal{g},\hbar I)]^2\end{aligned}$$

Thus *N*(*H*) is left regular.

Now suppose (ii)⇒(i) It follows from neutrosophic properties (1), (2) and (3) so we have,

$$\begin{aligned}\{(g,\hbar I)\} &\in (\hbar,lI)o[(g,\hbar I)]^2\\ &=(\hbar,lI)o[(g,\hbar I)o[(5,5I)o(g,\hbar I)]]\\ &=(\hbar,lI)o[[(g,\hbar I)o(5,5I)]o(g,\hbar I)]\\ &=[(g,\hbar I)o(5,5I)]o[(\hbar,lI)o(g,\hbar I)]\\ &=[(g,\hbar I)o(\hbar,lI)]o[(5,5I)o(g,\hbar I)]\\ &=[(g,\hbar I)o(\hbar,lI)]o(g,\hbar I)\end{aligned}$$

Thus *N*(*H*) is weakly regular.

(i)⇒(iii) Let *N*(*H*) is weakly regular neutrosophic LA-semihypergroup with neutrosophic left identity then for any neutrosophic element $\{(g,\hbar I)\} \in N(H)$ there exist $(5,5I),(\hbar,lI) \in \langle H \cup I \rangle$ and assume $(5,5I)o(\hbar,lI)=(\hbar,lI)$. Such that

$$\begin{aligned}\{(g,\hbar I)\} &\in [(g,\hbar I)o(\hbar,lI)]o(g,\hbar I)\\ &=[(g,\hbar I)o(\hbar,lI)]o[(5,5I)o(g,\hbar I)]\\ &=[(g,\hbar I)o(5,5I)]o[(\hbar,lI)o(g,\hbar I)]\\ &=[[(\hbar,lI)o(g,\hbar I)]o(5,5I)]o(g,\hbar I)\\ &=[(\hbar,lI)o(g,\hbar I)]o[(5,5I)o(g,\hbar I)]\\ &=[(\hbar,lI)o(g,\hbar I)]o(g,\hbar I)\\ &=[(g,\hbar I)o(g,\hbar I)]o(\hbar,lI)\\ &=[(g,\hbar I)]^2 o(\hbar,lI)\end{aligned}$$

Implies that *N*(*H*) is right regular neutrosophic LA-semihypergroup.

Now the proof of left regular neutrosophic LA-semihypergroup can follow from above (ii). Thus *N*(*H*) is right regular, left regular and weakly regular neutrosophic LA-semihypergroup so by definition consequently *N*(*H*) is completely regular.

(iii)⇒(i) is obvious.

Theorem 4.7: Every neutrosophic LA-semihypergroup *N*(*H*) with neutrosophic pure left identity (*e,eI*) is intra regular if *N*(*H*) is left (right) invertible neutrosophic LA-semihypergroup.

Proof: Assume that *N*(*H*) is left invertible neutrosophic LA-semihypergroup with neutrosophic pure left identity (*e,eI*) then for $(g,\hbar I) \in \langle H \cup I \rangle$ there exist $(\hbar,lI) \in \langle H \cup I \rangle$ such that

$$(5,5I)\in(\mathcal{g},\hbar I)o(\hbar,lI) \quad (43)$$

Now by using neutrosophic left invertive law and (1) we get

$$\begin{aligned}(\mathcal{g},\hbar I)&\in(5,5I)o(\mathcal{g},\hbar I)=(5,5I)o[(5,5I)o(\mathcal{g},\hbar I)]\\&\in[(\mathcal{g},\hbar I)o(\hbar,lI)]o[(5,5I)o(\mathcal{g},\hbar I)]\\&\in(5,5I)o[[(\mathcal{g},\hbar I)o(\hbar,lI)]o(\mathcal{g},\hbar I)]\\&\in N(H)o[[(\mathcal{g},\hbar I)oN(H)]o(\mathcal{g},\hbar I)]\\&\in[(\mathcal{g},\hbar I)oN(H)]o[N(H)o(\mathcal{g},\hbar I)]\end{aligned}$$

because $ko(ros)=ro(kos)$ for all $k,r,s\in N(H)$ with neutrosophoic pure left identity. Also $N(H)=N(H)oN(H)$. So

$$\begin{aligned}(\mathcal{g},\hbar I)&\in[(\mathcal{g},\hbar I)oN(H)]o[N(H)oN(H)]o(\mathcal{g},\hbar I)\\&\in[N(H)oN(H)]o[N(H)o(\mathcal{g},\hbar I)]o(\mathcal{g},\hbar I) \quad \text{by}(3)\\&\in[N(H)oN(H)]o[(\mathcal{g},\hbar I)o(\mathcal{g},\hbar I)]oN(H) \quad \text{by left invertive law}\\&=[N(H)o(\mathcal{g},\hbar I)^2]oN(H)\end{aligned}$$

This implies $N(H)$ that is intra regular. The case for neutrosophic right identity is follow in a similar way.

Theorem 4.8: For a left invertible neutrosophic LA-semihypergroup with neutrosophic pure left identity, the following conditions are equivalent:

(i) $N(H)$ is intra regular.
(ii) $N(\mathcal{R})\cap N(\mathcal{L})=N(\mathcal{R})oN(\mathcal{L})$, where $N(\mathcal{R})$ is neutrosophic right hyper ideal and $N(\mathcal{L})$ are neutrosophic left hyper ideals of $N(H)$.

Proof: (i)⇒(ii) Let us suppose that $N(H)$ is an intra regular neutrosophic LA-semihypergroup with neutrosophic pure left identity and assume $\{(\mathcal{g},\hbar I)\}\in N(H)$ then by definition there exist $(5,nI),(\hbar,lI)\in\langle H\cup I\rangle$ such that

$$\{(\mathcal{g},\hbar I)\}\in[(5,nI)o(\mathcal{g},\hbar I)^2]o(\hbar,lI)$$

Let $N(\mathcal{R})$ and $N(\mathcal{L})$ be any neutrosophic right and left hyper ideals of *N*(*H*) respectively. Then clearly

$$N(\mathcal{R}) \cap N(\mathcal{L}) \subseteq N(\mathcal{R})oN(\mathcal{L}) \tag{44}$$

Consider

$$\{(\mathcal{I},\ell I)\} \in N(\mathcal{R}) \cap N(\mathcal{L}) \Rightarrow \{(\mathcal{I},\ell I)\} \in N(\mathcal{R}) \text{ and } \{(\mathcal{I},\ell I)\} \in N(\mathcal{L})$$

By using (1) and neutrosophic left invertive law, we have

$$\begin{aligned}\{(\mathcal{I},\ell I)\} &\in \left[(e,nI)o(\mathcal{I},\ell I)^2\right]o(\hbar,lI)\\ &\in \left[N(H)o(\mathcal{I},\ell I)^2\right]oN(H)\\ &= \left[N(H)o\left[(\mathcal{I},\ell I)o(\mathcal{I},\ell I)\right]\right]oN(H)\end{aligned}$$

because $(\mathcal{I},\ell I)^2 = (\mathcal{I},\ell I)o(\mathcal{I},\ell I)$ with neutrosophic pure identity

$$= \left[(\mathcal{I},\ell I)o\left[N(H)o(\mathcal{I},\ell I)\right]\right]oN(H) \text{ by } NLI \text{ law}$$

$$= \left[(\mathcal{I},\ell I)o\left[N(H)o(\mathcal{I},\ell I)\right]\right]o\left[N(H)oN(H)\right]$$

$$= \left[(\mathcal{I},\ell I)oN(H)\right]o\left[\left[N(H)o(\mathcal{I},\ell I)\right]oN(H)\right] \text{ by } NM \text{ law}$$

$$= \left[N(H)o(\mathcal{I},\ell I)\right]o\left[\left[(\mathcal{I},\ell I)oN(H)\right]oN(H)\right] \text{ by } NLI \text{ law}$$

$$= \left[N(H)o(\mathcal{I},\ell I)\right]o\left[\left[N(H)oN(H)\right]o(\mathcal{I},\ell I)\right] \text{ by } NLI \text{ law}$$

$$= \left[N(H)o(\mathcal{I},\ell I)\right]o\left[N(H)o(\mathcal{I},\ell I)\right]$$

Because *N*(*H*) = *N*(*H*)*oN*(*H*)

$$= \left[\left[N(H)oN(H)\right]o(\mathcal{I},\ell I)\right]o\left[N(H)o(\mathcal{I},\ell I)\right]$$

$$=\left[\left[(\mathcal{I},\ell I)oN(H)\right]oN(H)\right]o\left[N(H)o(\mathcal{I},\ell I)\right] \quad \text{by } NLI \text{ law}$$

$$\subseteq\left[\left[N(\mathcal{R})oN(H)\right]oN(H)\right]o\left[N(H)oN(\mathcal{L})\right]$$

$$\subseteq\left[N(\mathcal{R})oN(H)\right]o\left[N(H)oN(\mathcal{L})\right]$$

$$\subseteq N(\mathcal{R})oN(\mathcal{L})$$

$$\Rightarrow N(\mathcal{R})\cap N(\mathcal{L})\subseteq N(\mathcal{R})oN(\mathcal{L}) \tag{45}$$

From (44) and (45) we get $N(\mathcal{R})\cap N(\mathcal{L})=N(\mathcal{R})oN(\mathcal{L})$

(ii)⇒(i) Let $N(\mathcal{R})$ and $N(\mathcal{L})$ are any neutrosophic right and left hyper ideals of *N*(*H*) respectively. Then $\{(g,\hbar I)\}\in N(H)$ there exist $(e,eI),(\hbar,lI)\in\langle H\cup I\rangle$ such that

As $(g,\hbar I)^2 oN(H)$ is neutrosophic hyper ideal of *N*(*H*) so $(g,\hbar I)\in(g,\hbar I)^2 oN(H)$ Therefore by applying neutrosophic left inventive law, given statement and (1) we have

$$(g,\hbar I)\in\left[(g,\hbar I)^2 oN(H)\right]\cap\left[(g,\hbar I)^2 oN(H)\right]$$

$$=\left[(g,\hbar I)^2 oN(H)\right]o\left[(g,\hbar I)^2 oN(H)\right]$$

$$=(g,\hbar I)^2 o\left[\left[(g,\hbar I)^2 oN(H)\right]oN(H)\right]$$

$$=(g,\hbar I)^2 o\left[\left[N(H)oN(H)\right]o(g,\hbar I)^2\right]$$

$$=\left[(g,\hbar I)o(g,\hbar I)\right]o\left[\left[N(H)oN(H)\right]o(g,\hbar I)^2\right]$$

$$=\left[(g,\hbar I)o(g,\hbar I)\right]o\left[N(H)o(g,\hbar I)^2\right]$$

$$=\left[\left[N(\mathcal{H})o(g,\hbar I)^2\right]o(g,\hbar I)\right]o(g,\hbar I)$$

$$\Rightarrow(g,\hbar I)\in\left[\left[(\hbar,lI)o(g,\hbar I)^2\right]o(g,\hbar I)\right]o(g,\hbar I) \quad \text{for some } (\hbar,lI)\in N(H)$$

Now by using neutrosophic left inventive law, we have

$$\begin{aligned}\left[(g,\hbar I)o(g,\hbar I)\right]o(\hbar,lI) &\in \left[\left[\left[\left[(\hbar,lI)o(g,\hbar I)^2\right]o(g,\hbar I)\right]o(g,\hbar I)\right]o(g,\hbar I)\right]o(\hbar,lI)\\ &\in \left[(\hbar,lI)o(g,\hbar I)\right]o\left[\left[\left[(\hbar,lI)o(g,\hbar I)^2\right]o(g,\hbar I)\right]o(g,\hbar I)\right]\end{aligned}$$

And

$$\begin{aligned}\left[(\hbar,lI)o(g,\hbar I)\right]o(g,\hbar I) &\in \left[(\hbar,lI)o(g,\hbar I)\right]o\left[\left[\left[(\hbar,lI)o(g,\hbar I)^2\right]o(g,\hbar I)\right]o(g,\hbar I)\right]\\ &= \left[\left[(\hbar,lI)o(g,\hbar I)^2\right]o(g,\hbar I)\right]o\left[\left[(\hbar,lI)o(g,\hbar I)\right]o(g,\hbar I)\right]\\ &= \left[\left[(\hbar,lI)o(g,\hbar I)^2\right]o(g,\hbar I)\right]o\left[\left[(g,\hbar I)o(g,\hbar I)\right]o(\hbar,lI)\right]\\ &= \left[\left[(\hbar,lI)o(g,\hbar I)^2\right]o(g,\hbar I)\right]o\left[(g,\hbar I)^2 o(\hbar,lI)\right]\\ &= \left[(\hbar,lI)o(g,\hbar I)^2\right]o\left[(g,\hbar I)o\left[(\hbar,lI)o(g,\hbar I)^2\right]\right]\\ &= \left[(\hbar,lI)o(g,\hbar I)^2\right]o\left[(\hbar,lI)o\left[(g,\hbar I)o(g,\hbar I)^2\right]\right]\\ &= \left[(\hbar,lI)o(g,\hbar I)^2\right]o(\hbar,lI)\\ &\Rightarrow (g,\hbar I)\in\left[(\hbar,lI)o(g,\hbar I)^2\right]o(\hbar,lI)\end{aligned}$$

Hence *N(H)* is intra regular neutrosophic LA-semihypergroup.

CONCLUSION

The main aim of this chapter is to provide a platform for characterizing neutrosophic LA-semihypergroups and its extended form regular and intra-regular neutrosophic LA-semihypergroups. Also, some basic results and examples with different neutrosophic identities are discussed here.

REFERENCES

Abdullah, S., Aslam, M., & Anwar, T. (2011). A note on M-hyper systems and N-hypersystems in Γ-semihypergroups. *Quasi groups and Related Systems*, *19*, 169-172.

Cutland, N. J. (1983). Non-standard measure theory and its applications. *Bulletin of the London Mathematical Society*, *15*(6), 529–589. doi:10.1112/blms/15.6.529

Davvaz, B., & Fotea, V. L. (2007). *Hyper ring theory and applications*. Gulf Breeze, FL: International Academic Pres.

Davvaz, B., & Poursalavati, N. S. (2000). Semihypergroups and S-hyper systems. *Pure Mathematics and Applications*, *11*, 43–49.

Gulistan, M., Khalaf, M., Yaqoob, N., & Rashid, U. (2019). An introduction to non-associative neutrosophic semihypergroups. *Periodica Mathematica Hungarica*. (Manuscript submitted for publication)

Hasankhani, A. (1999). Ideals in a semihypergroup and greens relations ratio mathematica. *Journal of Applied Mathematics*, *13*, 19–36.

Hila, K., Davvaz, B., & Naka, K. (2011). On quasi-hyper ideals in semihypergroups. *Journal of Communications in Algebra*, *39*(11), 4183–4194. doi:10.1080/00927872.2010.521932

Hila, K., & Dine, J. (2011). On hyper ideal s in left almost semihypergroups: *ISRN Algebra. Article ID, 953124*(8), 68–76.

Howie., (1976). *Introduction to semigroup theory*. Academic Press.

Kazim, M. A., & Naseerudden, M. (1972). On almost semi groups. *The Aligarh Bulletin of Mathematics*, *2*, 1–7.

Khan, M., & Ahmad, N. (2010). Characterizations of left almost semi groups by their ideals. *Journal of Advanced Research in Pure Mathematics*, *2*, 61–71.

Khan, M., & Asif, T. (2010). Intra-regular left almost semi groups characterized by their anti fuzzy ideals. *Journal of Mathematics Research*, *2*, 1–11.

Klir, G. J., & Yuan, B. (1995). *Fuzzy sets and fuzzy logic theory and applications*. PRT Prentice Hall.

Maji, P. K. (2013). Neutrosophic soft set. *Annuals of Fuzzy Mathematics and Informatics*, *5*, 157–168.

Mushtaq, Q., & Khan, M. (2009). Ideals in left almost semigroup. *Journal of Mathematics Research.*

Mushtaq, Q., & Khan, M. (2009). M-systems in LA-semigroups. *Southeast Asian Bulletin of Mathematics*, *33*, 321–327.

Onipchuk, S.V. (1992). Regular semihypergroups. *Mathematic he ski Sbornik, 183*, 43-54.

Smarandachi, F. (2005). Neutrosophic set. *International Journal of Pure and Applied Mathematic, 24*, 287-297.

Vougiouklis, T. (1994). *Hyper structures and their representations*. Palm Harbor, FL: Hadronic Press.

Yaqoob, N., Corsini, P., & Yousafzai, F. (2013). On Intra-regular LA-semihypergroups with Pure Left Identity. *Journal of Algebra, Number Theory, Advances and Applications, 10.*

Yousafzai, F., & Corsini, P. (2013). Some characterization problems in LA-semihypergroup. *Journal of Algebra, Number Theory. Advances and Applications*, *10*, 41–55.

Zadeh, L. A. (1965). Fuzzy sets. *Information and Control*, *8*(3), 338–353. doi:10.1016/S0019-9958(65)90241-X

Chapter 18

Series of Aggregation Operators for Picture Fuzzy Environments and Their Applications:

Aggregation Operators for Picture Fuzzy Sets

Saleem Abdullah
Abdul Wali Khan University, Pakistan

Shahzaib Ashraf
https://orcid.org/0000-0002-8616-8829
Abdul Wali Khan University, Pakistan

ABSTRACT

The main objective of the chapter is to introduce a series of picture fuzzy weighted averaging and geometric aggregation operators by using t-norm and t-conorm. In this chapter, they discussed generalized form of weighted averaging and geometric aggregation operator for picture fuzzy information. Further, the proposed aggregation operators of picture fuzzy number are applied to multi-attribute group decision making problems. To implement the proposed models, they provide some numerical applications of group decision making problems. Also compared with previous model, they conclude that the proposed technique is more effective and reliable.

1. INTRODUCTION

Multiple-criteria decision making (MCDM) means that the best alternative is selected from the limited alternatives set according to the multiple criteria, which can be regarded as cognitive processing. MCDM is a vital division of the decision-making theory which has been widely used in human activities. Because the real decision-making problems were frequently produced from the complicated environment, the evaluation information is usually fuzzy. In general, the fuzzy information takes two forms: one is

DOI: 10.4018/978-1-7998-0190-0.ch018

quantitative and other one is qualitative. The quantitative fuzzy information can be expressed by fuzzy set (FS) (Zadeh, 1965), intuitionistic fuzzy set (IFS) (Atanassov, 1986), Pythagorean fuzzy set (PyFS) [45] picture fuzzy set (PFS) (Beliakov, Bustince, Gosawmi, Mukherjee, & Pal, 2011), spherical fuzzy set (SFS) (Ashraf & Abdullah, 2018) and so on. FS theory proposed by Zadeh (Zadeh, 1965) has been used to describe fuzzy quantitative information which contains only a membership degree. Based on this, Atanassov (1986) presented IFS, which consists of membership and non-membership degrees that fulfills the constraint form that the sum of their degrees is less or equal to 1. For more study about IFS, we refer (Ejegwa, Tyoakaa, & Ayenge, 2016; Ejegwa & Modom, 2015; Ejegwa, 2015; Ejegwa, Akubo, & Joshua, 2014; Ejegwa, Onoja, & Emmanuel, 2014). However, some-times the two degrees don't meet the constraint, but the sum of squares of the two degrees is less or equal to 1. Yager and Abbasov (2014) introduced the PyFS in which the square sum of membership and non-membership degrees is equal to or less than 1.

Atanassov's structure of IFSs and Yager's structure of PyFSs discourses only satisfaction and dissatisfaction degree of elements in a set which is quite insufficient as human nature has some sort of abstain and refusal issues too. The notion of the picture fuzzy set was proposed Cuong (2014), basically (PFS) is the generalization of the intuitionistic fuzzy set (IFS). Outstanding qualities of (PFS) is that it allows to each element the degree of membership, neutral membership and non-membership. Due to its capacity of considerate hesitancy in human decision processes, PFSs have been generally enforced to the field of decision making. PFSs can be applied to the direction that involving human assessment like as, "yes", "abstain", "no" and "refusal". For example, voting is a good example in this position, the voters may be divided in four groups, "vote for", "abstain", vote against", "refusal of voting". Picture fuzzy obtain thought of many authors in this area. In 2015, Singh gave an interrelationship coefficient for the (PFSs). Son (2016) gave a generalized picture distance measure and applied it to deal with clustering problem under the picture fuzzy surroundings. Wei (2017) conferred a decision-making art depend on the picture fuzzy weighted cross-entropy and apply it to rank the alternatives. In 2017, wang, Zhou, Tu, and Tao used the picture fuzzy information to check the multi-criteria decision-making problems, also introduced some picture fuzzy geometric operators and discuss their elemental properties. Phong and Cuong (2014) also studied multi-attribute group decision making problem using picture linguistic numbers. Ashraf, Mahmood, Abdullaj, and Khan (2018) propose some aggregation operators for picture fuzzy information.

In this article, we present some aggregation operators of the PFS based on the triangular norms. We study some properties of PFS. We define score and accuracy function of the PFNs for the comparison of picture fuzzy numbers. In this study, we have proposed series of aggregation operators namely, Picture fuzzy weighted averaging, ordered weighted averaging, hybrid ordered weighted averaging, generalized weighted averaging, geometric weighted, geometric ordered weighted, hybrid geometric weighted and generalized geometric weighted operators based on triangular norms to utilized to aggregate the picture fuzzy information's. Furthermore, these operators are used for decision-making problems in which specialists offer their preferences in the picture fuzzy information to show the practicality and effectiveness of the proposed approach. The remainder of the study is structured as follows:

The rest of this study is designed as follows. Section II briefly introduces the basic knowledge of extension of fuzzy sets. The novel operations "picture fuzzy Set and their operations" is presented in Section III. The series of aggregation operator based on triangular norms for PFNs are proposed in section IV. Section V makes some discussions on application of proposed method. Discussion of the essential conclusions are drawn in last Section.

2. PRELIMINARIES

The section of this study gives reassess the basic notations and concept which are helpful for proposing novel operations for fuzzy extensions.

Definition 1: (Zadeh, 1965) Let the universe set R. Then μ is known to be an FS of R, if

$$\mu = \{< Þ_{\mu}(r_1) > | r_1 \in R\},$$

where $Þ_{\mu}$: $R \rightarrow [0,1]$ is said to be positive-membership degree of r_1 in R.

Definition 2: (Atanassov, 1986) Let the universe set R. Then μ is known to be an IFS of R, if

$$\mu = \{< Þ_{\mu}(r_1), g_{\mu}(r_1) > | r_1 \in R\},$$

where $Þ_{\mu}$: $R \rightarrow [0,1]$ and g_{μ}:$R \rightarrow [0,1]$ are said to be positive-membership degree of r_1 in $\boldsymbol{R}$ and negative-membership degree of r_1 in R respectively. Also, $Þ_{\mu}$ and g_{μ} fulfil the following condition,

$$(0 \leq Þ_{\mu}(r_1) + g_{\mu}(r_1) \leq 1), \ (\forall r_1 \in R).$$

Definition 3: (Yager & Abbasov, 2014) Let the universe set R. Then μ is known to be an PyFS of R, if

$$\mu = \{< Þ_{\mu}(r_1), g_{\mu}(r_1) > | r_1 \in R\},$$

where $Þ_{\mu}$: $R \rightarrow [0,1]$ and g_{μ}:$R \rightarrow [0,1]$ are said to be positive-membership degree of r_1 in R and negative-membership degree of r_1 in R respectively. Also, $Þ_{\mu}$ and g_{μ} fulfil the following condition,

$$(0 \leq Þ^2_{\mu}(r_1) + g^2_{\mu}(r_1) \leq 1), \ (\forall r_1 \in R).$$

3. PICTURE FUZZY SETS AND ITS OPERATIONS

In this section, operations of picture fuzzy number are introduced on the basis of algebraic t-conorm and t-norm. Operations for PFSs can defined as:

Definition 4: (Cuong, 2014) Let the universe set R. Then μ is known to be PFS of R, if

$$\mu = \{< Þ_{\mu}(r_1), /_{\mu}(r_1), g_{\mu}(r_1) > | r_1 \in R\},$$

where Þ $_{\mu}$: $R \rightarrow [0,1]$, ị $_{\mu}$: $p \rightarrow [0,1]$ and g_{μ}: $p \rightarrow [0,1]$ are said to be degree of positive-membership of r_1 in R, neutral-membership degree of r_1 in R and negative-membership degree of r_1 in R respectively. Also, Þ$_{\mu}$, ị$_{\mu}$ and g_{μ} satisfy the following condition:

$$(\forall r_1 \in R)\ \ (0 \leq Þ_{\mu}(r_1) + /_{\mu}(r_1) + g_{\mu}(r_1) \leq 1).$$

and $\gamma_p(r_1) = 1 - (Þ_{\mu}(r_1) + ị_{\mu}(r_1) + g_{\mu}(r_1))$ said to be refusal degree of r_1 in R.

Definition 5: (Wei, Alsaadi, Hayat, & Alsaedi, 2016) Let $\mu_1 = \left\langle Þ_{\mu_1}, /_{\mu_1}, g_{\mu_1} \right\rangle$ and $\mu_2 = \left\langle Þ_{\mu_2}, /_{\mu_2}, g_{\mu_2} \right\rangle$ be two PFNs, and v>0 . Then the operations for PFNs are defined as:

$$(1)\quad \mu_1 \otimes \mu_2 = \left\langle \left(Þ_{\mu_1} + /_{\mu_1}\right)\left(Þ_{\mu_2} + /_{\mu_2}\right) - /_{\mu_1} \cdot /_{\mu_2},\ /_{\mu_1} \cdot /_{\mu_2}, 1 - \left(1 - g_{\mu_1}\right)\left(1 - g_{\mu_2}\right) \right\rangle;$$

$$(2)\quad \mu_1^{v} = \left\langle (Þ_{\mu_1} + /_{\mu_1})^{v} - (/_{\mu_1})^{v},\ (/_{\mu_1})^{v},\ 1 - (1 - g_{\mu_1})^{v} \right\rangle.$$

Definition 6: (Wei, 2017) Let $\mu_1 = \left\langle Þ_{\mu_1}, /_{\mu_1}, g_{\mu_1} \right\rangle$ and $\mu_2 = \left\langle Þ_{\mu_2}, /_{\mu_2}, g_{\mu_2} \right\rangle$ be two PFNs, and v>0 . Then the operations for PFNs are defined as:

$$(1)\quad \mu_1 \otimes \mu_2 = \left\langle Þ_{\mu_1} . Þ_{\mu_2}, /_{\mu_1} + /_{\mu_2} - /_{\mu_1} \cdot /_{\mu_2}, g_{\mu_1} + g_{\mu_2} - g_{\mu_1} \cdot g_{\mu_2} \right\rangle;$$

$$(2)\quad \mu_1^{v} = \left\langle (Þ_{\mu_1})^{v},\ 1 - (1 - /_{\mu_1})^{v},\ 1 - (1 - g_{\mu_1})^{v} \right\rangle.$$

Substantially, it is easy to find that neither the additions for picture fuzzy numbers [30,31] nor the multiplication for picture fuzzy numbers [30,31] is universal to all situations. The limitations will be remove by proposing novel operations for picture fuzzy information's that satisfies the condition that sum of his membership degrees are less or equal to 1.

Definition 7: Let $\mu_1 = \left\langle Þ_{\mu_1}, /_{\mu_1}, g_{\mu_1} \right\rangle$ and $\mu_2 = \left\langle Þ_{\mu_2}, /_{\mu_2}, g_{\mu_2} \right\rangle$ are PFNs in finite universe and ß>0 . Then the operations for PFNs, constructed from norms are defined as:

$$(1)\quad \mu_1 \oplus \mu_2 = \left\langle c^{-1}\left(c\left(Þ_{\mu_1}\right) + c\left(Þ_{\mu_2}\right)\right), \_^{-1}(\_(/_{\mu_1}) + \_(/_{\mu_2})), \_^{-1}(\_(g_{\mu_1}) + \_(g_{\mu_2})) \right\rangle;$$

$$(2)\quad \mu_1 \otimes \mu_2 = \left\langle \_^{-1}(\_(Þ_{\mu_1}) + \_(Þ_{\mu_2})), \_^{-1}(\_(/_{\mu_1}) + \_(/_{\mu_2})), c^{-1}(c(g_{\mu_1}) + c(g_{\mu_2})) \right\rangle;$$

$$(3)\quad ß\mu_1 = \left\langle c^{-1}(ßc(Þ_{\mu_1})), \_^{-1}(ß\_(/_{\mu_1})), \_^{-1}(ß\_(g_{\mu_1})) \right\rangle;$$

$$(4)\quad \mu_1^{ß} = \left\langle \_^{-1}(ß\_(Þ_{\mu_1})), \_^{-1}(ß\_(/_{\mu_1})), c^{-1}(ßc(g_{\mu_1})) \right\rangle.$$

allocate ş (r_1)= -log(r_1) to generator ş, operations of PFNs will attain the form,

$$(5)\quad \mu_1 \oplus \mu_2 = \left\langle Þ_{\mu_1} + Þ_{\mu_2} - Þ_{\mu_1} \cdot Þ_{\mu_2},\ /_{\mu_1} \cdot /_{\mu_2},\ g_{\mu_1} \cdot g_{\mu_2} \right\rangle;$$

$$(6)\quad \mu_1 \otimes \mu_2 = \left\langle Þ_{\mu_1} \cdot Þ_{\mu_2},\ /_{\mu_1} \cdot /_{\mu_2},\ g_{\mu_1} + g_{\mu_2} - g_{\mu_1} \cdot g_{\mu_2} \right\rangle;$$

(7) $ß\mu_1 = \left\langle 1-(1-\text{Þ}_{\mu_1})^{ß},\ (/_{\mu_1})^{ß},\ (\text{g}_{\mu_1})^{ß} \right\rangle$;

(8) $\mu_1^{ß} = \left\langle (\text{Þ}_{\mu_1})^{ß},\ (/_{\mu_1})^{ß},\ 1-(1-\text{g}_{\mu_1})^{ß} \right\rangle$.

Example: Assume that $\mu_1 = \langle 0.6, 0.1, 0.2\rangle$, $\mu_2 = \langle 0.5, 0.2, 0.2\rangle$ and ß =2 When $ş(r_1)$= -log(r_1), then

(1) $\mu_1 \oplus \mu_2 = \langle 0.8, 0.02, 0.04\rangle$;

(2) $\mu_1 \otimes \mu_2 = \langle 0.3, 0.02, 0.36\rangle$;

(3) $2\cdot\mu_1 = \langle 0.84, 0.01, 0.04\rangle$;

(4) $\mu_1^2 = \langle 0.36, 0.01, 0.36\rangle$.

Theorem 1: Let three PFNs $\mu_1 = \langle \text{Þ}_{\mu_1}, /_{\mu_1}, \text{g}_{\mu_1}\rangle$, $\mu_2 = \langle \text{Þ}_{\mu_2}, /_{\mu_2}, \text{g}_{\mu_2}\rangle$ and $\mu_3 = \langle \text{Þ}_{\mu_3}, /_{\mu_3}, \text{g}_{\mu_3}\rangle$. Then the given identities are satisfies:

(1) $\mu_1 \oplus \mu_2 = \mu_2 \oplus \mu_1$;

(2) $\mu_1 \otimes \mu_2 = \mu_2 \otimes \mu_1$;

(3) $ß(\mu_1 \oplus \mu_2) = ß\mu_1 \oplus ß\mu_2$, ß>0;

(4) $(\mu_1 \otimes \mu_2)^{ß} = \mu_1^{ß} \otimes \mu_2^{ß}$, ß >0;

(5) $ß_1\mu_1 \oplus ß_2\mu_1 = (ß_1 + ß_2)\mu_1$, $ß_1>0$, $ß_2>0$;

(6) $\mu_1^{ß_1} \otimes \mu_1^{ß_2} = \mu_1^{(ß_1+ß_2)}$, $ß_1>0$, $ß_2>0$;

(7) $(\mu_1 \oplus \mu_2) \oplus \mu_3 = \mu_1 \oplus (\mu_2 \oplus \mu_3)$;

(8) $(\mu_1 \otimes \mu_2) \otimes \mu_3 = \mu_1 \otimes (\mu_2 \otimes \mu_3)$.

Proof (1), (2), (7) and (8) are understandable; thus, we prove the others:

(3) *L.H.S*=$ß(\mu_1 \oplus \mu_2)$

$$= ß\cdot\left\langle \text{c}^{-1}(\text{ɑ}(\text{Þ}_{\mu_1}) + \text{ɑ}(\text{Þ}_{\mu_2})), \_^{-1}(\_(/_{\mu_1}) + \_(/_{\mu_2})), \_^{-1}(\_(\text{g}_{\mu_1}) + \_(\text{g}_{\mu_2}))\right\rangle$$
$$= \left\langle \text{c}^{-1}[ß\text{ɑ}(\text{c}^{-1}(\text{ɑ}(\text{Þ}_{\mu_1}) + \text{ɑ}(\text{Þ}_{\mu_2})))], \_^{-1}[ß\_(\_^{-1}(\_(/_{\mu_1}) + \_(/_{\mu_2})))], \_^{-1}[ß\_(\_^{-1}(\_(\text{g}_{\mu_1}) + \_(\text{g}_{\mu_2})))]\right\rangle$$
$$= \left\langle \text{c}^{-1}(ß\text{ɑ}(\text{Þ}_{\mu_1}) + ß\text{ɑ}(\text{Þ}_{\mu_2})), \_^{-1}(ß\_(/_{\mu_1}) + ß\_(/_{\mu_2})), \_^{-1}(ß\_(\text{g}_{\mu_1}) + ß\_(\text{g}_{\mu_2}))\right\rangle$$

Now *R.H.S*=$ß\mu_1 \oplus ß\mu_2$

$$
\begin{aligned}
&= \left\langle c^{-1}(ßɗ(Þ_{\mu_1})), \_^{-1}(ß\_(/_{\mu_1})), \_^{-1}(ß\_(g_{\mu_1}))\right\rangle \oplus \left\langle c^{-1}(ßɗ(Þ_{\mu_2})), \_^{-1}(ß\_(/_{\mu_2})), \_^{-1}(ß\_(g_{\mu_2}))\right\rangle \\
&= \Big\langle c^{-1}\left[c\left(c^{-1}(ßɗ(Þ_{\mu_1}))\right) + ɗ(c^{-1}(ßɗ(Þ_{\mu_2})))\right], \_^{-1}[\_(\_^{-1}(ß(/_{\mu_1}))) \\
&\quad + \_(\_^{-1}(ß\_(/_{\mu_2})))], \_^{-1}[\_(\_^{-1}(ß\_(g_{\mu_1}))) + \_(\_^{-1}(ß\_(g_{\mu_2})))]\Big\rangle \\
&= \left\langle c^{-1}(ßɗ(Þ_{\mu_1}) + ßɗ(Þ_{\mu_2})), \_^{-1}(ß\_(/_{\mu_1}) + ß\_(/_{\mu_2})), \_^{-1}(ß\_(g_{\mu_1}) + ß\_(g_{\mu_2}))\right\rangle
\end{aligned}
$$

Then, $ß(\mu_1 \oplus \mu_2) = ß\mu_1 \oplus ß\mu_2$, for every $ß > 0$.

(4) $L.H.S = (\mu_1 \otimes \mu_2)^{ß}$

$$
\begin{aligned}
&= \left\{\left\langle \_^{-1}(\_(Þ_{\mu_1}) + \_(Þ_{\mu_2})), \_^{-1}(\_(/_{\mu_1}) + \_(/_{\mu_2})), c^{-1}(ɗ(g_{\mu_1}) + ɗ(g_{\mu_2}))\right\rangle\right\}^{ß} \\
&= \left\langle \_^{-1}[ß\_(\_^{-1}(\_(Þ_{\mu_1}) + \_(Þ_{\mu_2})))], \_^{-1}[ß\_(\_^{-1}(\_(/_{\mu_1}) + \_(/_{\mu_2})))], c^{-1}[ßɗ(c^{-1}(ɗ(g_{\mu_1}) + ɗ(g_{\mu_2})))]\right\rangle \\
&= \left\langle \_^{-1}(ß\_(Þ_{\mu_1}) + ß\_(Þ_{\mu_2})), \_^{-1}(ß\_(/_{\mu_1}) + ß\_(/_{\mu_2})), c^{-1}(ßɗ(g_{\mu_1}) + ßɗ(g_{\mu_2}))\right\rangle
\end{aligned}
$$

Now $R.H.S = \mu_1 \otimes \mu_2$

$$
\begin{aligned}
&= \left\langle \_^{-1}(ß\_(Þ_{\mu_1})), \_^{-1}(ß\_(/_{\mu_1})), c^{-1}(ßɗ(g_{\mu_1}))\right\rangle \otimes \left\langle \_^{-1}(ß\_(Þ_{\mu_2})), \_^{-1}(ß\_(/_{\mu_2})), c^{-1}(ßɗ(g_{\mu_2}))\right\rangle \\
&= \Big\langle \_^{-1}[\_(\_^{-1}(ß\_(Þ_{\mu_1}))) + \_(\_^{-1}(ß\_(Þ_{\mu_2})))], \_^{-1}[\_(\_^{-1}(ß(/_{\mu_1}))) \\
&\quad + \_(\_^{-1}(ß\_(/_{\mu_2})))], c^{-1}[ɗ(c^{-1}(ßɗ(g_{\mu_1}))) + ɗ(c^{-1}(ßɗ(g_{\mu_2})))]\Big\rangle \\
&= \left\langle \_^{-1}(ß\_(Þ_{\mu_1}) + ß\_(Þ_{\mu_2})), \_^{-1}(ß\_(/_{\mu_1}) + ß\_(/_{\mu_2})), c^{-1}(ßɗ(g_{\mu_1}) + ßɗ(g_{\mu_2}))\right\rangle
\end{aligned}
$$

Hence, $(\mu_1 \otimes \mu_2)^{ß} = \mu_1 \otimes \mu_2$, for every $ß > 0$.

(5) $\text{L.H.S} = ß_1\mu_1 \oplus ß_2\mu_1$

$$
\begin{aligned}
&= \left\langle c^{-1}(ß_1ɗ(Þ_{\mu_1})), \_^{-1}(ß_1\_(/_{\mu_1})), \_^{-1}(ß_1\_(g_{\mu_1}))\right\rangle \oplus \left\langle c^{-1}(ß_2ɗ(Þ_{\mu_1})), \_^{-1}(ß_2\_(/_{\mu_1})), \_^{-1}(ß_2\_(g_{\mu_1}))\right\rangle \\
&= \Big\langle c^{-1}[ɗ(c^{-1}(ß_1ɗ(Þ_{\mu_1}))) + ɗ(c^{-1}(ß_2ɗ(Þ_{\mu_1})))], \_^{-1}[\_(\_^{-1}(ß_1(/_{\mu_1}))) \\
&\quad + \_(\_^{-1}(ß_2\_(/_{\mu_1})))], \_^{-1}[\_(\_^{-1}(ß_1\_(g_{\mu_1}))) + \_(\_^{-1}(ß_2\_(g_{\mu_1})))]\Big\rangle \\
&= \left\langle c^{-1}(ß_1ɗ(Þ_{\mu_1}) + ß_2ɗ(Þ_{\mu_1})), \_^{-1}(ß_1\_(/_{\mu_1}) + ß_2\_(/_{\mu_1})), \_^{-1}(ß_1\_(g_{\mu_1}) + ß_2\_(g_{\mu_1}))\right\rangle \\
&= \left\langle c^{-1}((ß_1 + ß_2)ɗ(Þ_{\mu_1})), \_^{-1}((ß_1 + ß_2)\_(/_{\mu_1})), \_^{-1}((ß_1 + ß_2)\_(g_{\mu_1}))\right\rangle \\
&= (ß_1 + ß_2)\mu_1 = R.H.\_
\end{aligned}
$$

Hence, $ß_1\mu_1 \oplus ß_2\mu_1 = (ß_1 + ß_2)\mu_1$, for every $ß_1 > 0, ß_2 > 0$.

(6) $L.H.S = \mu_1^{ß_1} \otimes \mu_1^{ß_2}$

$$= \left\langle \_^{-1}(ß_1\_(Þ_{\mu_1})), \_^{-1}(ß_1\_(/_{\mu_1})), c^{-1}(ß_1 ɑ(g_{\mu_1})) \right\rangle \otimes \left\langle \_^{-1}(ß_2\_(Þ_{\mu_1})), \_^{-1}(ß_2\_(/_{\mu_1})), c^{-1}(ß_2 ɑ(g_{\mu_1})) \right\rangle$$
$$= \left\langle \_^{-1}[\_(\_^{-1}(ß_1\_(Þ_{\mu_1}))) + \_(\_^{-1}(ß_2\_(Þ_{\mu_1})))], \_^{-1}[\_(\_^{-1}(ß_1(/_{\mu_1})))\right.$$
$$\left. + \_(\_^{-1}(ß_2\_(/_{\mu_1})))], c^{-1}[ɑ(c^{-1}(ß_1 ɑ(g_{\mu_1}))) + ɑ(c^{-1}(ß_2 ɑ(g_{\mu_1})))] \right\rangle$$
$$= \left\langle \_^{-1}(ß_1\_(Þ_{\mu_1}) + ß_2\_(Þ_{\mu_1})), \_^{-1}(ß_1\_(/_{\mu_1}) + ß_2\_(/_{\mu_1})), c^{-1}(ß_1 ɑ(g_{\mu_1}) + ß_2 ɑ(g_{\mu_1})) \right\rangle$$
$$= \left\langle \_^{-1}((ß_1 + ß_2)\_(Þ_{\mu_1})), \_^{-1}((ß_1 + ß_2)\_(/_{\mu_1})), c^{-1}((ß_1 + ß_2) ɑ(g_{\mu_1})) \right\rangle$$
$$= \mu_1^{(ß_1+ß_2)} = L.H.S$$

Hence $\mu_1^{ß_1} \otimes \mu_1^{ß_2} = \mu_1^{(ß_1+ß_2)}, forevery_1 > 0, ß_2 > 0.$ which done the proof.

Comparison Rules

To compare the PFNs, the following comparison techniques are used as:

Definition 8: [34] Let $\mu_k = \left\langle Þ_{\mu_k}, /_{\mu_k}, g_{\mu_k} \right\rangle$ $k \in N$ be the PFNs. Then

(1) $so(\mu_k) = (Þ_{\mu_k} - g_{\mu_k})$,

(2) $au(\mu_k) = Þ_{\mu_k} + ɨ_{\mu_k} + g_{\mu_k}$,

where $so \in [-1,1]$ and $au \in [0,1]$, $so(\mu_k)$ & $au(\mu_k)$ are said to be score and accuracy functions of the PFNs μ_k.

Definition 9: Let $\mu_j = \left\langle Þ_{\mu_j}, /_{\mu_j}, g_{\mu_j} \right\rangle$ and $\mu_k = \left\langle Þ_{\mu_k}, /_{\mu_k}, g_{\mu_k} \right\rangle$ be any two PFNs. Then by using the Definition 8, comparison technique can be described as,

(a) If $so(\mu_j) > so(\mu_k)$, then $\mu_j > \mu_k$.
(b) If $so(\mu_j) = so(\mu_k)$ and $au(\mu_j) > au(\mu_k)$, then $\mu_j > \mu_k$.
(c) If $so(\mu_j) = so(\mu_k)$ and $au(\mu_j) = au(\mu_k)$ then $\mu_j = \mu_k$.

4. AGGREGATION OPERATORS FOR PFNS

4.1. Averaging Aggregation Operators

Definition 10: Let $\mu_j = \left\langle Þ_{\mu_j}, /_{\mu_j}, g_{\mu_j} \right\rangle$ ($j \in N$) be a collection of PFNs, and *PFNWA*: $PFN^n \rightarrow PFN$,

$PFNWA_w(\mu_1, \mu_2, \dots \mu_n) = \Sigma_{j=1}^{n} \vartheta_j \mu_j$, and then the PFNWA operator is said to be picture fuzzy number weighted averaging operator having the dimensions n, where the weight vector of μ_j is $\vartheta=\{\vartheta_1,\vartheta_2,\ldots,\vartheta_g\}$ such that ($j\in N$), with $\vartheta_j\geq 0$ ($j\in N$) and $\Sigma_{j=1}^{n}\vartheta_j=1$.

Theorem 2: Let $\mu_j=\left\langle Þ_{\mu_j}, /_{\mu_j}, g_{\mu_j}\right\rangle$ ($j\in N$) be a collection of PFNs, and the weight vector of μ_j is $\vartheta=\{\vartheta_1,\vartheta_2,\ldots,\vartheta_n\}$ such that ($j\in N$), with $\vartheta_j\geq 0$ ($j\in N$) and $\Sigma_{j=1}^{n}\vartheta_j=1$. Then their accumulated outcome utilizing the PFNWA operator is again PFN, and

$$PFNWA_w(\mu_1,\mu_2,\ldots\mu_n)=\left\langle c^{-1}(\Sigma_{j=1}^{n}\vartheta_j ɖ(Þ_{\mu_j})), \_^{-1}(\Sigma_{j=1}^{n}\vartheta_j \_(/_{\mu_j})), \_^{-1}(\Sigma_{j=1}^{n}\vartheta_j \_(g_{\mu_j}))\right\rangle$$

If additive generator $ş(r_1)$= -log(r_1), then the Theorem 2 take the form,

$$PFNWA_w(\mu_1,\mu_2,\ldots\mu_n)=\left\langle 1-\Pi_{j=1}^{n}(1-Þ_{\mu_j})^{\vartheta_j}, \Pi_{j=1}^{n}/_{\mu_j}^{\vartheta_j}, \Pi_{j=1}^{n}g_{\mu_j}^{\vartheta_j}\right\rangle.$$

Proof: By utilized the technique of mathematical induction on n . Therefore, we have the following.

(1) let n=2, since $\vartheta_1\mu_1\oplus\vartheta_2\mu_2$

$$\begin{aligned}
&=\left\langle c^{-1}(\vartheta_1 ɖ(Þ_{\mu_1})), _^{-1}(\vartheta_1_(/_{\mu_1})), _^{-1}(\vartheta_1_(g_{\mu_1}))\right\rangle\oplus\left\langle c^{-1}(\vartheta_2 ɖ(Þ_{\mu_2})), _^{-1}(\vartheta_2_(/_{\mu_2})), _^{-1}(\vartheta_2_(g_{\mu_2}))\right\rangle\\
&=\left\langle c^{-1}[ɖ(c^{-1}(\vartheta_1 ɖ(Þ_{\mu_1})))+ɖ(c^{-1}(\vartheta_2 ɖ(Þ_{\mu_2})))], _^{-1}[_(_^{-1}(\vartheta_1(/_{\mu_1})))\right.\\
&\quad\left.+_(_^{-1}(\vartheta_2_(/_{\mu_2})))], _^{-1}[_(_^{-1}(\vartheta_1_(g_{\mu_1})))+_(_^{-1}(\vartheta_2_(g_{\mu_2})))]\right\rangle\\
&=\left\langle c^{-1}(\vartheta_1 ɖ(Þ_{\mu_1})+\vartheta_2 ɖ(Þ_{\mu_2})), _^{-1}(\vartheta_1_(/_{\mu_1})+\vartheta_2_(/_{\mu_2})), _^{-1}(\vartheta_1_(g_{\mu_1})+\vartheta_2_(g_{\mu_2}))\right\rangle,
\end{aligned}$$

Let, it is true for $n=k$, then equation take the form $PFNWA_w(\mu_1,\mu_2,\ldots,\mu_k)$

$$\begin{aligned}
&=\vartheta_1\mu_1\oplus\vartheta_2\mu_2\oplus\ldots\oplus\vartheta_k\mu_k\\
&=\left\langle c^{-1}(\Sigma_{j=1}^{k}\vartheta_j ɖ(Þ_{\mu_j})), _^{-1}(\Sigma_{j=1}^{k}\vartheta_j_(/_{\mu_j})), _^{-1}(\Sigma_{j=1}^{k}\vartheta_j_(g_{\mu_j}))\right\rangle.
\end{aligned}$$

for $n=k+1$, then $PFNWA_w(\mu_1,\mu_2,\ldots,\mu_k,\mu_{k+1})$

$$= \vartheta_1\mu_1 \oplus \vartheta_2\mu_2 \oplus \ldots \oplus \vartheta_{k+1}\mu_{k+1}$$
$$= \left\langle c^{-1}(\Sigma_{j=1}^{k}\vartheta_j ɋ(Þ_{\mu_j})), \_^{-1}(\Sigma_{j=1}^{k}\vartheta_j \_(/_{\mu_j})), \_^{-1}(\Sigma_{j=1}^{k}\vartheta_j \_(g_{\mu_j}))\right\rangle$$
$$\oplus \left\langle c^{-1}(\vartheta_{k+1} ɋ(Þ_{e_{k+1}})), \_^{-1}(\vartheta_{k+1} \_(/_{e_{k+1}})), \_^{-1}(\vartheta_{k+1} \_(g_{e_{k+1}}))\right\rangle$$
$$= \left\langle c^{-1}[ɋ(c^{-1}(\Sigma_{j=1}^{k}\vartheta_j ɋ(Þ_{\mu_j}))) + ɋ(c^{-1}(\vartheta_{k+1} ɋ(Þ_{e_{k+1}})))], \_^{-1}[\_(\_^{-1}(\Sigma_{j=1}^{k}\vartheta_j \_(/_{\mu_j})))\right.$$
$$\left. + \_(\_^{-1}(\vartheta_{k+1} \_(/_{e_{k+1}})))], \_^{-1}[\_(\_^{-1}(\Sigma_{j=1}^{k}\vartheta_j \_(g_{\mu_j}))) + \_(\_^{-1}(\vartheta_{k+1} \_(g_{e_{k+1}})))]\right\rangle$$
$$= \left\langle c^{-1}(\Sigma_{j=1}^{k+1}\vartheta_j ɋ(Þ_{\mu_j})), \_^{-1}(\Sigma_{j=1}^{k+1}\vartheta_j \_(/_{\mu_j})), \_^{-1}(\Sigma_{j=1}^{k+1}\vartheta_j \_(g_{\mu_j}))\right\rangle$$

i.e., Equation (eq1) satisfy for $n=k+1$. Thus, Equation (eq1) satisfied for whole n. Therefore

$$PFNWA_w(\mu_1,\mu_2,\ldots,\mu_n) = \left\langle c^{-1}(\Sigma_{j=1}^{n}\vartheta_j ɋ(Þ_{\mu_j})), \_^{-1}(\Sigma_{j=1}^{n}\vartheta_j \_(/_{\mu_j})), \_^{-1}(\Sigma_{j=1}^{n}\vartheta_j \_(g_{\mu_j}))\right\rangle.$$

which done the proof.

Properties

There are some properties which are PFNWA operator has satisfies obviously.

Let $\mu_j = \left\langle Þ_{\mu_j}, /_{\mu_j}, g_{\mu_j}\right\rangle$ $(j\varepsilon N)$ be a collection of PFNs, therefore

(1) **Idempotency:** Let $\mu_j(j\in N)$ be a collection of PFNs. If all of $\mu_j(j\in N)$ are identical, i.e., $\mu_j=\mu$, for whole $(j\in N)$, therefore $PFNWA_w(\mu_1,\mu_2,\ldots,\mu_n)=\mu$.

(2) **Boundedness:** Assuming that

$$\mu_j^- = \left\langle \min_j Þ_{\mu_j}, \min_j /_{\mu_j}, \max_j g_{\mu_j}\right\rangle \text{ and } \mu_j^+ = \left\langle \max_j Þ_{\mu_j}, \min_j /_{\mu_j}, \min_j g_{\mu_j}\right\rangle$$

for all $(j\in N)$, therefore $\mu_j^- \subseteq PFNWA_w(\mu_1,\mu_2,\ldots,\mu^n) \subseteq \mu_j^+$.

(3) **Monotonically:** Suppose that $\mu_j = \left\langle Þ_{\mu_j}^{\#}, /_{\mu_j}^{\#}, g_{\mu_j}^{\#}\right\rangle$ $(j\in N)$ be a collection of PFNs. If it satisfies that $\mu_j \subseteq \mu_j^{\#}$, for whole, $j\in\{1,2,\ldots,n\}$, then

$$PFNWA_w(\mu_1,\mu_2,\ldots,\mu_n) \subseteq PFNWA_w(\mu_1^{\#},\mu_2^{\#},\ldots,\mu_n^{\#}).$$

Definition 11: Let $\mu_j = \left\langle Þ_{\mu_j}, /_{\mu_j}, g_{\mu_j}\right\rangle$ $(j\in N)$ be a collection of PFNs, and $PFNOWA$: $PFN^n\to PFN$,

$$PFNOWA_w(\mu_1,\mu_2,\ldots\mu_n) = \sum_{j=1}^{n}\vartheta_j\mu_{\eta(j)},$$

and then the *PFNOWA* operator is said to be a picture fuzzy number ordered weighted averaging operator having the dimensions *n*, where $\mu_{\eta(j)}$ is the *j*th biggest value consequently by total order $\mu_{\eta(1)} \geq \mu_{\eta(2)} \geq \ldots \geq \mu_{\eta(n)}$. $\vartheta=\{\vartheta_1,\vartheta_2,\ldots,\vartheta_n\}$ are the aggregation-associated weight vectors such that $\vartheta_j \geq 0$ ($j \in N$) and $\Sigma_{j=1}^{n} \vartheta_j = 1$.

Theorem 3: Suppose that $\mu_j = \langle Þ_{\mu_j}, /_{\mu_j}, g_{\mu_j} \rangle$ ($j \in N$) be a collection of PFNs, and $\vartheta=\{\vartheta_1,\vartheta_2,\ldots,\vartheta_n\}$ is the weight vector such that $\vartheta_j \geq 0$ ($j \in N$) and $\Sigma_{j=1}^{n} \vartheta_j = 1$. Then their accumulated outcome utilizing the PFNOWA operator is again PFN, and

$$PFNOWA_w(\mu_1,\mu_2,\ldots\mu_n) = \left\langle c^{-1}\left(\Sigma_{j=1}^{n} \vartheta_j c\left(Þ_{\mu_{\eta(j)}}\right)\right), \_^{-1}\left(\Sigma_{j=1}^{n} \vartheta_j \_\left(/_{\mu_{\eta(j)}}\right)\right), \_^{-1}\left(\Sigma_{j=1}^{n} \vartheta_j \_\left(g_{\mu_{\eta(j)}}\right)\right)\right\rangle$$

If additive generator $§(r_1)= -\log(r_1)$, then the Theorem 3 take the form,

$$PFNOWA_w(\mu_1,\mu_2,\ldots\mu_n) = \left\langle 1 - \Pi_{j=1}^{n}(1 - Þ_{\mu_{\eta(j)}})^{\vartheta_j}, \Pi_{j=1}^{n} /_{\mu_{\eta(j)}}^{\vartheta_j}, \Pi_{j=1}^{n} g_{\mu_{\eta(j)}}^{\vartheta_j} \right\rangle.$$

where $\mu_{\eta(j)}$ is *j*th largest value consequently by total order $\mu_{\eta(1)} \geq \mu_{\eta(2)} \geq \ldots \geq \mu_{\eta(n)}$.

Proof: Procedure is similar as Theorem 1.

Properties

There are some properties which are PFNOWA operator has satisfies obviously.

Suppose that $\mu_j = \langle Þ_{\mu_j}, /_{\mu_j}, g_{\mu_j} \rangle$ ($j \in N$) be a collection of PFNs, therefore

(1) **Idempotency:** Let μ_j($j \varepsilon N$) be a collection of PFNs. If all of μ_j($j \varepsilon N$) are identical, i.e., $\mu_j=\mu$, for all ($j \in N$), then $PFNOWA_w(\mu_1,\mu_2,\ldots,\mu_n)=\mu$.

(2) **Boundedness:** Assuming that

$$\mu_j^- = \left\langle \min_j Þ_{\mu_j}, \min_j /_{\mu_j}, \max_j g_{\mu_j} \right\rangle \text{ and } \mu_j^+ = \left\langle \max_j Þ_{\mu_j}, \min_j /_{\mu_j}, \min_j g_{\mu_j} \right\rangle$$

for all ($j \in N$), then $\mu_1^- \subseteq PFNOWA_w(\mu_1,\mu_2,\ldots,\mu_n) \subseteq \mu_1^+$.

(3) **Monotonically:** Suppose that $\mu_j = \left\langle Þ_{\mu_j^\#}, /_{\mu_j^\#}, g_{\mu_j^\#} \right\rangle$ ($j \in N$) be a collection of PFNs, if it satisfy that $\mu_j \subseteq \mu_j^\#$, for all $j \in N$, then

$$PFNOWA_w(\mu_1,\mu_2,\ldots,\mu_n) \subseteq PFNOWA_w(\mu^\#{}_1,\mu^\#{}_2,\ldots,\mu_n^\#).$$

Definition 12: Let $\mu_j = \left\langle Þ_{\mu_j}, /_{\mu_j}, g_{\mu_j} \right\rangle$ (*jεN*) be a collection of PFNs, and *PFNHWA*: *PFN*^{n}→*PFN*,

$$PFNHWA_w(\mu_1, \mu_2, \ldots \mu_n) = \sum_{j=1}^{n} \vartheta_j \mu'_{\eta(j)},$$

and then the *PFNHWA* operator is said to be a picture fuzzy number hybrid weighted averaging operator having the dimensions *n*, where *j*th biggest weighted value is $\mu_{\eta(j)}$ and $\mu_j \left(\mu_j = n\vartheta_j \mu_j,\ j\varepsilon N\right)$, the weight vector of μ_j is $\vartheta=\{\vartheta_1,\vartheta_2,\ldots,\vartheta_n\}$ such that (*jεN*) with $\vartheta_j \geq 0$ (*jεN*) and $\Sigma_{j=1}^{n} \vartheta_j = 1$. $\omega=\{\omega_1,\omega_2,\ldots,\omega_n\}$ is the aggregation-associated vectors such that $\omega_j \geq 0$ (*jεN*) and $\Sigma_{j=1}^{n} \omega_j = 1$, and balancing coefficient is *n*.

Theorem 4: Suppose that $\mu_j = \left\langle Þ_{\mu_j}, /_{\mu_j}, g_{\mu_j} \right\rangle$ (*jεN*) be a collection of PFNs, and the weight vector of μ_j is $\vartheta=\{\vartheta_1,\vartheta_2,\ldots,\vartheta_n\}$ such that (*jεN*) with $\vartheta_j \geq 0$ (*jεN*), $\Sigma_{j=1}^{n} \vartheta_j = 1$ and $\omega=\{\omega_1,\omega_2,\ldots,\omega_n\}$ is the aggregation-associated vectors such that $\omega_j \geq 0$ (*jεN*) and $\Sigma_{j=1}^{n} \omega_j = 1$. Then their accumulated outcome utilizing the PFNHWA operator is again PFN, and

$$PFNHWA_w(\mu_1, \mu_2, \ldots \mu_n) = \left\langle c^{-1}\left(\Sigma_{j=1}^{n} \omega_j c\left(Þ_{\mu_{\eta(j)}}\right)\right), \_^{-1}\left(\Sigma_{j=1}^{n} \omega_j \_\left(/_{\mu_{\eta(j)}}\right)\right), \_^{-1}\left(\Sigma_{j=1}^{n} \omega_j \_\left(g_{\mu_{\eta(j)}}\right)\right)\right\rangle.$$

If additive generator ş(r_1)= -log(r_1), then the Theorem 4 take the form,

$$PFNHWA_w(\mu_1, \mu_2, \ldots \mu_n) = \left\langle 1 - \Pi_{j=1}^{n}(1 - Þ_{\mu_{\eta(j)}})^{\omega_j}, \Pi_{j=1}^{n} /_{\mu_{\eta(j)}}^{\omega_j}, \Pi_{j=1}^{n} g_{\mu_{\eta(j)}}^{\omega_j} \right\rangle.$$

where *j*th biggest weighted value is $\mu_{\eta(j)}$ and $\mu_j \left(\mu_j = n\vartheta_j \mu_j,\ j\varepsilon N\right)$ consequently by total order $\mu'_{\eta(1)} \geq \mu'_{\eta(2)} \geq \ldots \geq \mu'_{\eta(n)}$.

Proof: Procedure is similar as Theorem 1.

Properties

There are some properties which are PFNHWA operator has satisfies obviously.

Let $\mu_j = \left\langle Þ_{\mu_j}, /_{\mu_j}, g_{\mu_j} \right\rangle$ (*jεN*) be a collection of PFNs, therefore

(1) **Idempotency:** Let μ_j(*jεN*) be a collection of PFNs. If all of μ_j(*jεN*) are identical, i.e., $\mu_j=\mu$, for all (*jεN*), then $PFNHWA_w(\mu_1,\mu_2,\ldots,\mu_n)=\mu$
(2) **Boundedness:** Assuming that

$$\mu_j^- = \left\langle \min_j \text{Þ}_{\mu_j}, \min_j /_{\mu_j}, \max_j \text{g}_{\mu_j} \right\rangle \text{ and } \mu_j^+ = \left\langle \max_j \text{Þ}_{\mu_j}, \min_j /_{\mu_j}, \min_j \text{g}_{\mu_j} \right\rangle$$

for all ($j\varepsilon N$), then $\mu_j^- \subseteq PFNHWA_w(\mu_1, \mu_2, ..., \mu_n) \subseteq \mu_j^+$.

(3) **Monotonically:** Suppose that $\mu_j = \left\langle \text{Þ}_{\mu_j^\#}, /_{\mu_j^\#}, \text{g}_{\mu_j^\#} \right\rangle$ ($j\varepsilon N$) be a collection of PFNs, if it satisfy that $\mu_j \subseteq \mu_j^\#$, for whole $j \in N$, then

$$PFNHWA_w(\mu_1, \mu_2, ..., \mu_n) \subseteq PFNHWA_w(\mu_1^\#, \mu_2^\#, ..., \mu_n^\#).$$

Definition 13: Let $\mu_j = \left\langle \text{Þ}_{\mu_j}, /_{\mu_j}, \text{g}_{\mu_j} \right\rangle$ ($j\varepsilon N$) be a collection of PFNs, and $GPFNWA$: $PFN^n \rightarrow PFN$,

$$GPFNWA_w(\mu_1, \mu_2, ...\mu_n) = \left(\Sigma_{j=1}^n \vartheta_j \left(\mu_j\right)^{\text{ß}} \right)^{\frac{1}{\text{ß}}},$$

and then the GPFNWA operator is said to be generalized picture fuzzy number weighted averaging operator having the dimensions n, where the weight vector of μ_j is $\vartheta=\{\vartheta_1,\vartheta_2,\ldots,\vartheta_n\}$ such that ($j\varepsilon N$), with $\vartheta_j \geq 0$ ($j\varepsilon N$) and $\Sigma_{j=1}^n \vartheta_j = 1$, $\text{ß} \geq 0$.

Theorem 5: Suppose that $\mu_j = \left\langle \text{Þ}_{\mu_j}, /_{\mu_j}, \text{g}_{\mu_j} \right\rangle$ ($j\varepsilon N$) be a collection of PFNs, and the weight vector of μ_j is $\vartheta=\{\vartheta_1,\vartheta_2,\ldots,\vartheta_n\}$ such that ($j\varepsilon N$), with $\text{ß}>0$, $\vartheta_j \geq 0$ ($j\varepsilon N$) and $\Sigma_{j=1}^n \vartheta_j = 1$. Then their accumulated outcome utilizing the GPFNWA operator is again PFN, and

$$GPFNWA_w(\mu_1, \mu_2, ...\mu_n)$$
$$= \left\langle \_^{-1}\left(\tfrac{1}{\text{ß}}\_\left(\text{c}^{-1}\left(\Sigma_{j=1}^{\text{g}} \vartheta_j \text{c}\left(\_^{-1}(\text{ß}\_(\text{Þ}_{\mu_j}))\right)\right)\right)\right), \text{c}^{-1}\left(\tfrac{1}{\text{ß}}\text{c}\left(\_^{-1}\left(\Sigma_{j=1}^{\text{g}} \vartheta_j \_(\text{c}^{-1}(\text{ß}\text{c}(/_{\mu_j})))\right)\right)\right), \text{c}^{-1}\left(\tfrac{1}{\text{ß}}\text{c}\left(\_^{-1}\left(\Sigma_{j=1}^{\text{g}} \vartheta_j \_(\text{c}^{-1}(\text{ß}\text{c}(\text{g}_{\mu_j})))\right)\right)\right) \right\rangle.$$

If additive generator $\text{ş}(r_1)$= -log(r_1), then the Theorem 5 take the form,

$$GPFNWA_w(\mu_1, \mu_2, ...\mu_n)$$
$$= \left\langle \left(\alpha, 1 - \Pi_{j=1}^{\text{g}} (1 - \left(\text{Þ}_{\mu_j}\right)^{\text{ß}})^{\vartheta_j} \right)^{\frac{1}{\text{ß}}}, 1 - (1 - \Pi_{j=1}^{\text{g}} (1 - (1 - /_{\mu_j})^{\text{ß}})^{\vartheta_j})^{\frac{1}{\text{ß}}}, 1 - (1 - \Pi_{j=1}^{\text{g}} (1 - (1 - \text{g}_{\mu_j})^{\text{ß}})^{\vartheta_j})^{\frac{1}{\text{ß}}} \right\rangle.$$

Apparently, if we use $\text{ß}=1$, then GPFNWA operator becomes into PFNWA operator.

Proof: Procedure is similar as Theorem 1.

4.1. Geometric Aggregation Operators

Definition 14: let $\mu_j = \left\langle Þ_{\mu_j}, /_{\mu_j}, g_{\mu_j} \right\rangle$ (*jεN*) be a collection of PFNs, and *PFWG*: *PFN*^{n}→*PFN*,

$$PFWG_w(\mu_1, \mu_2, \ldots \mu_n) = \Pi_{j=1}^{n} \mu_j^{\vartheta_j},$$

and then the PFWG operator is said to be a picture fuzzy weighted geometric operator having the dimensions *n*, where the weight vector of μ_j is $\vartheta=\{\vartheta_1,\vartheta_2,\ldots,\vartheta_n\}$ such that (*jεN*), with $\vartheta_j \geq 0$ ($j \in \{1,2,\ldots,n\}$) and $\Sigma_{j=1}^{g} \vartheta_j = 1$.

Theorem 6: Suppose that $\mu_j = \left\langle Þ_{\mu_j}, /_{\mu_j}, g_{\mu_j} \right\rangle$ (*jεN*) be a collection of PFNs, and the weight vector of μ_j is $\vartheta=\{\vartheta_1,\vartheta_2,\ldots,\vartheta_n\}$ such that (*jεN*), with $\vartheta_j \in [0,1]$ and $\Sigma_{j=1}^{n} \vartheta_j = 1$. Then their accumulated outcome utilizing the PFWG operator is again PFN, and

$$PFWG_w(\mu_1, \mu_2, \ldots \mu_n) = \left\langle \_^{-1}\left(\Sigma_{j=1}^{n} \vartheta_j \_(Þ_{\mu_j})\right), \_^{-1}\left(\Sigma_{j=1}^{n} \vartheta_j \_(/_{\mu_j})\right), c^{-1}\left(\Sigma_{j=1}^{n} \vartheta_j c(g_{\mu_j})\right) \right\rangle$$

If additive generator $ş(r_1)= -\log(r_1)$, then the Theorem 6 take the form,

$$PFWG_w(\mu_1, \mu_2, \ldots \mu_n) = \left\langle \Pi_{j=1}^{n} \left(Þ_{\mu_j}\right)^{\vartheta_j}, \ \Pi_{j=1}^{n} (/_{\mu_j})^{\vartheta_j}, \ 1 - \Pi_{j=1}^{n} (1 - g_{\mu_j})^{\vartheta_j} \right\rangle.$$

Proof: By utilized the technique of mathematical induction on *n*. Therefore, we have the following.

(1) let *n*=2, since $\mu_1^{\vartheta_1} \otimes \mu_2^{\vartheta_2}$

$$= \left\langle \_^{-1}(\vartheta_1 \_(Þ_{e_1})), \ \_^{-1}(\vartheta_1 \_(/_{e_1})), \ c^{-1}(\vartheta_1 c(g_{e_1})) \right\rangle \otimes \left\langle \_^{-1}(\vartheta_2 \_(Þ_{e_2})), \ \_^{-1}(\vartheta_2 \_(/_{e_2})), \ c^{-1}(\vartheta_2 c(g_{e_2})) \right\rangle$$
$$= \left\langle \_^{-1}\left[\_\left(\_^{-1}(\vartheta_1 \_(Þ_{e_1}))\right) + \_\left(\_^{-1}(\vartheta_2 \_(Þ_{e_2}))\right)\right], \_^{-1}\left[\_\left(\_^{-1}(\vartheta_1 \_(/_{e_1}))\right) + \_\left(\_^{-1}(\vartheta_2 \_(/_{e_2}))\right)\right], \right.$$
$$\left. c^{-1}[c(c^{-1}(\vartheta_1 c(g_{e_1}))) + c(c^{-1}(\vartheta_2 c(g_{e_2})))] \right\rangle$$
$$= \left\langle \_^{-1}\left[\left(\vartheta_1 \_(Þ_{e_1})\right) + \left(\vartheta_2 \_(Þ_{e_2})\right)\right], \_^{-1}\left[\left(\vartheta_1 \_(/_{e_1})\right) + \left(\vartheta_2 \_(/_{e_2})\right)\right], c^{-1}[(\vartheta_1 c(g_{e_1})) + (\vartheta_2 c(g_{e_2}))] \right\rangle$$
$$= \left\langle \_^{-1}\left(\Sigma_{j=1}^{2} \vartheta_j \_(Þ_{\mu_j})\right), \ \_^{-1}\left(\Sigma_{j=1}^{2} \vartheta_j \_(/_{\mu_j})\right), \ c^{-1}\left(\Sigma_{j=1}^{2} \vartheta_j c(g_{\mu_j})\right) \right\rangle$$

Then

$$PFWG_w(\mu_1,\mu_2) = \mu_1^{\vartheta_1} \otimes \mu_2^{\vartheta_2}$$
$$= \left\langle \_^{-1}\left(\Sigma_{j=1}^{2}\vartheta_j\_(Þ_{\mu_j})\right),\ \_^{-1}\left(\Sigma_{j=1}^{2}\vartheta_j\_(/_{\mu_j})\right),\ c^{-1}\left(\Sigma_{j=1}^{2}\vartheta_j c(g_{\mu_j})\right)\right\rangle$$

Let, it is true for $n=k$, then equation take the form

$$PFWG_w(\mu_1,\mu_2,\ldots,\mu_k) = \mu_1^{\vartheta_1} \otimes \mu_2^{\vartheta_2} \otimes \ldots \otimes \mu_k^{\vartheta_k}$$
$$= \left\langle \_^{-1}\left(\Sigma_{j=1}^{k}\vartheta_j\_(Þ_{\mu_j})\right),\ \_^{-1}\left(\Sigma_{j=1}^{k}\vartheta_j\_(/_{\mu_j})\right),\ c^{-1}\left(\Sigma_{j=1}^{k}\vartheta_j c(g_{\mu_j})\right)\right\rangle$$

We have to prove for $n=k+1$, then $PFWG_w(\mu_1,\mu_2,\ldots,\mu_k,\mu_{k+1})$

$$= \mu_1^{\vartheta_1} \otimes \mu_2^{\vartheta_2} \otimes \ldots \otimes \mu_k^{\vartheta_k} \otimes \mu_{k+1}^{\vartheta_{k+1}}$$
$$= \left\langle \_^{-1}\left(\Sigma_{j=1}^{k}\vartheta_j\_(Þ_{\mu_j})\right),\_^{-1}\left(\Sigma_{j=1}^{k}\vartheta_j\_(/_{\mu_j})\right),\ c^{-1}\left(\Sigma_{j=1}^{k}\vartheta_j c(g_{\mu_j})\right)\right\rangle$$
$$\otimes\left\langle \_^{-1}(\vartheta_{k+1}\_(Þ_{e_{k+1}})),\_^{-1}(\vartheta_{k+1}\_(/_{e_{k+1}})),c^{-1}(\vartheta_{k+1}c(g_{e_{k+1}}))\right\rangle$$
$$= \left\langle \_^{-1}\left[\_\left(\_^{-1}\left(\Sigma_{j=1}^{k}\vartheta_j\_(Þ_{\mu_j})\right)\right)+\_\left(\_^{-1}(\vartheta_{k+1}\_(Þ_{e_{k+1}}))\right)\right],\_^{-1}\left[\_\left(\_^{-1}\left(\Sigma_{j=1}^{k}\vartheta_j\_(/_{\mu_j})\right)\right)+\_\left(\_^{-1}(\vartheta_{k+1}\_(/_{e_{k+1}}))\right)\right],\right.$$
$$\left. c^{-1}\left[c\left(c^{-1}\left(\Sigma_{j=1}^{k}\vartheta_j c(g_{\mu_j})\right)\right)+c\left(c^{-1}\left(\vartheta_{k+1}c(g_{e_{k+1}})\right)\right)\right]\right\rangle$$
$$= \left\langle \_^{-1}\left(\Sigma_{j=1}^{k+1}\vartheta_j\_(Þ_{\mu_j})\right),\_^{-1}\left(\Sigma_{j=1}^{k+1}\vartheta_j\_(/_{\mu_j})\right),c^{-1}\left(\Sigma_{j=1}^{k+1}\vartheta_j c(g_{\mu_j})\right)\right\rangle$$

i.e., Equation satisfy for $n=k+1$. Thus, Equation satisfied for whole n. Therefore

$$PFWG_w(\mu_1,\mu_2,\ldots,\mu_n) = \left\langle \_^{-1}\left(\Sigma_{j=1}^{n}\vartheta_j\_(Þ_{\mu_j})\right),\_^{-1}\left(\Sigma_{j=1}^{n}\vartheta_j\_(/_{\mu_j})\right),c^{-1}\left(\Sigma_{j=1}^{n}\vartheta_j c(g_{\mu_j})\right)\right\rangle.$$

which done the proof.

Properties

There are some properties which are PFWG operator has satisfies obviously.

Let $\mu_j = \left\langle Þ_{\mu_j}, /_{\mu_j}, g_{\mu_j}\right\rangle$ $(j\varepsilon N)$ be a collection of PFNs, therefore

(1) **Idempotency:** Let $\mu_j(j\varepsilon N)$ be a collection of PFNs. If all of $\mu_j(j\varepsilon N)$ are identical, i.e., $\mu_j=\mu$, for all $j\varepsilon N$, then $PFWG_w(\mu_1,\mu_2,\ldots,\mu_n)=\mu$.

(2) **Boundedness:** Assuming that

$$\mu_j^- = \left\langle \min_j Þ_{\mu_j}, \min_j /_{\mu_j}, \max_j g_{\mu_j}\right\rangle \quad\text{and}\quad \mu_j^+ = \left\langle \max_j Þ_{\mu_j}, \min_j /_{\mu_j}, \min_j g_{\mu_j}\right\rangle$$

for all $j\varepsilon N$, then $\mu_j^- \subseteq PFWG_w(\mu_1,\mu_2,\ldots,\mu_n) \subseteq \mu_j^+$.

(3) **Monotonically:** Suppose that $\mu_j = \left\langle Þ^{\#}_{\mu_j}, /^{\#}_{\mu_j}, g^{\#}_{\mu_j} \right\rangle$ $(j\varepsilon N)$ be a collection of PFNs if it is satisfy that $\mu_j \subseteq \mu_j^{\#}$, for $j\in\{1,2,\ldots,n\}$ then

$$PFWG_w(\mu_1,\mu_2,\ldots,\mu_n) \subseteq PFWG_w(\mu_1^{\#},\mu_2^{\#},\ldots,\mu_n^{\#}).$$

Definition 15: Let $\mu_j = \left\langle Þ_{\mu_j}, /_{\mu_j}, g_{\mu_j} \right\rangle$ $(j\varepsilon N)$ be a collection of PFNs, and *PFOWG*: $PFN^n \rightarrow PFN$,

$$PFOWG_w(\mu_1,\mu_2,\ldots,\mu_n) = \Pi_{j=1}^{n} \mu_{\eta(j)}^{\vartheta_j},$$

and then the PFOWG operator is said to be picture fuzzy ordered weighted geometric operator having the dimensions g, where *j*th biggest weighted value is $\mu_{\eta(j)}$ consequently by total order $\mu_{\eta(1)} \geq \mu_{\eta(2)} \geq \ldots \geq \mu_{\eta(n)}$. $\vartheta=\{\vartheta_1,\vartheta_2,\ldots,\vartheta_n\}$ is the aggregation-associated vectors such that $\vartheta_j \geq 0$ $(j\varepsilon N)$ and $\Sigma_{j=1}^{n} \vartheta_j = 1$.

Theorem 7: Suppose that $\mu_j = \left\langle Þ_{\mu_j}, /_{\mu_j}, g_{\mu_j} \right\rangle$ $(j\varepsilon N)$ be a collection of PFNs, and $\vartheta=\{\vartheta_1,\vartheta_2,\ldots,\vartheta_n\}$ is the aggregation-associated vectors such that $\vartheta_j \geq 0$ $(j\varepsilon N)$ and $\Sigma_{j=1}^{n} \vartheta_j = 1$. Then their accumulated outcome utilizing the PFOWG operator is again PFN, and

$$PFOWG_w(\mu_1,\mu_2,\ldots,\mu_n) = \left\langle \_^{-1}\left(\Sigma_{j=1}^{g} \vartheta_j \_(Þ_{\mu_{\eta(j)}})\right), \_^{-1}\left(\Sigma_{j=1}^{g} \vartheta_j \_(/_{\mu_{\eta(j)}})\right), c^{-1}\left(\Sigma_{j=1}^{g} \vartheta_j ç(g_{\mu_{\eta(j)}})\right) \right\rangle$$

If additive generator $ş(r_1)= -\log(r_1)$, then the Theorem 7 take the form,

$$PFOWG_w(\mu_1,\mu_2,\ldots,\mu_n) = \left\langle \Pi_{j=1}^{g} \left(Þ_{\mu_{\eta(j)}}\right)^{\vartheta_j}, \Pi_{j=1}^{g} (/_{\mu_{\eta(j)}})^{\vartheta_j}, 1-\left(\Pi_{j=1}^{g}(1-g_{\mu_{\eta(j)}})^{\vartheta_j}\right) \right\rangle$$

where $\mu_{\eta(j)}$ is *j*th largest value consequently by total order $\mu_{\eta(1)} \geq \mu_{\eta(2)} \geq \ldots \geq \mu_{\eta(g)}$.

Proof: Procedure is similar as Theorem 6.

Properties

There are some properties which are PFOWG operator has satisfies obviously.

Suppose that $\mu_j = \left\langle Þ_{\mu_j}, /_{\mu_j}, g_{\mu_j} \right\rangle$ $(j\varepsilon N)$ be a collection of PFNs, therefore

(1) **Idempotency:** Let $\mu_j(j\varepsilon N)$ be a collection of PFNs. If all of $\mu_j(j\varepsilon N)$ are identical, i.e., $\mu_j=\mu$, for all $(j\varepsilon N)$, then $PFOWG_w(\mu_1,\mu_2,\ldots,\mu_n)=\mu$.

(2) **Boundedness:** Assuming that

$$\mu_j^- = \left\langle \min_j Þ_{\mu_j}, \min_j /_{\mu_j}, \max_j g_{\mu_j} \right\rangle \text{ and } \mu_j^+ = \left\langle \max_j Þ_{\mu_j}, \min_j /_{\mu_j}, \min_j g_{\mu_j} \right\rangle$$

for all (*jεN*), then $\mu_j^- \subseteq PFOWG_w(\mu_1, \mu_2, ..., \mu_n) \subseteq \mu_j^+$

(3) **Monotonically:** Suppose that $\mu_j = \left\langle Þ^{\#}_{\mu_j}, /^{\#}_{\mu_j}, g^{\#}_{\mu_j} \right\rangle$ (*jεN*) be a collection of PFNs, if it satisfy that $\mu_j \subseteq \mu_j^{\#}$, for all *j*∈*N*, then

$$PFOWG_w(\mu_1, \mu_2, ..., \mu_n) \subseteq PFOWG_w(\mu_1^{\#}, \mu_2^{\#}, ..., \mu_g^{\#}).$$

Definition 16: Let $\mu_j = \left\langle Þ_{\mu_j}, /_{\mu_j}, g_{\mu_j} \right\rangle$ (*j*∈*g*) be a collection of PFNs and *PFHWG*: *PFN^n→PFN,*

$$PFHWG_w(\mu_1, \mu_2, ..., \mu_n) = \Pi_{j=1}^{n} \mu_{\eta(j)}^{\vartheta_j{}'},$$

and then the *PFHWG* operator is said to be a picture fuzzy hybrid weighted geometric operator having the dimensions *n*, where *j*th biggest weighted value is $\mu_{\eta(j)}$ and $\mu_j\left(\mu_j = n\vartheta_j\mu_j,\ j\varepsilon N\right)$, the weight vector of μ_j is $\vartheta=\{\vartheta_1,\vartheta_2,\ldots,\vartheta_n\}$ such that (*jεN*) with $\vartheta_j \geq 0\ (j \in \{1,2,...n\})$ and $\Sigma_{j=1}^{n}\vartheta_j = 1$. Also $\omega=\{\omega_1,\omega_2,\ldots,\omega_n\}$ is the aggregation-associated vector such that $\omega_j \geq 0\ (j \in \{1,2,...n\})$ and $\Sigma_{j=1}^{n}\omega_j = 1$, and balancing coefficient is *n*.

Theorem 8: Suppose that $\mu_j = \left\langle Þ_{\mu_j}, /_{\mu_j}, g_{\mu_j} \right\rangle$ (*j*∈*N*) be a collection of PFNs, the weight vector of μ_j is $\vartheta=\{\vartheta_1,\vartheta_2,\ldots,\vartheta_n\}$ such that (*jεN*) with $\vartheta_j \geq 0$ and $\Sigma_{j=1}^{n}\vartheta_j = 1$, $\omega=\{\omega_1,\omega_2,\ldots,\omega_n\}$ is the aggregation-associated vectors such that $\omega_j \geq 0$ (*j*∈{1,2,…,*n*}) and $\Sigma_{j=1}^{n}\omega_j = 1$. Then their accumulated outcome utilizing the PFHWG operator is again PFN, and

$$PFHWG_w(\mu_1, \mu_2, ..., \mu_n) = \left\langle \_^{-1}\left(\Sigma_{j=1}^{g}\omega_j \_(Þ_{\mu_{\eta(j)}})\right), \_^{-1}\left(\Sigma_{j=1}^{g}\omega_j \_(/_{\mu_{\eta(j)}})\right), c^{-1}\left(\Sigma_{j=1}^{g}\omega_j c(g_{\mu_{\eta(j)}})\right) \right\rangle$$

If additive generator ş(r_1)= -log(r_1), then the Theorem 8 take the form,

$$PFHWG_w(\mu_1, \mu_2, ..., \mu_n) = \left\langle \Pi_{j=1}^{g}\left(Þ_{\mu_{\eta(j)}}\right)^{\omega_j}, \Pi_{j=1}^{g}\left(/_{\mu_{\eta(j)}}\right)^{\omega_j}, 1 - \Pi_{j=1}^{g}\left(1 - g_{\mu_{\eta(j)}}\right)^{\omega_j} \right\rangle$$

where $\mu_{\eta(j)}$ is *j*th biggest value consequently by total order $\mu_{\eta(1)} \geq \mu_{\eta(2)} \geq ... \geq \mu_{\eta(n)}$ and *j*th biggest weighted value is $\mu_{\eta(j)}$, $\mu_j\left(\mu_j = n\vartheta_j\mu_j,\ j\varepsilon N\right)$.

Proof: Procedure is similar as Theorem 6.

Properties

There are some properties which are PFHWG operator has satisfies obviously.

Let $\mu_j = \langle Þ_{\mu_j}, /_{\mu_j}, g_{\mu_j} \rangle$ ($j\varepsilon N$) be a collection of PFNs, therefore

(1) **Idempotency:** Let μ_j($j\varepsilon N$) be a collection of PFNs. If all of μ_j($j\varepsilon N$) are identical, i.e., μ_j=μ, for all ($j\varepsilon N$), then $PFHWG_w(\mu_1,\mu_2,\ldots,\mu_n)=\mu$.

(2) **Boundedness:** Assuming that

$$\mu_j^- = \langle \min_j Þ_{\mu_j}, \min_j /_{\mu_j}, \max_j g_{\mu_j} \rangle \text{ and } \mu_j^+ = \langle \max_j Þ_{\mu_j}, \min_j /_{\mu_j}, \min_j g_{\mu_j} \rangle$$

for all ($j\varepsilon N$), then $\mu_j^- \subseteq PFHWG_w(\mu_1, \mu_2, \ldots, \mu_n) \subseteq \mu_j^+$.

(3) **Monotonically:** Suppose that $\mu_j = \langle Þ^{\#}_{\mu_j}, /^{\#}_{\mu_j}, g^{\#}_{\mu_j} \rangle$ ($j\varepsilon N$) be a collection of PFNs, if it satisfy that $\mu_j \subseteq \mu_j^{\#}$, for all $j\in N$, then

$$PFHWG_w(\mu_1, \mu_2, \ldots, \mu_n) \subseteq PFHWG_w(\mu_1^{\#}, \mu_2^{\#}, \ldots, \mu_n^{\#}).$$

Definition 17: Let $\mu_j = \langle Þ_{\mu_j}, /_{\mu_j}, g_{\mu_j} \rangle$ ($j\in\{1,2,\ldots,n\}$) be a collection of PFNs, and $GPFWG$: $PFN^n \rightarrow PFN$,

$$GPFWG_w(\mu_1, \mu_2, \ldots, \mu_n) = \frac{1}{\mu}\left(\Pi_{j=1}^{n}\left(\mu e_j\right)^{\vartheta_j}\right),$$

and then the GPFWG operator is said to be generalized picture fuzzy weighted geometric operator having the dimensions n, where the weight vector of μ_j is $\vartheta=\{\vartheta_1,\vartheta_2,\ldots,\vartheta_n\}$ such that ($j\varepsilon N$), with $\vartheta_j\geq 0$ ($j\in N$) and $\Sigma_{j=1}^{n}\vartheta_j = 1$.

Theorem 9: Suppose that $\mu_j = \langle Þ_{\mu_j}, /_{\mu_j}, g_{\mu_j} \rangle$ ($j\in N$) be a collection of PFNs. Then their accumulated outcome utilizing the GPFWG operator is again PFN, and $GPFWG_w(\mu_1,\mu_2,\ldots,\mu_n)$

$$= \left\langle c^{-1}\left[\tfrac{1}{\mu} c\left(\_^{-1}\left(\Sigma_{j=1}^{g}\vartheta_j \_(c^{-1}(\mu t(Þ_{\mu_j})))\right)\right)\right], c^{-1}\left[\tfrac{1}{\mu} c\left(\_^{-1}\left(\Sigma_{j=1}^{g}\vartheta_j \_(c^{-1}(\mu t(/_{\mu_j})))\right)\right)\right], \_^{-1}\left(\tfrac{1}{\mu} \_\left(c^{-1}\left(\Sigma_{j=1}^{g}\vartheta_j c\left(\_^{-1}(\mu s(g_{\mu_j}))\right)\right)\right)\right). \right\rangle$$

If additive generator $ş(r_1)$= -log(r_1), then the Theorem 9 take the form,

$$GPFWG_w(\mu_1, \mu_2, \ldots, \mu_n) = \left\langle 1-\left[1-\Pi_{j=1}^{g}\left(1-\left(1-Þ_{\mu_j}\right)^{\mu}\right)^{\vartheta_j}\right]^{\frac{1}{\mu}}, 1-\left[1-\Pi_{j=1}^{g}\left(1-\left(1-/_{\mu_j}\right)^{\mu}\right)^{\vartheta_j}\right]^{\frac{1}{\mu}}, \left[1-\Pi_{j=1}^{g}\left(1-\left(g_{\mu_j}\right)^{\mu}\right)^{\vartheta_j}\right]^{\frac{1}{\mu}} \right\rangle.$$

Apparently, if we use μ=1 then the GPFWG operator is becomes into PFWG operator.

Proof: Procedure is similar as Theorem 6.

Similarly, we show that the GPFWG operators have same properties as the PFWG operator.

5. THE MCGDM METHOD UTILIZING AGGREGATION OPERATORS FOR PFNS

Let $H=(h_1,h_2,\ldots,h_m)$ be a distinct set of m probable alternatives and $Y=(y_1,y_2,\ldots,y_n)$ be a finite set of n criteria, where h_i indicate the ith alternatives and y_j indicate the jth criteria. Let $D=(d_1,d_2,\ldots,d_t)$ be a finite set of k experts, where d_k indicate the kth expert. The expert d_k supply her appraisal of an alternative h_i on an attribute y_j as a (PFN) r_{ij}^k $(i=1,2,\ldots,m; j=1,2,\ldots,n)$. The expert's information is represented by the picture fuzzy decision matrices $R_k=(r_{ij}^k)_{m\times n}(k=1,2,\ldots,p)$.

Assume that $\vartheta=\{\vartheta_1,\vartheta_2,\ldots,\vartheta_m\}$ is weight vector of the attribute y_j such that $0\leq\vartheta_j\leq1$, $\sum_{j=1}^{n}\vartheta_j=1$ and $\psi=\{\psi_1,\psi_2,\ldots,\psi_m\}$ is the weight vector of the decision makers d_j such that $\psi_j\leq1, \sum_{j=1}^{n}\psi_j=1$.

Step.1: Take the decision information from the given matrix R_k, and used the PFWG operator, the individual total picture fuzzy preference value r_i^k of the alternative h_i is derived as follows:

$$r_i^k = LPFWA_w(r_{i1}^k, r_{i2}^k, \ldots, r_{i3}^k), (i=1,2,\ldots,m; k=1,2,\ldots,t)$$

where $\vartheta=\{\vartheta_1,\vartheta_2,\ldots,\vartheta_m\}$is weight vector of the attribute.

Step.2: Due to the separate total picture fuzzy preference value w_i^k of alternative $h_i(i=1,2,\ldots,m; k=1,2,\ldots,t)$ Using PFHG/PFHA operator with has associated weight vector $\psi=(\psi_1,\psi_2,\ldots,\psi_m)^T$

$$r_i^k = LPFHA_{w,\psi}(r_{i1}^1, r_{i2}^2, \ldots, r_{i3}^t), (i=1,2,\ldots,m)$$

Collect the cumulative total picture fuzzy preference value $r_1$of the alternative h_i $(i=1,2,\ldots,m)$, where weight vector of the decision maker as $\vartheta=(\vartheta_1,\vartheta_2,\ldots,\vartheta_m)$.

Step.3: After finding the values of r_i then we can calculate the score and accuracy functions by Definition 8.
Step.4: According to Definition 9, we give the ranks to alternatives for choice the best one option.

A Descriptive Example

A construction company wants to find the outstanding universal supplier for one of its frequently critical section used in gathering operation. Assume that $SO_2(C_1)$, $NO_2(C_2)$, $PM_{10}(C_3)$, is the set of possible international suppliers (i.e., alternatives) under deliberation and y_1, y_2, y_3, y_4 is the set of criteria, represent for "total cost of the product", "quality of the product", "service performance of supplier", "suppliers profile", "risk factor", respectively. Decision makers and their weight vectors are $\vartheta=(0.314,$

0.355, 0.331). The numerical values which scrutinized under these indexes which describe in Tables 1, 2 and 3, and they can be shown by

$$\begin{pmatrix} \langle 0.265,0.150,0.385\rangle & \langle 0.330,0.190,0.280\rangle & \langle 0.245,0.175,0.380\rangle \\ \langle 0.345,0.145,0.310\rangle & \langle 0.430,0.190,0.180\rangle & \langle 0.245,0.275,0.310\rangle \\ \langle 0.365,0.150,0.335\rangle & \langle 0.480,0.310,0.105\rangle & \langle 0.340,0.310,0.190\rangle \\ \langle 0.430,0.210,0.170\rangle & \langle 0.460,0.145,0.165\rangle & \langle 0.310,0.430,0.070\rangle \end{pmatrix}$$

$$\begin{pmatrix} \langle 0.125,0.470,0.200\rangle & \langle 0.220,0.400,0.160\rangle & \langle 0.345,0.410,0.125\rangle \\ \langle 0.355,0.335,0.120\rangle & \langle 0.300,0.170,0.330\rangle & \langle 0.205,0.430,0.105\rangle \\ \langle 0.315,0.380,0.100\rangle & \langle 0.340,0.265,0.195\rangle & \langle 0.280,0.520,0.190\rangle \\ \langle 0.365,0.375,0.135\rangle & \langle 0.355,0.220,0.305\rangle & \langle 0.325,0.405,0.090\rangle \end{pmatrix}$$

$$\begin{pmatrix} \langle 0.260,0.075,0.395\rangle & \langle 0.220,0.414,0.160\rangle & \langle 0.255,0.370,0.275\rangle \\ \langle 0.270,0.160,0.360\rangle & \langle 0.320,0.015,0.465\rangle & \langle 0.135,0.575,0.090\rangle \\ \langle 0.245,0.365,0.290\rangle & \langle 0.250,0.570,0.110\rangle & \langle 0.175,0.330,0.165\rangle \\ \langle 0.390,0.340,0.160\rangle & \langle 0.305,0.435,0.120\rangle & \langle 0.465,0.425,0.076\rangle \end{pmatrix}$$

For computing the alternatives matrix A_{WG} we use GPFWG operator as:

$$A_{WG} = \begin{pmatrix} \langle 0.279,0.171,0.347\rangle & \langle 0.213,0.424,0.161\rangle & \langle 0.243,0.233,0.278\rangle \\ \langle 0.333,0.197,0.266\rangle & \langle 0.278,0.285,0.196\rangle & \langle 0.227,0.105,0.325\rangle \\ \langle 0.392,0.246,0.211\rangle & \langle 0.311,0.370,0.164\rangle & \langle 0.220,0.413,0.188\rangle \\ \langle 0.395,0.233,0.136\rangle & \langle 0.347,0.318,0.186\rangle & \langle 0.378,0.399,0.118\rangle \end{pmatrix}$$

Step 1: For calculating the inclusive value y_i of all alternatives $a_i(i=1,2,\ldots,n)$, we utilized PFHWG operator as:

$$\begin{pmatrix} e_1 = \langle 0.250,0.232,0.285\rangle \\ e_2 = \langle 0.276,0.165,0.277\rangle \\ e_3 = \langle 0.297,0.329,0.192\rangle \\ e_4 = \langle 0.378,0.307,0.139\rangle \end{pmatrix}$$

Step 2: Calculate score and accuracy function by using Definition 8, of H_{WG} is expressed as:

$$\begin{pmatrix} sc(e_1) = -0.035 & ac(e_1) = 0.768 \\ sc(e_2) = -0.001 & ac(e_2) = 0.719 \\ sc(e_3) = 0.104 & ac(e_3) = 0.820 \\ sc(e_4) = 0.239 & ac(e_4) = 0.826 \end{pmatrix}$$

Note that if we get the different values of the score function then the accuracy function not need to be calculated.

Step 3: Calculate values for ranking the alternatives and selected the finest one. According to Definition 9 and the outcome in Step 3, for H_{WG}, the final ranking is $e_4>e_3>e_2>e_1$. And the best alternative is e_4.

CONCLUSION

In this paper, we have introduced the notion of picture fuzzy weighted geometric operator, picture fuzzy order weighted geometric operator, picture fuzzy hybrid weighted geometric operator and generalized picture fuzzy weighted geometric operator. Also establish the technique to handling the MAGDM problems for picture fuzzy information. Finally, based on these operators, a decision-making method has been

Figure 1. Scores got by GPFNWG operator, 0≤λ≤10

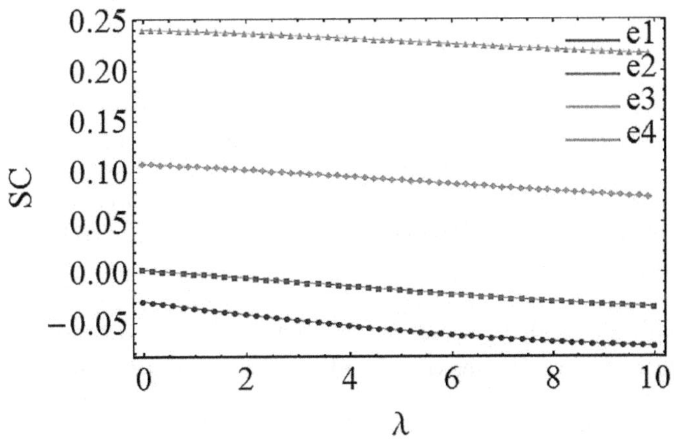

Figure 2. Accuracy got by FPFNWG operator, 0≤λ≤10

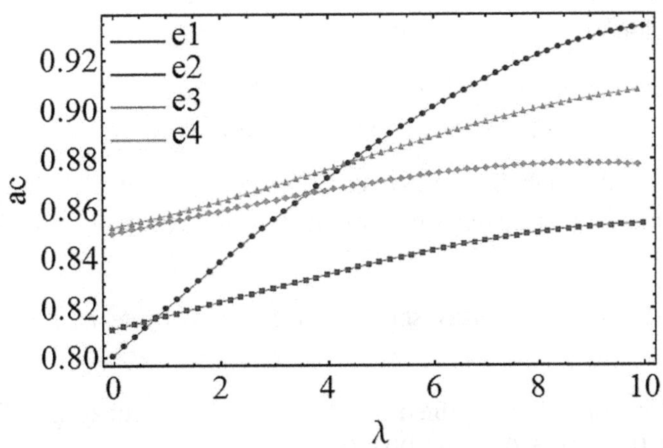

established for ranking the dissimilar alternatives by utilizing picture fuzzy environment. The suggested techniques have been demonstrated with a descriptive example for viewing their effectiveness as well as reliability. A test checks the reliability and validity has also been conducted for viewing the supremacy of the suggested techniques. Thus, the suggested operators provide a new direction to the information measure theory and give a new easier track to grip the uncertainties throughout the decision-making procedure. As picture fuzzy set is generalized concept of fuzzy set, intuitionistic fuzzy set and Pythagorean fuzzy set to deals with uncertainty more effectively and reliability. Also gave the more opportunity to deal with more situation in decision making problems.

ACKNOWLEDGMENT

The authors declare that they have no conflict of interest.

Ethical approval: This article does not contain any studies with human participants or animals performed by any of the authors.

REFERENCES

Ashraf, S., & Abdullah, S. (2018). Spherical aggregation operators and their application in multi attribute group decision-making. *International Journal of Intelligent Systems*, doi:10.1002/int.22062

Ashraf, S., Abdullah, S., & Mahmood, T. (n.d.). GRA method based on spherical linguistic fuzzy Choquet integral environment and its application in multi-attribute decision-making problems. *Mathematical Sciences*. doi:10.100740096-018-0266-0

Ashraf, S., Abdullah, S., Mahmood, T., Ghani, F., & Mahmood, T. (n.d.). Spherical fuzzy sets and their applications in multi-attribute decision making problems. *Journal of Intelligent & Fuzzy Systems*, 1-16.

Ashraf, S., Abdullah, S., & Qadir, A. (2018). Novel concept of cubic picture fuzzy sets. *J. New Theory*, *24*, 59–72.

Ashraf, S., Abdullah, S., Smarandache, F., & Amin, N. (2019). Logarithmic Hybrid Aggregation Operators Based on Single Valued Neutrosophic Sets and Their Applications in Decision Support Systems. *Symmetry*, *11*(3), 364. doi:10.3390ym11030364

Ashraf, S., Mahmood, T., Abdullah, S., & Khan, Q. (2018). Picture fuzzy linguistic sets and their applications for multi-attribute group decision making problems. *Nucleus*, *55*(2), 66–73.

Ashraf, S., Mahmood, T., Abdullah, S., & Khan, Q. (2018). Different approaches to multi-criteria group decision making problems for picture fuzzy environment. *Bull. Braz. Math. Soc. New Ser.* 4-018-0103-y doi:10.10070057

Atanassov, K. (1986). Intuitionistic fuzzy sets. *Fuzzy Sets and Systems*, *20*(1), 87–96. doi:10.1016/S0165-0114(86)80034-3

Atanassov, K. (1994). New operations defined over the intuitionistic fuzzy sets. *Fuzzy Sets and Systems*, *61*(2), 137–142. doi:10.1016/0165-0114(94)90229-1

Atanassov, K. (1999). *Intuitionistic Fuzzy Sets*. Heidelberg, Germany: Springer. doi:10.1007/978-3-7908-1870-3

Atanassov, K. (2007). Remark on intuitionistic fuzzy numbers. *Notes Intuit. Fuzzy Sets*, *13*, 29–32.

Atanassov, K., & Gargov, G. (1989). Interval valued intuitionistic fuzzy sets. *Fuzzy Sets and Systems*, *31*(3), 343–349. doi:10.1016/0165-0114(89)90205-4

Beliakov, G., Bustince, H., Goswami, D. P., Mukherjee, U. K., & Pal, N. R. (2011). On averaging operators for Atanassov's intuitionistic fuzzy sets. *Inf. Sci.*, *181*(6), 1116–1124. doi:10.1016/j.ins.2010.11.024

Cuong, B. C. (2014). Picture fuzzy sets. *J. Comput. Sci. Cybern.*, *30*(4), 409–420.

Cuong, B. C. (2013). Picture fuzzy sets first results. part 1. Seminar Neuro-Fuzzy Systems with Applications. Institute of Mathematics, Hanoi.

Cuong, B. C., Kreinovitch, V., & Ngan, R. T. (2016). A classification of representable t-norm operators for picture fuzzy sets. *Eighth International Conference on Knowledge and Systems Engineering*. 10.1109/KSE.2016.7758023

Cuong, B. C., & Phan, V. H. (2015). Some fuzzy logic operations for picture fuzzy sets. *Proceedings of Seventh International Conference on Knowledge and System Engineering*.

Ejegwa, P. A. (2015). Intuitionistic fuzzy sets approach in appointment of positions in an organization via max-min-max rule. *Global Journal of Science Frontier Research: Mathematics & Decision Sciences*, *15*(6), 1–5.

Ejegwa, P. A. (2018). *Distance and similarity measures for Pythagorean fuzzy sets*. Granular Computing. doi:10.100741066-018-00149-z

Ejegwa, P. A. (2019). *Improved composite relation for Pythagorean fuzzy sets and its application to medical diagnosis*. Granular Computing. doi:10.100741066-019-00156-8

Ejegwa, P. A. (2019). *Pythagorean fuzzy set and its application in career placements based on academic performance using max-min-max composition*. Complex and Intelligent Systems. doi:10.100740747-019-0091-6

Ejegwa, P. A., Akubo, A. J., & Joshua, O. M. (2014). Intuitionistic fuzzy set and its application in career determination via normalized Euclidean distance method. *European Scientific Journal*, *10*(15), 529–536.

Ejegwa, P. A., & Modom, E. S. (2015). Diagnosis of viral hepatitis using new distance measure of intuitionistic fuzzy sets. *International Journal of Fuzzy Mathematical Archive*, *8*(1), 1–7.

Ejegwa, P. A., Onoja, A. M., & Emmanuel, I. T. (2014). A note on some models of intuitionistic fuzzy sets in real life situations. *Journal of Global Research in Mathematical Archives*, *2*(5), 42–50.

Ejegwa, P. A., Tyoakaa, G. U., & Ayenge, A. M. (2016). Application of intuitionistic fuzzy sets in electoral system. *International Journal of Fuzzy Mathematical Archive*, *10*(1), 35–41.

Farhadinia, B. (2013). A theoretical development on the entropy of interval-valued fuzzy sets based on the intuitionistic distance and its relationship with similarity measure. *Knowledge-Based Systems, 39*, 79–84. doi:10.1016/j.knosys.2012.10.006

Garg, H. (2017). Some picture fuzzy aggregation operators and their applications to mul-ticriteria decision-making. *Arabian Journal for Science and Engineering, 42*(12), 5275–5290. doi:10.100713369-017-2625-9

Gautam, S. S. (2016). TOPSIS for multi criteria decision making in intuitionistic fuzzy environment. *International Journal of Computers and Applications, 156*(8), 42–49. doi:10.5120/ijca2016912514

Liu, H. W., & Wang, G. J. (2007). Multi-criteria methods based on intuitionistic fuzzy sets. *European Journal of Operational Research, 179*(1), 220–233. doi:10.1016/j.ejor.2006.04.009

Liu, P., Chu, Y., Li, Y., & Chen, Y. (2014). Some generalized neutrosophic number Hamacher aggregation operators and their application to group decision making. *International Journal of Fuzzy Systems, 16*(2).

Phong, P. H., Hieu, D. T., Ngan, R. H., & Them, P. T. (2014). Some composition of picture fuzzy relations. *Conference Paper.*

Singh, P. (2015). Correlation coefficients for picture fuzzy sets. *Journal of Intelligent & Fuzzy Systems, 28*, 591–604.

Son, L. H. (2016). Generalized picture distance measure and applications to picture fuzzy clustering. *Applied Soft Computing, 46*, 284–295. doi:10.1016/j.asoc.2016.05.009

Tan, C. (2011). A multi-criteria interval-valued intuitionistic fuzzy group decision making with Choquet integral based TOPSIS. *Expert Systems with Applications, 38*(4), 3023–3033. doi:10.1016/j.eswa.2010.08.092

Wang, C., Zhou, X., Tu, H., & Tao, S. (2017). Some geometric aggregaton operators based on picture fuzzy sets and their application in multiple attribute decision making. *International Journal of Pure and Applied Mathematics, 37*, 477–492.

Wang, H., & Zhang, Y. Q. (2013). Multi-criteria decision-making approach based on Atanassov's intuitionistic fuzzy sets with incomplete certain information on weights. *IEEE Transactions on Fuzzy Systems, 21*(3), 510–515. doi:10.1109/TFUZZ.2012.2210427

Wang, L., Peng, J., & Wang, J. (2018). A multi-criteria decision-making framework for risk ranking of energy performance contracting project under picture fuzzy environment. *Journal of Cleaner Production, 191*, 105–118. doi:10.1016/j.jclepro.2018.04.169

Wang, L., Zhang, H., Wang, J., & Li, L. (2018). Picture fuzzy normalized projection-based VIKOR method for the risk evaluation of construction project. *Applied Soft Computing, 64*, 216–226. doi:10.1016/j.asoc.2017.12.014

Wei, G. (2017). Picture fuzzy aggregation operators and their application to multiple attribute decision making. *Journal of Intelligent & Fuzzy Systems, 33*(2), 713–724. doi:10.3233/JIFS-161798

Wei, G. W. (2017). Picture 2-tuple linguistic Bonferroni mean operators and their application to multiple attribute decision making. *International Journal of Fuzzy Systems*, *19*(4), 997–1010. doi:10.100740815-016-0266-x

Wei, G. W., Alsaadi, F. E., Hayat, T., & Alsaedi, A. (2016). Picture 2-tuple linguistic aggregation operators in multiple attribute decision making. *Soft Computing*.

Xu, Z. S. (2007). Intuitionistic fuzzy aggregation operations. *IEEE Transactions on Fuzzy Systems*, *15*(6), 1179–1187. doi:10.1109/TFUZZ.2006.890678

Xu, Z. S. (2011). Approaches to multiple attribute group decision making based on intuitionistic fuzzy power aggregation operators. *Knowledge-Based Systems*, *24*(6), 749–760. doi:10.1016/j.knosys.2011.01.011

Xu, Z. S., & Yager, R. R. (2006). Some geometric aggregation operators based on intuitionistic fuzzy sets. *International Journal of General Systems*, *35*(4), 417–433. doi:10.1080/03081070600574353

Yager, R. R., & Abbasov, A. M. (2014). Pythagorean membership grades, complex numbers, and decision making. *International Journal of Intelligent Systems*, *28*(5), 436–452. doi:10.1002/int.21584

Yue, Z. L. (2011). Deriving decision maker's weights based on distance measure for interval-valued intuitionistic fuzzy group decision making. *Expert Systems with Applications*, *38*(9), 11665–11670. doi:10.1016/j.eswa.2011.03.046

Zadeh, L. A. (1965). Fuzzy sets. *Information and Control*, *8*(3), 338–356. doi:10.1016/S0019-9958(65)90241-X

Zeng, S., Asharf, S., Arif, M., & Abdullah, S. (2019). Application of Exponential Jensen Picture Fuzzy Divergence Measure in Multi-Criteria Group Decision Making. *Mathematics*, *7*(2), 191. doi:10.3390/math7020191

Chapter 19
Novel Fuzzy Preview Control With Granular Computing for Underactuated Systems:
Underactuated System Case Studies

Srikanth Kavirayani
https://orcid.org/0000-0002-5058-410X
Gayatri Vidya Parishad College of Engineering (Autonomous), India

ABSTRACT

This chapter provides an elaborate discussion on the use of fuzzy preview control designed with the help of look up tables and granular computing for the studies on under actuated systems which are dynamic systems which have lesser degrees of freedom than the actuating input signals. It is an important design consideration for most systems to be categorized under this section which makes them the essential component for analysis. The chapter provides an in depth analysis taking the case studies of various control system test beds such as the family of the inverted pendulum systems which make the prototype models for validation before applying the developed logic to control of high end complex dynamic systems in terms of stability, robustness, speed of operation and more importantly to see the fuzzy structure that controls these systems in terms of lookup tables.

1. INTRODUCTION

Under actuated systems are electromechanical systems which have less degrees of freedom than the actuating signals. Mathematical modeling of these systems is important due to difficulties associated with physical model based experiments. The studies of these systems is of paramount importance due to the extremely important nature of functionality associated with the machinery which operates as under actuated. It is very well known that most systems which are simple in design and do complex tasks such as robots for instance are under actuated. This chapter gives a glimpse of such systems from a fuzzy view point in line with the advantages of fuzzy structures. Granular computing is one additional focal

DOI: 10.4018/978-1-7998-0190-0.ch019

dimension added in this chapter which highlights the necessity for speeding up computations in the fuzzy domain. (Yang et al. 2013) discusses about algebraic structure of knowledge distances in establishing a framework for granular computation and its importance in fuzzy logic. (Lai and Chiang 2002) links how constraint based fuzzy modeling is key by taking the example of a rotary cement klin.(Pottem and Nigam 2014) elaborates on rule explosion by taking granular computing into fuzzy for speeding up processing of rules for a faster rule explosion. (Chen and Su 2008) elaborates on models with unbalanced data processing used in granular computing. Many Authors have discussed about distinct aspects and advantages of using granular computing applied to typical problems of data processing.

A good controller is needed for the better control of a under actuated system often taking into consideration design characterized by flexibility, efficiency and cost involved. Various types of controllers such as classes of combinations of P, I, and D, PID, LQR, LQG, H-Infinity, Kalman filter based observer design have been applied by many researchers and few are listed here (Reyhanoglul et al., 1996; Welch and Bishop, 2006; Ordaz et al., 2017; Rigatos et al., 2017; Su, 2017; Trinh and Huong, 2018). A simple PID controller will not suffix the requirements in all cases as it cannot be physically realized for all values. (Paliwal, Chopra, and Singla 2016) discusses the idea of a fuzzy PID controller with two loops for angle and position control which is one preliminary discussion of a test system on a mobile inverted pendulum environment. (Paliwal, Chopra, and Singla 2016) discusses the idea of a fuzzy PID controller with two loops for angle and position control which is one preliminary discussion of a test system on a mobile inverted pendulum environment.

Additionally, Meta-heuristic methods are naturally inspired and are being used worldwide in the context of the numerous benefits that the system would have in terms of energy usage(Bashiri and Karimi 2012). In this context, it is required to understand the basic advantages of fuzzy logic in comparison to naturally inspired algorithms. It is a very well known fact that naturally inspired algorithms are chosen and are mimicked for the sole reason that most species of animals from which humans learn are energy efficient when compared with other techniques that does not involve mimicking nature. A naturally inspired algorithm such as particle swarm optimization shows how it is more efficient when compared with other optimization techniques that are practiced by researchers. However, inspite of all the advantages due to meta-heuristics still the focus of this chapter remains with fuzzy logic due to the complexity of real time systems with degree of uncertainty being very high in systems which are under actuated. (N.P.Padhy 2014) indicates the hardware design is one advantage which necessitates the need for using fuzzy structures in system design.

2. RESEARCH METHODOLOGY

This section outlines the basic methods that have been tested on various inverted pendulum systems that are existing in the market by various vendors. The simulation studies are the basis by which one can extrapolate and test the systems. The systems that are tested are double, triple and rotary inverted pendulum systems which are highly under actuated and unstable when the pendulums are in the upright equilibrium positions. (N.P.Padhy 2014) indicates that the extension principal is a mathematical tool for extending crisp mathematical notions and operations to the milieu of fuzziness and it can be generalized for several variables from different universes of discourse. The work using lookup tables as a traditional method is available in the literature however work from the perspective of granular computing addition is not there in literature.

System Definition:

The family of inverted pendulum systems was considered in this work as the test bed. Generally, systems are under actuated and are of the form as given below which represents systems with less degree of freedom. The generalized representation of such a system is obtained from the Euler Lagrangian equations. The system generalized equations are given by the following

$$\frac{d}{dt}\left(\frac{\partial T}{\partial \dot{q}_i}\right) - \frac{\partial T}{\partial \dot{q}_i} + \frac{\partial U}{\partial \dot{q}_i} = Q_i$$

where

T: kinetic energy
U: potential energy
Qi: generalized forces not taken into account in T,U
qi: generalized coordinates

If the system equations are converted into state space form taking the generalized coordinates of a inverted pendulum on cart taking the variables of the cart position, pendulum angles will yield the functional aspect of such a system. The inverted pendulum system are given as below for the double and triple inverted pendulum systems as follows with varying degrees of freedom.

The block diagram realization of the double inverted pendulum system can be given as follows in figure 1.

The equations of such a system are represented as follows represented in figure 2 are as follows.

$$\dot{X}_4 = f_1(x_2, x_3, x_4) + B_{41}U$$

Figure 1. Block diagram of typical under actuated system on cart

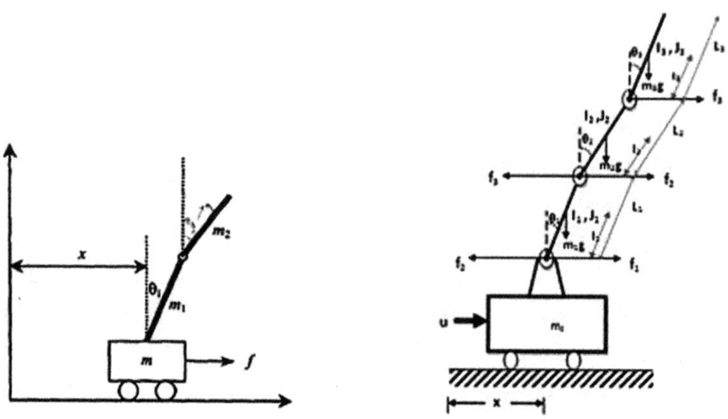

Figure 2. Block diagram of double inverted pendulum system on cart

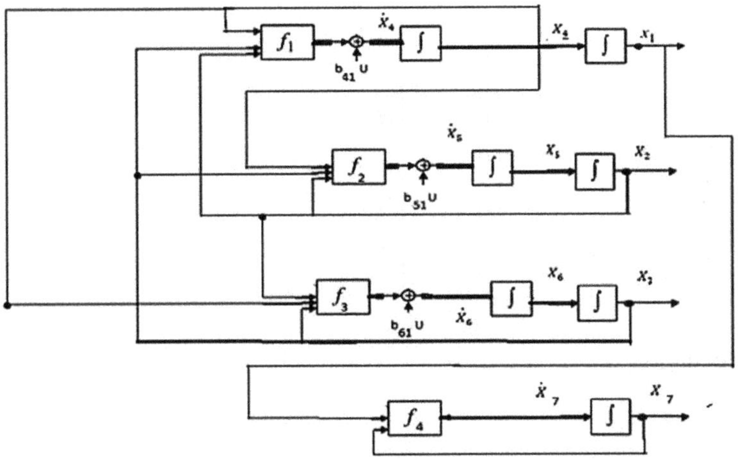

$$\dot{X}_5 = f_2(x_2, x_3, x_4) + B_{51}U$$

$$\dot{X}_6 = f_3(x_2, x_3, x_4) + B_{61}U$$

$$\dot{X}_7 = f_4(x_1, x_7)$$

Functionally, the equations that got added due to the variable x7 are the variables that are integrated into the system due to the presence of the time delay component into the system definition. This could be generalized as follows:

$$\dot{X}_i = f_j(x_l, x_m, x_n) + B_{ij}U$$

Where i,j are indices of the system equation and l,m,n are indices for the state variables. U is the input to the system.

If the system represented in figure 2 is generalized and a controller is to be designed due to the highly unstable nature of the system. The definition would yield the following,

The generalized state equation would be of the form

$$\begin{bmatrix} \dot{x}_1 \\ \dot{x}_2 \\ \cdots \\ \cdots \\ \vdots \\ \vdots \\ \dot{x}_n \end{bmatrix} = \begin{bmatrix} 0 & 0 & 0 & 1 & 0 & 0 & 0 \\ 0 & 0 & 0 & 0 & 1 & 0 & 0 \\ 0 & 0 & 0 & 0 & 0 & 1 & 0 \\ 0 & A_{42} & A_{43} & A_{44} & 0 & 0 & 0 \\ 0 & A_{52} & A_{53} & A_{54} & 0 & 0 & 0 \\ 0 & A_{62} & A_{63} & A_{64} & 0 & 0 & 0 \\ \frac{-4}{\Gamma} & 0 & 0 & 0 & 0 & 0 & \frac{2}{\Gamma} \end{bmatrix} \begin{bmatrix} x_1 \\ x_2 \\ \vdots \\ \vdots \\ \vdots \\ \vdots \\ x_n \end{bmatrix} + \begin{bmatrix} 0 \\ 0 \\ 0 \\ B_{41} \\ B_{51} \\ B_{61} \\ 0 \end{bmatrix} u$$

Thus the system equations have been generalized and a time delay has been integrated into the system definitions.

3. GRANULAR COMPUTING

Granular computing is a method of defining computation by dividing the tasks into granular level. It involves considering information granules for computation. (Pedrycz 2001) has discussed about the need for splitting a larger task into smaller task and making it granular which indicates a need to comprehend the problem for further insight into the problem. Granular computing as opposed to numeric computing is knowledge-oriented which makes it much suitable in the context of using it with fuzzy logic design. (Xu and Li 2014)propose a two-way machine learning system in fuzzy datasets and discuss fuzzy information granules as another method by taking the example of loans given to developing countries. For a better analysis of system dynamics, it is important to assemble lower level granules into higher or divide higher level granules into low grannies which paves way to interesting aspects of control. (Yao, Vasilakos, and Pedrycz 2013) have discussed the various applications of granular computing and how it is useful for interdisciplinary research. Fundamentally, the concept of fuzzy granular computing is done by differentiating the error that propagates into the system in the form of feedback after a control is applied to the plant. Then, because of the differentiation we are granulizing the error which makes the rule explosion faster in the process for looking up for the values in the process of evaluation for the input that needs to be passed to the plant.

A rule base composed by rules where the rule-antecedent contains the state variables X= (x1, x2, x3, x4,x5, x6) represents the controller input variables. The consequent output contains a single control output U,. The process states variables X can take n1 and n2 linguistic values respectively. Under these conditions, the maximum number of rule base is n1* n2 . The region of study for the analysis of the preview based fuzzy controller in the state space is normally bounded by some finite values(min,max) for Xi=1 to 6

A crisp element of X= (x1, x2, x3, x4,x5, x6) of the state space belong to the subspace of the partition associated with rule j, if it holds that

$$\forall_j \neq k\forall\left(x_1, x_2, x_3, x_4, x_5, x_6\right) : \mu_{R_j}\left(x_1, x_2, x_3, x_4, x_5, x_6\right) \geq \mu_{R_k}\left(x_1, x_2, x_3, x_4, x_5, x_6\right)$$

where $\mu_{R_j}\left(x_1, x_2, x_3, x_4, x_5, x_6\right)$ and $\mu_{R_k}\left(x_1, x_2, x_3, x_4, x_5, x_6\right)$ are the fuzzy relations representing the meaning of the rule-antecedents of rules j and k respectively, where,

Rule j**:** if x_1 is $LX_1(j)$ and x_2 is $LX_2(j)$ then U is $LU(j)$
Rule k**:** if x_1 is $LX_1(k)$ and x_2 is $LX_2(k)$ then U is $LU(k)$

Then μ_{R_j} and μ_{R_k} in the case of mamdani type implication are defined as follows:

$$\forall x_1, x_2, x_3, x_4, x_5, x_6 : \mu_{R_j}\left(x_1, x_2, x_3, x_4, x_5, x_6\right) = \min\left(\mu_{LX_1}(j), \mu_{LX_2}(j)\right)$$

$$\forall x_1, x_2, x_3, x_4, x_5, x_6 : \mu_{R_k}\left(x_1, x_2, x_3, x_4, x_5, x_6\right) = \min\left(\mu_{LX_1}(k), \mu_{LX_2}(k)\right)$$

A Type-1 Fuzzy Logic Controller has been designed with 2 inputs – error and rate of error, and 1 output – F. The controller uses traditional lookup tables which are defined based on fuzzy members for the fuzzy membership values defined based on the possible inputs. The rate of error is additionally differentiated for further granular computing. The analysis is done by taking the time derivative of error and error itself and passing on to the fuzzy lookup table where the lookup table has data which is a set of possible fuzzy numbers defined for each of the adjectives of distortion of the pendulum from the unstable equilibrium position which decides the deviation. The output of the fuzzy lookup table is given as the input to the plant model as shown in figure and then a feedback gain matrix defined is used for controlling the plant for a given reference cart position. The model that is developed can be considered for the preview of stepping that happens during a stair case stepping case of the humanoid robot. The preview data is looked up using the fuzzy controller and is passed on to the plant model and is analyzed. The analysis of the controller is shown in figure 3 and table 1. Typical responses with overshoots and undershoots in control efforts is seen in figures 3, 4 and 5.

Figure 3. Double pendulum performance with fuzzy controller

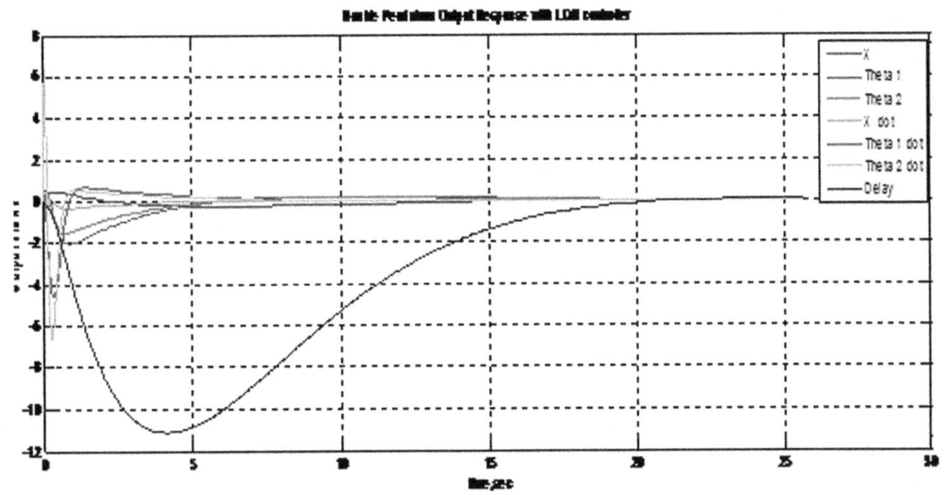

Table 1. Fuzzy analysis with variations

Case ID	Error gain value	Error rate gain	Output Gain	Feedback gain	Fuzzy Analysis
1	20	0.002	1	G = [-51.8604 -35.5576 44.2448 16.7295 -7.4139 2.8860 1.3030]	Possible
2	20	0.002	100	G = [-51.8604 -35.5576 44.2448 16.7295 -7.4139 2.8860 1.3030]	Difficult
3	20	0.002	10	G = [-51.8604 -35.5576 44.2448 16.7295 -7.4139 2.8860 1.3030]	Difficult
4	2	0.002	1	G = [-51.8604 -35.5576 44.2448 16.7295 -7.4139 2.8860 1.3030]	Possible

Figure 4. Case ID #1 from table 2

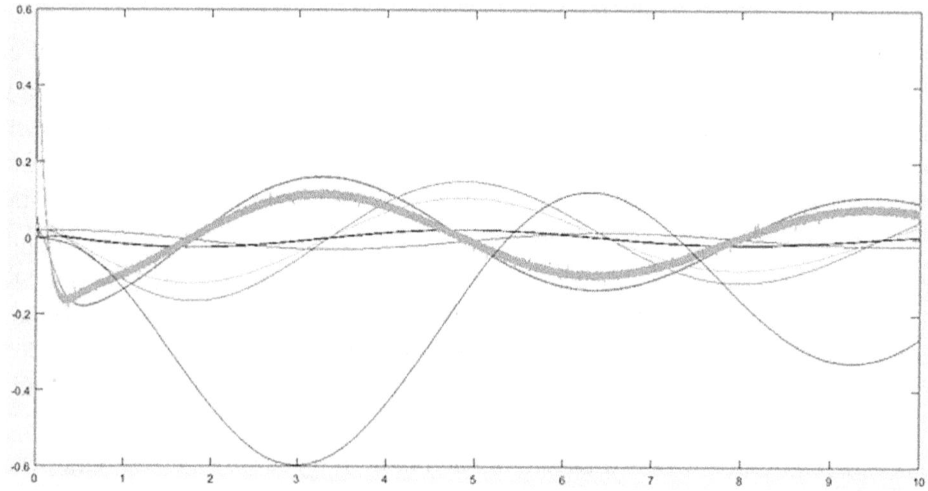

Figure 5. Case ID #4 from table 2

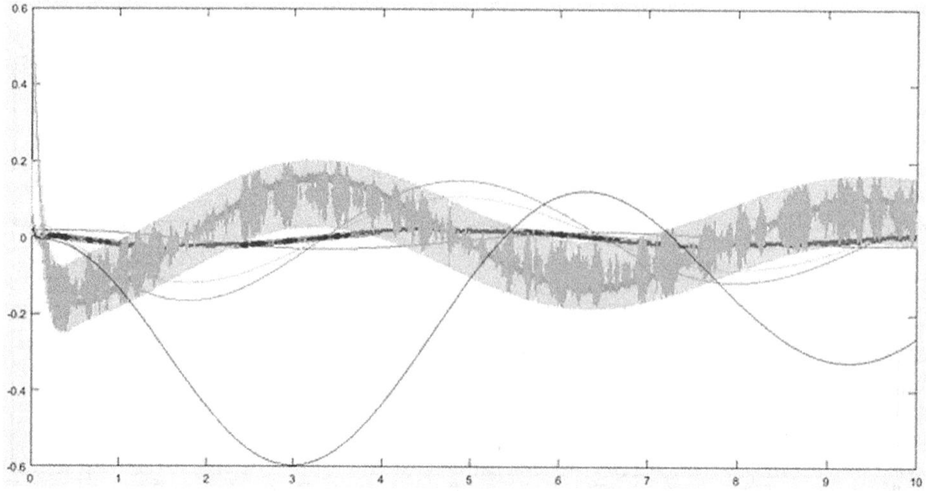

Figure 6. System model with fuzzy control

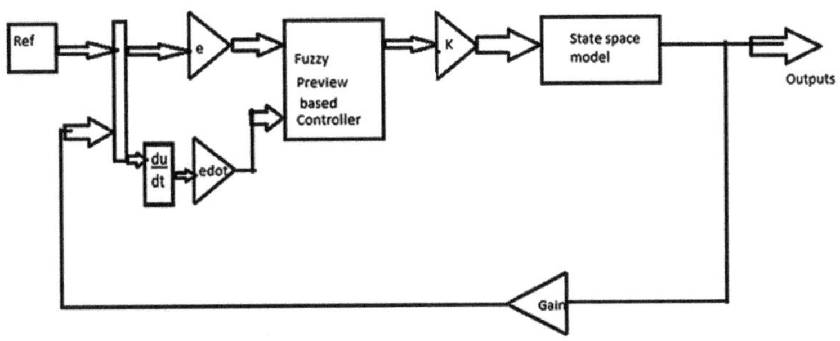

3.1 Fuzzy Preview Control With Granular Computing

The fuzzy control block diagram model that is proposed is represented in figure 6.

As shown in figure 4, the important parameters that are playing a key role are the error and the error rate which represent two inputs to the fuzzy preview controller which lookups the values based on which a decision is made and the output is passed on to the controller for amplification of the signal which is then passed as input to the plant. The feedback path has a LQR controller which does the state feedback control. The fuzzy granular based preview controller has a faster rule explosion which results in efficient control.

A weighted average method is adopted to calculate the hierarchical fuzzy controller with type-1 fuzzy controller and type -2 fuzzy controller.

u=wtype1*utype1fuzzy+wtype2*utype2fuzzy

As given in equation 9, where granulation is causing refinement in the way the control effort is used for the smoother control of the plant model. The incremental control action is not only a function of error and rate of error but also the time delay component which makes it a nonlinear controller which is more efficient than a liberalized LQR controller in terms of the control effort for reduced oscillations.

3.2 Results From Fuzzy Preview Control

The system matrices representing the model of the integrated rotary inverted pendulum model that were taken during the process are given as follows taken from [2] are as follows

$$A = \begin{bmatrix} 0 & 0 & 1 & 0 & 0 \\ 0 & 0 & 0 & 1 & 0 \\ 0 & 39.2 & -14.52 & 0 & 0 \\ 0 & 81.78 & -13.98 & 0 & 0 \\ -4/Tuo & 2/Tuo & 0 & 0 & 0 \end{bmatrix}$$

$$B = \begin{bmatrix} 0 \\ 0 \\ 25.54 \\ 25.59 \\ 0 \end{bmatrix}$$

C is an identity matrix and D is a null matrix for the model of the Rotary pendulum taken from [2]. The system definitions were obtained by values taken from Appendix A. The system is as such obtained from the dynamic model of the rotary inverted pendulum with an integration of the time delay component added to the system as a fifth state other than the four states which define the vertical and horizontal motions of the plant. The plant model has two rotations (α, θ) which are varied in order to achieve the control. The velocity components of the same are as well considered in the system. The system model dynamics are developed from a standard Quanser product of the rotary inverted pendulum as taken in [2].

The gain values that have been calculated from the classical linear quadratic controller are given as

Gain Matrix = [-101.0896 175.5642 -20.1725 23.4626 1.3030]

The initial conditions for all the cases is taken as

Xo=[pi/6 0 0 0 0]

It is assumed that the initial conditions for the pendulum assign the pendulum to a position where it is in the vicinity of the upright equilibrium position ensuring that the case studies only the stabilization about the upright equilibrium position and not the control problem of swing up from the downward equilibrium position which is another case study found in literature.

The various test cases tested as per the block diagram represented in figure 1 is shown below where the analysis is done for cases 1 to 12. The error values variations resulted in control action where the pendulum was stabilized about the upright equilibrium position.

The various plots that have been obtained are given below in figures 2 and 3. In figure 2,the control effort was successful in the stabilization of the pendulum and figure 3 we present a counter example case where the failure of the stabilization occurs which is indicative that the failure is also possible if the constraints on the system definition limits are violated. The fuzzy lookup table is as shown below in figure 7 and failure case in figure 8 for the lookup table classification in figure 9.

It can be observed clearly the status of control for the system from swing up to upright equilibrium in figures 10, 11 and 12 where the switching happens between stable and unstable regions for the laboratory model of the Quanser rotary inverted pendulum which is an actual experiment conducted on the test bed.

Table 2. Test case of error variations

Case ID	E	Edot	Standard Ref	Control Action
1	20	.002	.001	Success
2	2	.002	.001	Success
3	2	0.2	.001	Fails
4	2	.002	.001	Fails
5	2	.2	.001	Fails
6	2	.02	1	Fails
7	2	.002	1	Success
8	.002	.002	1	Fails
9	.002	.00002	1	Fails
10	.002	.00002	.001	Fails
11	20	20	.001	Fails
12	2000	20	.001	Fails
13	2000	2000	1	Fails

Figure 7. Case1 output states

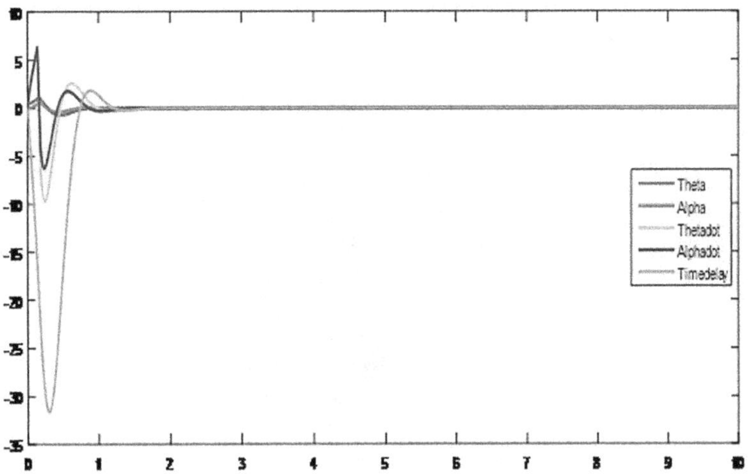

4. RESULTS AND DISCUSSION OF TRIPLE INVERTED PENDULUM

The output obtained with a basic linear quadratic controller with the integrated model are given in figure 13 wherein the integrated plant model is taken and peak overshoot is seen for the delay component of the plant dynamics. Table 3 indicates the various case studies performed on the triple inverted pendulum system. Figure 13 to 18 to indicates the response of the simulation for the plant model analysed using MATLAB simulink for the preview control with variations in error and rate of error. The analysis was performed to understand the significance of error influences on the dynamic response of the system. The analysis is done as per table I where figures 14 indicate the case wherein the stability is difficult to

Figure 8. Case 9 of output states (unstable)

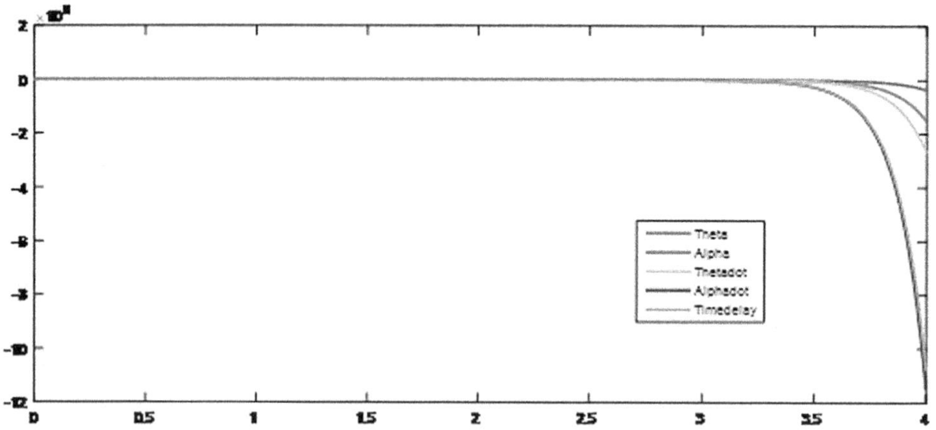

Figure 9. Lookup table classification

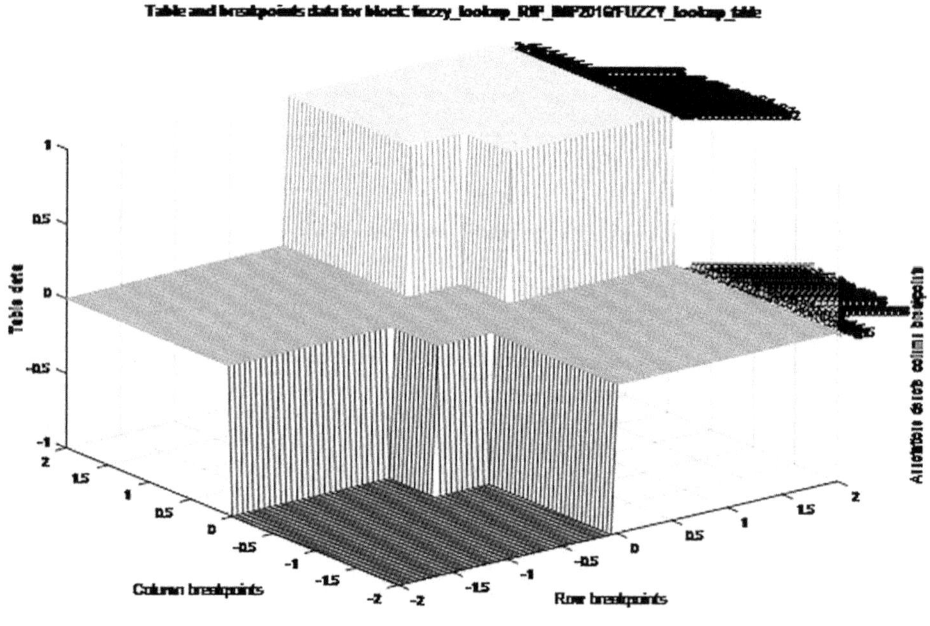

achieve with fuzzy preview control. This indicates the practical case wherein it is virtually impossible to obtain zero error and also the case when the error is in the extremity of being at a very high value. In figures 13 to 18 it is clearly seen that incorporating the delay as a separate state has yielded a condition wherein maximum effort is required for stabilizing the state representing and corresponding to the time delay component. This categorically implies that the effort that is required in terms of the input for stabilization of other states is less. The other important observation is that the variations indicate that the system being highly unstable would oscillate continuously which indicates the difficulty involved in stabilization a system such as the triple inverted pendulum plant.

Figure 10. Typical real time performance of RIP with offset

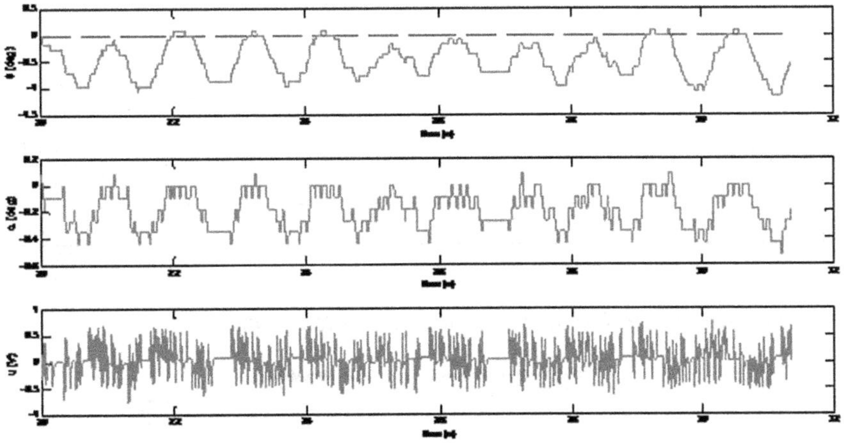

Figure 11. Typical stabilization failure case in real time

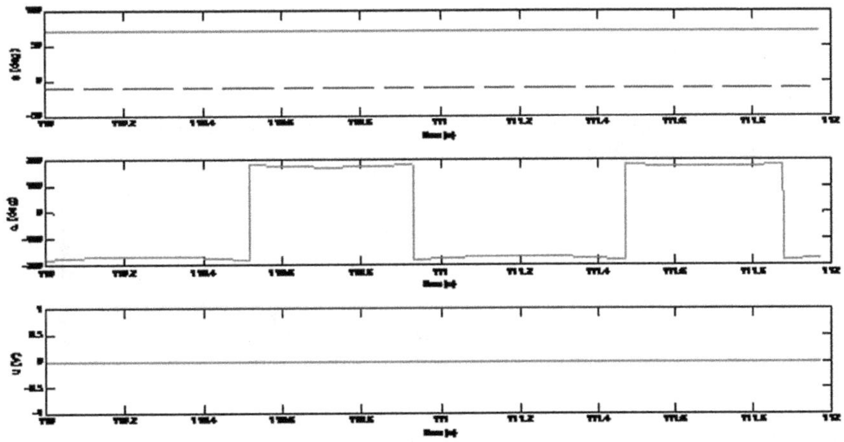

Figure 12. Swing up and stabilization in real time

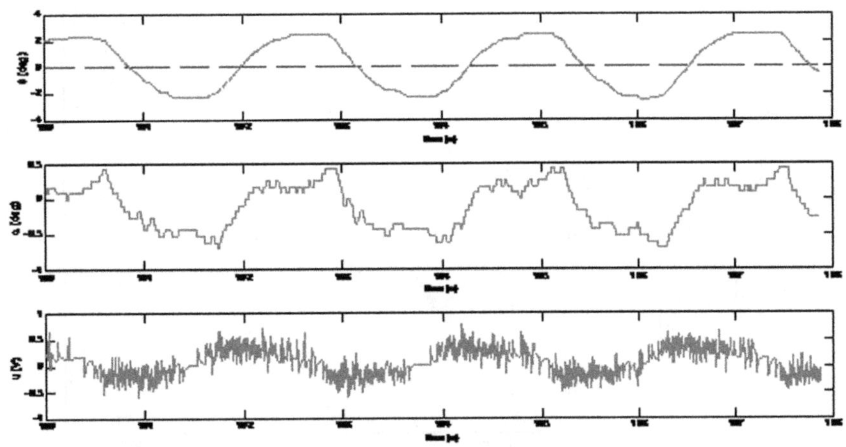

Table 3. Case studies for preview control

CaseID	Error gain	Error rate gain	Control Status
1	20	0.02	Stable
2	2	0.002	Stable
3	0	0	UnStable
4	10	0.001	Stable
5	100	0	Unstable

Figure 13. Linear quadratic controller response for integrated plant

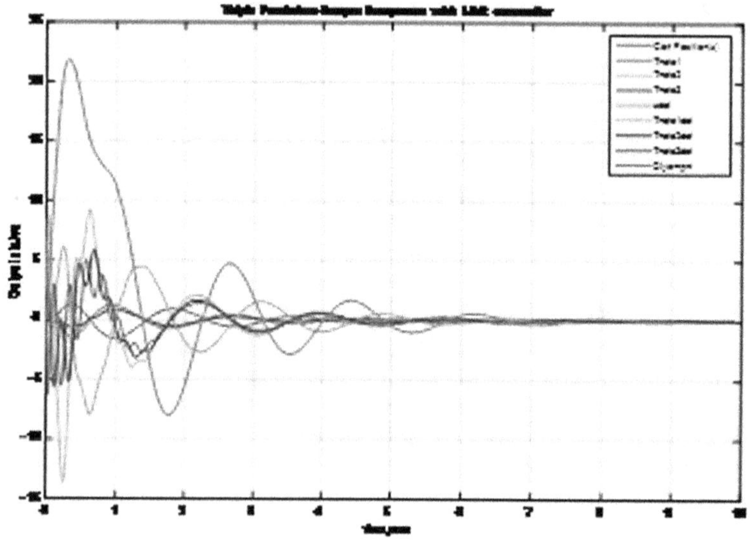

Figure 14. Caseid #1, lower error gain and lower error rate gain

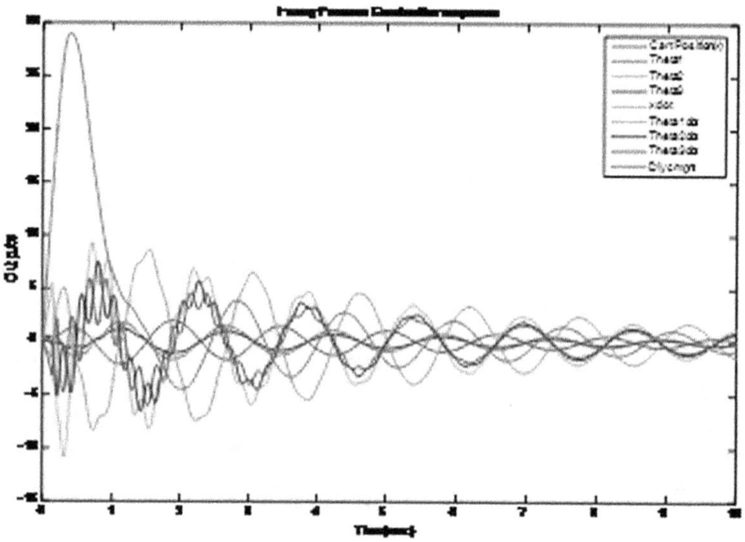

Figure 15. Case #2, lower error rate and nominal error gain

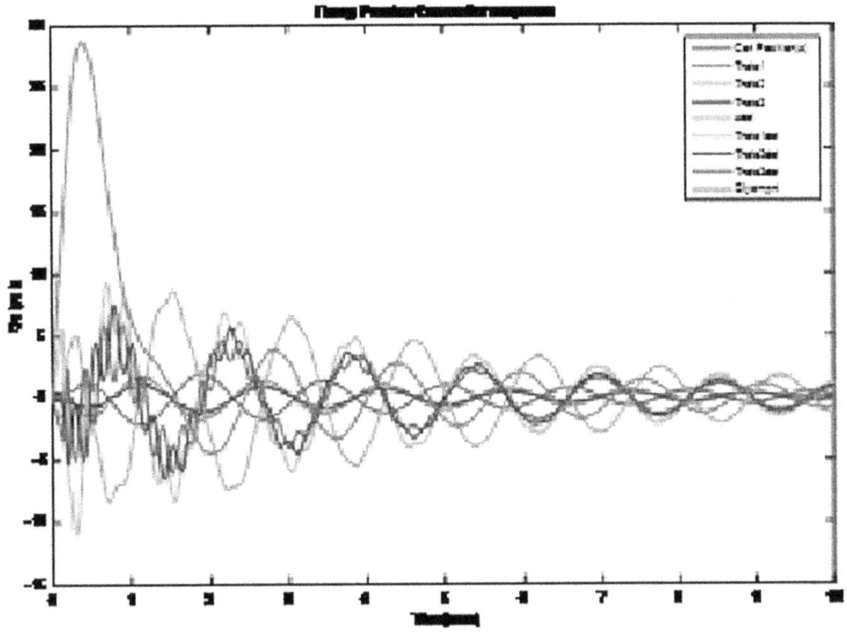

Figure 16. Case#3, zero error gain and zero error rate gain

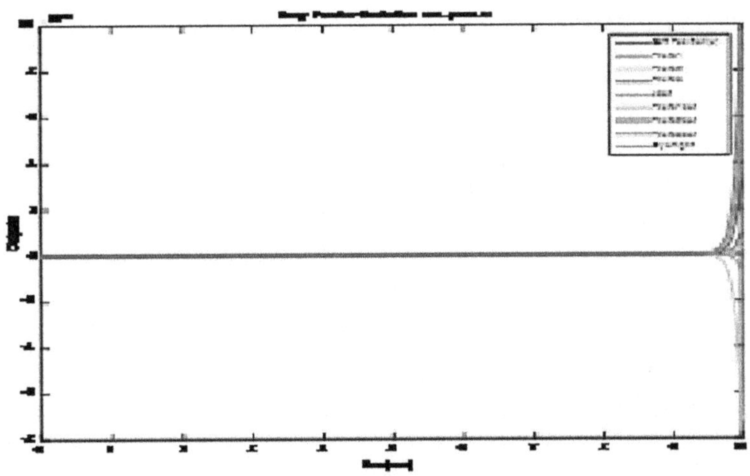

Figure 17 . Case#5, Very high error gain and zero error rate gain

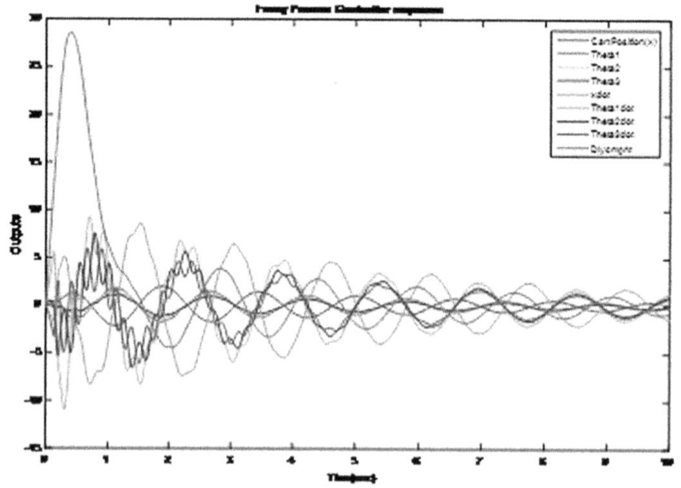

Figure 18. Relative analysis for error gain and error rate

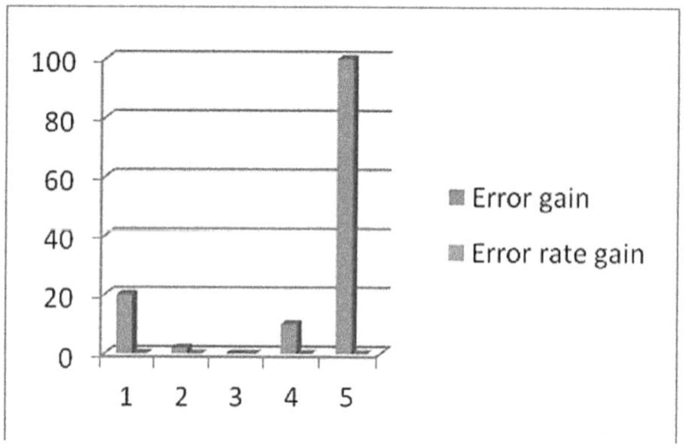

The relative importance of error gain variation with error rate gain which act as the inputs for the fuzzy lookup tables are projected in figure 18. The variations are as per table 3 which indicates that the variation of error gain with error rate gain should be relatively less in order to better stabilize a system smoothly without much variation.

4. CONCLUSION

Granular computing essentially was done using an additional differentiation (granulation of error) which resulted in splitting the error involved from the feedback and was used for speeding up the control process and smoothening the response curve using fuzzy look up tables and preview control. Parseval's analysis has indicated that the energy is infinite for non casual systems with the power in finite limit for various underactuated systems. The analysis can be extended to high end dynamic large scale systems where delays can be incorporated as time varying; the realization of the fuzzy preview control can be achieved with the help of cheap integrated circuits. The methodology developed, tested and validated can be used in various fields starting from entertainment, military, and other systems linked with intelligent system design.

The following conclusions can be drawn from the analysis

1. Fuzzy granular computing is faster than traditional computing with less logical steps
2. The precision of the system is improved with fuzzy granular computing
3. The system performance improves with fuzzy granular computing with preview control
4. Lookup tables help faster resolution of the problem within the domain.

ACKNOWLEDGMENT

The authors declare that they have no conflict of interest.

Ethical approval: This article does not contain any studies with human participants or animals performed by any of the authors.

REFERENCES

Bashiri, M., & Karimi, H. (2012). Effective Heuristics and Meta-Heuristics for the Quadratic Assignment Problem with Tuned Parameters and Analytical Comparisons. *Journal of Industrial Engineering International*, *8*(1), 1–9. Retrieved from http://www.jiei-tsb.com/content/8/1/6

Chen, L. S., & Su, C. T. (2008). Using Granular Computing Model to Induce Scheduling Knowledge in Dynamic Manufacturing Environments. *International Journal of Computer Integrated Manufacturing*, *21*(5), 569–583.

Lai, R., & Chiang, D. (2002). Constraint-Based Granular Computing for Fuzzy Modeling. *2002 IEEE World Congress on Computational Intelligence. 2002 IEEE International Conference on Fuzzy Systems. FUZZ-IEEE'02. Proceedings (Cat. No.02CH37291), 1*, 584–89.

Ordaz, P., Santos-Sánchez, O. J., Rodríguez-Guerrero, L., & González-Facundo, A. (2017). Nonlinear Stabilization for a Class of Time Delay Systems via Inverse Optimality Approach. *ISA Transactions*, *67*, 1–8. doi:10.1016/j.isatra.2016.11.015 PMID:27939564

Padhy, N. P. (2014). *Artificial Intelligence and Intelligent Systems* (13th ed.). Oxford University Press.

Paliwal, S., Chopra, V., & Singla, S. K. (2016). Stabilization of Mobile Inverted Pendulum Using Fuzzy PID Controllers. *2016 7th India International Conference on Power Electronics (IICPE)*, 1–4.

Pedrycz, W. (2001). Granular Computing: An Introduction. *Proceedings Joint 9th IFSA World Congress and 20th NAFIPS International Conference (Cat. No. 01TH8569)*, 1349–54. Retrieved from http://ieeexplore.ieee.org/document/943745/

Pottem & Nigam. (2014). Hierarchical Fuzzy PID Controller Using Granular Computing for Inverted Pendulum. *International Journal of Electrical, Electronics and Data Communication, 1*(4), 31–35.

Reyhanoglul, Van Der Schaft, McClamroch, & Kolmanovsky. (1996). Nonlinear Control of a Class of Underactuated Systems. *Proceedings of the 35th Conference on Decision and Control*, 1682–87.

Rigatos, G., Siano, P., Abbaszadeh, M., & Ademi, S. (2017). Nonlinear H-Infinity Control for the Rotary Pendulum. *2017 11th International Workshop on Robot Motion and Control (RoMoCo)*, 217–22.

Su. (2017). Stabilization of Double Inverted Pendulum on Cart: LQR Approach. *International Journal of Mechanical and Production Engineering, 5*(2), 149–53.

Trinh, H., & Huong, D. C. (2018). A New Method for Designing Distributed Reduced-Order Functional Observers of Interconnected Time-Delay Systems. *Journal of the Franklin Institute*. Retrieved from http://linkinghub.elsevier.com/retrieve/pii/S0016003218300383

Welch, G., & Bishop, G. (2006). An Introduction to the Kalman Filter. *In Practice, 7*(1), 1–16. Retrieved from http://citeseerx.ist.psu.edu/viewdoc/download?doi=10.1.1.79.6578&rep=rep1&type=pdf

Xu & Li. (2014). Granular Computing Approach to Two-Way Learning Based on Formal Concept Analysis in Fuzzy Datasets. *IEEE Trans. Cybern., 46*, 366–79.

Yang, X., Song, X., She, Y., & Yang, J. (2013). Hierarchy on Multigranulation Structures: A Knowledge Distance Approach. *International Journal of General Systems, 42*(7), 754–773.

Yao, Vasilakos, & Pedrycz. (2013). Granular Computing: Perspectives and Challenges. *IEEE Transactions on Cybernetics, 43*(6), 1977–89. Retrieved from http://www.ncbi.nlm.nih.gov/pubmed/23757594

Chapter 20
Fuzzy Similarity Relations in Decision Making

Mohamed El Alaoui
https://orcid.org/0000-0001-5342-4964
National High School for the Arts and Professions, Meknès, Morocco

Khalid El Yassini
Faculty of Sciences, Moulay Ismail University, Meknès, Morocco

ABSTRACT

Similarity is an ambiguous term that can be interpreted differently depending on the context of use. In this chapter, the authors review some of its uses before focusing on decision making. Ignoring the uncertainty of human knowledge would be denying a major attribute. Thus, they linked it to the fuzzy context. However, even taking this aspect into consideration, the opinion itself must be relevant. Therefore, the more a decision is similar to other opinions, the more it is coherent. Hence, there is a need to measure the similarity between each couple of expressed opinions.

INTRODUCTION

An ambiguous term is a word that can have many interpretations (Klir 1987), and similarity fits perfectly this definition. A criminal can be arrested on the basis of the similarity of his fingerprints to those found at the scene (Dhivya and Sridevi 2018; Keatley and Clarke 2019). In mechanics, the laws of similarity and similitude are a necessary step linking the essay prototype in the laboratory, to the real model on field (Lalanne 2014). In the musical field, measuring the similarity of a new submitted melody to the previous existing ones will permit verifying authenticity (Velankar, Sahasrabuddhe, and Kulkarni 2015). The musical example, already illustrate the notion complexity, similarities between texts (Schedl et al. 2011), melodies (Velankar et al. 2015) tone (Krasser et al. 2012) etc. … (Bello 2011; Knees and Schedl 2013). Other uses could be found in psychology (Wan et al. 2018), mathematics (Bluman and Cole 1974), chemistry (Carbó-Dorca and Mezey 2001), biology (Klinkenberg 2001), marketing (Bèzes

DOI: 10.4018/978-1-7998-0190-0.ch020

and Mercanti-Guérin 2017), and even between science fields (Deriglazov 2011). The followers of the aforementioned disciplines try to give a mathematical aspect to their field, thus the majority of the works consider similarity as a distance duel, also called dissimilarity.

In addition to ambiguity, similarity is also vague, in the absence of a clearly established boundary, what are the criteria to confirm with no doubt that two objects or concepts are similar. If a teacher finds doubtful similarities between the works of two of his students. What is the necessary amount to arouse his suspicions, and what is the sufficient amount to assert cheating. The question is all the more troubling since, using different plagiarism detection software, we achieve significantly different results.

Decision-making is a complex process whose result is paramount in different disciplines, where the similarity between opinions can be regarded as a criterion of validity and coherence (El Alaoui, El Yassini, and Ben-azza 2018), with various applications ranging from agriculture (El Alaoui, Ben-azza, and El Yassini 2019b) to education (El Alaoui, El Yassini, and Ben-azza 2019). In the sequel, we review similarity properties from a mathematical point of view and what it is meant to represent in the chapter 'what is similarity!' Human knowledge is highly tainted with uncertainty, what needs to be taken into account from the outset. Fuzzy logic happens to be a powerful tool in such situations.

In the chapter 'the need to include fuzzy logic', the required mathematical background from fuzzy logic is presented. The chapter 'similarity in a fuzzy context' shows the evolution of the similarity notion in the fuzzy context according to what it must reflect. Then, in the 'comparative example', the extended similarity and distance studied are compared in order to prove their merits.

WHAT IS SIMILARITY?

Evaluating the similarity between two objects, could be realized by measuring the relationship that links them (Teng et al. 2014). From a mathematical point of view, generally, a relation R is defined between two objects $X(x_1,\ldots,x_n)$ and $Y(y_1,\ldots,y_m)$ by the Cartesian product: R: $X{\times}Y \rightarrow \{0,1\}$ where the elements x_i and y_j are either completely related, i.e $R(x_i,y_j)=1$ or not at all i.e $R(x_i,y_j)=0$. In such situation, containing two objects only, we talk about a binary relation. If the relation consists of three, four or five objects; then the relation is called ternary, quaternary or quinary (Sivanandam, Sumathi, and Deepa 2007).

A relation is reflexive if for all x_i: $R(x_i,x_i)=1$. A relation is symmetric if for all x_i,x_j: $R(x_i,x_j) = R(x_j,x_i)$. A relation is max-min transitive if for all x_i,x_j,x_k

$$R\left(x_i, x_j\right) \geq \max_{x_k \in X}\left(\min\left(R\left(x_i, x_k\right); R\left(x_k, x_j\right)\right)\right).$$

A relation which verify the three conditions is called a similarity S (Zadeh 1971).

While humans try to give a definition to all concepts for a better understanding and correct use. Similarity is difficult to surround. There is therefore no universal definition, giving way to definitions adapted to the domain of use.

A widely used approach is to define it from what it is not. More precisely, from what can be called its inverse, a distance, or a dissimilarity D (Szmidt 2014). Which gives rise to the first category of similarity approaches, and the frequently used ones (El Alaoui, Ben-azza, and El Yassini 2018b, 2019a), called similarity based distance, or proximity based measures. Given its simplicity, the most used formula is the one proposed in (Santini and Jain 1997): S=1 - D. Other formulas could be found in the literature S=1/(1+D) (Kóczy T. and Tikk 2000), $S=e^{-\alpha D}$ where α is a steepness measure (Williams and Steele 2002). In (Zhou et al. 2015) the authors divided similarities into distance metric based methods and score function based methods. In addition to the distance based similarity methods, referred to as proximity based methods, authors in (Cross and Sudkamp 2002) added set-theoretic measures and logic-based measures. Another classification in the datamining context divided similarities in accordance to the data computed as follows (Teng et al. 2014): similarity between numerical objects, non-numerical objects and mixed ones.

The existence of words meaning roughly the same, complicates more the apprehension of similarity, such as: similitude, fit, congruence, compatibility. In the mechanical framework, similarity and similitude can be used interchangeably (Zohuri 2015). According to (Thibaut 1997), similitude requires a perfect matching, while similarity means near resemblance. Compatibility is an ideal that we look for with our coworkers, friends, or partners. Awkwardly often mistaken for similarity. While some authors consider compatibility as a generalization of similarity (Cross and Sudkamp 2002), we believe that neither of them involves the other. A discussion of closely related terms was presented in (Bèzes and Mercanti-Guérin 2017).

Even if one tends to have a single and unique definition the plurality and versatility of explanations must be seen as an opportunity to encompass the multiplicity of disciplines and uses (Cross and Sudkamp 2002; Gregson 1975). What cannot be done otherwise (Ismat and Samina 2009).

What the similarity is supposed to reflect, is not the subject of unanimity either. The earliest approach was proposed by Jaccard (Jaccard 1908), in which the similarity between A and B noted $S(A,B)$ is meant to reflect commonalities, which is a function of their intersection $A \cap B$ and union $A \cup B$ as follows:

$$S\left(A,B\right) = A \cap B \,/\, A \cup B\,.$$

Dubois and Parade (1992) sustained that a similarity must reflect symmetrical difference, what can be represented in classical logic by decomposing the union as follows:

$$A \cup B = \left(A \cap B\right) \cup \left(A \cap \bar{B}\right) \cup \left(\bar{A} \cap B\right).$$

where $\bar{A}$ and $\bar{B}$ represent respectively A and B complements.

Hence, the similarity could be written in function of three entities:

$$S\left(A,B\right) = f\left(A \cap B, A \cap \bar{B}, \bar{A} \cap B\right).$$

In addition to positive matches, Sokal and Sneath (Sokal and Sneath 1963) maintained that a similarity measure should take also into consideration negative matches. Hence, they included common differences $\bar{A} \cap \bar{B}$ what added an extra parameter to consider as follows:

$$S\left(A,B\right) = f\left(A \cap B, A \cap \bar{B}, \bar{A} \cap B, \bar{A} \cap \bar{B}\right).$$

Assume the universe considered is very large, in that case, for any two objects; the negative matches would be considerable, leading to large similarity values. In the presence of a ranking order, the similarity could be considered in function of the negative matches encompassed between the compared objects (El Alaoui, Ben-azza, and Zahi 2016):

$$S\left(A,B\right)=f\left(A\cap B,A\cap\bar{B},\bar{A}\cap B,\overline{A\cap B}\right)$$

such that $A\leq\overline{A\cap B}\leq B$ and $\overline{A\cap B}$ represents $A\cap B$ complement.

Further applications were presented in (El Alaoui, El Yassini, et al. 2018)

THE NEED TO INCLUDE FUZZY LOGIC

Our knowledge of the world is not perfect. Hence, the acceptance of some form of uncertainty is trivial. Without going into the details of the most appropriate form to model the uncertainties (Zadeh 2006). We believe that fuzzy logic (Zadeh 1965), permits a coherent perception, closing the gap between natural human language's and machine's one (Gupta, Jain, and Joshi 2018). In this situation, the relation between two elements is no longer limited to two possible choices, but to a panoply between the values 0 and 1. A first attempt to define fuzzy similarity relation was proposed in (Zadeh 1971) in which Zadeh, the founder of fuzzy logic, defined it as a generalization of equivalence relation. Here we present the necessary background from fuzzy logic.

Let X be a universal set and F a fuzzy subset in X. F is defined as follows (El Alaoui 2018):

$$F=\{\left\langle x,\mu_F\left(x\right)\right\rangle \mid x\in X\}$$

where $\mu_F(x)$ is the degree of membership of x in F in the unity interval μ_F: $X\rightarrow[0,1]$.

A fuzzy set is convex if and only if:

$$\mu_F\left(\lambda x_1+\left(1-\lambda\right)x_2\right)\geq\min\left(\mu_F\left(x_1\right),\mu_F\left(x_2\right)\right)$$

for all x_1,x_2∈X and λ∈[0,1]. A fuzzy set F is normalized if $\sup(\mu_F)=1$ otherwise it is called subnormal.

A Fuzzy Number (FN) is a convex normalized fuzzy set of the real line R^1 whose membership function is piecewise continuous. A FN that does not fulfill normality condition is called a Fuzzy Sub Number FSN.

A FN N is called positive, denoted by $N>0$ if its membership function $\mu_N(x)=0$ for all $x<0$.

In order to achieve the compromise between simple representation and complexity (Grzegorzewski and Mrówka 2005, 2007), the most used notation is based on General Trapezoidal Fuzzy Numbers (GTFN), where a GTFN $\tilde{A}$ is denoted by (a^1,a^2,a^3,a^4,h) as follows (Javanmard and Nehi 2018):

$$\mu_{\tilde{A}}(x) = \begin{cases} \dfrac{x - a^1}{a^2 - a^1} h, & a^1 \leq x \leq a^2 \\ h, & a^2 \leq x \leq a^3 \\ \dfrac{x - a^4}{a^3 - a^4} h, & a^3 \leq x \leq a^4 \\ 0, & otherwise \end{cases}.$$

If h=1 the GTFN is called Trapezoidal FN (TFN). Otherwise, (i.e. h<1 it is called a Trapezoidal Fuzzy Sub Number (TFSN).

If $a^2=a^3$ then the GTFN is called a General Triangular FN (GTrFN). Similarly a GTrFN is called a Triangular FN (TrFN) if h=1 and a Triangular Fuzzy Sub Number otherwise (TrFSN).

If $a^1=a^2=a^3=a^4$ then we have a classical crisp number.

For any two positive GTFNs $\tilde{A}\left(a^1, a^2, a^3, a^4, h_A\right)$ and $\tilde{B}\left(b^1, b^2, b^3, b^4, h_B\right)$ the fuzzy addition, subtraction, and multiplication are respectively defined as follows:

$$\tilde{A} \oplus \tilde{B} = \left(a^1 + b^1, a^2 + b^2, a^3 + b^3, a^4 + b^4, \min\left\{h_A, h_B\right\}\right) \tag{1}$$

$$\tilde{A} \ominus \tilde{B} = \left(a^1 - b^4, a^2 - b^3, a^3 - b^2, a^4 - b^1, \min\left\{h_A, h_B\right\}\right) \tag{2}$$

$$\tilde{A} \otimes \tilde{B} = \left(a^1 * b^1, a^2 * b^2, a^3 * b^3, a^4 * b^4, \min\left\{h_A, h_B\right\}\right) \tag{3}$$

According to (Zadeh 1975), it is counterintuitive to ask for an exact membership function, while the purpose is to permit fuzziness. Hence, the previously mentioned fuzzy sets will be referred as type 1 fuzzy sets.

An Interval Type 2 Fuzzy Set (IT2FS) $\tilde{\tilde{A}}$ is bounded by two type 1 fuzzy sets $\tilde{A}^L$ and $\tilde{A}^U$ such that $\tilde{A}^L \subset \tilde{A}^U$ which means that $\mu_{\tilde{A}^L} \leq \mu_{\tilde{A}^U}$, therefore $\tilde{\tilde{A}} = \left[\tilde{A}^L, \tilde{A}^U\right]$.

Three types of IT2FS could be found in the literature (Javanmard and Nehi 2018; Mendel 2017):

- An IT2FS is called: Perfectly normal Interval Type 2 FN (PIT2FN), if $\tilde{A}^L$ and $\tilde{A}^U$ are FN.
- An IT2FS is called: Sub normal IT2FN or simply an Interval Type 2 FN (IT2FN), if $\tilde{A}^L$ is a FSN and $\tilde{A}^U$ is a FN.
- An IT2FS is called: Generalized IT2FN (GIT2FN), if $\tilde{A}^L$ and $\tilde{A}^U$ are FSN.

According to (Mendel 2017), only PIT2FN, are worth being called IT2FN.

Suppose $\tilde{\tilde{A}} = \left[\tilde{A}^L, \tilde{A}^U\right]$ and $\tilde{\tilde{B}} = \left[\tilde{B}^L, \tilde{B}^U\right]$ are two IT2FNs. The extended arithmetic operations on IT2FNs are defined as follows (Javanmard and Nehi 2018):

$$\tilde{\tilde{A}} \oplus \tilde{\tilde{B}} = \left[\tilde{A}^L \oplus \tilde{B}^L, \tilde{A}^U \oplus \tilde{B}^U\right] \quad (4)$$

$$\tilde{\tilde{A}} \ominus \tilde{\tilde{B}} = \left[\tilde{A}^L \ominus \tilde{B}^L, \tilde{A}^U \ominus \tilde{B}^U\right] \quad (5)$$

$$\tilde{\tilde{A}} \otimes \tilde{\tilde{B}} = \left[\tilde{A}^L \otimes \tilde{B}^L, \tilde{A}^U \otimes \tilde{B}^U\right] \quad (6)$$

If $\tilde{A}^L$ and $\tilde{A}^U$ are both GTFN, then the IT2FN $\tilde{\tilde{A}} = \left[\tilde{A}^L, \tilde{A}^U\right]$ is called a General Interval Type 2 TFN (GIT2TFN).

If $\tilde{A}^L$ and $\tilde{A}^U$ are both TFN, then the IT2FN $\tilde{\tilde{A}} = \left[\tilde{A}^L, \tilde{A}^U\right]$ is called an Interval Type 2 TFN (IT2TFN).

SIMILARITY IN A FUZZY CONTEXT

Similarity definition in a fuzzy context is all the more debatable (Baccour, Alimi, and John 2014), and can also be adapted to the relative domain. Whereas some authors (Cross and Sudkamp 2002; Tversky 1977) see that similarity is not necessarily symmetric; we can say that a son's attitude or face is similar to his father's, but the inverse is barely acceptable which suggests the existence of a reference; others maintain the usefulness of the three conditions: reflexivity, symmetry and max-min transitivity. In the fuzzy context, the max-min transitivity condition becomes (Szmidt 2014):

For three fuzzy subsets $\tilde{A}$, $\tilde{B}$, and $\tilde{C}$ such that $\tilde{A} \subset \tilde{B} \subset \tilde{C}$ then:

$$\min\left(S\left(\tilde{A}, \tilde{B}\right); S\left(\tilde{B}, \tilde{C}\right)\right) \geq S\left(\tilde{A}, \tilde{C}\right).$$

In addition to the axiomatic divergences, and without denying the influx of fuzzy logic, the latter generates new problematics peculiar to it, which have no equivalent in the classic logic. Namely, the interpretation of conjunctions and disjunctions commonly referred to as unions and intersections. These two concepts were modeled using T-norms and T-conorms. However, their use cannot be carried out indifferently from the field of use (Harmse 1996), reinforcing the necessity of a plurality of definitions.

Zwick et al. (Zwick, Carlstein, and Budescu 1987) extended Jaccard's index (Jaccard 1908) to the fuzzy context, representing the intersection and the union respectively by the min and max operator as follows:

$$S\left(\tilde{A}, \tilde{B}\right) = \frac{\int \min\left\{\mu_{\tilde{A}}, \mu_{\tilde{B}}\right\}}{\int \max\left\{\mu_{\tilde{A}}, \mu_{\tilde{B}}\right\}}$$

where $\mu_{\tilde{A}}$ and $\mu_{\tilde{B}}$ represent the membership function of $\tilde{A}$ and $\tilde{B}$ respectively.

Lu et al. (Lu, Lan, and Wang 2006) proposed two major improvements to Zwick's similarity. First, the similarity measures the intersection areas without distinguishing the degree of certainty of each area concerned, the larger the intersection support is, the larger the similarity should be. Hence, the similarity between two opinions $\tilde{R}_i$ and $\tilde{R}_j$ becomes:

$$S_w\left(\tilde{R}_i,\tilde{R}_j\right)=\frac{\int\left(\min\left\{\mu_{\tilde{R}_i},\mu_{\tilde{R}_j}\right\}\right)^2}{\int\left(\max\left\{\mu_{\tilde{R}_i},\mu_{\tilde{R}_j}\right\}\right)^2} \tag{7}$$

Second, even if the expressed opinions have no common intersection, it does not imply that they are dissimilar. Suppose three decision makers assess an alternative respectively as, excellent good and bad, it is obvious that the first and second opinions are similar to each other than to the third. Hence Lu et al. (Lu et al. 2006) maintained that what they call consistency between opinions should take into account both similarity and distance:

$$r\left(\tilde{R}_i,\tilde{R}_j\right)=\beta S_w\left(\tilde{R}_i,\tilde{R}_j\right)+\left(1-\beta\right)\left(1-d\left(\tilde{R}_i,\tilde{R}_j\right)\right).$$

However, the measure proposed was sensitive to the other considered opinions, due to the distance used (El Alaoui, Ben-azza, and El Yassini 2018a). In addition, the additive aggregation operator used to merge distance and similarity imposes a careful weight choosing, $0\leq\beta\leq1$, otherwise it can over valuate an aspect over the other.

While coherency (Lane 1972) and consistency are both linked to logic (Carnielli and Coniglio 2016; Carnielli and Malinowski 2018), an opinion's consistency, in our point of view, reflects its ability to encompass others in order to provide solid evidence. The coherence proposed in (El Alaoui et al. 2016) (Eq. 4) allows simultaneous consideration of all aspects constituting a similarity, in which the multiplicative aggregation permits creating a class of quasi-equivalent similarities (Lesot and Rifqi 2012). This cannot be achieved by choosing an additive weighting method, which imposes arbitrations between the aspects dealt with.

$$r_p\left(\tilde{R}_i,\tilde{R}_j\right)=\frac{p+S_w\left(\tilde{R}_i,\tilde{R}_j\right)}{p+1}*\left(1-\frac{D\left(\tilde{R}_i,\tilde{R}_j\right)}{M}\right) \tag{8}$$

where:

$$D\left(\tilde{R}_i,\tilde{R}_j\right)=\frac{1}{2}\int_{a^i}^{d^i}\mu_{\tilde{R}_i}-\min\left(\mu_{\tilde{R}_i},\mu_{\tilde{R}_j}\right)+\frac{1}{2}\int_{a^j}^{d^j}\mu_{\tilde{R}_j}-\min\left(\mu_{\tilde{R}_i},\mu_{\tilde{R}_j}\right)+\int_{\min\left(c^i,c^j\right)}^{\max\left(b^i,b^j\right)}1-\max\left(\mu_{\tilde{R}_i},\mu_{\tilde{R}_j}\right)^2$$

M: the width of the rating scale and p: positive real.

Generalization of the weighted product aggregation were proposed in (El Alaoui and Ben-azza 2017b, 2017a).

While type 1 fuzzy sets represents words, type 2 fuzzy sets can model perceptions (Singh 2014). To able to compute the coherence between perceptions, the similarity (Eq. 3) and the distance (Eq. 4) can be extended as follows:

$$S_w\left(\tilde{\tilde{R}}_i,\tilde{\tilde{R}}_j\right)=\frac{S_w\left(\tilde{R}_i^{\,L},\tilde{R}_i^{\,L}\right)+S_w\left(\tilde{R}_i^{\,U},\tilde{R}_i^{\,U}\right)}{2} \tag{9}$$

$$D\left(\tilde{\tilde{R}}_i,\tilde{\tilde{R}}_j\right)=\frac{D\left(\tilde{R}_i^{\,L},\tilde{R}_i^{\,L}\right)+D\left(\tilde{R}_i^{\,U},\tilde{R}_i^{\,U}\right)}{2} \tag{10}$$

Then the coherence could be computed as in (Eq. 4) by:

$$r_p\left(\tilde{\tilde{R}}_i,\tilde{\tilde{R}}_j\right)=\frac{p+S_w\left(\tilde{\tilde{R}}_i,\tilde{\tilde{R}}_j\right)}{p+1}*\left(1-\frac{D\left(\tilde{\tilde{R}}_i,\tilde{\tilde{R}}_j\right)}{M}\right) \tag{11}$$

COMPARATIVE EXAMPLE

To able comparison, we use the example presented in (Wu and Mendel 2009), in which 32 IT2FS were compared.

Before going further, the proposed classification is not unanimous. As there is no complete order for fuzzy entities, three approaches exist in the literature (Wang and Kerre 2001):

In the first, each fuzzy set $\tilde{\tilde{A}}$ .is associated to a real number $\mathcal{R}\left(\tilde{\tilde{A}}\right)$.such that the comparison occurs between the associated crisp numbers.

$$\tilde{\tilde{A}}\preceq\tilde{\tilde{B}} \text{ if } \mathcal{R}\left(\tilde{\tilde{A}}\right)\leq\mathcal{R}\left(\tilde{\tilde{B}}\right) \tag{12}$$

$$\tilde{\tilde{A}}\cong\tilde{\tilde{B}} \text{ if } \mathcal{R}\left(\tilde{\tilde{A}}\right)=\mathcal{R}\left(\tilde{\tilde{B}}\right) \tag{13}$$

$$\tilde{\tilde{A}}\succeq\tilde{\tilde{B}} \text{ if } \mathcal{R}\left(\tilde{\tilde{A}}\right)\geq\mathcal{R}\left(\tilde{\tilde{B}}\right) \tag{14}$$

where $\succeq$, $\cong$, and $\preceq$, denotes the fuzzy versions of $\geq$, $=$, and $\leq$, respectively.

Suppose $\tilde{\tilde{A}}\left[\left(a^{1L},a^{2L},a^{3L},a^{4L},h_A^L\right)\left(a^{1U},a^{2U},a^{3U},a^{4U},h_A^U\right)\right]$ is a GIT2TFN, the ranking value associated is (Javanmard and Nehi 2018):

Table 1. Compared IT2FS

Rank	word	LMF	UMF
1	None to very	(0,0, 0.05, 0.66 ; 1)	(0,0, 0.14, 1.97 ; 1)
2	Teeny-weeny	(0,0, 0.01, 0.13 ; 1)	(0,0, 0.14, 1.97 ; 1)
3	A smidgen	(0,0, 0.05, 0.63 ; 1)	(0,0, 0.26, 2.63 ; 1)
4	Tiny	(0,0, 0.05, 0.63 ; 1)	(0,0, 0.36, 2.63 ; 1)
5	Very small	(0,0, 0.10, 1.16 ; 1)	(0,0, 0.64, 2.47 ; 1)
6	Very little	(0,0, 0.09, 0.99 ; 1)	(0,0, 0.64, 2.63 ; 1)
7	A bit	(0.79, 1.68, 1.68, 2.21 ; 0.74)	(0.59, 1.50, 2.00, 3.41 ; 1)
8	Little	(1.09, 1.83, 1.83, 2.21 ; 0.53)	(0.38, 1.50, 2.50, 4.62 ; 1)
9	Low amount	(1.67, 1.92, 1.92, 2.21 ; 0.30)	(0.09, 1.25, 2.50, 4.62 ; 1)
10	Small	(1.79, 2.28, 2.28, 2.81 ; 0.40)	(0.09, 1.50, 3.00, 4.62 ; 1)
11	Somewhat small	(2.29, 2.70, 2.70, 3.21 ; 0.42)	(0.59, 2.00, 3.25, 4.41 ; 1)
12	Some	(2.88, 3.61, 3.61, 4.21 ; 0.35)	(0.38, 2.50, 5.00, 7.83 ; 1)
13	Some to moderate	(4.09, 4.65, 4.65, 5.41 ; 0.40)	(1.17, 3.50, 5.50, 7.83 ; 1)
14	Moderate amount	(4.29, 4.75, 4.75, 5.21 ; 0.38)	(2.59, 4.00, 5.50, 7.62 ; 1)
15	Fair amount	(4.79, 5.29, 5.29, 6.02 ; 0.41)	(2.17, 4.25, 6.00, 7.83 ; 1)
16	Medium	(4.86, 5.03, 5.05, 5.14 ; 0.27)	(3.59, 4.75, 5.50, 6.91 ; 1)
17	Modest amount	(4.79, 5.30, 5.30, 5.71 ; 0.42)	(3.59, 4.75, 6.00, 7.41 ; 1)
18	Good amount	(5.79, 6.50, 6.50, 7.21 ; 0.41)	(3.38, 5.50, 7.50, 9.62 ; 1)
19	Sizeable	(6.79, 7.38, 7.38, 8.21 ; 0.49)	(4.38, 6.50, 8.00, 9.41 ; 1)
20	Quite a bit	(6.79, 7.38, 7.38, 8.21 ; 0.49)	(4.38, 6.50, 8.00, 9.41 ; 1)
21	Considerable amount	(7.19, 7.58, 7.58, 8.21 ; 0.37)	(4.38, 6.50, 8.25, 9.62 ; 1)
22	Substantial amount	(7.79, 8.22, 8.22, 8.81 ; 0.45)	(5.38, 7.50, 8.75, 9.81 ; 1)
23	A lot	(7.69, 8.19, 8.19, 8.81 ; 0.47)	(5.38, 7.50, 8.75, 9.83 ; 1)
24	High amount	(7.79, 8.30, 8.30, 9.21 ; 0.53)	(5.38, 7.50, 8.75, 9.81 ; 1)
25	Very sizable	(8.29, 8.56, 8.56, 9.21 ; 0.38)	(5.38, 7.50, 9.00, 9.81 ; 1)
26	Large	(8.03, 8.36, 8.36, 9.17 ; 0.57)	(5.98, 7.75, 8.60, 9.52 ; 1)
27	Very large	(8.72, 9.91, 10, 10 ; 1)	(7.37, 9.41, 10, 10 ; 1)
28	Humongous amount	(8.95, 9.93, 10, 10 ; 1)	(7.37, 9.59, 10, 10 ; 1)
29	Huge amount	(9.34, 9.95, 10, 10 ; 1)	(7.37, 9.73, 10, 10 ; 1)
30	Very high amount	(9.37, 9.95, 10, 10 ; 1)	(7.37, 9.82, 10, 10 ; 1)
31	Extreme amount	(9.74, 9.98, 10, 10 ; 1)	(7.37, 9.82, 10, 10 ; 1)
32	Maximum amount	(9.61, 9.97, 10, 10 ; 1)	(8.68, 9.91, 10, 10 ; 1)

$$\mathcal{R}\left(\tilde{\tilde{A}}\right) = h_{\tilde{A}}^{L}\frac{a^{1L} + a^{2L} + a^{3L} + a^{4L}}{4} + h_{\tilde{A}}^{U}\frac{a^{1U} + a^{2U} + a^{3U} + a^{4U}}{4} \tag{15}$$

In the second, each fuzzy quantity is compared to a reference value. In a q-imensional space IR^q with q=8 for GIT2TFN, the normalized Minkowski distances, could be written as follows (Figueroa-García, Chalco-Cano, and Román-Flores 2015; Grzegorzewski 2004):

$$d\left(\tilde{\tilde{A}}, \tilde{\tilde{B}}\right) = \left(\frac{1}{q}\sum_{i=1}^{q}\left|a^{i} - b^{i}\right|^{m}\right)^{\frac{1}{m}} \tag{16}$$

where m=1 for the Hamming distance, and m=2 for the Euclidean distance. Thus, the 32 IT2FS could be ranked by computing their distances to the zero [(0,0,0,0,1) (0,0,0,0,1)] (Table. 2).

While in the third approach, pairwise comparisons permit ordering each couple. The latter is cumbersome and requires $n(n - 1)/2 = 32*31/2 = 496$ comparison. Where n denotes the number of the compared quantities.

The ranking proposed in (Wu and Mendel 2009) that was assumed right in (Gong, Yang, and Dai 2018) seems to be counterintuitive especially for extremal compared IT2TFS. 'Teeny weeny' have the smallest tuples values, thus it should be ranked first. However, (Wu and Mendel 2009) ranked it in the second place, while all other methods sort it in the first place. Otherwise, its sorting capacity is acceptable.

The Euclidean and the Hamming metric achieves the same ranking except the words 'very small' and 'very little' that have an inverted ranking.

The proposed ranking is similar to the one extracted from the Hamming distance with a single reversed ranking between 'low amount' and 'little'.

In general, the ranking proposed and the ones given using the Hamming distance or the Euclidean distance could be perceived as identical. Whereas the classification proposed in (Javanmard and Nehi 2018) is the furthest from all compared methods. Table 3 resumes Kendall correlation coefficients between each ranking.

Returning back to similarity, and using (Wu and Mendel 2009) ranking to ease comparison. Table 4 resumes the similarities between 'none to very' and the other words.

Methods in (Mitchell 2006; Wu and Mendel 2009) failed to capture any similarity between the first word and all words ranging from the 14th to the 32nd position.

The method in (Zeng and Li 2006) able computing similarity between all compared words. However, the results achieved are all superior to 0.5. Furthermore, the variations are inconsistent with words ranking.

The two similarities proposed in (Gong et al. 2018) obtain a more consistent value distribution. However, they are extremely sensitive the lower bound of the compared IT2FS, comparing 'quite a bite' and 'considerable amount', the first tuples are smaller, thus should be more similar to 'none to very'. The same could be said comparing 'high amount' are 'very sizable' similarities to the reference word.

The coherence proposed was computed with different parameter values to verify their consistency p=1, p=2, and p=10. That unlike the method presented in (Gong et al. 2018) preserves the same order.

Table 2. Compared ranking for IT2FS

(Javanmard and Nehi 2018)		hamming		Euclidean		Proposed distance		word
value	ranking	value	ranking	value	ranking	value	ranking	
0.353	2	0.353	2	0.736	2	0.352	2	None to very
0.281	1	0.281	1	0.7	1	0.281	1	Teeny-weeny
0.446	3	0.446	3	0.961	3	0.446	3	A smidgen
0.459	4	0.459	4	0.965	4	0.459	4	Tiny
0.546	6	0.546	6	0.992	5	0.546	6	Very small
0.544	5	0.544	5	1.019	6	0.544	5	Very little
1.526	8	1.733	7	1.916	7	1.903	7	A bit
1.586	9	1.995	8	2.312	8	2.173	9	Little
1.347	7	2.023	9	2.351	9	2.131	8	Low amount
1.609	10	2.296	10	2.6	10	2.447	10	Small
1.854	11	2.644	11	2.839	11	2.785	11	Somewhat small
2.59	12	3.753	12	4.254	12	3.989	12	Some
3.19	13	4.6	13	4.93	13	4.820	13	Some to moderate
3.366	15	4.839	14	5.021	14	4.989	14	Moderate amount
3.627	16	5.205	16	5.423	16	5.399	16	Fair amount
3.271	14	5.101	15	5.173	15	5.213	15	Medium
3.827	17	5.356	17	5.455	17	5.499	17	Modest amount
4.583	18	6.5	18	6.713	18	6.730	18	Good amount
5.359	21	7.256	20	7.386	20	7.466	20	Sizeable
5.359	21	7.256	20	7.386	20	7.466	20	Quite a bit
5.007	19	7.414	21	7.55	21	7.599	21	Considerable amount
5.789	23	8.06	23	8.15	23	8.260	23	Substantial amount
5.864	24	8.043	22	8.134	22	8.252	22	A lot
6.156	25	8.13	24	8.226	24	8.327	24	High amount
5.606	22	8.289	26	8.385	26	8.455	26	Very sizable
6.398	26	8.221	25	8.282	25	8.372	25	Large
9.426	27	9.426	27	9.468	27	9.695	27	Very large
9.48	28	9.48	28	9.52	28	9.747	28	Humongous amount
9.549	29	9.549	29	9.587	29	9.796	29	Huge amount
9.564	30	9.564	30	9.602	30	9.816	30	Very high amount
9.614	31	9.614	31	9.652	31	9.838	31	Extreme amount
9.771	32	9.771	32	9.781	32	9.904	32	Maximum amount

Table 3. Kendall correlation coefficients between compared ranking methods

Correlation	(Wu and Mendel 2009)	(Javanmard and Nehi 2018)	Euclidean	Hamming	Proposed
(Wu and Mendel 2009)	1	0.955	0.983	0.979	0.975
(Javanmard and Nehi 2018)	0.955	1	0.956	0.960	0.964
Euclidean	0.983	0.956	1	0.996	0.992
Hamming	0.979	0.960	0.996	1	0.996
Proposed	0.975	0.964	0.992	0.996	1

Table 4. Compared IT2FNs

Compared words	Jaccard (Wu and Mendel 2009)	(Mitchell 2006)	(Zeng and Li 2006)	(Gong et al. 2018)		Proposed coherence		
				S_1	S_2	r_1	r_2	r_{10}
(1,2)	0.8	0.57	0.93	0.9841	0.9841	0.7936	0.8601	0.9567
(1,3)	0.77	0.62	0.92	0.9016	0.9016	0.9094	0.9362	0.9752
(1,4)	0.75	0.61	0.91	0.8881	0.8881	0.8977	0.928	0.9721
(1,5	0.64	0.6	0.84	0.8480	0.8480	0.7744	0.8431	0.9431
(1,6)	0.65	0.6	0.86	0.8393	0.8393	0.7908	0.8542	0.9463
(1,7)	0.11	0.11	0.58	0.5460	0.4491	0.4308	0.5659	0.7624
(1,8)	0.11	0.09	0.63	0.5040	0.4257	0.4173	0.5487	0.7397
(1,9)	0.16	0.13	0.66	0.5254	0.4597	0.4273	0.5579	0.7478
(1,10)	0.13	0.11	0.62	0.4829	0.4136	0.4071	0.5337	0.7179
(1,11)	0.08	0.06	0.6	0.4561	0.3739	0.3791	0.5017	0.6801
(1,12)	0.05	0.04	0.62	0.3356	0.2814	0.3167	0.4204	0.5712
(1,13)	0.01	0.01	0.63	0.3043	0.2425	0.2684	0.3577	0.4876
(1,14)	0	0	0.69	0.2970	0.2299	0.258	0.344	0.4691
(1,15)	0	0	0.66	0.2812	0.2179	0.2375	0.3167	0.4319
(1,16)	0	0	0.75	0.2956	0.2245	0.2468	0.3291	0.4488
(1,17)	0	0	0.72	0.2736	0.2076	0.2325	0.31	0.4228
(1,18)	0	0	0.7	0.2325	0.1772	0.171	0.228	0.3109
(1,19)	0	0	0.74	0.2109	0.1578	0.1341	0.1789	0.2439
(1,20)	0	0	0.74	0.2109	0.1578	0.1341	0.1789	0.2439
(1,21)	0	0	0.74	0.2155	0.1619	0.1275	0.17	0.2319
(1,22)	0	0	0.78	0.1959	0.1453	0.0945	0.126	0.1718
(1,23)	0	0	0.77	0.1946	0.1444	0.0949	0.1265	0.1725
(1,24)	0	0	0.77	0.1898	0.1406	0.0911	0.1215	0.1657
(1,25)	0	0	0.77	0.1975	0.1468	0.0847	0.1129	0.154
(1,26)	0	0	0.8	0.1858	0.1370	0.0889	0.1185	0.1616
(1,27)	0	0	0.82	0.1342	0.0976	0.0227	0.0303	0.0413
(1,28)	0	0	0.86	0.1336	0.0971	0.0201	0.0268	0.0366
(1,29)	0	0	0.83	0.1329	0.0966	0.0176	0.0235	0.0321
(1,30)	0	0	0.84	0.1327	0.0964	0.0166	0.0222	0.0303
(1,31)	0	0	0.85	0.1322	0.0961	0.0156	0.0208	0.0283
(1,32)	0	0	0.89	0.1310	0.0951	0.0123	0.0164	0.0223

CONCLUSION

In this work, we reviewed similarity interpretations from a mathematical perspective and used fuzzy logic to encompass the inherent uncertainty. We also extended a fuzzy distance metric that we used also as a ranking tool. An example proving its sorting capacity was provided. Then, a successful comparison of the distance multiplicative fusion with similarity, called coherence, was presented.

REFERENCES

Baccour, L., Alimi, A. M., & John, R. I. (2014). Some Notes on Fuzzy Similarity Measures and Application to Classification of Shapes. *Recognition of Arabic Sentences and Mosaic.*, *41*(2), 81–90.

Bello, J. P. (2011). Measuring Structural Similarity in Music. *IEEE Transactions on Audio, Speech, and Language Processing*, *19*(7), 2013–2025. doi:10.1109/TASL.2011.2108287

Bèzes, C., & Mercanti-Guérin, M. (2017). Similarity in Marketing: Scope, Measurement, and Fields of Application. [English Edition]. *Recherche et Applications en Marketing*, *32*(1), 83–105. doi:10.1177/2051570716676258

Bluman, G. W., & Cole, J. D. (1974). *Similarity Methods for Differential Equations* (1st ed.). New York: Springer-Verlag. doi:10.1007/978-1-4612-6394-4

Carbó-Dorca, R., & Mezey, P. G. (Eds.). (2001). *Fundamentals of Molecular Similarity*. New York: Springer US. doi:10.1007/978-1-4757-3273-3

Carnielli, W., & Coniglio, M. E. (2016). *Paraconsistent Logic: Consistency, Contradiction and Negation*. Springer International Publishing.

Carnielli, W., & Malinowski, J. (Eds.). (2018). *Contradictions, from Consistency to Inconsistency*. Springer International Publishing. doi:10.1007/978-3-319-98797-2

Cross, V. V., & Sudkamp, T. A. (2002). *Similarity and Compatibility in Fuzzy Set Theory: Assessment and Applications* (1st ed.). Berlin: Physica-Verlag Heidelberg. doi:10.1007/978-3-7908-1793-5

Deriglazov, A. A. (2011). Formal Similarity Between Mathematical Structures of Electrodynamics and Quantum Mechanics. *International Journal of Theoretical Physics*, *50*(3), 654–662. doi:10.100710773-010-0589-6

Dhivya, J., & Sridevi, B. (2018). Intuitionistic Fuzzy Similarity Measure for Generalized Fuzzy Numbers and Its Application in Fingerprint Matching. *Journal of the Institution of Electronics and Telecommunication Engineers*, *0*(0), 1–12.

Dubois, D., & Prade, H. (1992). Gradual Inference Rules in Approximate Reasoning. *Information Sciences*, *61*(1), 103–122. doi:10.1016/0020-0255(92)90035-7

El Alaoui & Ben-azza. (2017b). Generalization of the Weighted Product Aggregation Applied to Data Fusion of Intuitionistic Fuzzy Quantities. *2017 Intelligent Systems and Computer Vision (ISCV)*.

El Alaoui, M. (2018). SMART Grid Evaluation Using Fuzzy Numbers and TOPSIS. *IOP Conference Series. Materials Science and Engineering*, *353*(1), 012019. doi:10.1088/1757-899X/353/1/012019

El Alaoui, M., & Ben-azza, H. (2017a). Aggregation of Performance Indicators for Supply Chain and Fuzzy Logic Extensions Applied to Green Supply Chain. *2017 International Colloquium on Logistics and Supply Chain Management (LOGISTIQUA)*. 10.1109/LOGISTIQUA.2017.7962870

El Alaoui, M., Ben-azza, H., & El Yassini, K. (2018a). Optimal Weighting Method for Fuzzy Opinions. Paris, France: Academic Press.

El Alaoui, M., Ben-azza, H., & El Yassini, K. (2018b). Optimal Weighting Method for Interval-Valued Intuitionistic Fuzzy Opinions. *Notes on Intuitionistic Fuzzy Sets*, *24*(3), 106–110. doi:10.7546/nifs.2018.24.3.106-110

El Alaoui, M., Ben-azza, H., & El Yassini, K. (2019a). Achieving Consensus in Interval Valued Intuitionistic Fuzzy Environment. *Procedia Computer Science*, *148*, 218–225. doi:10.1016/j.procs.2019.01.064

El Alaoui, M., Ben-azza, H., & El Yassini, K. (2019b). Fuzzy TOPSIS with Coherent Measure: Applied to a Closed Loop Agriculture Supply Chain. In M. Ezziyyani (Ed.), *Advanced Intelligent Systems for Sustainable Development (AI2SD'2018), Advances in Intelligent Systems and Computing* (pp. 106–117). Springer International Publishing. doi:10.1007/978-3-030-11878-5_12

El Alaoui, M., Ben-azza, H., & Zahi, A. (2016). New Multi-Criteria Decision-Making Based on Fuzzy Similarity, Distance and Ranking. In *Proceedings of the Third International Afro-European Conference for Industrial Advancement — AECIA 2016, Advances in Intelligent Systems and Computing*. Springer.

El Alaoui, M., El Yassini, K., & Ben-azza, H. (2018). Enhancing MOOCs Peer Reviews Validity and Reliability by a Fuzzy Coherence Measure. In *Proceedings of the 3rd International Conference on Smart City Applications, SCA '18*. New York: ACM. 10.1145/3286606.3286834

El Alaoui, M., El Yassini, K., & Ben-azza, H. (2019). Peer Assessment Improvement Using Fuzzy Logic. In Innovations in Smart Cities Applications Edition 2, Lecture Notes in Intelligent Transportation and Infrastructure. Springer International Publishing. doi:10.1007/978-3-030-11196-0_35

Figueroa-García, J. C., Chalco-Cano, Y., & Román-Flores, H. (2015). Distance Measures for Interval Type-2 Fuzzy Numbers. *Discrete Applied Mathematics*, *197*, 93–102. doi:10.1016/j.dam.2014.11.016

Gong, Y., Yang, S., & Dai, L. (2018). Some New Signed Distances and Similarity Measures of Interval Type-2 Trapezoidal Fuzzy Numbers and Comparative Study. *Journal of Intelligent & Fuzzy Systems*, *35*(3), 3465–3475. doi:10.3233/JIFS-17671

Gregson, R. A. M. (1975). *Psychometrics of Similarity*. New York: Academic Press.

Grzegorzewski, P. (2004). Distances between Intuitionistic Fuzzy Sets and/or Interval-Valued Fuzzy Sets Based on the Hausdorff Metric. *Fuzzy Sets and Systems*, *148*(2), 319–328. doi:10.1016/j.fss.2003.08.005

Grzegorzewski, P., & Mrówka, E. (2005). Trapezoidal Approximations of Fuzzy Numbers. *Fuzzy Sets and Systems*, *153*(1), 115–135. doi:10.1016/j.fss.2004.02.015

Grzegorzewski, P., & Mrówka, E. (2007). Trapezoidal Approximations of Fuzzy Numbers—Revisited. *Fuzzy Sets and Systems*, *158*(7), 757–768. doi:10.1016/j.fss.2006.11.015

Gupta, C., Jain, A., & Joshi, N. (2018). Fuzzy Logic in Natural Language Processing – A Closer View. *Procedia Computer Science*, *132*, 1375–1384. doi:10.1016/j.procs.2018.05.052

Harmse, J. (1996). Continuous Fuzzy Conjunctions and Disjunctions. *IEEE Transactions on Fuzzy Systems*, *4*(3), 295–314. doi:10.1109/91.531772

Ismat, B., & Samina, A. (2009). Similarity Measures for Fuzzy Sets. *Applied and Computational Mathematics*, *8*(2), 192–202.

Jaccard, P. (1908). Nouvelles Recherches Sur La Distribution Florale. *Bulletin de la Société Vaudoise des Sciences Naturelles*, *44*, 223–270.

Javanmard & Nehi. (2018). Rankings and Operations for Interval Type-2 Fuzzy Numbers: A Review and Some New Methods. *Journal of Applied Mathematics and Computing*.

Keatley, D. A., & Clarke, D. D. (2019). Crime Linkage: Finding a Behavioral Fingerprint Using the 'Path Similarity Metric.'. *Journal of Police and Criminal Psychology*. doi:10.100711896-019-9315-3

Klinkenberg, B. (2001). Similarity of the Flora of the Erie Islands: Implications for Conservation Biology. *The Canadian Geographer. Geographe Canadien*, *45*(3), 414–426. doi:10.1111/j.1541-0064.2001.tb01191.x

Klir, G. J. (1987). Where Do We Stand on Measures of Uncertainty, Ambiguity, Fuzziness, and the Like? *Fuzzy Sets and Systems*, *24*(2), 141–160. doi:10.1016/0165-0114(87)90087-X

Knees & Schedl. (2013). A Survey of Music Similarity and Recommendation from Music Context Data. *ACM Trans. Multimedia Comput. Commun. Appl.*, *10*(1), 2:1–2:21.

Krasser, J., Abeßer, J., Großmann, H., Dittmar, C., & Cano, E. (2012). Improved Music Similarity Computation Based on Tone Objects. In *Proceedings of the 7th Audio Mostly Conference: A Conference on Interaction with Sound, AM '12*. New York, NY: ACM. 10.1145/2371456.2371464

Lalanne, C. (2014). Appendix: Similitude in Mechanics. In Mechanical Shock. John Wiley & Sons, Ltd.

Lane, S. M. (Ed.). (1972). *Coherence in Categories*. Berlin: Springer-Verlag.

László & Tikk. (2000). Fuzzy rendszerek. Budapest: Academic Press.

Lesot, M.-J., & Rifqi, M. (2012). Degrés d'équivalence de Mesures de Comparaison Pour Données Binaires et Pour Données Numériques. *Revue Des Nouvelles Technologies de l'Information RNTI-E*, *22*, 145–164.

Lu, C., Lan, J., & Wang, Z. (2006). Aggregation of Fuzzy Opinions Under Group Decision-Making Based on Similarity and Distance. *Journal of Systems Science and Complexity*, *19*(1), 63–71. doi:10.100711424-006-0063-y

Mendel, J. M. (2017). *Uncertain Rule-Based Fuzzy Systems: Introduction and New Directions* (2nd ed.). Springer International Publishing. doi:10.1007/978-3-319-51370-6

Mitchell, H. B. (2006). Ranking Type-2 Fuzzy Numbers. *IEEE Transactions on Fuzzy Systems*, *14*(2), 287–294. doi:10.1109/TFUZZ.2005.864078

Santini, S., & Jain, R. (1997). Similarity Is a Geometer. *Multimedia Tools and Applications*, *5*(3), 277–306. doi:10.1023/A:1009651725256

Schedl, Pohle, Knees, & Widmer. (2011). Exploring the Music Similarity Space on the Web. *ACM Trans. Inf. Syst.*, *29*(3), 14:1–14:24.

Singh, P. (2014). Some New Distance Measures for Type-2 Fuzzy Sets and Distance Measure Based Ranking for Group Decision Making Problems. *Frontiers of Computer Science*, *8*(5), 741–752. doi:10.100711704-014-3323-3

Sivanandam, S. N., Sumathi, S., & Deepa, S. N. (2007). *Introduction to Fuzzy Logic Using MATLAB*. Berlin: Springer-Verlag. doi:10.1007/978-3-540-35781-0

Sokal, R. R., & Peter, H. A. S. (1963). *Principles of Numerical Taxonomy*. W. H. Freeman.

Szmidt, E. (2014). *Distances and Similarities in Intuitionistic Fuzzy Sets*. Springer International Publishing. doi:10.1007/978-3-319-01640-5

Teng, S., Li, J., Li, R., & Zhang, W. (2014). The Calculation of Similarity and Its Application in Data Mining. In Pervasive Computing and the Networked World, Lecture Notes in Computer Science. Springer International Publishing. doi:10.1007/978-3-319-09265-2_57

Thibaut, J.-P. (1997). Similarité et catégorisation. *L'Année Psychologique*, *97*(4), 701–736. doi:10.3406/psy.1997.28989

Tversky, A. (1977). Features of Similarity. *Psychological Review*, *84*(4), 327–352. doi:10.1037/0033-295X.84.4.327

Velankar, M. R., Sahasrabuddhe, H. V., & Kulkarni, P. A. (2015). Modeling Melody Similarity Using Music Synthesis and Perception. *Procedia Computer Science*, *45*, 728–735. doi:10.1016/j.procs.2015.03.141

Wan, Yu., Chen, Y., Shi, X., & Zhou, C. (2018). Constructing and Validating Word Similarity Datasets by Integrating Methods from Psychology, Brain Science and Computational Linguistics. *Soft Computing*, *22*(21), 6967–6979. doi:10.100700500-018-3174-1

Wang, X., & Kerre, E. E. (2001). Reasonable Properties for the Ordering of Fuzzy Quantities (I). *Fuzzy Sets and Systems*, *118*(3), 375–385. doi:10.1016/S0165-0114(99)00062-7

Williams, J., & Steele, N. (2002). Difference, Distance and Similarity as a Basis for Fuzzy Decision Support Based on Prototypical Decision Classes. *Fuzzy Sets and Systems*, *131*(1), 35–46. doi:10.1016/S0165-0114(01)00253-6

Wu, D., & Mendel, J. M. (2009). A Comparative Study of Ranking Methods, Similarity Measures and Uncertainty Measures for Interval Type-2 Fuzzy Sets. *Information Sciences*, *179*(8), 1169–1192. doi:10.1016/j.ins.2008.12.010

Zadeh, L. A. (1965). Fuzzy Sets. *Information and Control*, *8*(3), 338–353. doi:10.1016/S0019-9958(65)90241-X

Zadeh, L. A. (1971). Similarity Relations and Fuzzy Orderings. *Information Sciences*, *3*(2), 177–200. doi:10.1016/S0020-0255(71)80005-1

Zadeh, L. A. (1975). The Concept of a Linguistic Variable and Its Application to Approximate Reasoning—I. *Information Sciences*, *8*(3), 199–249. doi:10.1016/0020-0255(75)90036-5

Zadeh, L. A. (2006). Generalized Theory of Uncertainty (GTU)—Principal Concepts and Ideas. *Computational Statistics & Data Analysis*, *51*(1), 15–46. doi:10.1016/j.csda.2006.04.029

Zeng, W., & Li, H. (2006). Relationship between Similarity Measure and Entropy of Interval Valued Fuzzy Sets. *Fuzzy Sets and Systems*, *157*(11), 1477–1484. doi:10.1016/j.fss.2005.11.020

Zhou, Z., Zhang, R., Lin, Y., & Wang, R. (2015). A Comparison of Similarity Measures of Intuitionistic Fuzzy Sets. In Z. Zhang, Z. M. Shen, J. Zhang, & R. Zhang (Eds.), *LISS 2014* (pp. 1237–1242). Springer Berlin Heidelberg. doi:10.1007/978-3-662-43871-8_179

Zohuri, B. (2015). Similitude Theory and Applications. In B. Zohuri (Ed.), *Dimensional Analysis and Self-Similarity Methods for Engineers and Scientists* (pp. 93–193). Cham: Springer International Publishing.

Zwick, R., Carlstein, E., & Budescu, D. V. (1987). Measures of Similarity among Fuzzy Concepts: A Comparative Analysis. *International Journal of Approximate Reasoning*, *1*(2), 221–242. doi:10.1016/0888-613X(87)90015-6

Chapter 21
Application of Fuzzy Logic in Investment-Intensive Decision Making

Prateek Pandey
Jaypee University of Engineering and Technology, Guna, India

Shishir Kumar
Jaypee University of Engineering and Technology, Guna, India

Sandeep Shrivastava
Jaypee University of Engineering and Technology, Guna. India

ABSTRACT

Fuzzy logic has been serving the industry for decades by resolving the ambiguities that appear as a result of imprecise environment. High-stake decision-making processes require inputs from various stakeholders to incorporate. If the risk is high, as in the case of high investment decision making, a robust system of incorporating opinions from multiple stakeholders must be set in place in order to avoid any inconsistency or bad decisions. Fuzzy matrices and arithmetic can play a rescuer in such situations. In this chapter, the authors demonstrate a decision-making framework incorporating the use of fuzzy numbers and arithmetic to make critical decisions in strategic marketing and new product development. Forecasting in the domains of new products is an utmost complex and critical process because no relevant history is available owing to the product's 'one-of-its-kind' nature. In such cases, computation via analogy is an interesting paradigm, which is also discussed in the chapter.

1. INTRODUCTION

Human has always been trying to model the behaviour of future events scientifically. A scientific approach to forecasting is based on the supporting facts and reasoning to come to a conclusion. However, the mantra behind every scientific approach remains the same and that is to reduce the uncertainty as

DOI: 10.4018/978-1-7998-0190-0.ch021

much as possible. There exist a number of forecasting methods to deal with uncertainty in information. Usually, the choice of a forecasting method is determined by nature of the domain in which forecast has to be made and availability of the prior information or evidence. For example, there are different preferable methods for technological forecasting, financial forecasting and product sales forecasting. Also, the situations—when the evidences or the historical data is sufficiently available; when the historical data is absolutely unavailable, and; when the historical data is available for short periods only—influence the choice of a forecasting method.

Suppose an organization is ready with a design or a prototype of the innovation that it has produced, and huge amount of money and its image is at stake. Now, the organization decides that "it would be much beneficial, in implementing a promotional plan and deciding the production capacity of the upcoming setup, if an estimate, although imprecise, of the initial demand pattern is available." This is that daunting forecasting problem where no prior data or historical observations are available to perform any analysis. It is not needed to mention that time series or regression analysis cannot be performed as no historical data is available. A special class of methods called qualitative methods is applicable in this context. Qualitative methods make use of expert opinion in forecasting and are thus subject to a number of biases and imprecision. Some techniques that can be used in qualitative forecasting are Delphi method, forecasting by analogy, scenario forecasting, sales force composite, executive opinion, and customer intentions (Hyndman & Athanasopoulos, 2013).

The authors propose an analogy detection methodology where a number of seemingly analogous analogy products are considered. It is considered a good practice to diversify the investments, as it minimizes the risk. Analogically similar practice is followed in the methodology. Instead of counting on a single analogy for forecasting, a set of analogies is often preferred. A few possible reasons behind using multiple analogies may be:

1. Decision makers often recall multiple resemblances, not one.
2. It is riskier to put the entire stake on a single analogy.
3. Different analogies may resemble in different aspects.

Because, every analogy cannot possess equal resemblance, there is a need to identify the analogies with significant resemblance. Identifying such analogies is not a trivial task. The process of selecting a set of good analogies, in fact, requires complex decision making ability at the end of decision maker (DM). Thus, a well structured approach for deciding the appropriate analogies is warranted at this juncture.

2. SUPPLEMENTARY TECHNIQUES FOR FORECASTING IN ABSOLUTE ABSENCE OF DATA

Forecasting is indispensable in an organization because crucial decisions are often required to be made that are based on the decision makers belief in forecasting results. A reverse version of the said statement is also valid sometimes i.e. decision making sometimes become essential for generating forecasts. Decisions are made to be taken in the presence of multiple, often conflicting, criteria; had there been a single criterion, no decision would have to be taken at all (Shi, Wang, Kou & Wallenius, 2011). Decision making is not a trivial process. It requires tremendous efforts from the side of a decision maker. Therefore, a structured way of decision making is being practiced and studied independently under the name of Multiple Criteria Decision Making (MCDM). A number of approaches and models exist in MCDM literature that make the decision making structured and the outcomes more reasonable. MCDM can broadly be classified into two categories: Multiple Attribute Decision Making (MADM) and Multiple Objective Decision Making (MODM). MADM deals with rating of the alternatives to form a rank, based on preferences (Kenevissi, 2014). MODM, on the other hand, provides a mathematical framework for designing a set of decision alternatives.

3. NEW PRODUCT, INNOVATION AND FORECASTING

The words 'new product' and 'innovation' are sometimes used synonymously in the forecasting fraternity and the same trend is also followed in this thesis. The interpretation of the term 'innovation', considered in this thesis, is warranted at this place and is deduced from the definitions given by White, Braczyk, Ghobadian & Niebuhr (1988) and Rogers (2003).

A firm could be counted as innovative even if it had made only a marginal product improvement (White et al., 1988). This improvement may result in:

1. A modified version of an existing product range.
2. A new model in the existing product range.
3. A new product outside the existing range but in a similar field of technology.
4. A totally new product in a new field of technology.

Rogers (2003) described innovation as an idea, practice, or an object perceived as new by an individual or other unit of adoption. This definition of innovation allows using the words new product and innovation interchangeably. Thus, a new product may be an innovation that is totally idiosyncratic of its kind or it might be a slight improvement over an existing product.

New product forecasting (NPF) is the science of predicting the degree of success a new product will enjoy in the marketplace. It is perhaps the most difficult and thankless of all forecasting endeavours. Organizations commit significant resources to new product development and release, and sometimes even "bet the company" on promising new ideas (Gilliland, 2012). New products affect every aspect of the life of individuals, communities, countries, and economies. Thus there is a need to improve the quality of forecasting for new products.

Kahn (2006) provides a discussion on a variety of estimates that are possible with new product forecasting. These include market potential, sales potential, market forecast, and sales forecast. Potential estimates represent the maximum attainable under a given set of conditions, with market pertaining to all companies within a given industry marketplace and sales pertaining to only the respective focal company. Forecasts represent likely or reasonable attainable estimates. Based on the said information about estimates, interpretation of different types of estimates can be formed:

1. **Market potential estimate** A market potential estimate is a prediction of maximum total market volume under a given set of conditions.
2. **Sales Potential Estimate** A sales potential estimate is a prediction of maximum company sales within the given market under a given set of conditions.

3. **Market Forecast** A market forecast is therefore a reasonable estimate of total market volume to be attained by all firms in that market under a given set of conditions.
4. **Sales Forecast** A sales forecast is a reasonable estimate of attainable sales volume for a focal company within a given market under a given set of conditions.

4. ANALOGY BASED FORECASTING

Sometimes, conventional analysis cannot be performed to obtain forecasts due to the scarcity of historical data. In such cases, forecasts can be derived by doing the analysis, mapping, and projections on analogies. Thus, the accuracy of forecasts is a function of the quality of analogies used. A typical example of the situation when a forecaster encounters 'total-data-absence' is the launch of a new product, service or an innovation by an organization. In such cases, forecasting has to be performed under the maximum uncertain conditions, as the data pertaining to the past performance of the new product or the service cannot be made available. An obvious technique in such circumstances that one can count on is to perform forecasting by analogies.

Forecasting by analogy is not a new concept. Thomas (1985) offered a systematic modelling approach for estimating demand growth of a new consumer durable products and services by incorporating diffusion models and concept of analogy. Ignoring the fact that the human opinions are intrinsically imprecise, the methodology; however, requires consumers' perception in estimating the diffusion parameters using analogies. Easingwood (1989) hypothesized that "some new products may exhibit diffusions that are characteristic of a particular group of products". Author presented a novel approach of diffusion maps to locate the diffusion parameters estimates for a new product; however, a good anticipation of the location on diffusion map is desired for a good approximation. Green and Armstrong (2007) proposed a structured judgmental procedure for forecasting using analogies. Authors argue that the people or the organizations often rely on forecasting using analogies in an unstructured manner, which often results in inaccurate forecasts. Authors suggest a minimum of five experts' opinions to generate a forecast. Despite knowing its importance for the industries, new product sales forecasting literature has received relatively little attention (Kahn, 2006).

Clearly, the success of an analogy based forecasting technique depends on the quality of the analogies used, a structured and convenient procedure to assess analogies, and of course a way to handle the imprecision and ambiguity in human judgments.

In section 8, a framework for the assessment and selection of analogies is proposed by pandey et al. (2014, 2015), following which a methodology based on the proposed framework is also discussed to order the set of analogies according to their appropriateness to a new product. The framework also takes care of the intrinsic ambiguities associated with the human judgments.

Considering that the selection of appropriate analogies is indeed a complex decision making process, which will often be taken in the presence of multiple, often conflicting, criteria, a recap of the prominent techniques of multiple criteria decision making is presented in the section 6 which is followed by an overview of the fuzzy sets and fuzzy numbers.

5. MCDM TECHNIQUES AND THEIR USAGE

MCDM is required, when decision making involves multiple, often conflicting, criteria. There are basically two approaches to multiple criteria decision making. One involves selection of a number of alternatives—on the basis of suitability of those alternatives against a set of attributes—to achieve an overall result. This approach is commonly known as multiple attribute decision making (MADM). Other involves synthesis of attributes on the basis of prioritized objectives. This approach is commonly known as multiple objective decision making (MODM). In the MODM approach, contrary to the MADM approach, the decision alternatives are not given. The proposed framework is based on the MADM approaches to decision making. Hwang and Yoon (1981) suggest two models for MADM: Compensatory, where analysis involves tradeoffs between high and low performances and; Non Compensatory, where no such tradeoffs exist. Hwang and Yoon (1981) further divide the methods for MADM into three categories:

1. When no preference information is given by DM
2. When information on attributes is known or given by DM
3. When information on alternatives is given by DM

Dominance, Maximin, and Maximax are the examples of the common techniques that can be used in the absence of preference information. Another class of MADM methods requires a DM to provide the weights for the attributes. In literature, such an attribute preference is called cardinal preference (Hwang & Yoon, 1981). SAW is one such method.

TOPSIS suggests that a better alternative is one whose distance is minimum from the positive ideal solution and maximum from the negative ideal solution (Chen, 2000). A good estimate of positive ideal solution and negative ideal solution can be obtained by selecting maximum and minimum values across all alternatives with respect to each criterion (Olson, 2004). Distance can be found out by measuring the Euclidian distance for each alternative from the positive ideal solution and negative ideal solution.

Apart from the methods discussed so far, two families of methods namely, MAUT and outranking methods are found in (Fülöp, 2005 and Kahraman, 2008) and a separate discussion, apart from the basic models discussed in Hwang and Yoon (1981), is often provided on these two classes of MADM methods. SAW and WPM are the examples of MAUT techniques, while ELECTRE and PROMETHEE are the examples of outranking methods.

When the concerned information is either ambiguous or not known, the classic MADM methods often proved to be inefficient. In such cases, fuzzy flavour of MADM methods provides reliable solutions. The next section summarizes the concepts that are useful in fuzzy assessment of the problem.

6. FUZZY SETS AND FUZZY NUMBERS

Zadeh (1965) introduced the concepts of fuzzy sets and then a number of applications were introduced where fuzzy sets and numbers can be used effectively (Kumar & Pandey, 2010; Pandey, Kumar & Shrivastav, 2013a, 2013b, 2017; and Pandey, Litoriya & Tiwari, 2018)

Sections 6.1, 6.2 and 6.3 define fuzzy sets, fuzzy numbers, and their operations which have been defined in pandey et al. (2015) and are reproduced here for convenience of the readers. Readers are advised to refer the mentioned article for more details.

6.1 Definitions

If Ω is some set, then a fuzzy subset $\bar{A}$ of Ω is defined by its membership function, written $\bar{A}$ x), which provides values in [0, 1] for all x in Ω.

A fuzzy set $\bar{A}$ is called normalized iff atleast one of its elements attain the maximum possible membership grade.

An α- cut of a fuzzy set $\bar{A}$, designated as $\bar{A}$ α], is a crisp set that contains all the elements of the universal set of U that have a membership grade in $\bar{A}$ greater than or equal to the specified value of α . Thus, $\bar{A}$ α] = {x ∈.Ω | $\bar{A}$ (x) ≥ α }. Therefore, for all 0<α≤ 1, $\bar{A}$ 0] is defined separately and is called support of $\bar{A}$.

A fuzzy number is a convex, normalized fuzzy set $\bar{A} \subseteq \mathbb{R}$ whose membership is at least segmentally continuous and has the function value $\mu_{\bar{A}}$ (x) =1 at precisely one element.

A triangular fuzzy number $\bar{A}$ is defined by three numbers a<b<c where the base of the triangle is the interval [a, c] and its vertex is at x=b. Triangular fuzzy numbers will be written as $\bar{A}$ = (a/b/c).

6.2 Operations on Fuzzy Numbers

There are two ways to perform operations on fuzzy numbers: First, using α. cut and interval arithmetic; Second, using extension principal (Buckley, 1985). Since, representing a fuzzy number into corresponding α- cut notation and performing interval arithmetic is simple and computer friendly, with a little compromise over accuracy (Buckley, 1985), the demonstrated analysis resort to this method only.

For any fuzzy number $\bar{A}$, $\bar{A}$ [α] is a closed and bounded interval, for 0<α<1. $\bar{A}$ [α] = [a_1(α), a_2(α)] where a_1(α), a_2(α) is an increasing (decreasing) function of α with $a_1(1) = a_2(1)$.

Thus for any two fuzzy numbers $\bar{A}$ and $\bar{B}$

if $\bar{C} = \bar{A}\bar{B}$ then $\bar{C}\alpha] = \bar{A}\alpha] + \bar{B}\alpha]$,
if $\bar{C} = \bar{A}\bar{B}$. then $\bar{C}\alpha] = \bar{A}\alpha] - \bar{B}\alpha]$,
if $\bar{C} = \bar{A}\bar{B}$ then $\bar{C}\alpha] = \bar{A}\alpha] * \bar{B}\alpha]$, and
if $\bar{C} = \bar{A}\bar{B}$. then $\bar{C}\alpha] = \bar{A}\alpha] / \bar{B}\alpha]$, provided that zero does not belong to $\bar{B}$ α] for all α.

6.3 Fuzzy Synthetic Extent Analysis

Assume that X={x_1, x_2,…,x_n) is an object set, and U={u_1, u_2,…,u_m} is a goal set, and μ_i^j (i=1,2,…, n; j=1, 2, …,m) are fuzzy numbers representing the performance of the object x_i with regard to each goal u_j.

By using fuzzy synthetic extent analysis as discussed in Deng (1999), the value of fuzzy synthetic extent with respect to the ith object x_i (i=1, 2,…, n) that represents the overall performance of the object across all goals involved can be determined by

$$S_i = \frac{\sum_{j=1}^{m} \mu_i^j}{\sum_{i=1}^{n} \sum_{j=1}^{m} \mu_i^j}, \; i=1, 2 \ldots n \quad (1)$$

There exist two schools of thought when subjective opinions, from an expert, are collected and computations are required to come to a decision. One advocates quantization of the qualitative judgments as pure quantities, other emphasizes in representing the judgments as fuzzy quantities. Fuzzy numbers are capable to explain the imprecision in judgments by the experts (Laarhoven & Pedrycz, 1983 and Buckley, 1985). Therefore, instead of crisp values, fuzzy numbers are used to record the assessments of experts in this work. Fuzzy numbers and their arithmetic has also been used in a variety of disciplines (M Pandey, Litoriya, & Pandey, 2016; Mamta Pandey, Litoriya, & Pandey, 2018a, 2018b, 2019c, 2019b, 2019d, 2019e, 2019a; P. Pandey & Litoriya, 2019c, 2019b, 2019a)

7. FRAMEWORK

Before starting an analysis to find good analogies, users or managers should well be equipped with three raw requirements:

1. Object of forecasting: The object can be a product, a service, an idea or something else.
2. List of similar objects: This provides an initial set of analogies on which to perform the analysis.
3. Subject experts: This is essential as analogy assessment is subjective, and any verdict should reasonably be a collective decision.

Having equipped with these preliminary requirements, the next and possibly the most crucial steps are to understand the nature of the object and to identify the factors that govern the penetration of the object in its target market. Once the nature is understood and the factors are identified, the managers can identify a list of surrogate objects. A potential surrogate object should be selected, however, on the basis of the closeness in comparable attributes between the object and the surrogate objects. The next step is to identify a panel of subject experts (in case the object is a new product, the subject experts can be replaced with consumers). As stated earlier, a panel contains no less than five members can be expected to generate a reliable forecast. Assuming the problem of finding an appropriate analogy for the object as a goal and factors governing the object as the attributes and the surrogate objects as the alternatives, a typical environment for MADM is set in place. Depending upon the nature of the problem, desired outcomes, and the information at hand, managers or the users can decide an appropriate MADM technique. The results obtained on application of a MADM technique are the desired analogies that may be used further for forecasting.

8. UNDERSTANDING THE NEW PRODUCT AND INTERPRETATION OF THE VARIABLES

In this context, obviously, the object of forecasting is the new product itself. A total of five or more marketing experts can be involved in this experiment whose opinions about the nature of new product will be taken into account. To understand the nature of the object, the first need is to understand the scope of the term 'new product'. A new product may be an innovation that is totally idiosyncratic of its kind; it might be a slight improvement over an existing product range; it might be a new model in the existing product range, or it might be a new product outside the existing range but in the same field of technology (White et al., 1988). Unlike weather forecasting, which is based on the scientific study of the natural phenomenon, or stock forecasting, which is based on the understanding of complex economic processes and market sentiments, new product forecasting is a phenomenon that is often explained using the behavior of a social system. Rogers (2003) defined social system as a set of interrelated units that are engaged in joint problem solving to accomplish a common goal. When a new product or an innovation gets introduced to the market, it follows a penetration pattern that is a function of the gap between the ceiling on the total number of possible adoptions and the number of previous adopters at that time (Bass, 1969). This process of market penetration is called diffusion. Thus, innovation diffusion is the process of the market penetration of new products and services, which is driven by social influences (Peres, Muller & Mahajan, 2010).

Most of the variance in the rate of adoption of innovations, from 49 to 87 percent, is explained by five attributes: relative advantage, compatibility, complexity, trialability, and observability (Rogers, 2003). Other variables that are considered to have an impact on the rate of adoption are: the type of innovation decision, the nature of communication channels, the nature of the social system, and the extent of promotion efforts in diffusing the innovation. These variables are further used as criteria to assess the respective resemblance of analogical products to the new product. Therefore, a brief idea of these variables is warranted.

- **Relative Advantage**: Degree to which an innovation is perceived as being better than the idea it supersedes.
- **Compatibility**: Degree to which an innovation is perceived as consistent with the existing values, past experiences, and needs of potential adopters.
- **Complexity**: Degree to which an innovation is perceived as relatively difficult to understand and use.
- **Trialability**: Degree to which an innovation may be experimented with on a limited basis.
- **Observability**: Degree to which the results of an innovation are visible to others.
- **Innovation decision**: The more person involved in making an innovation decision, the slower the rate of adoption.
- **Nature of Communication Channels**: Mass media or interpersonal.
- **Nature of the Social System**: Social norms, degree of network interconnections.
- **Promotion Efforts:** Aggressive starting promotions lead to faster diffusion at early stages.

To obtain a list of similar products, a common and reasonable practice is to choose from the intra-domain products. For example, house hold electronic appliances like televisions, refrigerators, washing machines form a list of products from the same domain. Thus, with an object of forecasting, list of similar products (i.e. analogies), team of experts, and the factors that affect the penetration of a new product (relative advantage, compatibility, etc.) in our sleeves, any of the MADM model can be applied. Because the problem is well structured and requires tradeoff analysis, the recommendation is to use AHP by Saaty (2003) in this context. In most of the real world problems, where the concerned information is either ambiguous or not known completely, it is almost impossible to clearly define or represent a problem in crisp form. Therefore, a fuzzy flavor of AHP known as Fuzzy AHP is used in this methodology to rank the alternative analogical products.

9. METHODOLOGY

The methodology adopted can be explained as a divide and conquer approach. Here we only explain the method in details. To understand this methodology in-depth, readers are advised to refer to Pandey et al. (2015), where the methodology is explained with an example.

9.1 Dividing the Problem

First, the problem of identifying a suitable analogy (that mimic the diffusion pattern) for the new product, based on the specified attributes or criteria, is divided into two sub problems: to find the relative importance of each criteria over other with respect to the goal; and to find the relative importance of each alternative product over other alternative products on the basis of each criterion. These pair-wise comparisons are represented in the form of fuzzy judgment matrices. Pair-wise comparisons of criteria with respect to goal and pair-wise comparisons of alternatives with respect to criteria k can be represented as:

$$C2C = \begin{bmatrix} \overline{c}_{11} & \overline{c}_{12} \cdots \overline{c}_{1m} \\ \overline{c}_{21} & \overline{c}_{22} \cdots \overline{c}_{2m} \\ \cdots & \cdots\cdots \cdots \\ \overline{c}_{m1} & \overline{c}_{i2} \cdots \overline{c}_{mm} \end{bmatrix} \tag{2}$$

and,

$$A2C_k = \begin{bmatrix} \overline{a}_{11} & \overline{a}_{12} & \cdots \overline{a}_{1n} \\ \overline{a}_{21} & \overline{a}_{22} \cdots & \overline{a}_{2n} \\ \cdots & \cdots & \cdots \cdots \\ \overline{a}_{n1} & \overline{a}_{n2} & \cdots \overline{a}_{nn} \end{bmatrix} \tag{3}$$

where each $\overline{c}_{ij}$ i=1...m, j=1...m, a fuzzy number, represents the relative importance of criteria c_i. over criteria c_j.in predicting the diffusion of a new product (P) and each $\overline{a}_{ij}$ i=1...n, j=1...n represents the degree to which an alternative product a_i outperforms alternative product a_j.in resembling the new product P on the basis of a particular criteria C_k, k=1...m . Each $\overline{c}_{ij}$ is assessed by an expert or a team of experts using the fuzzified five point scale as shown in Table 1, where relative importance is characterized by five linguistic values. Each $\overline{a}_{ij}$ is also assessed using Table 1, but after identifying a suitable degree of resemblance for an alternative product with the new product, for a specified attribute, using Table 2 and Table 3. Degree of relative resemblance is the difference of the index values, assigned to each of the linguistic resemblance values (Table 3), that different competing analogous products obtain through the process.

Table 1. Linguistic values with their corresponding fuzzy numbers

Relative Importance	Degree of Relative Resemblance	Fuzzy Number
Equal Importance	0	(1, 1, 1+δ)
Weak Dominance	1	(3-δ, 3, 3+δ)
Strong Dominance	2	(5-δ, 5, 5+δ)
Demonstrated Dominance	3	(7-δ, 7, 7+δ)
Absolute Dominance	4	(9-δ, 9, 9)

δ is the fuzzy distance (0.5 ≤ δ ≤ 2)

Table 2. Resemblance of an alternative product and the new product

		New Product			
Alternative Products	**Attributes**	**High**	**Fairly High**	**Fairly Low**	**Low**
	High	So Close	Close	Fairly Distant	Distant
	Fairly High	Close	So Close	Separated	Fairly Distant
	Fairly Low	Fairly Distant	Separated	So Close	Close
	Low	Distant	Fairly Distant	Close	So Close

Table 3. Resemblance with their assigned indices

Resemblance Values	Index
So Close	1
Close	2
Separated	3
Fairly Distant	4
Distant	5

9.2 Consistency of a Fuzzy Judgment Matrix

In order to get the most reliable estimates, each of the fuzzy judgment matrices needs to be consistent. Csutora and Buckley (2001) analyzed the consistency of a fuzzy judgment matrix $\bar{A} = [\bar{a}_{ij}$, where $\bar{a}_{ij}$ is a fuzzy number of the form $\bar{a}_{ij}$ (α/β/γ). Authors further suggest that if A is consistent in A=[β_{ij}] then $\bar{A}$ will also be consistent. Thus, to check the consistency of a fuzzy judgment matrix $\bar{A}$, consistency of A is a sufficient and necessary condition.

9.3 Solving the Problems

Now, as we have divided the problem of finding a resembling analogy for a product in subject by looking at different criteria, the fitness of each criterion taking remaining criteria together can be calculated by applying fuzzy synthetic extent as discussed in (1). Corresponding weights of criteria can be determined as:

$$\overline{w}_i = \frac{\sum_{j=1}^{m} \overline{c}_{ij}}{\sum_{k=1}^{m} \sum_{l=1}^{m} \overline{c}_{kl}}, \tag{4}$$

where i=1…..m.

Each $\overline{w}_i$ is a fuzzy number and represents the importance of the criterion i over other criteria with respect to the goal. Hence, w_i will be an element of a weight vector W. Let the final criteria weight vector W= { $\overline{w}_1, \overline{w}_2, \ldots, \overline{w}_m$ }

Similarly, all the alternatives can be evaluated against each criteria in turn using fuzzy synthetic extent to produce decision matrix (7) as:

$$\overline{x}_{ij} = \frac{\sum_{k=1}^{n} \overline{a}_{ik}}{\sum_{p=1}^{n} \sum_{q=1}^{n} \overline{a}_{pq}} \tag{5}$$

where i=1…n, j=1…m and $\overline{x}_{ij}$ implies the importance of alternative i over other alternatives with respect to criteria j.

$$X = \begin{bmatrix} \overline{x}_{11} & \overline{x}_{12} & \cdots & \overline{x}_{1m} \\ \overline{x}_{21} & \overline{x}_{22} & \cdots & \overline{x}_{2m} \\ \cdots & \cdots & \cdots & \cdots \\ \overline{x}_{n1} & \overline{x}_{n2} & \cdots & \overline{x}_{nm} \end{bmatrix}. \tag{6}$$

9.4 Merging the Solutions

As proposed by Deng (1999), an overall performance of each alternative across all criteria is given by:

$$\text{A2C or } Z = \begin{bmatrix} \overline{w}_1\overline{x}_{11} & \overline{w}_2\overline{x}_{12} & \cdots & \overline{w}_m\overline{x}_{1m} \\ \overline{w}_1\overline{x}_{21} & \overline{w}_2\overline{x}_{22} & \cdots & \overline{w}_m\overline{x}_{2m} \\ \cdots & \cdots & \cdots & \cdots \\ \overline{w}_1\overline{x}_{n1} & \overline{w}_2\overline{x}_{n2} & \cdots & \overline{w}_m\overline{x}_{nm} \end{bmatrix} \tag{7}$$

In order to calculate the relative performances of alternative analogous products, the decision matrix A2C or Z must be defuzzified.

9.5 Defuzzification of the Decision Matrix

By representing performance matrix (7) into its α-cut form, an interval performance matrix can be derived as in (8), where $0 \leq \alpha \leq 1$. The chosen value of α is the reflection of the decision maker's confidence in his or her judgments. If α is small, the tendency of the assessments will deviate from the middle of the triangular fuzzy numbers (a/b/c). If α is large, the tendency will converge towards the middle.

$$Z_\alpha = \begin{bmatrix} \left[z_{11l}^{\alpha}, z_{11r}^{\alpha}\right] & \left[z_{12l}^{\alpha}, z_{12r}^{\alpha}\right] & \cdots & \left[z_{1ml}^{\alpha}, z_{1mr}^{\alpha}\right] \\ \left[z_{21l}^{\alpha}, z_{21r}^{\alpha}\right] & \left[z_{22l}^{\alpha}, z_{22r}^{\alpha}\right] & \cdots & \left[z_{2ml}^{\alpha}, z_{2mr}^{\alpha}\right] \\ \cdots & \cdots & \cdots & \cdots \\ \left[z_{n1l}^{\alpha}, z_{n1r}^{\alpha}\right] & \left[z_{n2l}^{\alpha}, z_{n2r}^{\alpha}\right] & \cdots & \left[z_{nml}^{\alpha}, z_{nmr}^{\alpha}\right] \end{bmatrix} \tag{8}$$

Crisp performance matrix is calculated as in (9), using the equation $z_{ij\alpha}^{\lambda'}, \lambda z_{ijr}^{\alpha}$ (1-λ) z_{ijl}^{α} . λ ∈.[0, 1], where λ represents the expert's opinion about the overall trend of the fuzzy assessments made using triangular fuzzy numbers. λ=0.5 signifies a neutral trend (moderate DM), λ=0 can be interpreted as a left trend (pessimistic DM) and λ=1 indicates a trend towards the right (optimistic DM).

$$Z_{\alpha}^{\lambda'} = \begin{bmatrix} z_{11\alpha}^{\lambda'} & z_{12\alpha}^{\lambda'} & \cdots & z_{1m\alpha}^{\lambda'} \\ z_{21\alpha}^{\lambda'} & z_{22\alpha}^{\lambda'} & \cdots & z_{2m\alpha}^{\lambda'} \\ \cdots & \cdots & \cdots & \cdots \\ z_{n1\alpha}^{\lambda'} & z_{n2\alpha}^{\lambda'} & \cdots & z_{nm\alpha}^{\lambda'} \end{bmatrix} \tag{9}$$

To facilitate the TOPSIS, normalization of (9) with regard to each criterion is required, and will be performed using (10) to get the normalized decision matrix (11).

$$z_{ij\alpha}^{\lambda} = \frac{z_{ij\alpha}^{\lambda'}}{\sqrt{\sum_{i=1}^{n}\left(z_{ij\alpha}^{\lambda'}\right)^2}}, \tag{10}$$

$$\begin{bmatrix} z_{11\alpha}^{\lambda} & z_{12\alpha}^{\lambda} & \cdots & z_{1m\alpha}^{\lambda} \\ z_{21\alpha}^{\lambda} & z_{22\alpha}^{\lambda} & \cdots & z_{2m\alpha}^{\lambda} \\ \cdots & \cdots & \cdots & \cdots \\ z_{n1\alpha}^{\lambda} & z_{n2\alpha}^{\lambda} & \cdots & z_{nm\alpha}^{\lambda} \end{bmatrix}. \tag{11}$$

9.6 Ranking Procedure

TOPSIS suggests that a better alternative is one whose distance is minimum from the positive ideal solution and maximum from the negative ideal solution (Chen, 2000). Distance can be found out by measuring the Euclidian distance for each alternative from the positive ideal solution and negative ideal solution. TOPSIS is used as a method for ranking, because unlike simple weighted summation methods, TOPSIS is heuristic, understandable, and simple. Also, its performance is only partially affected by the number of alternatives (Chen, 2000).

A good estimate of positive ideal solution and negative ideal solution can be obtained by selecting maximum and minimum values across all alternatives with respect to each criterion (Olson, 2004).

$$A_{\alpha}^{\lambda+} = \left(z_{1\alpha}^{\lambda+}, z_{2\alpha}^{\lambda+}, \ldots, z_{m\alpha}^{\lambda+}\right). \tag{12}$$

$$A_{\alpha}^{\lambda-} = \left(z_{1\alpha}^{\lambda-}, z_{2\alpha}^{\lambda-}, \ldots z_{m\alpha}^{\lambda-}\right). \tag{13}$$

where:

$$z_{j\alpha}^{\lambda+} \begin{matrix} = \max(z_{1j\alpha}^{\lambda}, z_{2j\alpha}^{\lambda}, \ldots, z_{nj\alpha}^{\lambda} & \text{if } j \in J \\ = \min(z_{1j\alpha}^{\lambda}, z_{2j\alpha}^{\lambda}, \ldots, z_{nj\alpha}^{\lambda} & \text{if } j \in J' \end{matrix} \tag{14}$$

$$z_{j\alpha}^{\lambda-} \begin{matrix} = \min(z_{1j\alpha}^{\lambda}, z_{2j\alpha}^{\lambda}, \ldots, z_{nj\alpha}^{\lambda} & \text{if } j \in J \\ = \max(z_{1j\alpha}^{\lambda}, z_{2j\alpha}^{\lambda}, \ldots, z_{nj\alpha}^{\lambda} & \text{if } j \in J' \end{matrix} \tag{15}$$

where J is a benefit attribute and J' is a cost attribute. A point to take a note here is that, in this particular case where analogies are analyzed on the basis of their resemblance to the new product, cost attributes and benefit attributes converge. For example, the attribute 'complexity' appears to be a cost attribute, as a new product diffuses better if its complexity is minimum, but, however, the goal of this analysis is to obtain analogies and analogies must resemble the subject in every aspect, whether costs or benefits.

Euclidean distances of each alternative from positive ideal solution and negative ideal solution can be given by

$$S_{i\alpha}^{\lambda+} = \sqrt{\sum_{j=1}^{m}\left(z_{j\alpha}^{\lambda+} - z_{ij\alpha}^{\lambda}\right)^2}, i = 1\ldots n \tag{16}$$

$$S_{i\alpha}^{\lambda-} = \sqrt{\sum_{j=1}^{m}\left(z_{j\alpha}^{\lambda-} - z_{ij\alpha}^{\lambda}\right)^2}, i = 1\ldots n \tag{17}$$

Therefore an overall performance index for each alternative, with the decision maker's alpha level of confidence in his fuzzy assessments and λ opinion about overall trend can be determined by (18) (Olson, 2004).

$$P^{\lambda}_{\alpha i} = \frac{S^{\lambda -}_{i\alpha}}{S^{\lambda +}_{i\alpha} + S^{\lambda -}_{i\alpha}}. \tag{18}$$

The larger the index value, the more preferred the alternative.

10. Example

In order to check the plausibility of the method, we take a real life example of four consumer durable products, viz. Television sets, refrigerators, washing machines, and microwaves in the Indian market. Of the four products, we choose television sets (TV), refrigerators (REF), and washing machines (WM) as the competing analogous products to forecast the diffusion of microwaves in the Indian market. In this example, we consider relative advantage, compatibility, and complexity as the criteria on which to decide the relative ranking of the competing analogous products. Remaining criteria are intentionally omitted to ease understandability. The opinions of five experts from marketing discipline in our own university are considered to populate different matrices. Initial matrices and the final defuzzified matrix formed during the test are represented as Table 4, Table 5, Table 6, Table 7, and Table 8.

Table 4. Criteria-to-Criteria matrix

C2C	RAD	COM	COX
RAD	$\overline{1}$	$\overline{5}$	$\overline{7}$
COM	$\overline{1/5}$	$\overline{1}$	$\overline{3}$
COX	$\overline{1/7}$	$\overline{1/3}$	$\overline{1}$

Table 5. Alternative-to-Alternative (A2A) matrix #1

RAD	TV	REF	WM
TV	$\overline{1}$	$\overline{3}$	$\overline{1/5}$
REF	$\overline{1/3}$	$\overline{1}$	$\overline{1/7}$
WM	$\overline{5}$	$\overline{7}$	$\overline{1}$

Table 6. A2A matrix #2

COM	TV	REF	WM
TV	$\overline{1}$	$\overline{5}$	$\overline{3}$
REF	$\overline{1/5}$	$\overline{1}$	$\overline{1/3}$
WM	$\overline{1/3}$	$\overline{3}$	$\overline{1}$

Table 7. A2A matrix #3

COX	TV	REF	WM
TV	$\overline{1}$	$\overline{3}$	$\overline{1}$
REF	$\overline{1/3}$	$\overline{1}$	$\overline{1/3}$
WM	$\overline{1}$	$\overline{3}$	$\overline{1}$

Table 8. Alternative-to-Criteria (A2C) matrix

A2C	RAD	COM	COX
TV	0.3663	0.8441	0.6751
REF	0.1723	0.2250	0.2973
WM	0.9144	0.4867	0.6751

Table 9. Performance of different competing products

Analogies	Performance Index
Television Sets	0.6
Refrigerators	0.0001
Washing Machines	0.7095

Considering α=0.5, λ=0.5 and applying TOPSIS on Table 8, the performance matrix obtained is as shown in Table 9.

Clearly, washing machines proved to be a better analogy for forecasting the diffusion of microwave ovens in India than television sets and refrigerators . In real practice, however, no single analogy is used to estimate the parameters of the new product. As a result of the proposed methodology, multiple analogies can be considered for parameter estimation with appropriate weightage given to them as per their performance indices.

11. COMPARISON WITH EXISTING METHODS

Applying Saaty's traditional AHP (Saaty, 1980; Saaty, 1995) on the given problem, for the sake of comparison, we get Table 10, Table 11, Table 12, and Table 13 as the corresponding eigen vectors for the matrices denoted by Table 4, Table 5, Table 6, and Table 7. Table 14 represents the overall performance of different consumer durables in matching the closeness to the new product. By looking at Table 14, it is clear that washing machines are the winners followed by tv sets and refrigerators.

While applying traditional AHP, rankings remain the same as they are in fuzzy AHP, however; performances using fuzzy AHP are relatively close as compared to traditional AHP. This difference is due to the uncertainty accommodated by fuzzy AHP method. Similarity in the results by both the methods gives a reasonable level of confidence in deciding the appropriate analogy/analogies for the new product.

On applying VIKOR to the decision matrix given in Table 8 by assuming group utility value as 0.5, three ranking lists (S, R, Q) are obtained and are shown in Table 15

Since the conditions of 'Acceptable advantage' and 'Acceptable stability in decision making' are satisfied, it can be concluded that in terms of resemblance to the new product ' microwave ovens' washing machines perform much better than refrigerators while television sets lie somewhere in between.

Table 10. Eigen Vectors for C2C

C2C	Eigen Vector
RAD	0.7306
COM	0.1884
COX	0.0810

Table 11. Eigen Vectors for RAD

RAD	Eigen Vector
TV	0.1884
REF	0.0810
WM	0.7306

Table 12. Eigen Vectors for COM

COM	Eigen Vector
TV	0.6370
REF	0.1047
WM	0.2583

Table 13. Eigen Vectors for COX

COX	Eigen Vector
TV	0.4286
REF	0.1428
WM	0.4286

Table 14. A2C Matrix (traditional AHP)

A2C	RAD	COM	COX	Performance
TV	0.1376	0.1200	0.0347	0.2923
REF	0.0592	0.0197	0.0116	0.0905
WM	0.5338	0.0487	0.0347	0.6172

Table 15. Three ranking lists obtained on applying VIKOR on Table 10

VIKOR	S	R	Q
TV	0.5170	0.5170	0.5703
REF	1	0.7	1
WM	0.1155	0.1155	0

12. CONCLUSION

The chapter provides a framework to find suitable analogies out of a set of predetermined analogies. The applicability of for finding analogies is of utmost important when a new product is launched into the market and its market acceptance pattern is to be projected. Since introducing a new product is a investment intensive decision, a structured framework is necessary. Believing that the selection of suitable analogies is a complex decision making process, the framework takes into consideration the MADM practices and methodologies. Further, a methodology is proposed to rank the competing products on the basis of their resemblance in diffusion to the new product. The method takes into account the vagueness and uncertainties intrinsic to the qualitative judgments made by the human experts. Further, a structured approach based on divide and conquer strategy to successfully rank the potential analogous products is suggested.

NOTE

Part of this work has earlier been submitted by the authors as a part of PhD thesis in Jaypee University of Engineering and Technology, Guna (MP) in the year 2015.

REFERENCES

Bass, F. M. (1969). A new product growth model for consumer durables. *Management Science*, *15*(5), 215–227. doi:10.1287/mnsc.15.5.215

Buckley, J. J. (1985). Ranking alternatives using fuzzy numbers. *Fuzzy Sets and Systems*, *15*(1), 21–31. doi:10.1016/0165-0114(85)90013-2

Chen, C. T. (2000). Extensions of the TOPSİS for group decision making under fuzzy environment. *Fuzzy Sets and Systems, 114*(1), 1–9. doi:10.1016/S0165-0114(97)00377-1

Csutora, R., & Buckley, J. J. (2001). Fuzzy hierarchical analysis: The lambda-max method. *Fuzzy Sets and Systems, 120*(2), 181–195. doi:10.1016/S0165-0114(99)00155-4

Deng, H. (1999). Multicriteria analysis with fuzzy pairwise comparison. *International Journal of Approximate Reasoning, 21*(3), 215–231. doi:10.1016/S0888-613X(99)00025-0

Easingwood, C. J. (1989). An analogical approach to the long term forecasting of major new product sales. *International Journal of Forecasting, 5*(1), 69–82. doi:10.1016/0169-2070(89)90065-4

Fülöp. (2005). Introduction to decision making methods. Laboratory of operations research and decision systems: computer and automation institute. *Hungarian Academy of Sciences*. Retrieved: http://academic.evergreen.edu/projects/bdei/documents/ decision making methods.pdf

Gilliland, M. (2012). Worst practices in new product forecasting. *The Journal of Business Forecasting, 31*(4), 31.

Hwang, C.-L., & Yoon, K. (1981). *Multiple attribute decision Mmaking, Lecture notes in economics and mathematical systems*. Heidelberg, Germany: Springer-Verlag. doi:10.1007/978-3-642-48318-9

Hyndman, R. J., & Athanasopoulos, G. (2013). Forecasting: principles and practice. *OTexts*. Retrieved: https://www.otexts.org/fpp

Kahn, K. B. (2006). *New product forecasting: an applied approach*. Armonk, NY: ME Sharpe.

Kahraman, C. (2008). Fuzzy multi-criteria decision making: theory and applications with recent developments. Springer Science+Business Media, LLC.

KenevissiF. (2014). Retrieved: http://www.edc.ncl.ac.uk/highlight/rhdecember2006.php

Kumar, S., & Pandey, P. (2010). Dynamic Performers using Fuzzy logic in Business process management. *Pragyan: Journal of Information Technology, 8*(12), 71–75.

Laarhoven, P. J. M., & Pedrycz, W. (1983). A extension of Saaty's priority theory. *Fuzzy Sets and Systems, 11*(1-3), 229–241. doi:10.1016/S0165-0114(83)80082-7

Olson, D. L. (2004). Comparison of weights in TOPSIS models. *Mathematical and Computer Modelling, 40*(7-8), 721–727. doi:10.1016/j.mcm.2004.10.003

Pandey, M., Litoriya, R., & Pandey, P. (2018b). Mobile App development based on agility function. *Ingénierie Des Systèmes d'information RSTI Série ISI, 23*(6), 19–44.

Pandey, M., Litoriya, R., & Pandey, P. (2019a). Application of Fuzzy Dematel Approach in Analyzing Mobile App Issues. *Programming and Computer Software, 45*(5), 268–287. doi:10.1134/S0361768819050050

Pandey, M., Litoriya, R., & Pandey, P. (2019b). Empirical Analysis of Defects in Handheld Device Applications. In M. Singh, P. K. Gupta, V. Tyagi, J. Flusser, T. Ören, & R. Kashyap (Eds.), Advances in Computing and Data Sciences (pp. 103–113). Singapore: Springer Singapore.

Pandey, M., Litoriya, R., & Pandey, P. (2019c). Identifying Causal Relationships in Mobile App Issues: An Interval Type-2 Fuzzy DEMATEL Approach. *Wireless Personal Communications*, 1–28.

Pandey, M., Litoriya, R., & Pandey, P. (2019e). Perception-Based Classification of Mobile Apps: A Critical Review. In A. K. Luhach, K. B. G. Hawari, I. C. Mihai, P.-A. Hsiung, & R. B. Mishra (Eds.), Smart Computational Strategies: Theoretical and Practical Aspects (pp. 121–133). Academic Press. doi:10.1007/978-981-13-6295-8_11

Pandey, M., Litoriya, R., & Pandey, P. (2016). Mobile applications in context of big data: A survey. *2016 Symposium on Colossal Data Analysis and Networking (CDAN)*, 1–5. 10.1109/CDAN.2016.7570942

Pandey, M., Litoriya, R., & Pandey, P. (2018a). An ISM Approach for Modeling the Issues and Factors of Mobile App Development. *International Journal of Software Engineering and Knowledge Engineering*, *28*(07), 937–953. doi:10.1142/S0218194018400119

Pandey, M., Litoriya, R., & Pandey, P. (2019d). Novel Approach for Mobile Based App Development Incorporating MAAF. *Wireless Personal Communications*, *107*(4), 1687–1708. doi:10.100711277-019-06351-9

Pandey, P., Kumar, S., & Shrivastava, S. (2013a). A Critical Evaluation of Computational Methods of Forecasting Based on Fuzzy Time Series. *International Journal of Decision Support System Technology*, *5*(1), 25–40. doi:10.4018/jdsst.2013010102

Pandey, P., Kumar, S., & Shrivastava, S. (2013b). Forecasting using Fuzzy Time Series for Diffusion of Innovation: Case of Tata Nano Car in India. *National Academy Science Letters*, *36*(3), 299–309. doi:10.100740009-013-0140-4

Pandey, P., Kumar, S., & Shrivastava, S. (2014). A unified strategy for forecasting of a new product. *Decision (Washington, D.C.)*, *41*(4), 411–424.

Pandey, P., Kumar, S., & Shrivastava, S. (2015). A Fuzzy Decision Making Approach for Analogy Detection in New Product Forecasting. *Journal of Intelligent & Fuzzy Systems*, *28*(5), 2047–2057. doi:10.3233/IFS-141483

Pandey, P., Kumar, S., & Shrivastava, S. (2017). An efficient time series forecasting method exploiting fuzziness and turbulence in data. *International Journal of Fuzzy System Applications*, *6*(4), 83–98. doi:10.4018/IJFSA.2017100106

Pandey, P., & Litoriya, R. (2019a). A Predictive Fuzzy Expert System for Crop Disease Diagnostic and Decision Support. In Fuzzy Expert Systems and Applications in Agricultural Diagnosis (pp. 175–194). Academic Press. doi:10.4018/978-1-5225-9175-7.ch010

Pandey, P., & Litoriya, R. (2019b). An activity vigilance system for elderly based on fuzzy probability transformations. *Journal of Intelligent & Fuzzy Systems*, *36*(3), 2481–2494. doi:10.3233/JIFS-181146

Pandey, P., & Litoriya, R. (2019c). Legal/Regulatory Issues for MMBD in IoT. In S. Tanwar, S. Tyagi, & N. Kumar (Eds.), *Multimedia Big Data Computing for IoT Applications Concepts, Paradigms and Solutions* (pp. 367–388). Springer Nature Singapore Ltd.

Pandey, P., Litoriya, R., & Tiwari, A. (2018). A framework for fuzzy modelling in agricultural diagnostics. *Journal Européen des Systèmes Automatisés*, *51*(4-6), 203–223. doi:10.3166/jesa.51.203-223

Peres, R., Muller, E., & Mahajan, V. (2010). Innovation diffusion and new product growth models: A critical review and research directions. *International Journal of Research in Marketing*, *27*(2), 91–106. doi:10.1016/j.ijresmar.2009.12.012

Rogers, E. M. (2003). *Diffusion of Innovations*. New York: Free Press.

Saaty, T. L. (2003). *The Analytic Hierarchy Process*. New York: McGra Hill.

Shi, Y., Wang, S., Kou, G., & Wallenius, J. (2011). New state of MCDM in the 21st century: Selected Papers of the 20th International Conference on Multiple Criteria Decision Making 2009. *Lecture Notes in Economics and Mathematical Systems*.

Thomas, R. J. (1985). Estimating market growth for new products: An analogical diffusion model approach. *Journal of Product Innovation Management*, *2*(1), 45–55. doi:10.1111/1540-5885.210045

White, M., Braczyk, H. J., Ghobadian, A., & Niebuhr, J. (1988). Small firms' innovation: why regions differ. *Policy Studies Institute*. Retrieved from: http://www.psi.org.uk/site/publication_detail/735

Zadeh, L. A. (1965). Fuzzy sets. *Information and Control*, *8*(3), 338–353. doi:10.1016/S0019-9958(65)90241-X

Compilation of References

Abbott, J. C. (1967). Semi-bolean algebra. *Matem. Vestnik*, *4*, 177-198. http://eudml.org/doc/258960

Abdullah, S., Aslam, M., & Anwar, T. (2011). A note on M-hyper systems and N-hypersystems in Γ-semihypergroups. *Quasi groups and Related Systems*, *19*, 169-172.

Abdullah, S., Naeem, M., & Imran, M. (2013). On intuitionistic fuzzy Γ-ideals of Γ-LA-semigroups. *Annals of Fuzzy Mathematics and Informatics*, *6*(1), 17–31.

Abou-Zaid, A. (1989). On fuzzy subnear-ring. *Fuzzy Sets and Systems*, *81*, 383–393.

Adak, A. K., Bhowmik, M., & Pal, M. (2012d). Semiring of Generalized Interval-valued Intuitionistic Fuzzy Matrices. *World Applied Sciences Journal*, *16*, 7-16.

Adak, A. K., & Bhowmik, M. (2011a). Interval cut-set of interval-valued intuitionistic fuzzy sets. *African Journal of Mathematics and Computer Sciences*, *4*(4), 192–200.

Adak, A. K., Bhowmik, M., & Pal, M. (2011b). Application of generalized intuitionistic fuzzy matrix in multi-criteria decision making problem. *Journal of Mathematical and Computational Science*, *1*(1), 19–31.

Adak, A. K., Bhowmik, M., & Pal, M. (2011c). Semiring of interval-valued intuitionistic fuzzy matrices. *Global Journal of Computer Application and Technology*, *1*(3), 340–347.

Adak, A. K., Bhowmik, M., & Pal, M. (2012a). Interval cut-set of generalized interval-valued intuitionistic fuzzy sets. *International Journal of Fuzzy System Applications*, *2*(3), 35–50.

Adak, A. K., Bhowmik, M., & Pal, M. (2012b). Intuitionistic fuzzy block matrix and its some properties. *Annals of Pure and Applied Mathematics*, *1*(1), 13–31.

Adak, A. K., Bhowmik, M., & Pal, M. (2012c). Some properties of generalized intuitionistic fuzzy nilpotent matrices over distributive lattice. *International Journal of Fuzzy Information and Engineering*, *4*(4), 371–387.

Adak, A. K., Bhowmik, M., & Pal, M. (2013). Distributive Lattice over Intuitionistic Fuzzy Matrices. *The Journal of Fuzzy Mathematics*, *21*(2), 401–416.

Adnan Abdul-Ghani, Mahmood Khalil, Abd Ulrazaq, & Firas Muhammad Jawad Al-Musawi. (2018). New Branch of Intuitionistic Fuzzification in Algebras with Their Applications. *International Journal of Mathematics and Mathematical Sciences*.

Ajmal, N., & Prajapati, A. S. (1992). Fuzzy cosets and fuzzy normal subgroups. *Information Sciences*, *64*(1-2), 17–25. doi:10.1016/0020-0255(92)90107-J

Akram, M. (2011). Bipolar fuzzy graphs. *Information Sciences*, *181*(24), 5548–5564. doi:10.1016/j.ins.2011.07.037

Aktas, H., & Cagman, N. (2007). Soft sets and soft groups. *Information Science*, *177*, 2726–2735.

Ali Abbas, N.M., Khalil, S.M. & Sadiq, H.J. (2019). A Note on the New Closed Classes and GP-Regular Classes in Soft. *Journal of Southwest Jiaotong University, 54*(1). doi:10. 3969/j.issn.0258-2724.2019.155

Ali, M. I., Feng, F., & Shabir, M. A. (2012). Note on (∈,∈∨q) -fuzzy equivalence relations and indistinguishability operators. *Hacettepe Journal of Mathematics and Statistics*, *40*, 383–400.

Ali, M. I., & Shabir, M. (2009). Soft ideals and generalized fuzzy ideals in semigroups. *New Mathematics and Natural Computation*, *5*, 599–615.

Ali, T., & Ray, A. K. (2007). On product of fuzzy sublattices. *Journal of Fuzzy Mathematics*, *15*, 375–381.

Alkhatib, K., Najadat, H., Hmeidi, I., & Shatnawi, M. K. A. (2013). Stock Price Prediction Using K-Nearest Neighbor (kNN) Algorithm. *Int. J. Bus. Humanit. Technol.*, *3*(3), 32–44.

Allen, P. J. (1969). A fundamental theorem of homomorphism for semirings. *Proceedings of the American Mathematical Society*, *21*(2), 412–416. doi:10.1090/S0002-9939-1969-0237575-4

Ansari, M. A. A., & Chandramouleeswaran, M. (2013). Fuzzy β− subalgebras of β− algebras. *International Journal of Mathematical Sciences and Engineering applications*, *5*(7), 239-249.

Ansari, M. A. A., & Chandramouleeswaran, M. (2014). Fuzzy β−ideals of β− algebras. *International Journal of Mathematical Sciences and Engineering Applications, 8*(1), 1-10.

Ansari, M. A. A., & Chandramouleeswaran, M. (2014b). T- Fuzzy of β−subalgebras of β−algebras, *International Journal of Mathematical Sciences and Engineering Applications, 8*(1), 177-187.

Ansari, M. A. A., & Chandramouleeswaran, M. (2014a). Cartesian product of fuzzy β−ideals of β−algebras. *Global Journal of Pure and Applied Mathematics*, *10*(1), 35–42.

Ashraf, S., Abdullah, S., & Mahmood, T. (n.d.). GRA method based on spherical linguistic fuzzy Choquet integral environment and its application in multi-attribute decision-making problems. *Mathematical Sciences*. doi:10.100740096-018-0266-0

Ashraf, S., Abdullah, S., Mahmood, T., Ghani, F., & Mahmood, T. (n.d.). Spherical fuzzy sets and their applications in multi-attribute decision making problems. *Journal of Intelligent & Fuzzy Systems*, 1-16.

Ashraf, S., Mahmood, T., Abdullah, S., & Khan, Q. (2018). Different approaches to multi-criteria group decision making problems for picture fuzzy environment. *Bull. Braz. Math. Soc. New Ser.* 4-018-0103-y doi:10.10070057

Ashraf, S., & Abdullah, S. (2018). Spherical aggregation operators and their application in multi attribute group decision-making. *International Journal of Intelligent Systems*, doi:10.1002/int.22062

Ashraf, S., Abdullah, S., & Qadir, A. (2018). Novel concept of cubic picture fuzzy sets. *J. New Theory*, *24*, 59–72.

Ashraf, S., Abdullah, S., Smarandache, F., & Amin, N. (2019). Logarithmic Hybrid Aggregation Operators Based on Single Valued Neutrosophic Sets and Their Applications in Decision Support Systems. *Symmetry*, *11*(3), 364. doi:10.3390ym11030364

Ashraf, S., Mahmood, T., Abdullah, S., & Khan, Q. (2018). Picture fuzzy linguistic sets and their applications for multi-attribute group decision making problems. *Nucleus*, *55*(2), 66–73.

Atanassov, K. (1994). New operations defined over the intuitionistic fuzzy sets. *Fuzzy Sets and Systems*, *61*(2), 137–142. doi:10.1016/0165-0114(94)90229-1

Atanassov, K. (1994). Operations over interval valued fuzzy set. *Fuzzy Sets and Systems*, *64*, 159–174.

Atanassov, K. (2007). Remark on intuitionistic fuzzy numbers. *Notes Intuit. Fuzzy Sets*, *13*, 29–32.

Atanassov, K. T. (1986). Intuitionistic fuzzy sets. *Fuzzy Sets and Systems*, *20*(1), 87–96.

Atanassov, K. T. (1986). Intuitionistic fuzzy sets. *Fuzzy Sets and Systems*, *20*, 87–96.

Atanassov, K. T. (1999). *Intuitionistic Fuzzy Sets: Theory and Applications, Studies in Fuzziness and Soft Computing* (Vol. 35). Heidelberg, Germany: Physica-Verlag. doi:10.1007/978-3-7908-1870-3

Atanassov, K., & Gargo, G. (1989). Interval-valued intuitionistic fuzzy sets. *Fuzzy Sets and Systems*, *31*(1), 343–349.

Atanassov, K., & Gargov, G. (1989). Interval valued intuitionistic fuzzy sets. *Fuzzy Sets and Systems*, *31*(3), 343–349. doi:10.1016/0165-0114(89)90205-4

Awolola, J.A., & Ibrahim, A.M. (2016). Some results on multi-groups. *Quasigroups and Related Systems*, (24), 169-177.

Aygunoglu, A., & Aygun, H. (2009). Introduction to fuzzy soft groups. *Computers & Mathematics with Applications (Oxford, England)*, *58*, 1279–1286.

Baby, A., Shinoj, T. K., & Sunil, J. J. (2015). On abelian fuzzy multigroups and order of fuzzy multigroups. *J. New Theory*, *5*(2), 80–93.

Baccour, L., Alimi, A. M., & John, R. I. (2014). Some Notes on Fuzzy Similarity Measures and Application to Classification of Shapes. *Recognition of Arabic Sentences and Mosaic.*, *41*(2), 81–90.

Bandaru, R. K. (2017). On QI-algebras. *Discussiones Mathematicae. General Algebra and Applications*, *37*(2), 137–145. doi:10.7151/dmgaa.1269

Bandaru, R. K., Shum, K. P., & Rafi, N. (2017). Some types of ideals in distributive implication groupoids. *International Journal of Pure and Applied Mathematics*, *116*(2), 377–384. doi:10.12732/ijpam.v116i2.9

Barbhuiya, S. R., (2014). Doubt fuzzy ideals of *BF*-algebra. *IOSR journal of Mathematics, 10*(2), 65-70.

Bashiri, M., & Karimi, H. (2012). Effective Heuristics and Meta-Heuristics for the Quadratic Assignment Problem with Tuned Parameters and Analytical Comparisons. *Journal of Industrial Engineering International*, *8*(1), 1–9. Retrieved from http://www.jiei-tsb.com/content/8/1/6

Bass, F. M. (1969). A new product growth model for consumer durables. *Management Science*, *15*(5), 215–227. doi:10.1287/mnsc.15.5.215

Bej, T., & Pal, M. (2014a). Doubt intuitionistic fuzzy H-ideals in BCK/BCI-algebras. *Annals of Fuzzy Mathematics and Informatics*, *8*(4), 593–605.

Bej, T., & Pal, M. (2015). Doubt Atanassovs intuitionistic fuzzy Sub-implicative ideals in *BCI. International Journal of Computational Intelligent Systems*, *8*(2), 240–249. doi:10.1080/18756891.2015.1001948

Bej, T., & Pal, M. (2015). Doubt intuitionistic fuzzy ideals in *BCK/BCI. Int. J. Fuzzy Logic Systems*, *5*(1), 1–13. doi:10.5121/ijfls.2015.5101

Beliakov, G., Bustince, H., Goswami, D. P., Mukherjee, U. K., & Pal, N. R. (2011). On averaging operators for Atanassov's intuitionistic fuzzy sets. *Inf. Sci.*, *181*(6), 1116–1124. doi:10.1016/j.ins.2010.11.024

Bello, J. P. (2011). Measuring Structural Similarity in Music. *IEEE Transactions on Audio, Speech, and Language Processing*, *19*(7), 2013–2025. doi:10.1109/TASL.2011.2108287

Bèzes, C., & Mercanti-Guérin, M. (2017). Similarity in Marketing: Scope, Measurement, and Fields of Application. [English Edition]. *Recherche et Applications en Marketing*, *32*(1), 83–105. doi:10.1177/2051570716676258

Bhakat, S. K. (1999). -level subset. *Fuzzy Sets and Systems*, *103*(3), 529–533. $(\in\vee q)$ doi:10.1016/S0165-0114(97)00158-9

Bhakat, S. K. (2000). -fuzzy normal, quasinormal and maximal subgroups. *Fuzzy Sets and Systems*, *112*(2), 299–312. $(\in,\in\vee q)$ doi:10.1016/S0165-0114(98)00029-3

Bhakat, S. K., & Das, P. (1992). On the definition of a fuzzy subgroup. *Fuzzy Sets and Systems*, *51*(2), 235–241. doi:10.1016/0165-0114(92)90196-B

Bhakat, S. K., & Das, P. (1996a). -fuzzy subgroup. *Fuzzy Sets and Systems*, *80*(3), 359–368. $(\in,\in\vee q)$ doi:10.1016/0165-0114(95)00157-3

Bhakat, S. K., & Das, P. (1996b). Fuzzy subrings and ideals redefined. *Fuzzy Sets and Systems*, *81*(3), 383–393. doi:10.1016/0165-0114(95)00202-2

Bhattacharya, P., & Mukherjee, N. P. (1987). Fuzzy groups: Some group theoretic analogs II. *Information Sciences*, *41*(1), 77–91. doi:10.1016/0020-0255(87)90006-5

Bhowmik, M., & Pal, M. (2010). Generalized interval-valued intuitionistic fuzzy sets. *The Journal of Fuzzy Mathematics*, *18*(2), 357–371.

Bhowmik, M., Senapati, T., & Pal, M. (2014). Intuitionistic L BG. *Afrika Matematika*, *25*(3), 577–590. doi:10.100713370-013-0139-5

Birkhoff, G. (1984). *Lattice theory*. American Mathematical Society Providence.

Birkhoof, G. (2009). *Lattice Theory*. New York: American Mathematical Society.

Biswas, R. (1990). Fuzzy subgroups and antifuzzy subgroups. *Fuzzy Sets and Systems*, *35*(1), 121–124. doi:10.1016/0165-0114(90)90025-2

Biswas, R. (1994). Rosenfeld fuzzy subgroups with interval-valued membership function. *Fuzzy Sets and Systems*, *63*(1), 87–90. doi:10.1016/0165-0114(94)90148-1

Biswas, R. (1999). An application of Yager's bag theory in multicriteria based decision making problems. *International Journal of Intelligent Systems*, *14*(12), 1231–1238. doi:10.1002/(SICI)1098-111X(199912)14:12<1231::AID-INT4>3.0.CO;2-6

Blizard, W.D. (1991). The Developement of Multiset theory. *Modern Logic*, (1), 319-352.

Bluman, G. W., & Cole, J. D. (1974). *Similarity Methods for Differential Equations* (1st ed.). New York: Springer-Verlag. doi:10.1007/978-1-4612-6394-4

Borns, D. W., & Mack, J. M. (1975). *An algebraic Introduction to Mathematical Logic*. Berlin: Springer. doi:10.1007/978-1-4757-4489-7

Bourke, M. M., & Fisher, D. G. (1996). Convergence, eigen fuzzy sets and stability analysis of relational matrices. *Fuzzy Sets and Systems*, *81*(2), 227–234. doi:10.1016/0165-0114(95)00212-X

Bo, Y., & Wangming, W. (1990). Fuzzy ideals on a distributive lattice. *Fuzzy Sets and Systems*, *35*(2), 231–240. doi:10.1016/0165-0114(90)90196-D

Brink, C. (1988). Multiset and the Algebra of Relevance Logic. *Non-Classical Logic*, (5), 75-95.

Buckley, J. J. (1985). Ranking alternatives using fuzzy numbers. *Fuzzy Sets and Systems*, *15*(1), 21–31. doi:10.1016/0165-0114(85)90013-2

Buckley, J. J., & Qu, Y. (1991). Solving systems of linear fuzzy equations. *Fuzzy Sets and Systems*, *43*(1), 33–43. doi:10.1016/0165-0114(91)90019-M

Bustince, H., & Burillo, P. (1995). Intuitionistic fuzzy relation (part-1). *Mathware and Soft Computing*, *2*, 5–38.

Calude, C. S., Paun, G., Rozenberg, G., & Salomaa, A. (2001). Mathematics of Multisets. Springer Verlag.

Carbó-Dorca, R., & Mezey, P. G. (Eds.). (2001). *Fundamentals of Molecular Similarity*. New York: Springer US. doi:10.1007/978-1-4757-3273-3

Carnielli, W., & Coniglio, M. E. (2016). *Paraconsistent Logic: Consistency, Contradiction and Negation*. Springer International Publishing.

Carnielli, W., & Malinowski, J. (Eds.). (2018). *Contradictions, from Consistency to Inconsistency*. Springer International Publishing. doi:10.1007/978-3-319-98797-2

Chakraborty, D. (2001). Structural quantification of vagueness in linguistic expert opinion in an evaluation programme. *Fuzzy Sets and Systems*, *119*(1), 171–186. doi:10.1016/S0165-0114(99)00044-5

Chen, B. (2013). Soft semi-open sets and related properties in soft topological spaces. *Applied Mathematics & Information Sciences*, *7*, 287–294.

Chen, C. T. (2000). Extensions of the TOPSİS for group decision making under fuzzy environment. *Fuzzy Sets and Systems*, *114*(1), 1–9. doi:10.1016/S0165-0114(97)00377-1

Chen, L. S., & Su, C. T. (2008). Using Granular Computing Model to Induce Scheduling Knowledge in Dynamic Manufacturing Environments. *International Journal of Computer Integrated Manufacturing*, *21*(5), 569–583.

Chen, W. Y., & Oliveira, J. S. (1995). Implication algebras and the metropolis rota axioms for cubic lattices. *Journal of Algebra*, *171*(2), 383–396. doi:10.1006/jabr.1995.1017

Corinna, C., & Vapnik, V. (1995). Support vector networks. *Machine Learning*, *20*(3), 273–297. doi:10.1007/BF00994018

Cover, T. M., & Hart, P. E. (1967). Nearest neighbor pattern classification. *IEEE Transactions on Information Theory*, *13*(1), 21–27. doi:10.1109/TIT.1967.1053964

Cross, V. V., & Sudkamp, T. A. (2002). *Similarity and Compatibility in Fuzzy Set Theory: Assessment and Applications* (1st ed.). Berlin: Physica-Verlag Heidelberg. doi:10.1007/978-3-7908-1793-5

Csutora, R., & Buckley, J. J. (2001). Fuzzy hierarchical analysis: The lambda-max method. *Fuzzy Sets and Systems*, *120*(2), 181–195. doi:10.1016/S0165-0114(99)00155-4

Cuong, B. C. (2013). Picture fuzzy sets first results. part 1. Seminar Neuro-Fuzzy Systems with Applications. Institute of Mathematics, Hanoi.

Cuong, B. C. (2014). Picture fuzzy sets. *J. Comput. Sci. Cybern.*, *30*(4), 409–420.

Cuong, B. C., Kreinovitch, V., & Ngan, R. T. (2016). A classification of representable t-norm operators for picture fuzzy sets. *Eighth International Conference on Knowledge and Systems Engineering*. 10.1109/KSE.2016.7758023

Cuong, B. C., & Phan, V. H. (2015). Some fuzzy logic operations for picture fuzzy sets. *Proceedings of Seventh International Conference on Knowledge and System Engineering*.

Cutland, N. J. (1983). Non-standard measure theory and its applications. *Bulletin of the London Mathematical Society*, *15*(6), 529–589. doi:10.1112/blms/15.6.529

Davey, B. A., & Priestley, H. A. (1990). *Introduction to Lattices and Order*. Cambridge, UK: Cambridge University Press.

Davvaz, B. (2006). -fuzzy subnearrings and ideals. *Soft Computing*, *10*(3), 206–211. (∈,∈∨q) doi:10.100700500-005-0472-1

Davvaz, B., & Fotea, V. L. (2007). *Hyper ring theory and applications*. Gulf Breeze, FL: International Academic Pres.

Davvaz, B., & Poursalavati, N. S. (2000). Semihypergroups and S-hyper systems. *Pure Mathematics and Applications*, *11*, 43–49.

Deng, H. (1999). Multicriteria analysis with fuzzy pairwise comparison. *International Journal of Approximate Reasoning*, *21*(3), 215–231. doi:10.1016/S0888-613X(99)00025-0

Deriglazov, A. A. (2011). Formal Similarity Between Mathematical Structures of Electrodynamics and Quantum Mechanics. *International Journal of Theoretical Physics*, *50*(3), 654–662. doi:10.100710773-010-0589-6

Dhivya, J., & Sridevi, B. (2018). Intuitionistic Fuzzy Similarity Measure for Generalized Fuzzy Numbers and Its Application in Fingerprint Matching. *Journal of the Institution of Electronics and Telecommunication Engineers*, *0*(0), 1–12.

Dubois, D., & Prade, H. (1992). Gradual Inference Rules in Approximate Reasoning. *Information Sciences*, *61*(1), 103–122. doi:10.1016/0020-0255(92)90035-7

Dubois, D., & Prade, H. (1997). *Fuzzy sets and systems: Theory and applications*. Orlando, FL: Academic Press.

Dubois, D., & Prade, H. (2008). An introduction to bipolar representations of information and preference. *International Journal of Intelligent Systems*, *23*(8), 866–877. doi:10.1002/int.20297

Easingwood, C. J. (1989). An analogical approach to the long term forecasting of major new product sales. *International Journal of Forecasting*, *5*(1), 69–82. doi:10.1016/0169-2070(89)90065-4

Ebrahimnejad, A., Adak, A. K., & Jamkhaneh, E. B. (2019). Eigenvalue of Intuitionistic Fuzzy Matrices Over Distributive Lattice. *International Journal of Fuzzy System Applications*, *8*(1), 1–18.

Ejegwa, P.A. (2017). Upper and Lower Cuts of Multigroups. *Prajna International Journal of Mathematical Sciences and Applications,* (1), 19-26.

Ejegwa, P. A. (2014). Correspondence between fuzzy multisets and sequences. *Global J. Sci. Frontier Research: Math. Decision Sci.*, *14*(7), 61–66.

Ejegwa, P. A. (2015). Intuitionistic fuzzy sets approach in appointment of positions in an organization via max-min-max rule. *Global Journal of Science Frontier Research: Mathematics & Decision Sciences*, *15*(6), 1–5.

Ejegwa, P. A. (2018). *Distance and similarity measures for Pythagorean fuzzy sets*. Granular Computing. doi:10.100741066-018-00149-z

Ejegwa, P. A. (2018a). On fuzzy multigroups and fuzzy submultigroups. *J. Fuzzy Math.*, *26*(3), 641–654.

Ejegwa, P. A. (2018b). On abelian fuzzy multigroups. *J. Fuzzy Math.*, *26*(3), 655–668.

Ejegwa, P. A. (2018c). On normal fuzzy submultigroups of a fuzzy multigroup. *Theory Appl. Math. Comp. Sci.*, *8*(1), 64–80.

Ejegwa, P. A. (2018d). Homomorphism of fuzzy multigroups and some of its properties. *Applications and Applied Mathematics*, *13*(1), 114–129.

Ejegwa, P. A. (2019). *Improved composite relation for Pythagorean fuzzy sets and its application to medical diagnosis.* Granular Computing. doi:10.100741066-019-00156-8

Ejegwa, P. A. (2019). *Pythagorean fuzzy set and its application in career placements based on academic performance using max-min-max composition.* Complex and Intelligent Systems. doi:10.100740747-019-0091-6

Ejegwa, P. A., Akubo, A. J., & Joshua, O. M. (2014). Intuitionistic fuzzy set and its application in career determination via normalized Euclidean distance method. *European Scientific Journal, 10*(15), 529–536.

Ejegwa, P. A., & Modom, E. S. (2015). Diagnosis of viral hepatitis using new distance measure of intuitionistic fuzzy sets. *International Journal of Fuzzy Mathematical Archive, 8*(1), 1–7.

Ejegwa, P. A., Onoja, A. M., & Emmanuel, I. T. (2014). A note on some models of intuitionistic fuzzy sets in real life situations. *Journal of Global Research in Mathematical Archives, 2*(5), 42–50.

Ejegwa, P. A., Tyoakaa, G. U., & Ayenge, A. M. (2016). Application of intuitionistic fuzzy sets in electoral system. *International Journal of Fuzzy Mathematical Archive, 10*(1), 35–41.

Ejegwa, P. J., Akubo, A. J., & Joshua, O. M. (2014). Intuitionistic Fuzzy Sets and its Application in Career Determination Via Normalized Euclidean Distance Method. *European Scientific Journal, 10*, 529–536.

El Alaoui & Ben-azza. (2017b). Generalization of the Weighted Product Aggregation Applied to Data Fusion of Intuitionistic Fuzzy Quantities. *2017 Intelligent Systems and Computer Vision (ISCV).*

El Alaoui, M., & Ben-azza, H. (2017a). Aggregation of Performance Indicators for Supply Chain and Fuzzy Logic Extensions Applied to Green Supply Chain. *2017 International Colloquium on Logistics and Supply Chain Management (LOGISTIQUA).* 10.1109/LOGISTIQUA.2017.7962870

El Alaoui, M., Ben-azza, H., & El Yassini, K. (2018a). Optimal Weighting Method for Fuzzy Opinions. Paris, France: Academic Press.

El Alaoui, M., El Yassini, K., & Ben-azza, H. (2019). Peer Assessment Improvement Using Fuzzy Logic. In Innovations in Smart Cities Applications Edition 2, Lecture Notes in Intelligent Transportation and Infrastructure. Springer International Publishing. doi:10.1007/978-3-030-11196-0_35

El Alaoui, M. (2018). SMART Grid Evaluation Using Fuzzy Numbers and TOPSIS. *IOP Conference Series. Materials Science and Engineering, 353*(1), 012019. doi:10.1088/1757-899X/353/1/012019

El Alaoui, M., Ben-azza, H., & El Yassini, K. (2018b). Optimal Weighting Method for Interval-Valued Intuitionistic Fuzzy Opinions. *Notes on Intuitionistic Fuzzy Sets, 24*(3), 106–110. doi:10.7546/nifs.2018.24.3.106-110

El Alaoui, M., Ben-azza, H., & El Yassini, K. (2019a). Achieving Consensus in Interval Valued Intuitionistic Fuzzy Environment. *Procedia Computer Science, 148*, 218–225. doi:10.1016/j.procs.2019.01.064

El Alaoui, M., Ben-azza, H., & El Yassini, K. (2019b). Fuzzy TOPSIS with Coherent Measure: Applied to a Closed Loop Agriculture Supply Chain. In M. Ezziyyani (Ed.), *Advanced Intelligent Systems for Sustainable Development (AI2SD'2018), Advances in Intelligent Systems and Computing* (pp. 106–117). Springer International Publishing. doi:10.1007/978-3-030-11878-5_12

El Alaoui, M., Ben-azza, H., & Zahi, A. (2016). New Multi-Criteria Decision-Making Based on Fuzzy Similarity, Distance and Ranking. In *Proceedings of the Third International Afro-European Conference for Industrial Advancement — AECIA 2016, Advances in Intelligent Systems and Computing*. Springer.

El Alaoui, M., El Yassini, K., & Ben-azza, H. (2018). Enhancing MOOCs Peer Reviews Validity and Reliability by a Fuzzy Coherence Measure. In *Proceedings of the 3rd International Conference on Smart City Applications, SCA '18*. New York: ACM. 10.1145/3286606.3286834

Elsner, L., & van den Driessche, P. (1999). On the power method in max algebra. *Linear Algebra and Its Applications, 302–303*, 17–32. doi:10.1016/S0024-3795(98)10171-4

Fan, Z. T., & Liu, D.-F. (1997). Convergence of the power sequence of a nearly monotone increasing fuzzy matrix. *Fuzzy Sets and Systems, 88*(3), 363–372. doi:10.1016/S0165-0114(96)00040-1

Fan, Z.-T. (1999). A note on the power sequence of a fuzzy matrix. *Fuzzy Sets and Systems, 102*(2), 281–286. doi:10.1016/S0165-0114(97)00118-8

Fan, Z.-T., & Liu, D.-F. (1998a). A detailed study on the power sequence of matrices of commonly used types. *Fuzzy Sets and Systems, 99*(2), 197–203. doi:10.1016/S0165-0114(97)00025-0

Fan, Z.-T., & Liu, D.-F. (1998b). On the oscillating power sequence of a fuzzy matrix. *Fuzzy Sets and Systems, 93*(1), 75–85. doi:10.1016/S0165-0114(96)00194-7

Farhadinia, B. (2013). A theoretical development on the entropy of interval-valued fuzzy sets based on the intuitionistic distance and its relationship with similarity measure. *Knowledge-Based Systems, 39*, 79–84. doi:10.1016/j.knosys.2012.10.006

Feng, Y., Duan, H., & Zeng, Q. (2010). $\left(\lambda,\hat{\gamma}\right)$-fuzzy sublattices and $\left(\lambda,\hat{\gamma}\right)$-fuzzy subhyperlattices. *Fuzzy Information and Engineering, 78*, 17-26.

Fenga, F., Jun, Y. B., & Zhao, X. (2008). Soft semirings. *Computers & Mathematics with Applications (Oxford, England), 56*, 2621–2628.

Feng, Y., Zeng, Q., Zou, L., & Zhang, B. (2010). On Cartesian product of $\left(», \hat{\gamma}\right)$. *Fuzzy Information and Engineering, 4*(4), 413–422. doi:10.100712543-010-0059-0

Feng, Y., & Zou, L. (2010). On product of fuzzy subhyperlattice and interval-valued fuzzy subhyperlattice. *Discrete Mathematics, Algorithms, and Applications, 2*(02), 239–246. doi:10.1142/S1793830910000619

Figueroa-García, J. C., Chalco-Cano, Y., & Román-Flores, H. (2015). Distance Measures for Interval Type-2 Fuzzy Numbers. *Discrete Applied Mathematics, 197*, 93–102. doi:10.1016/j.dam.2014.11.016

Fix, E., & Hodges, J. L. (1951). *Discriminatory analysis, Nonparametric discrimination; consistency properties.* Tech. Rep. 4, USAF School of Aviation Medicine, Randolph Field, TX.

Fülöp. (2005). Introduction to decision making methods. Laboratory of operations research and decision systems: computer and automation institute. *Hungarian Academy of Sciences*. Retrieved: http://academic.evergreen.edu/projects/bdei/documents/ decision making methods.pdf

Garg, H. (2017). Some picture fuzzy aggregation operators and their applications to mul-ticriteria decision-making. *Arabian Journal for Science and Engineering, 42*(12), 5275–5290. doi:10.100713369-017-2625-9

Gautam, S. S. (2016). TOPSIS for multi criteria decision making in intuitionistic fuzzy environment. *International Journal of Computers and Applications, 156*(8), 42–49. doi:10.5120/ijca2016912514

Gilliland, M. (2012). Worst practices in new product forecasting. *The Journal of Business Forecasting, 31*(4), 31.

Girish, K. P., & John, S. J. (2009). Relations and functions in multiset context. Information Sciences, (179), 758–768.

Girish, K. P., & John, S. J. (2011). Rough Multiset and its Multiset Topology. Lecture Notes in Computer Science, 6600, 62-80.

Girish, K. P., & John, S. J. (2012). Multiset Topologies induced by multiset relations. Information Sciences, (188), 298–313.

Girish, K.P., & John, S.J. (2012). On multiset Topologies. *Theory and Applications of Mathematics and Computer Science*, (2), 37-52.

Goguen, J. A. (1985). Fuzzy Sets and Systems: Theory and Applications (Didier Dubois and Henri Prade). *SIAM Review*, *27*(2), 270–274. doi:10.1137/1027081

Gong, Y., Yang, S., & Dai, L. (2018). Some New Signed Distances and Similarity Measures of Interval Type-2 Trapezoidal Fuzzy Numbers and Comparative Study. *Journal of Intelligent & Fuzzy Systems*, *35*(3), 3465–3475. doi:10.3233/JIFS-17671

Good, R. A., & Hughes, D. R. (1952). Associated groups for a semigroup. *Bulletin of the American Mathematical Society*, *58*, 624–625.

Gregson, R. A. M. (1975). *Psychometrics of Similarity*. New York: Academic Press.

Grzegorzewski, P. (2002). Nearest interval approximation of a fuzzy number. *Fuzzy Sets and Systems*, *1320*, 321–330.

Grzegorzewski, P. (2004). Distances between Intuitionistic Fuzzy Sets and/or Interval-Valued Fuzzy Sets Based on the Hausdorff Metric. *Fuzzy Sets and Systems*, *148*(2), 319–328. doi:10.1016/j.fss.2003.08.005

Grzegorzewski, P., & Mrówka, E. (2005). Trapezoidal Approximations of Fuzzy Numbers. *Fuzzy Sets and Systems*, *153*(1), 115–135. doi:10.1016/j.fss.2004.02.015

Grzegorzewski, P., & Mrówka, E. (2007). Trapezoidal Approximations of Fuzzy Numbers—Revisited. *Fuzzy Sets and Systems*, *158*(7), 757–768. doi:10.1016/j.fss.2006.11.015

Gulistan, M., Khalaf, M., Yaqoob, N., & Rashid, U. (2019). An introduction to non-associative neutrosophic semihypergroups. *Periodica Mathematica Hungarica*. (Manuscript submitted for publication)

Gupta, C., Jain, A., & Joshi, N. (2018). Fuzzy Logic in Natural Language Processing – A Closer View. *Procedia Computer Science*, *132*, 1375–1384. doi:10.1016/j.procs.2018.05.052

Guu, S.-M., Chen, H.-H., & Pang, C.-T. (2001). Convergence of products of fuzzy matrices. *Fuzzy Sets and Systems*, *121*(2), 203–207. doi:10.1016/S0165-0114(00)00006-3

Hall, P., Park, B. U., & Samworth, R. J. (2008). Choice of neighbor order in nearest-neighbor classification. *Annals of Statistics*, *36*(5), 2135–2152. doi:10.1214/07-AOS537

Hansen, E. (1980). Global optimization using interval analysis? the multi-dimensional case. *Numerische Mathematik*, *34*(3), 247–270. doi:10.1007/BF01396702

Hansen, E. R., & Walster, G. W. (2004). *Global optimization using interval analysis*. New York: Marcel Dekker Inc.

Harmse, J. (1996). Continuous Fuzzy Conjunctions and Disjunctions. *IEEE Transactions on Fuzzy Systems*, *4*(3), 295–314. doi:10.1109/91.531772

Hasankhani, A. (1999). Ideals in a semihypergroup and greens relations ratio mathematica. *Journal of Applied Mathematics*, *13*, 19–36.

Hashimoto, H. (1983a). Canonical form of a transitive fuzzy matrix. *Fuzzy Sets and Systems*, *11*(1–3), 157–162. doi:10.1016/S0165-0114(83)80076-1

Hashimoto, H. (1983b). Convergence of powers of a fuzzy transitive matrix. *Fuzzy Sets and Systems*, *9*(1–3), 153–160. doi:10.1016/S0165-0114(83)80015-3

Hashimoto, H. (1984). Subinverses of fuzzy matrices. *Fuzzy Sets and Systems*, *12*(2), 155–168. doi:10.1016/0165-0114(84)90035-6

Hashimoto, H. (1985). Transitivity of generalized fuzzy matrices. *Fuzzy Sets and Systems*, *17*(1), 83–90. doi:10.1016/0165-0114(85)90009-0

Hemasinha, R., Pal, N. R., & Bezdek, J. C. (1993). Iterates of fuzzy circulant matrices. *Fuzzy Sets and Systems*, *60*(2), 199–206. doi:10.1016/0165-0114(93)90346-J

Hemavathi, P., Muralikrishna, P., & Palanivel, K. (2017). ivf β--Ideals of β-Algebras. Materials Science and Engineering Conference Series, 263(4), 1-10.

Hemavathi, P., Muralikrishna, P., & Palanivel, K. (2015). A note on interval valued fuzzy β-subalgebras. *Global Journal of Pure and Applied Mathematics*, *11*(4), 2553–2560.

Henriksen, M. (1958). Ideals in semirings with commutative addition. *Amer. Math. Soc. Notices*, *5*, 321.

He, P., Xin, X., & Zhan, J. (2014). (Fuzzy) hyperlattices and fuzzy preordered lattices. *Journal of Intelligent & Fuzzy Systems*, *26*, 2369–2381.

He, P., Xin, X., & Zhan, J. (2016). On derivations and their fixed point sets in residuated lattices. *Fuzzy Sets and Systems*, *303*, 97–113. doi:10.1016/j.fss.2016.01.006

Hila, K., Davvaz, B., & Naka, K. (2011). On quasi-hyper ideals in semihypergroups. *Journal of Communications in Algebra*, *39*(11), 4183–4194. doi:10.1080/00927872.2010.521932

Hila, K., & Dine, J. (2011). On hyper ideal s in left almost semihypergroups: *ISRN Algebra. Article ID*, *953124*(8), 68–76.

Hong, S. M., Jun, Y. B., Kim, S. J., & Kim, G. I. (1993).. . *Fuzzy BCI-Subalgebras with Interval Valued Membership Functions Math Japonica*, *40*(2), 199–202.

Howie., (1976). *Introduction to semigroup theory*. Academic Press.

Huang, Y. (2006). BCI-algebra. Science Press.

Huang, Y. S. (2006). BCI-algebra. Science Press.

Hu, Q. P., & Iseki, K. (1980). On BCI-algebra satisfying (x*y)*z=x*(y*z). *Math Sem Notes*, *8*, 553–555.

Hwang, C.-L., & Yoon, K. (1981). *Multiple attribute decision Mmaking, Lecture notes in economics and mathematical systems*. Heidelberg, Germany: Springer-Verlag. doi:10.1007/978-3-642-48318-9

Hyndman, R. J., & Athanasopoulos, G. (2013). Forecasting: principles and practice. *OTexts*. Retrieved: https://www.otexts.org/fpp

Ibrahim, A.M. & Ajegwa, P.A. (2016). A Survey on the Concept of Multigroups. *Journal of the Nigerian Association of Mathematical Physics*, (38), 1-8.

Imai, Y., & Iseki, K. (1966). On axiom system of propositional calculi. *XIV Proc. Japan Academy*, *42*, 19-22.

Imai, H., Okahara, Y., & Miyakoshi, M. (2000). The period of powers of a fuzzy matrix. *Fuzzy Sets and Systems*, *109*(3), 405–408. doi:10.1016/S0165-0114(98)00080-3

Imai, Y., & Iseki, K. (1966). On axiom system of propositional calculi. XIV. *Proc. of the Japan Aceademy*, *42*(1), 19–22. doi:10.3792/pja/1195522169

Imandoust, S. B., & Bolandraftar, M. (2013). Application of K-Nearest Neighbor (KNN) Approach for Predicting Economic Events: Theoretical Background. *Int. J. Eng. Res. Appl.*, *3*(5), 605–610.

Iseki, K. (1980). On BCI-algebras. *Mathematics Seminar Notes*, *8*, 125-130. http://www.math.kobe-u.ac.jp/jmsj/kjm/

Iseki, K. (1956). Ideal theory of semiring. *Proceedings of the Japan Academy*, *32*(8), 554–559. doi:10.3792/pja/1195525272

Iseki, K. (1958). Ideal in semirings. *Proceedings of the Japan Academy*, *34*(1), 29–31. doi:10.3792/pja/1195524845

Iseki, K. (1958). Quasi-ideals in semirings without zero. *Proceedings of the Japan Academy*, *34*(2), 79–84. doi:10.3792/pja/1195524783

Iseki, K. (1966). An algebra related with a propositional calculus. *Proceedings of the Japan Academy*, *42*(1), 26–29. doi:10.3792/pja/1195522171

Iseki, K., & Tanaka, S. (1978). An introduction to the theory of BCK-algebras. *Math. Japon*, *23*(1), 1–26.

Ismat, B., & Samina, A. (2009). Similarity Measures for Fuzzy Sets. *Applied and Computational Mathematics*, *8*(2), 192–202.

Jaccard, P. (1908). Nouvelles Recherches Sur La Distribution Florale. *Bulletin de la Société Vaudoise des Sciences Naturelles*, *44*, 223–270.

Jagatap, R. D., & Pawar, Y. S. (2009). Quasi-ideals and minimal quasi-ideals in Γ-semirings. *Novi Sad J. Math.*, *39*(2), 79–87.

Jahantigh, M.A.,Khezerloo, S., & Khezerloo, M. (2010). Complex fuzzy Linear Systems. *Int. Journal of Industrial Mathematics*, *2*(1), 21-28.

Jana, C., & Pal, M. (2016a). Applications of (α,β) *BCK/BCI*. *International Journal of Intelligent Systems Technologies and Applications*, *16*, 269–288. doi:10.1504/IJISTA.2017.085362 PMID:27160752

Jana, C., & Pal, M. (2017). On(∈α,∈α∨qβ) BCI. *Missouri Journal of Mathematical Sciences*, *29*(2), 197–215. doi:10.35834/mjms/1513306831

Jana, C., Pal, M., & Saied, A. B. (2017). -bipolar fuzzy (∈,∈∨q)BCK/BCI-algebras. *Missouri J. of Math. Sci.*, *29*(2), 139–160.

Jana, C., Pal, M., & Saied, A. B. (2017a). -bipolar fuzzy BCK/BCI-algebras. *Missouri Journal of Mathematical Sciences*, *29*(2), 139–160. (∈,∈∨q) doi:10.35834/mjms/1513306827

Jana, C., Pal, M., Senapati, T., & Bhowmik, M. (2015a). Atanassov's intutionistic 𝔏-fuzzy G-subalgebras of G-algebras. *Journal of Fuzzy Mathematics*, *23*, 195–209.

Jana, C., Senapati, T., Bhowmik, M., & Pal, M. (2015). On intuitionistic fuzzyG. *Fuzzy Information and Engineering*, *7*(2), 195–209. doi:10.1016/j.fiae.2015.05.005

Jana, C., Senapati, T., & Pal, M. (2016). (∈,∈∨q)-intuitionistic fuzzy BCI. *Journal of Intelligent & Fuzzy Systems*, *31*(1), 613–621. doi:10.3233/IFS-162175

Javanmard & Nehi. (2018). Rankings and Operations for Interval Type-2 Fuzzy Numbers: A Review and Some New Methods. *Journal of Applied Mathematics and Computing*.

Jian-Xin, L. (1992). Controllable fuzzy matrices. *Fuzzy Sets and Systems*, *45*(3), 313–319. doi:10.1016/0165-0114(92)90149-X

Jian-Xin, L. (1994). Convergence of powers of controllable fuzzy matrices. *Fuzzy Sets and Systems*, *62*(1), 83–88. doi:10.1016/0165-0114(94)90074-4

Johnson, R. A., & Wichern, D. W. (2008). *Applied Multivariate Statistical Analysis*. Pearson.

Jun, Y. B., & Kim. (2012). β−subalgebras and related topics. *Commun. Korean. Math. Soc*, *27*(2), 243-255.

Jun, Y. B., Ahn, S. S., & Muhiuddin, G. (2014). Hesitant fuzzy soft subalgebras and ideals in BCK/BCI-algebras. The Scientific World Journal.

Jun, Y. B. (1993). Closed fuzzy ideals in BCI-algebras. *Math Japonica*, *38*, 199–202.

Jun, Y. B. (1993). Closed fuzzy ideals in *BCI*-algebras. *Math. Japonica*, *38*, 199–202.

Jun, Y. B. (1994). Doubt fuzzy BCK/BCI-algebras. *Soochow Journal of Mathematics*, *20*(3), 351–358.

Jun, Y. B. (2000). Interval-valued fuzzy subalgebras/ideals in BCK-algebras. *Scientiae Mathematicae*, *3*(3), 435–444.

Jun, Y. B. (2002). Interval-valued fuzzy R-subgroups of near-rings. *Indian Journal of Pure and Applied Mathematics*, *33*(1), 71–80.

Jun, Y. B. (2005). On (α,β) *BCK/BCI*. *Bulletin of the Korean Mathematical Society*, *42*(4), 703–711. doi:10.4134/BKMS.2005.42.4.703

Jun, Y. B. (2007). Fuzzy subalgebras with thresholds in *BCK/BCI*. *Commun. Korean Math. Soc.*, *22*(2), 173–181. doi:10.4134/CKMS.2007.22.2.173

Jun, Y. B. (2009). Generalizations of(∈,∈∨q). *Computers & Mathematics with Applications (Oxford, England)*, *58*(7), 1383–1390. doi:10.1016/j.camwa.2009.07.043

Jun, Y. B., Kim, H. S., & Lee, K. J. (2009). Bipolar fuzzy translation in BCK/BCI-algebra. Journal of the Chungcheong Math. *Soc.*, *22*(3), 399–408.

Jun, Y. B., & Kim, K. H. (2000). Intuitionistic fuzzy ideals in *BCK*. *International Journal of Mathematics and Mathematical Sciences*, *24*(12), 839–849. doi:10.1155/S0161171200004610

Jun, Y. B., Kim, K. H., & Yon, Y. H. (1999). Intuitionistic fuzzy ideals of near-rings. *J. Inst. Math. Comp. Sci.*, *12*(3), 221–228.

Jun, Y. B., Öztürk, M. A., & Muhiuddin, G. (2016). A generalization of(∈,∈∨q). *International Journal of Algebra and Statistics*, *5*(1), 7–18. doi:10.20454/ijas.2016.1041

Jun, Y. B., & Song, S. Z. (2006). Generalized fuzzy interior ideals in semigroups. *Information Science*, *176*(20), 3079–3093. doi:10.1016/j.ins.2005.09.002

Jun, Y. B., & Song, S. Z. (2014). Hesitant fuzzy set theory applied to filters in MTL-algebras. *Honam Math. J.*, *36*(4), 813–830. doi:10.5831/HMJ.2014.36.4.813

Jun, Y. B., & Song, S. Z. (2015). Hesitant fuzzy prefilters and filters of EQ-algebras. *Applied Mathematical Sciences*, *9*, 515–532. doi:10.12988/ams.2015.411962

Jun, Y. B., Song, S. Z., & Muhiuddin, G. (2016). Hesitant fuzzy semigroups with a frontier. *Journal of Intelligent & Fuzzy Systems*, *30*(3), 1613–1618. doi:10.3233/IFS-151869

Kahn, K. B. (2006). *New product forecasting: an applied approach.* Armonk, NY: ME Sharpe.

Kahraman, C. (2008). Fuzzy multi-criteria decision making: theory and applications with recent developments. Springer Science+Business Media, LLC.

Kang, J. G. (2016). General types of(∈,∈qk). *Honam Mathematical J., 38*, 795–807. doi:10.5831/HMJ.2016.38.4.795

Kauffmann, A. (1965). *Introduction to the Theory of Fuzzy Subsets*. New York: Academic Press.

Kazanci, O., & Davvaz, B. (2010). More on fuzzy lattice. *Computers & Mathematics with Applications (Oxford, England), 64*(9), 2917–2925. doi:10.1016/j.camwa.2012.05.006

Kazanci, O., & Yamak, S. (2008). Generalized fuzzy bi-ideals of semigroup. *Soft Computing, 12*(11), 1119–1124. doi:10.100700500-008-0280-5

Kazim, M. A., & Naseerudden, M. (1972). On almost semi groups. *The Aligarh Bulletin of Mathematics, 2*, 1–7.

Kazim, M. A., & Naseerudin, M. (1972). On almost semigroups. *Alig. Bull. Math., 2*, 1–7.

Keatley, D. A., & Clarke, D. D. (2019). Crime Linkage: Finding a Behavioral Fingerprint Using the 'Path Similarity Metric.'. *Journal of Police and Criminal Psychology*. doi:10.100711896-019-9315-3

Keller, J. M., Gray, M. R., & Givens, J. A. (1985). A Fuzzy K-nearest Neighbor Algorithm. *IEEE Transactions on Systems, Man, and Cybernetics, SMC-15*(4), 580–585. doi:10.1109/TSMC.1985.6313426

KenevissiF. (2014). Retrieved: http://www.edc.ncl.ac.uk/highlight/rhdecember2006.php

Khalid, H. M., & Ahmad, B. (1999). Fuzzy H BCI. *Fuzzy Sets and Systems, 101*(1), 153–158. doi:10.1016/S0165-0114(97)00042-0

Khan, M.S.A., Abdullah, S., Ali, A., Amin, F., & Rahman, K. (2017). On generalized (∈, $\in\vee q_k$)-fuzzy quasi ideals in ordered semigroups. *Turkish Journal of Fuzzy Systems, 8*, 33-51.

Khan, F. M., Sarmin, N. H., Khan, A., & Khan, H. U. (2015). Some innovative types of fuzzy bi-ideals in ordered semigroups. *Journal of Advanced Mathematics and Applications, 4*(1), 24–36. doi:10.1166/jama.2015.1070

Khan, M. S. A., Abdullah, S. K., Rahman, K., & Abbass, S. Z. (2017). Characterization of regular semigroups by (∈, $\in\vee(k^*q_k)$. *Annals of Fuzzy Mathematics and Informatics, 13*(3), 403–419. doi:10.30948/afmi.2017.13.3.403

Khan, M. S. A., Abdullah, S., Jun, Y. B., & Rahman, K. (2017). More generalized fuzzy subsemigroups/ideals in semigroups. *Honam Mathematical J., 39*, 527–559.

Khan, M. S. A., Rahman, K. A., Fahmi, A., & Shakeel, M. (2018). Generalized (∈, $\in\vee q_k$) –fuzzy quasi-ideals in semigroups. *Punjab University Journal of Mathematics, 50*, 35–53.

Khan, M., & Ahmad, N. (2010). Characterizations of left almost semi groups by their ideals. *Journal of Advanced Research in Pure Mathematics, 2*, 61–71.

Khan, M., & Asif, T. (2010). Intra-regular left almost semi groups characterized by their anti fuzzy ideals. *Journal of Mathematics Research, 2*, 1–11.

Khan, M., Davvaz, B., Yaqoob, N., Gulistan, M., & Khalaf, M. M. (2015). On (∈,∈∨qk) -intuitionistic (fuzzy ideals, fuzzy soft ideals) of subtraction algebras. *Songklanakarin Journal of Science and Technology, 37*, 465–475.

Kim, J. B. (1988). Determinant theory for fuzzy and boolean matrices. *Congressus Numerantium*, 273–276.

Kim, D. S. (2003). Characteristic fuzzy groups. *Commun. Korean Math. Soc., 18*(1), 21–29. doi:10.4134/CKMS.2003.18.1.021

Kim, J. B., Baartmans, A., & Sahadin, N. S. (1989). Determinant theory for fuzzy matrices. *Fuzzy Sets and Systems, 29*(3), 349–356. doi:10.1016/0165-0114(89)90046-8

Kim, K. H. (1982). *Boolean matrix theory and applications*. New York: Dekker.

Kim, K. H., & Jun, Y. B. (2000). On fuzzy R-subgroups of near-rings. *J. Fuzzy Math., 8*, 549–558.

Kim, K. H., & Roush, F. W. (1980). Generalized fuzzy matrices. *Fuzzy Sets and Systems, 4*(3), 293–315. doi:10.1016/0165-0114(80)90016-0

Kim, K. H., & Roush, F. W. (1987). Fuzzy Matrix Theory. In *Analysis of Fuzzy Information*. CRC Press.

Klinkenberg, B. (2001). Similarity of the Flora of the Erie Islands: Implications for Conservation Biology. *The Canadian Geographer. Geographe Canadien, 45*(3), 414–426. doi:10.1111/j.1541-0064.2001.tb01191.x

Klir, G. J. (1987). Where Do We Stand on Measures of Uncertainty, Ambiguity, Fuzziness, and the Like? *Fuzzy Sets and Systems, 24*(2), 141–160. doi:10.1016/0165-0114(87)90087-X

Klir, G. J., & Yuan, B. (1995). *Fuzzy sets and fuzzy logic theory and applications*. PRT Prentice Hall.

Knees & Schedl. (2013). A Survey of Music Similarity and Recommendation from Music Context Data. *ACM Trans. Multimedia Comput. Commun. Appl., 10*(1), 2:1–2:21.

Knuth, D.E. (1981). The Art of Computer Programming. *Semi Numerical Algorithms, 2.*

Kołodziejczyk, W. (1987). Canonical form of a strongly transitive fuzzy matrix. *Fuzzy Sets and Systems, 22*(3), 297–302. doi:10.1016/0165-0114(87)90073-X

Kołodziejczyk, W. (1988). Convergence of powers of s-transitive fuzzy matrices. *Fuzzy Sets and Systems, 26*(1), 127–130. doi:10.1016/0165-0114(88)90011-5

Kordi, A., Moussavi, A., & Ahmadi, A. (2008). Fuzzy H-Ideals of BCI-Algebras with Interval Valued Membership Functions. *International Mathematical Forum, 3*(27), 1327-1338.

Krasser, J., Abeßer, J., Großmann, H., Dittmar, C., & Cano, E. (2012). Improved Music Similarity Computation Based on Tone Objects. In *Proceedings of the 7th Audio Mostly Conference: A Conference on Interaction with Sound, AM '12*. New York, NY: ACM. 10.1145/2371456.2371464

Kumar, R. (1992). Fuzzy characteristic subgroups of a group. *Fuzzy Sets and Systems, 45*(3), 397–398. doi:10.1016/0165-0114(92)90356-9

Kumar, S., & Pandey, P. (2010). Dynamic Performers using Fuzzy logic in Business process management. *Pragyan: Journal of Information Technology, 8*(12), 71–75.

Kuroki, N. (1979). Fuzzy bi-ideals in semigroups. *Comment. Math. Univ. St. Pauli, 28*, 17–21.

Kuroki, N. (1981). On fuzzy ideals and fuzzy bi-ideals in semigroups. *Fuzzy Sets and Systems*, 203–215.

Kuroki, N. (1991). On fuzzy semigroups. *Information Sciences, 53*(3), 203–236. doi:10.1016/0020-0255(91)90037-U

Laarhoven, P. J. M., & Pedrycz, W. (1983). A extension of Saaty's priority theory. *Fuzzy Sets and Systems, 11*(1-3), 229–241. doi:10.1016/S0165-0114(83)80082-7

Lai, R., & Chiang, D. (2002). Constraint-Based Granular Computing for Fuzzy Modeling. *2002 IEEE World Congress on Computational Intelligence. 2002 IEEE International Conference on Fuzzy Systems. FUZZ-IEEE'02. Proceedings (Cat. No.02CH37291), 1*, 584–89.

Lajos, S. (1969). On the bi-ideals in semigroups. *Proceedings of the Japan Academy*, *45*(8), 710–712. doi:10.3792/pja/1195520625

Lajos, S., & Szasz, F. A. (1970). On the bi-ideals in associative ring. *Proceedings of the Japan Academy*, *46*(6), 505–507. doi:10.3792/pja/1195520265

Lalanne, C. (2014). Appendix: Similitude in Mechanics. In Mechanical Shock. John Wiley & Sons, Ltd.

Lane, S. M. (Ed.). (1972). *Coherence in Categories*. Berlin: Springer-Verlag.

Larimi, M.A. (2013). On $(\in,\in\vee q_k)$ -intuitionistic fuzzy ideals of hemirings. *World Applied Sciences*, *21*, 54-67.

László & Tikk. (2000). Fuzzy rendszerek. Budapest: Academic Press.

Lee, H.-S. (2001). An optimal algorithm for computing the max–min transitive closure of a fuzzy similarity matrix. *Fuzzy Sets and Systems*, *123*(1), 129–136. doi:10.1016/S0165-0114(00)00062-2

Lee, K. J. (2009). Bipolar fuzzy subalgerbas and bipolar fuzzy ideals of BCK/BCI-algerbas. *Bulletin of the Malaysian Mathematical Sciences Society*, *32*(3), 361–373.

Lee, K. J., & Jun, Y. B. (2011). Bipolar fuzzy a BCI. *Communications of the Korean Mathematical Societ.*, *26*(4), 531–542. doi:10.4134/CKMS.2011.26.4.531

Lesot, M.-J., & Rifqi, M. (2012). Degrés d'équivalence de Mesures de Comparaison Pour Données Binaires et Pour Données Numériques. *Revue Des Nouvelles Technologies de l'Information RNTI-E*, *22*, 145–164.

Li, Y., Wang, X., & Yang, L. (2013). A Study of Fuzzy Subgroups. Journal of Applied Mathematics. doi:10.1155/2013/485768

Liao, H., & Xu, Z. (2017). *Hesitant Fuzzy Decision Making Methodologies and Applications*. Springer. doi:10.1007/978-981-10-3265-3

Li, J.-X. (1989). Some results on the nilpotent fuzzy matrices. *Fuzzy Systems Math.*, *1*, 52–55.

Liu, B. (2008). *Uncertainty Theory* (3rd ed.). Retrieved from http://orsc.edu.cn/liu/ut.pdf

Liu, F. A. (1982). Fuzzy invariant subgroups and fuzzy ideals. *Fuzzy Sets and Systems*, *8*(2), 133–139. doi:10.1016/0165-0114(82)90003-3

Liu, H. W., & Wang, G. J. (2007). Multi-criteria methods based on intuitionistic fuzzy sets. *European Journal of Operational Research*, *179*(1), 220–233. doi:10.1016/j.ejor.2006.04.009

Liu, P., Chu, Y., Li, Y., & Chen, Y. (2014). Some generalized neutrosophic number Hamacher aggregation operators and their application to group decision making. *International Journal of Fuzzy Systems*, *16*(2).

Liu, W., & Ji, Z. (2002). The periodicity of square fuzzy matrices based on minimal strong components. *Fuzzy Sets and Systems*, *126*(2), 233–240. doi:10.1016/S0165-0114(01)00033-1

Liu, Y. L., & Meng, J. (2001). Fuzzy ideals in BCI. *Fuzzy Sets and Systems*, *123*(2), 227–237. doi:10.1016/S0165-0114(00)00047-6

Lu, C., Lan, J., & Wang, Z. (2006). Aggregation of Fuzzy Opinions Under Group Decision-Making Based on Similarity and Distance. *Journal of Systems Science and Complexity*, *19*(1), 63–71. doi:10.100711424-006-0063-y

Luo, F., & Zhan, J. (2013). Fuzzy parameterized fuzzy soft lattice implication algebras. *Ars Combinatoria, 112*, 353–360.

Mahmood Khalil, S., Ulrazaq, M., & Abdul-Ghani, S. (2018). Abu Firas Al-Musawi, $\sigma-$Algebra and $\sigma-$Baire in Fuzzy Soft Setting. Advances in Fuzzy Systems.

Mahmood, S. (2014). Soft Regular Generalized b-Closed Sets in Soft Topological Spaces. *Journal of Linear and Topological Algebra, 3*, 195–204.

Mahmood, S. (2015). On intuitionistic fuzzy soft b - closed sets in intuitionistic fuzzy soft topological spaces. *Annals of Fuzzy Mathematics and Informatics, 10*, 221–233.

Mahmood, S. (2016). Dissimilarity Fuzzy Soft Points and their Applications. *Fuzzy Information and Engineering, 8*, 281–294. doi:10.1016/j.fiae.2016.11.003

Mahmood, S. (2017). Characteristics of Soft Tychonoff Spaces with New Soft Separation Axioms. *Journal of The College of Education, 2*, 1423–1446.

Mahmood, S. (2017). Soft Sequentially Absolutely Closed Spaces. *International Journal of Applications of Fuzzy Sets and Artificial Intelligence, 7*, 73–92.

Mahmood, S. (2017). Tychonoff Spaces in Soft Setting and their Basic Properties. *International Journal of Applications of Fuzzy Sets and Artificial Intelligence, 7*, 93–112.

Mahmood, S. (2018). S. and Abd Ulrazaq, M., Soft *BCH*-Algebras of the Power Sets. *American Journal of Mathematics and Statistics, 8*(1), 1–7.

Mahmood, S., & Abd Ulrazaq, M. (2018). New Category of Soft Topological Spaces. *American Journal of Computational and Applied Mathematics, 8*(1), 20–25.

Mahmood, S., & Alradha, M. A. (2017). Soft Edge $\rho-$Algebras of the power sets. *International Journal of Applications of Fuzzy Sets and Artificial Intelligence, 7*, 231–243.

Mahmood, S., & Hameed, F. (2018). An algorithm for generating permutation algebras using soft spaces. *Journal of Taibah University for Science, 12*(3), 299–308.

Mahmood, S., & Hameed, F. (2018). An algorithm for generating permutations in symmetric groups using soft spaces with general study and basic properties of permutations spaces. *J Theor Appl Inform Technol, 96*(9), 2445–2457.

Mahmood, S., & Muhamad, A. (2017). Soft BCL-Algebras of the Power Sets. *International Journal of Algebra, 11*(7), 329–341.

Maji, P. K. (2013). Neutrosophic soft set. *Annuals of Fuzzy Mathematics and Informatics, 5*, 157–168.

Maji, P. K., Biswas, R., & Roy, A. R. (2001). Intuitionistic fuzzy soft sets. *Journal of Fuzzy Mathematics, 9*, 677–692.

Maji, P. K., & Roy, A. R. (2004). On intuitionistic fuzzy soft sets. *Journal of Fuzzy Mathematics, 12*, 669–683.

Maji, P. K., Roy, A. R., & Biswas, R. (2001). Fuzzy soft sets. *Journal of Fuzzy Mathematics, 9*, 589–602.

Mandal, D. (2014). Fuzzy ideals and fuzzy interior ideals in ordered semirings. *Fuzzy Info. and Engg., 6*, 101-114.

Ma, X., Zhan, & J., Jun, Y.B. (2009). Some types of $(\in,\in\vee q)$ -interval-valued fuzzy ideals of BCI algebras. *Iranian Journal of Fuzzy Systems, 6*, 53–63.

Ma, X., & Zhan, J. (2009). On generalized fuzzy R-subgroups of near-rings. *International Workshop on Intelligent Systems and Applications*, 1724-1727.

Ma, X., Zhan, J., Davvaz, B., & Jun, Y. B. (2008). Some kinds of$(\in,\in\vee q)$ BCI. *Information Sciences*, *178*(19), 3738–3754. doi:10.1016/j.ins.2008.06.006

Ma, X., Zhan, J., & Jun, Y. B. (2011). Some kinds of$(\in\gamma,\in\gamma\vee q\delta)$ BCI. *Computers & Mathematics with Applications (Oxford, England)*, *61*(4), 1005–1015. doi:10.1016/j.camwa.2010.12.049 PMID:21542996

Ma, X., Zhan, J., & Jun, Y. B. (2012). New types of fuzzy ideals of *BCI*. *Neural Computing & Applications*, *21*(S1), S19–S27. doi:10.100700521-011-0558-x

Mendel, J. M. (2017). *Uncertain Rule-Based Fuzzy Systems: Introduction and New Directions* (2nd ed.). Springer International Publishing. doi:10.1007/978-3-319-51370-6

Meng, J., & Jun, Y. B. (1994). BCK-algebras. Kyung Moon Sa Co.

Meng, J., & Jun, Y. B. (1994). BCK-algebras. Kyungmoon Sa Co.

Meng, J. (1994). On ideals in BCK-algebras. *Math. Japonica*, *40*, 143–154.

Meng, J., & Guo, X. (2005). On fuzzy ideals in *BCK/BCI*. *Fuzzy Sets and Systems*, *149*(3), 509–525. doi:10.1016/j.fss.2003.11.014

Mishref, M. A., & Emam, E. G. (1992). Transitivity and subinverses in fuzzy matrices. *Fuzzy Sets and Systems*, *52*(3), 337–343. doi:10.1016/0165-0114(92)90241-U

Mitchell, H. B. (2006). Ranking Type-2 Fuzzy Numbers. *IEEE Transactions on Fuzzy Systems*, *14*(2), 287–294. doi:10.1109/TFUZZ.2005.864078

Mitra, S., & Pal, S. K. (2005). Fuzzy sets in pattern recognition and machine intelligence. *Fuzzy Sets and Systems*, *156*(3), 381–386. doi:10.1016/j.fss.2005.05.035

Miyamoto, S., & Mizutani, K. (2004). *Fuzzy multiset model and methods for nonlinear document clustering for information retrieval*. Springer-Verlag Berlin Heidelberg.

Miyamoto, S. (1996). Basic operations of fuzzy multisets. *J. Japan Soc. Fuzzy Theory Systems*, *8*(4), 639–645. doi:10.3156/jfuzzy.8.4_47

Mizutani, K., Inokuchi, R., & Miyamoto, S. (2008). Algorithms of nonlinear document clustering based on fuzzy multiset model. *International Journal of Intelligent Systems*, *23*(2), 176–198. doi:10.1002/int.20263

Molodtsov, D. (1999). Soft set theory–First results. *Computers & Mathematics with Applications (Oxford, England)*, *37*, 19–31.

Mondal, S., & Pal, M. (2013a). Similarity Relations, Invertibility and Eigenvalues of Intuitoinistic Fuzzy Matrix. *Fuzzy Information and Engineering*, *5*(4), 431–443. doi:10.100712543-013-0156-y

Mondal, S., & Pal, M. (2013b). Soft matrices. *Journal of Uncertain Systems*, *7*(4), 254–264.

Mondal, S., & Pal, M. (2014). Intuitionistic fuzzy incline matrix and determinant. *Annals of Fuzzy Mathematics and Informatics*, *8*(1), 19–32.

Mondal, S., & Pal, M. (2016a). Rank of interval-valued fuzzy matrices. *Afrika Matematika*, *27*(1–2), 97–114. doi:10.100713370-015-0325-8

Mondal, S., & Pal, M. (2016b). Similarity relations, eigenvalues and eigenvectors of bipolar fuzzy matrix. *Journal of Intelligent & Fuzzy Systems*, *30*(4), 2297–2307. doi:10.3233/IFS-152000

Moore, R. E., & Bierbaum, F. (1979). *Methods and applications of interval analysis*. Philadelphia: Siam.

Moore, R. E., Kearfott, R. B., & Cloud, M. J. (2009). *Introduction to interval analysis*. Philadelphia: SIAM.

Mordeson, J. N., & Malik, D. S. (1998). Fuzzy commutative algebra: World Scientific.

Mordeson, J. N., & Malik, D. S. (2002). *Fuzzy automata and languages: Theory and applications*. Chapman and Hall/CRC. doi:10.1016/0165-0114(81)90018-X

Mordeson, J. N., Malik, D. S., & Kuroki, N. N. (2003). Fuzzy semigroups. In Studies in Fuzziness and Soft Computing (p. 131). Springer-Verlag, Berlin.

Mordeson, J. M., Bhutani, K. R., & Rosenfeld, A. (2005). *Fuzzy group theory*. Heidelberg, Germany: Springer-Verlag Berlin. doi:10.1007/b12359

Mordeson, J. N., & Malik, D. S. (1998). *Fuzzy commutative algebra*. World scientific. doi:10.1142/3929

Muhiuddin, G. (2016). Hesitant fuzzy filters and hesitant fuzzy G-filters in residuated lattices. *Journal of Computational Analysis and Applications*, *21*(2), 394–404.

Muhiuddin, G., & Aldhafeeri, S. (2018). Subalgebras and ideals in *BCK/BCI*. *Eur. J. Pure Appl. Math*, *11*(2), 417–430. doi:10.29020/nybg.ejpam.v11i2.3246

Muhiuddin, G., & Al-roqi, A. (2015). Subalgebras of *BCK/BCI*-algebras based on (α,β) -type fuzzy sets. *Journal of Computational Analysis and Applications*, *18*, 1057–1064.

Muhiuddin, G., & Al-roqi, M., A. (2018). Regular hesitant fuzzy filters and MV -hesitant fuzzy filters of residuated lattices. *Journal of Computational Analysis and Applications*, *24*(6), 1133–1144.

Muhiuddin, G., Kim, H. S., Song, S. Z., & Jun, Y. B. (2017). Hesitant fuzzy translations and extensions of subalgebras and ideals in BCK/BCI-algebras. *Journal of Intelligent & Fuzzy Systems*, *32*(1), 43–48. doi:10.3233/JIFS-151031

Muhiuddin, G., Roh, E. H., Ahn, S. S., & Jun, Y. B. (2017). Hesitant fuzzy filters in lattice implication algebras. *Journal of Computational Analysis and Applications*, *22*(6), 1105–1113.

Mukherjee, N. P., & Bhattacharya, P. (1984). Fuzzy normal subgroups and fuzzy cosets. *Information Sciences*, *34*(3), 255–239. doi:10.1016/0020-0255(84)90050-1

Mukherjee, N. P., & Bhattacharya, P. (1986). Fuzzy groups: Some group theoretic analogs. *Information Sciences*, *39*(3), 247–268. doi:10.1016/0020-0255(86)90039-3

Murali Krishna Rao, M. (2017). Bi-quasi-ideals and fuzzy bi-quasi-ideals of Γ- semigroups. *Bull. Int. Math. Virtual Inst*, *7*(2), 231–242.

Murali Krishna Rao, M. (2017). fuzzy ideals in ordered *T*-Γ-. *Anl. of Fuzzy Math. and Info.*, *13*(2), 253–276. doi:10.30948/afmi.2017.13.2.253

Murali Krishna Rao, M. (2018). Bi –Ideals of semigroups. *Disc. Math. Gen. Alg. and Appl.*, *38*, 69–78.

Murali Krishna Rao, M. (2018). Fuzzy Bi-Interior Ideals of Semigroups, *Asia. Pacific Journal of Mathematics*, *5*, 208–218.

Murali Krishna Rao, M. (2018). Fuzzy left and right bi-quasi ideals of semirings. *Bull. Int. Math. Virtual Inst.*, *8*(3), 449–460.

Murali Krishna Rao, M. (2018). Left bi-quasi ideals of semirings. *Bull. Int. Math. Virtual Inst.*, *8*, 45–53.

Murali Krishna Rao, M., Venkateswarlu, B., & Adinaryana, Y. (1971). (commnicated). Bi-Interior Ideals of Semirings, Rosenfeld, A.(1971). Fuzzy groups. *Journal of Mathematical Analysis and Applications*, *35*, 512–517.

Muralikrishna, P., & Chandramouleeswaran, M. (2011). A Note on Intuitionistic L-Fuzzy H-ideals of BF-Algebras. *International Review of Fuzzy Mathematics*, *6*(2), 63–68.

Murali, V. (2004). Fuzzy points of equivalent fuzzy subsets. *Information Science*, *158*, 277–288. doi:10.1016/j.ins.2003.07.008

Mushtaq, Q., & Khan, M. (2009). Ideals in left almost semigroup. *Journal of Mathematics Research.*

Mushtaq, Q., & Kamran, M. S. (1989). On LA-semigroup with weak associative law. *Scientic Khyber*, *1*, 69–71.

Mushtaq, Q., & Khan, M. (2003). Ideals in Left Almost semigroups. *5th Proc. Int. Conf. Pur. Math.*, 23-29.

Mushtaq, Q., & Khan, M. (2008). Direct product of Abel Grassmann's Groupoids. *J. Interdiscip. Math.*, *11*, 461–467.

Mushtaq, Q., & Khan, M. (2009). M-System in LA-semigroups. *Southeast Asian Bulletin of Mathematics*, *33*, 321–327.

Mushtaq, Q., & Khan, M. (2009). M-systems in LA-semigroups. *Southeast Asian Bulletin of Mathematics*, *33*, 321–327.

Nazmul, S.K., Majumdar, P., & Samanta, S.K. (2013). On Multisets and Multi-groups. *Annals of Fuzzy Mathematics and Informatics,* (3), 643-656.

Neggers, J., & Kim, H. S. (2002). On β-algebras. *Mathematica Slovaca*, *52*(5), 517–530.

Neggers, J., & Kim, H. S. (2002a). On B-algebras. *Mat. Vesnik*, *54*, 21–29.

Olson, D. L. (2004). Comparison of weights in TOPSIS models. *Mathematical and Computer Modelling*, *40*(7-8), 721–727. doi:10.1016/j.mcm.2004.10.003

Onasanya, B. O., & Ilori, S. A. (2013). On fuzzy subgroups and fuzzy cosets. *International Journal of Computers and Applications*, *81*(14), 20–22. doi:10.5120/14184-2353

Onipchuk, S.V. (1992). Regular semihypergroups. *Mathematic he ski Sbornik, 183*, 43-54.

Ordaz, P., Santos-Sánchez, O. J., Rodríguez-Guerrero, L., & González-Facundo, A. (2017). Nonlinear Stabilization for a Class of Time Delay Systems via Inverse Optimality Approach. *ISA Transactions*, *67*, 1–8. doi:10.1016/j.isatra.2016.11.015 PMID:27939564

Otadi, M., & Mosleh, M. (2012). Solution of fuzzy matrix equation System. *International Journal of Mathematics and Mathematical Sciences*, *2012*, 1–8.

Padhy, N. P. (2014). *Artificial Intelligence and Intelligent Systems* (13th ed.). Oxford University Press.

Pal, Khan, & Shyamal. (2002). Intuitionistic fuzzy matrices. *Notes on Intuitionistic Fuzzy Sets*, *8*(2), 51–56.

Paliwal, S., Chopra, V., & Singla, S. K. (2016). Stabilization of Mobile Inverted Pendulum Using Fuzzy PID Controllers. *2016 7th India International Conference on Power Electronics (IICPE)*, 1–4.

Pal, M. (2001). Intuitionistic fuzzy determinant. *V. U. J. Physical Sciences*, *7*, 87–93.

Pal, M. (2015a). Fuzzy matrices with fuzzy rows and columns. *Journal of Intelligent & Fuzzy Systems*, *30*(1), 561–573. doi:10.3233/IFS-151780

Pal, M. (2015b). Interval-valued Fuzzy Matrices with Interval-valued Fuzzy Rows and Columns. *Fuzzy Information and Engineering*, *7*(3), 335–368. doi:10.1016/j.fiae.2015.09.006

Pal, M., & Mondal, S. (2019). Bipolar fuzzy matrices. *Soft Computing*. doi:10.100700500-019-03912-9

Pandey, M., Litoriya, R., & Pandey, P. (2016). Mobile applications in context of big data: A survey. *2016 Symposium on Colossal Data Analysis and Networking (CDAN)*, 1–5. 10.1109/CDAN.2016.7570942

Pandey, M., Litoriya, R., & Pandey, P. (2018b). Mobile App development based on agility function. *Ingénierie Des Systèmes d'information RSTI Série ISI, 23*(6), 19–44.

Pandey, M., Litoriya, R., & Pandey, P. (2019b). Empirical Analysis of Defects in Handheld Device Applications. In M. Singh, P. K. Gupta, V. Tyagi, J. Flusser, T. Ören, & R. Kashyap (Eds.), Advances in Computing and Data Sciences (pp. 103–113). Singapore: Springer Singapore.

Pandey, M., Litoriya, R., & Pandey, P. (2019e). Perception-Based Classification of Mobile Apps: A Critical Review. In A. K. Luhach, K. B. G. Hawari, I. C. Mihai, P.-A. Hsiung, & R. B. Mishra (Eds.), Smart Computational Strategies: Theoretical and Practical Aspects (pp. 121–133). Academic Press. doi:10.1007/978-981-13-6295-8_11

Pandey, P., & Litoriya, R. (2019a). A Predictive Fuzzy Expert System for Crop Disease Diagnostic and Decision Support. In Fuzzy Expert Systems and Applications in Agricultural Diagnosis (pp. 175–194). Academic Press. doi:10.4018/978-1-5225-9175-7.ch010

Pandey, M., Litoriya, R., & Pandey, P. (2019a). Application of Fuzzy Dematel Approach in Analyzing Mobile App Issues. *Programming and Computer Software, 45*(5), 268–287. doi:10.1134/S0361768819050050

Pandey, M., Litoriya, R., & Pandey, P. (2019c). Identifying Causal Relationships in Mobile App Issues: An Interval Type-2 Fuzzy DEMATEL Approach. *Wireless Personal Communications*, 1–28.

Pandey, M., Litoriya, R., & Pandey, P. (2018a). An ISM Approach for Modeling the Issues and Factors of Mobile App Development. *International Journal of Software Engineering and Knowledge Engineering, 28*(07), 937–953. doi:10.1142/S0218194018400119

Pandey, M., Litoriya, R., & Pandey, P. (2019d). Novel Approach for Mobile Based App Development Incorporating MAAF. *Wireless Personal Communications, 107*(4), 1687–1708. doi:10.100711277-019-06351-9

Pandey, P., Kumar, S., & Shrivastava, S. (2013a). A Critical Evaluation of Computational Methods of Forecasting Based on Fuzzy Time Series. *International Journal of Decision Support System Technology, 5*(1), 25–40. doi:10.4018/jdsst.2013010102

Pandey, P., Kumar, S., & Shrivastava, S. (2013b). Forecasting using Fuzzy Time Series for Diffusion of Innovation: Case of Tata Nano Car in India. *National Academy Science Letters, 36*(3), 299–309. doi:10.100740009-013-0140-4

Pandey, P., Kumar, S., & Shrivastava, S. (2014). A unified strategy for forecasting of a new product. *Decision (Washington, D.C.), 41*(4), 411–424.

Pandey, P., Kumar, S., & Shrivastava, S. (2015). A Fuzzy Decision Making Approach for Analogy Detection in New Product Forecasting. *Journal of Intelligent & Fuzzy Systems, 28*(5), 2047–2057. doi:10.3233/IFS-141483

Pandey, P., Kumar, S., & Shrivastava, S. (2017). An efficient time series forecasting method exploiting fuzziness and turbulence in data. *International Journal of Fuzzy System Applications, 6*(4), 83–98. doi:10.4018/IJFSA.2017100106

Pandey, P., & Litoriya, R. (2019b). An activity vigilance system for elderly based on fuzzy probability transformations. *Journal of Intelligent & Fuzzy Systems, 36*(3), 2481–2494. doi:10.3233/JIFS-181146

Pandey, P., & Litoriya, R. (2019c). Legal/Regulatory Issues for MMBD in IoT. In S. Tanwar, S. Tyagi, & N. Kumar (Eds.), *Multimedia Big Data Computing for IoT Applications Concepts, Paradigms and Solutions* (pp. 367–388). Springer Nature Singapore Ltd.

Pandey, P., Litoriya, R., & Tiwari, A. (2018). A framework for fuzzy modelling in agricultural diagnostics. *Journal Européen des Systèmes Automatisés, 51*(4-6), 203–223. doi:10.3166/jesa.51.203-223

Pedrycz, W. (2001). Granular Computing: An Introduction. *Proceedings Joint 9th IFSA World Congress and 20th NAFIPS International Conference (Cat. No. 01TH8569)*, 1349–54. Retrieved from http://ieeexplore.ieee.org/document/943745/

Peng, Z. X., & Iwamura, K. (2010). A Sufficient and Necessary Condition of Uncertainty Distribution. *Journal of Interdisciplinary Mathematics, 13*(3), 277–285. doi:10.1080/09720502.2010.10700701

Peres, R., Muller, E., & Mahajan, V. (2010). Innovation diffusion and new product growth models: A critical review and research directions. *International Journal of Research in Marketing, 27*(2), 91–106. doi:10.1016/j.ijresmar.2009.12.012

Phong, P. H., Hieu, D. T., Ngan, R. H., & Them, P. T. (2014). Some composition of picture fuzzy relations. *Conference Paper.*

Pottem & Nigam. (2014). Hierarchical Fuzzy PID Controller Using Granular Computing for Inverted Pendulum. *International Journal of Electrical, Electronics and Data Communication, 1*(4), 31–35.

Pradhan, R., & Pal, M. (2013). Convergence of maxgeneralized mean-min generalized mean powers of intuitionistic fuzzy matrices. *The Journal of Fuzzy Mathematics, 22*(2), 477–492.

Prince Williams, D. R., Latha, K. B., & Chandrasekar, E. (2013). -interval-valued T_G $(\in, \in, \vee q)$ -fuzzy subgroups. *Annals of Fuzzy Mathematics and Informatics, 5*(2), 283–300.

Pu, P., & Liu, Y. (1980). Fuzzy topology I. Neighbourhood structures of a fuzzy point and Moore-Smith convergence. *Journal of Mathematical Analysis and Applications, 76*(2), 571–579. doi:10.1016/0022-247X(80)90048-7

Ragab, M. Z., & Emam, E. G. (1995). On the min-max composition of fuzzy matrices. *Fuzzy Sets and Systems, 75*(1), 83–92. doi:10.1016/0165-0114(94)00332-2

Ragab, M. Z., & Eman, E. G. (1994). The determinant and adjoint of a square fuzzy matrix. *Fuzzy Sets and Systems, 61*(3), 297–307. doi:10.1016/0165-0114(94)90172-4

Rahgooy, T., Yazdi, H.S., & Monsefi, R. (2009). Fuzzy complex system of linear equations applied to circuit Analysis. *International Journal of Computer and Electrical Engineering, 1*(5), 535-541.

Rajam, K., & Chadramouleeswaran, M. (2015). L-Fuzzy T-ideals in β– Algebras. *Applied Mathematical Sciences, 9*(145), 7221–7228. doi:10.12988/ams.2015.59581

Rajarajeswari, R., & Uma, N. (2013). On Distance and Similarity Measures of Intuitionistic Fuzzy Multi Set. *IOSR Journal of Mathematics*, (4), 19-23.

Ray, A. K. (1999). On product of fuzzy subgroups. *Fuzzy Sets and Systems, 105*(1), 181–183. doi:10.1016/S0165-0114(98)00411-4

Reyhanoglul, Van Der Schaft, McClamroch, & Kolmanovsky. (1996). Nonlinear Control of a Class of Underactuated Systems. *Proceedings of the 35th Conference on Decision and Control*, 1682–87.

Rigatos, G., Siano, P., Abbaszadeh, M., & Ademi, S. (2017). Nonlinear H-Infinity Control for the Rotary Pendulum. *2017 11th International Workshop on Robot Motion and Control (RoMoCo)*, 217–22.

Rodriguez, M., Martinez, L., & Herrera, F. (2012). Hesitant fuzzy linguistic term sets for decision making. *IEEE Transactions on Fuzzy Systems*, *20*(1), 109–119. doi:10.1109/TFUZZ.2011.2170076

Rogers, E. M. (2003). *Diffusion of Innovations*. New York: Free Press.

Roh, E. H., Kim, K. H., & Lee, J. G. (2006). Intuitionistic Q-Fuzzy Subalgebras of BCK/BCI-Algebras. *International Mathematical Forum, 1*(24), 1167-1174.

Rosenfeld, A. (1971). Fuzzy Groups. Journal of Mathematical Analysis and Applications, (35), 512–517.

Rosenfeld, A. (1971). Fuzzy subgroups. *Journal of Mathematical Analysis and Applications*, *35*(3), 512–517. doi:10.1016/0022-247X(71)90199-5

Saaty, T. L. (2003). *The Analytic Hierarchy Process*. New York: McGra Hill.

Sabarinathan, S., Muralikrishna, P., & Kumar, D. C. (2017). Bipolar valued fuzzy H-ideals of BF-algebras. *International Journal of Pure and Applied Mathematics*, *112*(5), 87–92.

Saeid, A. B. (2011). Some results in Doubt fuzzy *BF*-algebras. *Analele Universitatii de Vest, Timi soara Seria Mathematica Informatica*, *1*, 125-134.

Saeid, A. B., & Rezvani, M. A. (2009). On fuzzy *BF*-algebras. *Int. Mathematical Forum*, *4*(1), 13-25.

Saeid, A. B., Kim, H. S., & Rezaei, A. (2017). On BI-algebras, *An.St. Univ.Ovidius Constana*, *25*(1), 177–194. doi:t doi:10.1515/auom-2017-0014

Saeid, A. B., Rezaei, A., & Borzooei, R. A. (2013). Some types of filters in BE-algebras. *Mathematics in Computer Science*, *7*(3), 341–352. doi:10.100711786-013-0157-6

Sahoo, L., Bhunia, A. K., & Kapur, P. K. (2012). Genetic algorithm based multi-objective reliability optimization in interval environment. *Computers & Industrial Engineering*, *62*(1), 152–160.

Sampath, S., & Senthil Kumar, R. (2014b). Clustering of Fuzzy Data using Credibilistic Expected and Critical Values. *IEEE Explore*, 176-181. 1 doi:0.1109/ICCCS.2014.7068189

Sampath, S., & Kalaivani, R. (2010). Clustering of Fuzzy data using credibilistic critical values. *IEEE International Conference on Signal and Image Processing*, 227–232. 10.1109/ICSIP.2010.5697474

Sampath, S., & Senthil Kumar, R. (2014a). Fuzzy Clustering using Credibilistic Critical Values. *International Journal of Computational Intelligence and Informatics*, *3*, 213–221.

Sampath, S., & Senthil Kumar, R. (2015a). Rough Clustering Using Credibilistic Critical Values. *International Journal of Data Mining and Emerging Technologies*, *5*(1), 24–30. doi:10.5958/2249-3220.2015.00005.1

Sampath, S., & Senthil Kumar, R. (2015b). Image Segmentation Using Credibilistic Critical Values. *Journal of Image Processing & Pattern Recognition Progress*, *2*, 46–55.

Santini, S., & Jain, R. (1997). Similarity Is a Geometer. *Multimedia Tools and Applications*, *5*(3), 277–306. doi:10.1023/A:1009651725256

Schedl, Pohle, Knees, & Widmer. (2011). Exploring the Music Similarity Space on the Web. *ACM Trans. Inf. Syst.*, *29*(3), 14:1–14:24.

Sebastian, S., & Ramakrishnan, T.V. (2011). Multi-fuzzy Sets: An Extension of Fuzzy Sets. *Fuzzy Information and Engineering,* (1), 35-43.

Sebastian, S., Jacob, M. K., Mary, V. M., & Daise, D. M. (2012). On Fuzzy Ideals and Rings. Journal of Computer and Mathematical Sciences, *3*(1), 115–120.

Senapati, T. (2015b). Bipolar fuzzy structure of BG-subalgebras. *J. Fuzzy Math.*, *23*(1), 209–220.

Senapati, T., Bhowmik, M., & Pal, M. (2012). Interval-valued intuitionistic fuzzy BG-subalgebras. *Journal of Fuzzy Mathematics*, *20*(3), 707–720.

Senapati, T., Bhowmik, M., & Pal, M. (2012c). Interval-valued intuitionistic fuzzy BG-subalgebras. *The Journal of Fuzzy Mathematics*, *20*(3), 707–720.

Senapati, T., Bhowmik, M., & Pal, M. (2013). Atanassov's intuitionistic fuzzy translations of intuitionistic fuzzy H-ideals in BCK/BCI-algebras. *Notes Intuition. Fuzzy Sets*, *19*(1), 32–47.

Senapati, T., Bhowmik, M., & Pal, M. (2013). Atanassov's intuitionistic fuzzy translations of intuitionistic fuzzy H-ideals in BCK/BCI-algebras. *Notes on Intuitionistic Fuzzy Sets*, *19*(1), 32–47.

Senapati, T., Bhowmik, M., & Pal, M. (2014). Fuzzy dot structure of BG. *Fuzzy Information & Engineering*, *6*(3), 315–329. doi:10.1016/j.fiae.2014.12.004

Senapati, T., Bhowmik, M., & Pal, M. (2016). Triangular norm based fuzzy BG-algebras. *Afrika Matematika*, *27*(1-2), 187–199. doi:10.100713370-015-0330-y

Senapati, T., Bhowmik, M., Pal, M., & Davvaz, B. (2015). Fuzzy translations of fuzzy H BCK/BCI. *Journal of the Indonesian Mathematical Society*, *21*(1), 45–58. doi:10.22342/jims.21.1.200.45-58

Senapati, T., Jana, C., Bhowmik, M., & Pal, M. (2015a). -fuzzy LGG. *Journal of the Egyptian Mathematical Society*, *23*(2), 219–223. doi:10.1016/j.joems.2014.05.010

Senapati, T., Kim, C. S., Bhowmik, M., & Pal, M. (2015c). Cubic subalgebras and cubic closed ideals of B. *Fuzzy Information & Engineering*, *7*(2), 129–149. doi:10.1016/j.fiae.2015.05.001

Sengupta, A., & Pal, T. K. (2000). On comparing interval numbers. *European Journal of Operational Research*, *127*(1), 28–43.

Seselja, B., Tepavcevic, A., & Vojvodic, G. (1993). 𝔏-fuzzy sets and codes. *Fuzzy Sets and Systems, 53,* 217-222.

Seselja, B., & Tepavcevic, A. (1994). Representation of lattices by fuzzy sets. *Information Sciences*, *79*(3-4), 171–180. doi:10.1016/0020-0255(94)90117-1

Seselja, B., & Tepavcevic, A. (1997). A note on fuzzy groups. *Yugoslav J. Operations Research*, *7*(1), 49–54.

Shabir, M. & Naz, M. (2011). On Soft topological spaces. *Comp. and Math. With Applications, 61*, 1786-1799.

Shabir, M., Jun, Y. B., & Nawaz, Y. (2010). Characterizations of regular semigroups by (α,β). *Computers & Mathematics with Applications (Oxford, England)*, *59*(1), 161–175. doi:10.1016/j.camwa.2009.07.062

Shabir, M., Jun, Y. B., & Nawaz, Y. (2010). Characterizations of regular semigroups by(∈,∈∨qk). *Computers & Mathematics with Applications (Oxford, England)*, *59*, 539–549. doi:10.1016/j.camwa.2009.06.014

Shabir, M., Jun, Y. B., & Nawaz, Y. (2010). Semigroups characterized by(∈,∈∨q). *Computers & Mathematics with Applications (Oxford, England)*, *60*(5), 1473–1493. doi:10.1016/j.camwa.2010.06.030

Shi, Y., Wang, S., Kou, G., & Wallenius, J. (2011). New state of MCDM in the 21st century: Selected Papers of the 20th International Conference on Multiple Criteria Decision Making 2009. *Lecture Notes in Economics and Mathematical Systems*.

Shinoj, T. K., Baby, A., & Sunil, J. J. (2015). On some algebraic structures of fuzzy multisets. *Ann. Fuzzy Math. Inform.*, *9*(1), 77–90.

Shouman, M., Turner, T. L., & Stocker, R. C. (2012). Applying k-nearest neighbour in diagnosing heart disease patients. *International Journal of Information and Education Technology (IJIET)*, *2*, 220–223. doi:10.7763/IJIET.2012.V2.114

Shyamal, A. K., & M. Pal. (2001). Distance between interval-valued intuitionistic fuzzy sets. *Busefal, 86*(1).

Shyamal, A. K., & Pal, M. (2004a). Matrices of interval numbers with applications. In *Proc. of National Seminar on ``Recent Advances in Applied Mathematics''*, (pp. 26–31). Midnapore, India: Vidyasagar University.

Shyamal, A. K., & Pal, M. (2004b). Triangular fuzzy matrices. In *Proc. `Fuzzy Logic and Its Applications in Technology and Management'*. Kharagpur, India: Indian Institute of Technology.

Shyamal, A. K., & Pal, M. (2002a). Distances between intuitionistics fuzzy matrices. *V. U. J. Physical Sciences*, *8*, 81–91.

Shyamal, A. K., & Pal, M. (2002b). Interval-valued fuzzy matrices. In *Proc. International Conference on 'Analysis and Discrete Structures'*, (pp. 22–24). Kharagpur, India: Indian Institute of Technology.

Shyamal, A. K., & Pal, M. (2004). Two new operators on fuzzy matrices. *Journal of Applied Mathematics and Computing*, *15*(1–2), 91–107. doi:10.1007/BF02935748

Shyamal, A. K., & Pal, M. (2005). Distance between fuzzy matrices and its applications-I. *Journal of Natural and Physical Sciences*, *19*(1), 38–51.

Sidky, F. I., & Mishref, M. A. A. (1991). Fully invariant, characteristic and S-fuzzy subgroups. *Information Sciences*, *55*(1-3), 27–33. doi:10.1016/0020-0255(91)90003-D

Singh, D., Ibrahim, A.M., Yohanna, T., & Singh, J.N. (2007). An overview of the application of multiset. *Novi Sad Journal of Math*ematics, (37), 73-92.

Singh, P. (2014). Some New Distance Measures for Type-2 Fuzzy Sets and Distance Measure Based Ranking for Group Decision Making Problems. *Frontiers of Computer Science*, *8*(5), 741–752. doi:10.100711704-014-3323-3

Singh, P. (2015). Correlation coefficients for picture fuzzy sets. *Journal of Intelligent & Fuzzy Systems*, *28*, 591–604.

Sivanandam, S. N., Sumathi, S., & Deepa, S. N. (2007). *Introduction to Fuzzy Logic Using MATLAB*. Berlin: Springer-Verlag. doi:10.1007/978-3-540-35781-0

Smarandachi, F. (2005). Neutrosophic set. *International Journal of Pure and Applied Mathematic, 24*, 287-297.

Sokal, R. R., & Peter, H. A. S. (1963). *Principles of Numerical Taxonomy*. W. H. Freeman.

Son, L. H. (2016). Generalized picture distance measure and applications to picture fuzzy clustering. *Applied Soft Computing*, *46*, 284–295. doi:10.1016/j.asoc.2016.05.009

Steinfeld, O. (1956). Uher die quasi ideals, *Von halbgruppend. Publicationes Mathematicae (Debrecen)*, *4*, 262–275.

Su. (2017). Stabilization of Double Inverted Pendulum on Cart: LQR Approach. *International Journal of Mechanical and Production Engineering, 5*(2), 149–53.

Sujatha, K., Chandramouleeswaran, M., & Muralikrishna, P. (2014). Product on Intuitionistic Fuzzy β-sub Algebras of β-algebras. *Indian Journal of Science and Technology*, *7*(3), 318–322.

Swamy, U. M., & Swamy, K. L. N. (1988). Fuzzy prime ideals of rings. *Journal of Mathematical Analysis and Applications*, *134*(1), 94–103. doi:10.1016/0022-247X(88)90009-1

Syropoulos, A. (2012). On generalized fuzzy multisets and their use in computation. *Iranian J. Fuzzy Syst.*, *9*(2), 113–125.

Szmidt, E. (2014). *Distances and Similarities in Intuitionistic Fuzzy Sets*. Springer International Publishing. doi:10.1007/978-3-319-01640-5

Szmidt, E., & Kacprzyk, J. (2000). Distances between intuitionistic fuzzy sets. *Fuzzy Sets and Systems*, *114*(3), 505–518. doi:10.1016/S0165-0114(98)00244-9

Tan, C. (2011). A multi-criteria interval-valued intuitionistic fuzzy group decision making with Choquet integral based TOPSIS. *Expert Systems with Applications*, *38*(4), 3023–3033. doi:10.1016/j.eswa.2010.08.092

Tan, P. N., Steinbach, M., & Kumar, V. (2006). *Introduction to Data Mining*. Pearson.

Tella, Y., & Daniel, S. (2013). A Study of Group Theory in the Context of Multiset Theory. International Journal of Science and Technology, *2*(8), 609–615.

Teng, S., Li, J., Li, R., & Zhang, W. (2014). The Calculation of Similarity and Its Application in Data Mining. In Pervasive Computing and the Networked World, Lecture Notes in Computer Science. Springer International Publishing. doi:10.1007/978-3-319-09265-2_57

Tepavcevic, A., & Trajkovski, G. (2001). -fuzzy lattices: An introduction. *Fuzzy Sets and Systems*, *123*(2), 209–216. L doi:10.1016/S0165-0114(00)00065-8

Thibaut, J.-P. (1997). Similarité et catégorisation. *L'Année Psychologique*, *97*(4), 701–736. doi:10.3406/psy.1997.28989

Thomason, M. G. (1977). Convergence of powers of a fuzzy matrix. *Journal of Mathematical Analysis and Applications*, *57*(2), 476–480. doi:10.1016/0022-247X(77)90274-8

Thomas, R. J. (1985). Estimating market growth for new products: An analogical diffusion model approach. *Journal of Product Innovation Management*, *2*(1), 45–55. doi:10.1111/1540-5885.210045

Torra, V. (2010). Hesitant fuzzy sets. *International Journal of Intelligent Systems*, *25*, 529–539.

Torra, V., & Narukawa, Y. (2009). On hesitant fuzzy sets and decision. *The 18th IEEE International Conference on Fuzzy Systems*, 1378–1382.

Trinh, H., & Huong, D. C. (2018). A New Method for Designing Distributed Reduced-Order Functional Observers of Interconnected Time-Delay Systems. *Journal of the Franklin Institute*. Retrieved from http://linkinghub.elsevier.com/retrieve/pii/S0016003218300383

Tsigkritis, T., Groumas, G., & Schneider, M. (2018). On the Use of k-NN in Anomaly Detection. *Journal of Information Security*, *9*(01), 70–84. doi:10.4236/jis.2018.91006

Ttripathy, B.C., Debnath, S., & Rakshit, D. (2018). On multiset group. *Proyecciones Journal of Mathematics*, (33), 479-489.

Tversky, A. (1977). Features of Similarity. *Psychological Review*, *84*(4), 327–352. doi:10.1037/0033-295X.84.4.327

Vandiver, H. S. (1934). Note on a simple type of algebra in which cancellation law of addition does not hold. *Bull. Amer. Math. Soc.(N.S.)*, *40*, 914--920.

Velankar, M. R., Sahasrabuddhe, H. V., & Kulkarni, P. A. (2015). Modeling Melody Similarity Using Music Synthesis and Perception. *Procedia Computer Science*, *45*, 728–735. doi:10.1016/j.procs.2015.03.141

Vougiouklis, T. (1994). *Hyper structures and their representations*. Palm Harbor, FL: Hadronic Press.

Walendziak, A. (2007). On BF-algebras. *Mathematica Slovaca*, *57*(2), 119–128.

Wang, C., Zhou, X., Tu, H., & Tao, S. (2017). Some geometric aggregaton operators based on picture fuzzy sets and their application in multiple attribute decision making. *International Journal of Pure and Applied Mathematics*, *37*, 477–492.

Wang, H., & Zhang, Y. Q. (2013). Multi-criteria decision-making approach based on Atanassov's intuitionistic fuzzy sets with incomplete certain information on weights. *IEEE Transactions on Fuzzy Systems*, *21*(3), 510–515. doi:10.1109/TFUZZ.2012.2210427

Wang, L., Peng, J., & Wang, J. (2018). A multi-criteria decision-making framework for risk ranking of energy performance contracting project under picture fuzzy environment. *Journal of Cleaner Production*, *191*, 105–118. doi:10.1016/j.jclepro.2018.04.169

Wang, L., Zhang, H., Wang, J., & Li, L. (2018). Picture fuzzy normalized projection-based VIKOR method for the risk evaluation of construction project. *Applied Soft Computing*, *64*, 216–226. doi:10.1016/j.asoc.2017.12.014

Wang, X., & Kerre, E. E. (2001). Reasonable Properties for the Ordering of Fuzzy Quantities (I). *Fuzzy Sets and Systems*, *118*(3), 375–385. doi:10.1016/S0165-0114(99)00062-7

Wang, Y. (2014). The theory of fuzzy arithmetic in the extended domain of fuzzy numbers. *Journal of Advanced Mathematics and Applications*, *3*(2), 165–175. doi:10.1166/jama.2014.1063

Wang, Y. (2015). Fuzzy probability algebra (FPA): A theory of fuzzy probability for fuzzy inference and computational intelligence. *Journal of Advanced Mathematics and Applications*, *4*(1), 37–54. doi:10.1166/jama.2015.1071

Wan, Yu., Chen, Y., Shi, X., & Zhou, C. (2018). Constructing and Validating Word Similarity Datasets by Integrating Methods from Psychology, Brain Science and Computational Linguistics. *Soft Computing*, *22*(21), 6967–6979. doi:10.100700500-018-3174-1

Wei, G. (2012). Hesitant fuzzy prioritized operators and their application to multiple attribute decision making. *Knowledge-Based Systems*, *31*, 176–182. doi:10.1016/j.knosys.2012.03.011

Wei, G. (2017). Picture fuzzy aggregation operators and their application to multiple attribute decision making. *Journal of Intelligent & Fuzzy Systems*, *33*(2), 713–724. doi:10.3233/JIFS-161798

Wei, G. W. (2017). Picture 2-tuple linguistic Bonferroni mean operators and their application to multiple attribute decision making. *International Journal of Fuzzy Systems*, *19*(4), 997–1010. doi:10.100740815-016-0266-x

Wei, G. W., Alsaadi, F. E., Hayat, T., & Alsaedi, A. (2016). Picture 2-tuple linguistic aggregation operators in multiple attribute decision making. *Soft Computing*.

Welch, G., & Bishop, G. (2006). An Introduction to the Kalman Filter. *In Practice*, *7*(1), 1–16. Retrieved from http://citeseerx.ist.psu.edu/viewdoc/download?doi=10.1.1.79.6578&rep=rep1&type=pdf

White, M., Braczyk, H. J., Ghobadian, A., & Niebuhr, J. (1988). Small firms' innovation: why regions differ. *Policy Studies Institute*. Retrieved from: http://www.psi.org.uk/site/publication_detail/735

Wildberger, N.J. (2003*). A New Look at Multisets*. School of Mathematics, UNSW.

Williams, J., & Steele, N. (2002). Difference, Distance and Similarity as a Basis for Fuzzy Decision Support Based on Prototypical Decision Classes. *Fuzzy Sets and Systems*, *131*(1), 35–46. doi:10.1016/S0165-0114(01)00253-6

Wu, D., & Mendel, J. M. (2009). A Comparative Study of Ranking Methods, Similarity Measures and Uncertainty Measures for Interval Type-2 Fuzzy Sets. *Information Sciences*, *179*(8), 1169–1192. doi:10.1016/j.ins.2008.12.010

Xia, X., & Xu, Z. S. (2011). Hesitant fuzzy information aggregation in decision making. *International Journal of Approximate Reasoning*, *52*(3), 395–407. doi:10.1016/j.ijar.2010.09.002

Xia, X., Xu, Z. S., & Chen, N. (2013). Some hesitant fuzzy aggregation operators with their application in group decision making. *Group Decision and Negotiation*, *22*(2), 259–279. doi:10.100710726-011-9261-7

Xi, O. (1991). Fuzzy BCK-algebras. *Math Japonica*, *36*, 935–942.

Xu & Li. (2014). Granular Computing Approach to Two-Way Learning Based on Formal Concept Analysis in Fuzzy Datasets. *IEEE Trans. Cybern.*, *46*, 366–79.

Xuecheng, L. (1992). Entropy, distance measure and similarity measure of fuzzy sets and their relations. *Fuzzy Sets and Systems*, *52*(3), 305–318.

Xu, Z. S. (2007). Intuitionistic fuzzy aggregation operations. *IEEE Transactions on Fuzzy Systems*, *15*(6), 1179–1187. doi:10.1109/TFUZZ.2006.890678

Xu, Z. S. (2011). Approaches to multiple attribute group decision making based on intuitionistic fuzzy power aggregation operators. *Knowledge-Based Systems*, *24*(6), 749–760. doi:10.1016/j.knosys.2011.01.011

Xu, Z. S. (2014). *Hesitant Fuzzy Sets Theory*. Springer. doi:10.1007/978-3-319-04711-9

Xu, Z. S., & Xia, X. (2011). Distance and similarity measures for hesitant fuzzy sets. *Inform. Sci.*, *181*(11), 2128–2138. doi:10.1016/j.ins.2011.01.028

Xu, Z. S., & Yager, R. R. (2006). Some geometric aggregation operators based on intuitionistic fuzzy sets. *International Journal of General Systems*, *35*(4), 417–433. doi:10.1080/03081070600574353

Yager, R. R. (1986). On the theory of bags. *International Journal of General Systems*, *13*(1), 23–37. doi:10.1080/03081078608934952

Yager, R. R., & Abbasov, A. M. (2014). Pythagorean membership grades, complex numbers, and decision making. *International Journal of Intelligent Systems*, *28*(5), 436–452. doi:10.1002/int.21584

Yang, C. (2008). A not on soft set theory. *Computers & Mathematics with Applications (Oxford, England)*, *56*, 1899–1900.

Yang, X., Song, X., She, Y., & Yang, J. (2013). Hierarchy on Multigranulation Structures: A Knowledge Distance Approach. *International Journal of General Systems*, *42*(7), 754–773.

Yao, Vasilakos, & Pedrycz. (2013). Granular Computing: Perspectives and Challenges. *IEEE Transactions on Cybernetics, 43*(6), 1977–89. Retrieved from http://www.ncbi.nlm.nih.gov/pubmed/23757594

Yaqoob, N., Aslam, M. & Hila, K. (2012). Rough Fuzzy Hyperideals in Ternary Semihypergroups. *Advances in Fuzzy Systems*.

Yaqoob, N., Corsini, P., & Yousafzai, F. (2013). On Intra-regular LA-semihypergroups with Pure Left Identity. *Journal of Algebra, Number Theory, Advances and Applications, 10.*

Yaqoob, N., Mostafa, S. M., & Ansari, M. A. (2013). On Cubic KU-Ideals of KU-Algebras. ISRN Algebra.

Yaqoob, N., Aslam, M., & Ansari, M. A. (2012). Structures of N-G-hyperideals in left almost G-semihypergroups. *World Applied Sciences Journal*, *17*, 1611–1617.

Yaqoob, N., Aslam, M., & Yousafzai, F. (2012). On soft Γ-hyperideals over left almost Γ-semihypergroups. *Journal of Advanced Research in Dynamical and Control Systems*, *4*, 1–2.

Yihua, L., & Vemuri, V. R. (2002). Use of K-NN for Intrusion Detection. *Computers & Society*, *21*(5), 439–448.

Young, B. J., & Min, H. S. (2001). Fuzzy subalgebras of BCK/BCI algebras redefined. *Scientiae. Mathematicae Japanicae, 4*, 769-775.

Yousafzai, F., & Corsini, P. (2013). Some characterization problems in LA-semihypergroup. *Journal of Algebra, Number Theory. Advances and Applications, 10*, 41–55.

Yuan, X. H., Li, H. X., & Lee, E. S. (2010). On the definition of the intuitionistic fuzzy subgroups. *Computers & Mathematics with Applications (Oxford, England), 59*(9), 3117–3129. doi:10.1016/j.camwa.2010.02.033

Yuan, X. H., Li, H. X., & Sun, K. B. (2009). The cut sets, decomposition theorems and representation theorems on intuitionistic fuzzy sets and interval-valued fuzzy sets [in Chinese]. *Science in China Series F: Information Sciences, 39*(9), 933–945.

Yue, Z. L. (2011). Deriving decision maker's weights based on distance measure for interval-valued intuitionistic fuzzy group decision making. *Expert Systems with Applications, 38*(9), 11665–11670. doi:10.1016/j.eswa.2011.03.046

Yuksel, S., Tozlu, N., & Guzel, Z. (2013). On Soft Generalized Closed Sets in Soft Topological Spaces. *Journal of Theoretical and Appl. Inform. Technology, 55*, 17–21.

Zadeh, Fu, Tanaka, & Shimura. (Eds.). (1975). Fuzzy sets and their applications to cognitive and decision processes: Papers. New York: Academic Press.

Zadeh, L. A. (1965). Fuzzy Sets. Information and Control, (8), 338–353.

Zadeh, L. A. (1965). Fuzzy sets. *Information and Control, 8*, 338-353.

Zadeh, L. A. (1965). Fuzzy sets. *Information and Control, 8*(3), 338–352.

Zadeh, L. A. (1965). Fuzzy Sets. *Information and Control, 8*(3), 338–353. doi:10.1016/S0019-9958(65)90241-X

Zadeh, L. A. (1965). Fuzzy sets. *Information and Control, 8*(5), 338–353.

Zadeh, L. A. (1965). Fuzzy sets. *Information and Control, 8*, 338-353.

Zadeh, L. A. (1971). Similarity Relations and Fuzzy Orderings. *Information Sciences, 3*(2), 177–200. doi:10.1016/S0020-0255(71)80005-1

Zadeh, L. A. (1971). Similarity relations and fuzzy orderings. *Information Sciences, 3*, 177–200.

Zadeh, L. A. (1975). The concept of a linguistic variable and its application to approximate reasoning-I. *Information Sciences, 8*(3), 199–249. doi:10.1016/0020-0255(75)90036-5

Zadeh, L. A. (1978). Fuzzy sets as a basis for a theory of possibility. *Fuzzy Sets and Systems, 1*(1), 3–28. doi:10.1016/0165-0114(78)90029-5

Zadeh, L. A. (2005). Toward a generalized theory of uncertainty (GTU)-an outline. *Inform. Sci., 172*(1-2), 1–40. doi:10.1016/j.ins.2005.01.017

Zadeh, L. A. (2006). Generalized Theory of Uncertainty (GTU)—Principal Concepts and Ideas. *Computational Statistics & Data Analysis, 51*(1), 15–46. doi:10.1016/j.csda.2006.04.029

Zadeh, L. A. (2008). Is there a need for fuzzy logic? *Inform. Sci., 178*(13), 2751–2779. doi:10.1016/j.ins.2008.02.012

Zadrozny, S., & Kacprzyk, J. (2012). Bipolar queries: An aggregation operator focused perspective. *Fuzzy Sets and Systems, 196*, 69–81. doi:10.1016/j.fss.2011.10.013

Zarandi, A., & Saeid, A. B. (2009). Interval-valued fuzzy BF-algebras. *Int. J. Contemp. Math. Sciences*, *4*(17), 803–814.

Zeng, S., Asharf, S., Arif, M., & Abdullah, S. (2019). Application of Exponential Jensen Picture Fuzzy Divergence Measure in Multi-Criteria Group Decision Making. *Mathematics*, *7*(2), 191. doi:10.3390/math7020191

Zeng, W., & Li, H. (2006). Relationship between Similarity Measure and Entropy of Interval Valued Fuzzy Sets. *Fuzzy Sets and Systems*, *157*(11), 1477–1484. doi:10.1016/j.fss.2005.11.020

Zhang, H., & Zhan, J. (2017). Rough soft lattice implication algebras and corresponding decision making methods. *International Journal of Machine Learning and Cybernetics*, *8*(4), 1301–1308. doi:10.100713042-016-0502-6

Zhang, W. R. (1994). Bipolar fuzzy sets and relations: a computational framework for cognitive modeling and multiagent decision analysis. *Proceedings of IEEE Conf.*, 305-309.

Zhang, W. R., & Zhang, L. (2004). YinYang bipolar logic and bipolar fuzzy logic. *Information Sciences*, *165*(3-4), 265–287. doi:10.1016/j.ins.2003.05.010

Zhan, J., & Tan, Z. (2004). Intuitionistic fuzzy *a*-ideals in BCI-algebras. *Soochow Math. J.*, *30*, 207–216.

Zhan, J., & Tan, Z. (2005). Fuzzy H-ideals in BCK-algebras. *Southeast Asian Bulletin of Mathematics*, *29*(6), 1165–1173.

Zhou, Z., Zhang, R., Lin, Y., & Wang, R. (2015). A Comparison of Similarity Measures of Intuitionistic Fuzzy Sets. In Z. Zhang, Z. M. Shen, J. Zhang, & R. Zhang (Eds.), *LISS 2014* (pp. 1237–1242). Springer Berlin Heidelberg. doi:10.1007/978-3-662-43871-8_179

Zhu, B., Xu, Z. S., & Xia, X. (2012). Hesitant fuzzy geometric Bonferroni means. *Inform. Sci.*, *205*, 72–85. doi:10.1016/j.ins.2012.01.048

Zhu, R. (1984). Convergence of powers of a fuzzy matrix. *Fuzzy Math.*, *3*, 45–49.

Zimmermann, H. J. (2001). Fuzzy Set Theory-and its Applications (4th ed.). Springer Science+ Business Media, LLC.

Zohuri, B. (2015). Similitude Theory and Applications. In B. Zohuri (Ed.), *Dimensional Analysis and Self-Similarity Methods for Engineers and Scientists* (pp. 93–193). Cham: Springer International Publishing.

Zwick, R., Carlstein, E., & Budescu, D. V. (1987). Measures of Similarity among Fuzzy Concepts: A Comparative Analysis. *International Journal of Approximate Reasoning*, *1*(2), 221–242. doi:10.1016/0888-613X(87)90015-6

About the Contributors

Chiranjibe Jana received his Bachelor of Science degree with honours in Mathematics in 2007 from Midnapore College, Pashim Medinipur, West Bengal, India and Master of Science degree in Mathematics in 2009 from Vidyasagar University, West Bengal, India. He received Ph.D. in Pure Mathematics with specialization of fuzzy BCK/BCI-algebra and related Algebras in 2018, Department of Applied Mathematics, Vidyasagar University, West Bengal, India. He has qualified NET in 2010. His current research interest is in the areas of multi-criteria decision-making, aggregation operators, similarity measure, neutrosophic set, fuzzy algebra, and soft algebraic structures. He has published 27 papers in the international reputed SCI journals.

Tapan Senapati received the B. Sc., M. Sc. and Ph.D. degrees in Mathematics all from the Vidyasagar University, India in 2006, 2008 and 2013 respectively. He has published one book and more than 50 articles in reputed international journals. He is a reviewer of several international journals. His main scientific interest concentrates on fuzzy sets and related algebraic systems.

Madhumangal Pal is currently a Professor of Applied Mathematics, Vidyasagar University. Prof. Pal has successfully guided 34 research scholars for Ph.D. degrees and has published more than 320 articles in international and national journals. His specializations include Algorithmic and Fuzzy Graph Theory, Fuzzy Matrices, Genetic and Parallel Algorithms, Soft Computing. Prof. Pal is the author of eight text books published from India and United Kingdom. He has published 17 book chapters. Prof. Pal completed four research project funded by UGC and DST, Govt. of India. Prof. Pal is the Editor-in-Chief of "Journal of Physical Sciences", "Annals of Pure and Applied Mathematics", section editor of "International Journal of Computational Intelligence Systems (SCI Index Journal)" and member of the editorial Boards of many journals. Also, he has visited China, Greece, London, Taiwan, Malaysia, Thailand, Hong Kong, Dubai and Bangladesh to participated, delivered invited talks and to chair conference event.

* * *

Saleem Abdullah received M.Phil. and Ph.D. degrees in mathematics from Quaid-i-Azam University, Islamabad, Pakistan. Currently he is working as Assistant Professor at Department of Mathematics Abdul Wali Khan University, Mardan, Pakistan. His research area is fuzzy logic, fuzzy set theory and decision making. He has published more than 150 papers in international journals.

Amal Kumar Adak received his M.Sc. degree in Mathematics and PhD from Vidyasagar University, West Bengal, India in 2006 and 2013, respectively. Currently, he is an Assistant Professor in Department of Mathematics of Ganesh Dutt College, Begusarai, India (Lalit Narayan Mithila University) since 2017. He is Associate-Editor of International Journal of End-User Computing and Development. He is a reviewer of several international journals. His research interest includes fuzzy and intuitionistic fuzzy sets, intuitionistic fuzzy matrices and operations research. His papers are published in several national and international journals. He has participated and presented papers in several national and international conferences.

Shahzaib Ashraf received MS degree in mathematics from International Islamic University, Islamabad, Pakistan. Currently he is a Ph.D. scholar at Department of Mathematics, Abdul Wali Khan University, Mardan, Pakistan. His main research interests are pure mathematics that is picture fuzzy set aggregation operators, Spherical fuzzy set and their aggregation operators. He has published more than 13 papers in international journals

Ravikumar Bandaru is working as an Assistant Professor in Department of Mathematics, GITAM(Deemed to be University), Hyderabad Campus. I have completed my Ph.D in 2010 from Andhra University, Visakhapatnam, Andhra Pradesh, India. I have published more than 40 research papers in various international journals. My area of research is Lattice theory, Generalizations of BCK-algebras.

Venkateswarlu Bollineni is working as an associate professor and HOD in the department of mathematics, GITAM(Deemed to be University), Bangalore campus. They have published 64 research papers in national and international peer reviewed journals. Their areas of research are Algebra and Complex analysis.

Guiyun Chen is a full Professor of School of Mathematics and Statistics, Southwest University, Beibei 400715, Chongqing, China.

Mohamed El Alaoui studied in his PhD thesis multi-objective optimization using fuzzy logic, in the Production and Industrial Engineering Department at Ecole National Supérieure des Arts et Metiers, ENSAM Meknes. He is an Industrial Engineer from the same institution.

Khalid El Yassini holds a PhD and MSc in Mathematics (University of Sherbrooke - Canada, 2000 and 1994). Previously, he had a Degree in Applied Mathematics (Statistics) at Abdelmalek Essaadi University in Morocco (1991). During his academic career, he has taught in several Canadian and Moroccan institutions such as UdeS, UQAR, St-Boniface University, International University of Rabat or Royal Military Academy. He has taught various courses in mathematics, computer science, operations research and logistics. His current research activities focus on mathematical programming, artificial intelligence, information systems, security in computer networks, telecommunications, intelligent systems, and logistics engineering with applications in hospital field or medication domain. Currently, he is a Professor in the Faculty of Sciences at Moulay Ismail University, Morocco.

Muhammad Gulistan got his MPhil degree from Quaide Azam University islamabad in 2011 and PhD degree from Hazara University in 2016. Now he is working as an assistant professor in the department of mathematics and statistics of Hazara university. His area of research is "Cubic sets and their generalizations, Non-associative hyperstructures, neutrosophic cubic sets, neutrosophic cubic graphs, decision making''. He published more than 50 research papers in different well reputed journals. He supervised many M.Phil. and PhD research students.

Miriam Kalpana is a teaching faculty member of the Department of Statistics, Madras Chrisitian College, Tambaram, Chennai, India. She has more than 15 years teaching experience.

Faruk Karaaslan was born March 25th, 1980, in Turkey. BSc in Mathematics from Erciyes University, Turkey (1999- 2003). Non-thesis master's program in Education of Mathematics from Yüzüncü Yıl University, Turkey (2003-2005). MSc in Mathematics (Topology) from Gaziosmanpaşa University, Tokat, Turkey (2005-2007). Ph.D in Mathematical Logic from Gaziosmanpaşa University, Tokat, Turkey (2009-2013). Currently serving as Associate Professor in Çankırı Karatekin University, Department of Mathematics, Turkey. Published more than 30 research papers in prestigious journals. Also reviewed international journals in different areas of algebraic structures, fuzzy sets and logics, neutrosophic sets and logics, rough sets. Refereeing Journals: Neural Computing and Applications (NCA), Autosoft journal Soft Computing and Automation Journal, Applied Mathematics & Information Sciences (AMIS), Applied Soft computing, Soft Computing, Informatica, Kuwait journal of Sciences, Artificial Intelligence in Medical Sciences, Annals of Fuzzy Mathematics and Informatics, Journal of New Theory, Annals of Fuzzy Mathematics and Informatics, Fuzzy Information and Engineering, IEEE Transaction on Fuzzy systems, Proceedings of the Pakistan Academy of Sciences.

Srikanth Kavirayani is an electrical engineering graduate from University of Missouri Columbia USA specializing in Intelligent systems. He has obtained his bachelors degree in electrical and electronics engineering from Jawahar lal Nehru Technological University, Hyderabad, India. He has to his credit over 20 journals and has participated in numerous conferences, workshops across the country. His areas of interest include robotics, computational intelligence, intelligent systems, control systems. He is currently working as an assistant professor in Gayatri Vidya Parishad College of Engineering(A), Visakhapatnam, India and is also a research scholar in GITAM (Deemed to be University), Visakhapatnam, India and has successfully submitted his PhD Thesis. He has received best paper award and editors choice award for his earlier works.

Muhammad Sajjad Ali Khan completed his PhD degree from Hazara University Mansehra, KPK, Pakistan. He published about 40 papers in well reported national and international journals.

Shishir Kumar is working as Professor & HOD. Dr. Shishir has completed his Ph. D. in 2005. He has joined this organization in July 2005 as Assistant Professor. He has ninteen years of teaching and research experience in various organizations of repute for PG & UG courses of Computer Science & IT. Many students of various UG and PG courses of Computer Science and IT had developed their projects under his guidance. He is also associated with many renowned International Journals as reviewer / editorial board member. He is also in a panel of reviewer for IEEE Communication Society. He is also working as honorary advisor to SETEPS-India for various academic and literary activities.

G. Muhiuddin is working at the Department of Mathematics, University of Tabuk, Saudi Arabia as an Associate Professor. He received his Ph.D. and M. Phil. in Pure Mathematics, more specifically, in Algebra with specialization in Category Theory. His mathematical research areas include Algebras related to logic (BCK, BCI, BCC-algebras, Hilbert algebras, and implication algebras), Fuzzy logical algebras, Theory of derivations in algebraic structures and Category theory. He serves as an expert referee for several research papers in these areas of his expertise. He also serves as reviewer for Mathematics Reviews of the American Math Society, USA (Mathematical Reviews/ MathSciNet Reviewer Number: 105900). He has published a number of research papers in internationally reputed mathematical journals. He is an active a member of the editorial boards of (i) Transnational Journal of Mathematical Analysis and Application and (ii) Annals of Fuzzy Mathematics and Informatics (iii) Financial Forum" (Singapore). Also, he serves as Editor-in-chief of the journal Annals of Communication in Mathematics (http://www.technoskypub.com/journal/acm/).

Prakasam Muralikrishna has 15 years of experience in Teaching and around 6 years in research. Published 50 articles in which nearly 20 indexed in SCOPUS.

Laxminarayan Sahoo is Assistant Professor at the Department of Mathematics, Raniganj Girls' College, West Bengal, India. He obtained his Ph.D. in Mathematics from The University of Burdwan and M.Sc. in Applied Mathematics from Vidyasagar University, India. His research interests include reliability optimization, computational optimization, artificial intelligence, soft computing, interval mathematics, interval ranking, and fuzzy mathematics. He has published several research papers in various national and international journals of repute. He has received MHRD fellowship form the Government of India and the Prof. M.N. Gopalan Award for the Best Ph.D. Thesis in Operations Research from the Operational Research Society of India.

Sampath Sundaram is currently holding the position of Professor of Data Science, Kalasalingam Academy of Higher Education, Srivilliputhur, India. He got his Ph.D. degree from the University of Madras. He has served in Loyola College, Chennai, Birla Institute of Science and Technology, Dubai, UAE and University of Madras before joining Kalasalingam Academy of Higher Education, Srivilliputhur, India. He has more than 35 years of experience in teaching and research. He has guided many scholars for their research degrees and currently four are working for their Ph.D. degree under his guidance. His main areas of research are Inference for Finite Populations, Credibility Theory, Testing and Estimation of Parameters in Imprecise Models and Data Mining. He has to his credit nearly 60 research articles in highly rated research journals. His book titled "Sampling Theory and Methods" published by CRC Press and Narosha Publishing house is a popular book in the academic circles.

Index

CPSIA information can be obtained
at www.ICGtesting.com
Printed in the USA
BVHW011318160320
575104BV00010B/51

9 781799 801900